Muscle Metabolism During Exercise

ADVANCES IN EXPERIMENTAL MEDICINE AND BIOLOGY

Editorial Board:

Nathan Back	*Chairman, Department of Biochemical Pharmacology, School of Pharmacy, State University of New York, Buffalo, New York*
N. R. Di Luzio	*Chairman, Department of Physiology, Tulane University School of Medicine, New Orleans, Louisiana*
Alfred Gellhorn	*University of Pennsylvania Medical School, Philadelphia, Pennsylvania*
Bernard Halpern	*Collège de France, Director of the Institute of Immuno-Biology, Paris, France*
Ephraim Katchalski	*Department of Biophysics, The Weizmann Institute of Science, Rehovoth, Israel*
David Kritchevsky	*Wistar Institute, Philadelphia, Pennsylvania*
Abel Lajtha	*New York State Research Institute for Neurochemistry and Drug Addiction, Ward's Island, New York*
Rodolfo Paoletti	*Institute of Pharmacology, University of Milan, Milan, Italy*

Volume 1
THE RETICULOENDOTHELIAL SYSTEM
AND ATHEROSCLEROSIS
 Edited by N. R. Di Luzio
 and R. Paoletti • 1967

Volume 2
PHARMACOLOGY OF HORMONAL
POLYPEPTIDES AND PROTEINS
 Edited by N. Back, L. Martini,
 and R. Paoletti • 1968

Volume 3
GERM-FREE BIOLOGY—Experimental
and Clinical Aspects
 Edited by E. A. Mirand and N. Back • 1969

Volume 4
DRUGS AFFECTING LIPID METABOLISM
 Edited by W. L. Holmes, L. A. Carlson,
 and R. Paoletti • 1969

Volume 5
LYMPHATIC TISSUE AND GERMINAL
CENTERS IN IMMUNE RESPONSE
 Edited by L. Fiore-Donati
 and M. G. Hanna, Jr. • 1969

Volume 6
RED CELL METABOLISM AND FUNCTION
 Edited by George J. Brewer • 1970

Volume 7
SURFACE CHEMISTRY OF
BIOLOGICAL SYSTEMS
 Edited by Martin Blank • 1970

Volume 8
BRADYKININ AND RELATED KININS:
Cardiovascular, Biochemical, and Neural Actions
 Edited by F. Sicuteri, M. Rocha e Silva,
 and N. Back • 1970

Volume 9
SHOCK: Biochemical, Pharmacological,
and Clinical Aspects
 Edited by A. Bertelli and N. Back • 1970

Volume 10
THE HUMAN TESTIS
 Edited by E. Rosemberg
 and C. A. Paulsen • 1970

Volume 11
MUSCLE METABOLISM DURING EXERCISE
 Edited by B. Pernow and B. Saltin • 1971

Volume 12
MORPHOLOGICAL AND FUNCTIONAL
ASPECTS OF IMMUNITY
 Edited by K. Lindahl-Kiessling, G. Alm, and
 M. G. Hanna, Jr. • 1971

Volume 13
CHEMISTRY AND BRAIN DEVELOPMENT
 Edited by R. Paoletti and A. N. Davison • 1971

Volume 14
MEMBRANE-BOUND ENZYMES
 Edited by G. Porcellati and F. di Jeso • 1971

Muscle Metabolism During Exercise

Proceedings of a Karolinska Institutet Symposium held in
Stockholm, Sweden, September 6–9, 1970

Honorary guest: E. Hohwü Christensen

Edited by

Bengt Pernow

Associate Professor and Chairman
Department of Clinical Physiology
Serafimerlasarettet, Karolinska Institutet
Stockholm, Sweden

and

Bengt Saltin

Associate Professor
Department of Physiology
Gymnastik & Idrottshögskolan
Stockholm, Sweden

PLENUM PRESS · NEW YORK-LONDON · 1971

Library of Congress Catalog Card Number 71-148821

SBN 306-39011-6

© 1971 Plenum Press, New York
A Division of Plenum Publishing Corporation
227 West 17th Street, New York, N.Y. 10011

United Kingdom edition published by Plenum Press, London
A Division of Plenum Publishing Company, Ltd.
Donington House, 30 Norfolk Street, London W.C.2, England

All rights reserved

No part of this publication may be reproduced in any form
without written permission from the publisher

Printed in the United States of America

LIST OF CONTRIBUTORS

Sten Adolfsson, Department of Physiology, University of Göteborg,
    Göteborg, Sweden

Gunvor Ahlborg, Department of Clinical Physiology, Karolinska
    Institutet at Serafimerlasarettet, Stockholm, Sweden

Kurt Ahrén, Department of Physiology, University of Göteborg,
    Göteborg, Sweden

Erling Asmussen, Gymnastikteoretisk Laboratory, University of
    Copenhagen, Copenhagen, Denmark

Kathryn Ballard, Department of Pharmacology, Karolinska Institutet,
    Stockholm, Sweden

Catherine Bauer, Department of Biochemistry, St. Louis University
    School of Medicine, St. Louis, Missouri

Jonas Bergström, Department of Clinical Chemistry, S:t Eriks
    Sjukhus, Stockholm, Sweden

Per Björntorp, First Medical Service, Sahlgrenska Sjukhuset,
    University of Göteborg, Göteborg, Sweden

George F. Cahill, Jr., Joslin Research Laboratory, Harvard Medical
    School, Boston, Massachusetts

Lars A. Carlson, King Gustaf V Research Institute, Karolinska
    Sjukhuset, Stockholm, Sweden

Erol Gerasi, Departments of Endocrinology and Metabolism,
    Karolinska Sjukhuset, Stockholm, Sweden

L. A. Cobb, Department of Medicine, University of Washington School
    of Medicine, Seattle, Washington

LIST OF CONTRIBUTORS

Jimmy C. Costin, Department of Medicine, Woodruff Medical Center of Emory University, Atlanta, Georgia

Richard E. Davies, School of Veterinary Medicine, University of Pennsylvania, Philadelphia, Pennsylvania

Peter E. di Prampero, Department of Physiology, University of Milano, Milan, Italy

I. J. Don, Department of Preventive Medicine, Washington University School of Medicine, St. Louis, Missouri

Lars Edström, Department of Neurology, Karolinska Sjukhuset, Stockholm, Sweden

Richard H. T. Edwards, Department of Clinical Chemistry, S:t Eriks Sjukhus, Stockholm, Sweden

L.-G. Ekelund, Department of Clinical Physiology, Karolinska Sjukhuset, Stockholm, Sweden

B. Essén, Department of Physiology, Gymnastik- och idrottshögskolan, Stockholm, Sweden

Philip Felig, Department of Medicine, Yale University School of Medicine, New Haven, Connecticut

Bertil B. Fredholm, Department of Pharmacology, Karolinska Institutet, Stockholm, Sweden

Sven O. Fröberg, King Gustaf V Research Institute, Karolinska Sjukhuset, Stockholm, Sweden

Philip D. Gollnick, Exercise Physiology Laboratory, Department of Physical Education for Men, and Electron Microscope Center, Washington State University, Pullman, Washington

Gunnar Grimby, Department of Clinical Physiology, University of Göteborg, Sahlgrenska Sjukhuset, Göteborg, Sweden

D. R. Gunn, Department of Medicine, University of Washington School of Medicine, Seattle, Washington

Lars Hagenfeldt, Department of Clinical Chemistry, Karolinska Institutet at Serafimerlasarettet, Stockholm, Sweden

Jan Häggendal, Department of Pharmacology, University of Göteborg, Göteborg, Sweden

LIST OF CONTRIBUTORS

R. C. Harris, Department of Clinical Chemistry, S:t Eriks Sjukhus, Stockholm, Sweden

Richard J. Havel, Cardiovascular Research Institute and Department of Medicine, University of California, San Francisco, California

Lars Hermansen, Institute of Work Physiology, Oslo, Norway

G. M. Hesser, Departments of Aviation and Naval Medicine, Karolinska Institutet, Stockholm, Sweden

John O. Holloszy, Department of Preventive Medicine, Washington University School of Medicine, St. Louis, Missouri

Olga Hudlicka, Department of Physiology, Medical School, University of Birmingham, England

Erik Hultman, Department of Clinical Chemistry, S:t Eriks Sjukhus, Stockholm, Sweden

C. David Ianuzzo, Exercise Physiology Laboratory, Department of Physical Education for Men, and Electron Microscope Center, Washington State University, Pullman, Washington

Lennart Jorfeldt, Department of Clinical Physiology, Karolinska Institutet at Serafimerlasarettet, Stockholm, Sweden

Lennart Kaijser, Department of Clinical Physiology, Karolinska Sjukhuset, Stockholm, Sweden

Jan Karlsson, Department of Physiology, Gymnastik- och idrotts-högskolan, Stockholm, Sweden

Joseph Keul, Medizinische Universitätsklinik, Frieburg in Breisgau, West Germany

Karl-Heinz Kiessling, Institute of Zoophysiology, University of Uppsala, Uppsala, Sweden

Douglas W. King, Exercise Physiology Laboratory, Department of Physical Education for Men, and Electron Microscope Center, Washington State University, Pullman, Washington

Howard G. Knuttgen, Department of Biology, Boston University, Boston, Massachusetts

D. Koh, Department of Clinical Chemistry, S:t Eriks Sjukhus, Stockholm, Sweden

LIST OF CONTRIBUTORS

Hans Kraus, Physiologisches Institut der Freien Universität Berlin, Berlin, West Germany

Eric Kugelberg, Department of Neurology, Karolinska Sjukhuset, Stockholm, Sweden

Ben W. Lassers, King Gustaf V Research Institute, Stockholm, Sweden

Rolf Luft, Departments of Endocrinology and Metabolism, Karolinska Sjukhuset, Stockholm, Sweden

Carl-Göran Lundquist, Institute of Zoophysiology, University of Uppsala, Uppsala, Sweden

A. Melcher, Departments of Aviation and Naval Medicine, Karolinska Institutet, Stockholm, Sweden

R. A. Melé, Department of Preventive Medicine, Washington University School of Medicine, St. Louis, Missouri

Thomas E. Morgan, Departments of Medicine, Pathology, and Orthopedics, University of Washington School of Medicine, Seattle, Washington

Albert S. Most, Cardiovascular Laboratory, Rhode Island Hospital, Providence, Rhode Island

L. H. Nilsson, Department of Clinical Chemistry, S:t Eriks Sjukhus, Stockholm, Sweden

L.-O. Nordesjö, Military Medical Examination Center, Karolinska Sjukhuset, Stockholm, Sweden

S. Nordlander, Departments of Endocrinology and Metabolism, Karolinska Sjukhuset, Stockholm, Sweden

Robert E. Olson, Department of Biochemistry, St. Louis University School of Medicine, St. Louis, Missouri

L. B. Oscai, Department of Preventive Medicine, Washington University School of Medicine, St. Louis, Missouri

Jan Östman, Departments of Endocrinology and Metabolism, Karolinska Sjukhuset, Stockholm, Sweden

Pavle Paul, Division of Research, Lankenau Hospital, Philadelphia, Pennsylvania

LIST OF CONTRIBUTORS

Bengt Pernow, Department of Clinical Physiology, Karolinska Institutet at Serafimerlasarettet, Stockholm, Sweden

Dirk Pette, Physiologisch-Chemisches Institut der Universität München, München, West Germany

Karin Piehl, Institute of Zoophysiology, University of Uppsala, Uppsala, Sweden

Esther D. R. Pruett, Institute of Work Physiology, Oslo, Norway

A. E. Roch Norlund, Department of Clinical Chemistry, S:t Eriks Sjukhus, Stockholm, Sweden

Flaviu C. A. Romanul, Neurological Unit, Boston City Hospital, and Department of Neurology, Harvard Medical School, Boston, Massachusetts

Sune Rosell, Department of Pharmacology, Karolinska Institutet, Stockholm, Sweden

R. Ross, Department of Medicine, University of Washington School of Medicine, Seattle, Washington

Loring B. Rowell, Departments of Physiology and Biophysics and of Medicine, University of Washington School of Medicine, Seattle, Washington

Bengt Saltin, Department of Physiology, Gymnastik- och idrottshögskolan, Stockholm, Sweden

F. A. Short, Department of Medicine, University of Washington School of Medicine, Seattle, Washington

N. Sheldon Skinner, Jr., Department of Medicine, Woodruff Medical Center of Emory University, Atlanta, Georgia

Göran Sterky, Department of Pediatrics at St. Göran's Children's Hospital, Stockholm, Sweden

Gösta Tibblin, Medical Service I, University of Göteborg, Sahlgrenska Sjukhuset, Göteborg, Sweden

G. Vastagh, Department of Medicine, Southwestern Medical School, Dallas, Texas

M. L. Wahlqvist, Department of Geriatrics, University of Uppsala, Uppsala, Sweden

John Wahren, Department of Clinical Physiology, Karolinska
    Institutet at Serafimerlasarettet, Stockholm, Sweden

O. Wigertz, Departments of Aviation and Naval Medicine, Karolinska
    Institutet, Karolinska Sjukhuset, Stockholm, Sweden

Lars Wilhelmsen, Medical Service I, University of Göteborg,
    Sahlgrenska Sjukhuset, Göteborg, Sweden

CONTENTS

Muscle Metabolism during Exercise in Man: A Historical
      Survey 1
  E. Asmussen

ADAPTIVE CHANGES IN MORPHOLOGY AND ENZYMES OF SKELETAL MUSCLES

Histochemical Effects of Contraction on Red and White Rat
      Muscle Fibres. Their Organization in the
      Motor Unit 13
  E. Kugelberg and L. Edström

Reversal of Enzymatic Profiles and Capillary Supply of
      Muscle Fibers in Fast and Slow Muscles After
      Cross Innervation 21
  F. C. A. Romanul

Metabolic Differentiation of Distinct Muscle Types at
      the Level of Enzymatic Organization 33
  D. Pette

Biochemical Adaptations to Endurance Exercise in
      Skeletal Muscle 51
  J. O. Holloszy, L. B. Oscai, P. A. Molé, and
  I. J. Don

Influence of Exercise on Electron Transport Capacity of
      Heart Mitochondria 63
  H. Kraus

Ultrastructural and Enzyme Changes in Muscles with Exercise   69
  P. D. Gollnick, C. D. Ianuzzo, and D. W. King

Effects of Long-Term Exercise on Human Muscle
        Mitochondria 87
  T. E. Morgan, L. A. Cobb, F. A. Short, R. Ross, and
  D. R. Gunn

Effect of Physical Training on Ultrastructural Features
        in Human Skeletal Muscle 97
  K.-H. Kiessling, K. Piehl, and C.-G. Lundquist

ENERGY STORES (MAGNITUDE, REGULATION AND INTERACTION)

Release of Substrates from Extramuscular Stores During Exercise

Metabolic Role of Muscle 103
  G. F. Cahill, Jr.

Adrenergic Neuro-Humoral Control of Lipolysis in
        Adipose Tissue 111
  S. Rosell and K. Ballard

Role of Circulating Noradrenaline and Adrenaline 119
  J. Häggendal

The Liver as an Energy Source in Man During Exercise 127
  L. B. Rowell

Liver Glycogen in Man. Effect of Different Diets and
        Muscular Exercise 143
  E. Hultman and L. H. Nilsson

Uptake and Oxidation of Energy Rich Compounds in the Muscle During Exercise

Metabolism of Free Fatty Acids and Ketone Bodies in
        Skeletal Muscle 153
  L. Hagenfeldt and J. Wahren

Plasma Insulin Levels During Prolonged Exercise 165
  E. D. R. Pruett

Glucose Uptake in Contracting, Isolated in Situ Dog
        Skeletal Muscle 177
  N. S. Skinner, Jr., J. C. Costin, B. Saltin, and
  G. Vastagh

Glucose Metabolism During Exercise in Man 189
    J. Wahren, G. Ahlborg, P. Felig, and L. Jorfeldt

Interrelationship Between Amino Acid and Carbohydrate
       Metabolism During Exercise: The Glucose
       Alanine-Cycle 205
    P. Felig and J. Wahren

Uptake of Substrates in Isolated Contracting Slow and
       Fast Muscles in situ in Relation to Fatigue 215
    O. Hudlicka

Uptake and Oxidation of Substrates in the Intact Animal
       During Exercise 225
    P. Paul

Fat Mobilization and Blood Lactate Concentration 249
    B. B. Fredholm

STORAGE AND USE OF INTRAMUSCULAR
SUBSTRATES DURING EXERCISE

Control Mechanisms for the Synthesis of Glycogen in
       Striated Muscle 257
    S. Adolfsson and K. Ahrén

Glycogen Storage in Human Skeletal Muscle 273
    E. Hultman, J. Bergström, and A. E. Roch-Norlund

Muscle Glycogen Utilization During Work of Different
       Intensities 289
    B. Saltin and J. Karlsson

The Effect of Propranolol on Glycogen Metabolism During
       Exercise 301
    R. C. Harris, J. Bergström, and E. Hultman

Local Lipid Stores and Exercise 307
    S. O. Fröberg, L. A. Carlson, and L.-G. Ekelund

Influence of Intensity and Duration of Exercise on
       Supply and Use of Fuels 315
    R. J. Havel

PHOSPHAGENS AND LACTATE METABOLISM

Energy-Rich Phosphagens 327
  R. E. Davies

Energy Rich Phosphagens in Dynamic and Static Work 341
  J. Bergström, R. C. Harris, E. Hultman, and
  L.-O. Nordesjö

Isometric Exercise — Factors Influencing Endurance
    and Fatigue 357
  R. H. T. Edwards, L.-O. Nordesjö, D. Koh,
  R. C. Harris, and E. Hultman

Lactate and Oxygen Debt: An Introduction 361
  H. G. Knuttgen

The Alactic Oxygen Debt: Its Power, Capacity, and
    Efficiency 371
  P. E. di Prampero

Muscle ATP, CP, and Lactate in Submaximal and Maximal
    Exercise 383
  J. Karlsson

Muscle ATP, CP, and Lactate During Exercise after
    Physical Conditioning 395
  B. Saltin and J. Karlsson

Lactate Production During Exercise 401
  L. Hermansen

Turnover of ^{14}C-L(+)-Lactate in Human Skeletal
    Muscle During Exercise 409
  L. Jorfeldt

Muscle Glycogen, Lactate, ATP, and CP in Intermittent
    Exercise 419
  B. Saltin and B. Essén

Blood Lactate Concentrations During Intermittent and
    Continuous Exercise with the Same Average
    Power Output 425
  R. H. T. Edwards, A. Melcher, C. M. Hesser, and
  O. Wigertz

CONTENTS

MYOCARDIAL METABOLISM

Utilization of Endogenous Lipids by the Isolated Perfused Rat Heart 429
   R. E. Olson and C. Bauer

Myocardial Metabolism in Athletes 447
   J. Keul

Myocardial Metabolism in Man at Rest and During Prolonged Exercise 457
   B. W. Lassers, L. Kaijser, M. L. Wahlqvist, and
   L. A. Carlson

HABITUAL PHYSICAL ACTIVITY; AEROBIC POWER AND BLOOD LIPIDS

Habitual Physical Activity: Aerobic Power and Blood Lipids 469
   G. Grimby, L. Wilhelmsen, P. Björntorp, B. Saltin, and G. Tibblin

METABOLIC STUDIES IN CLINICAL MATERIALS

Occlusive Artery Disease of the Heart and Skeletal Muscle

Myocardial Metabolism in Patients with Ischemic Heart Disease 483
   A. S. Most

Metabolism in Patients with Ischemic Heart Disease and Obesity After Training 493
   P. Björntorp

Metabolism of Free Fatty Acids During Exercise in Patients with Occlusive Arterial Disease of the Leg 505
   L. Hagenfeldt, B. Pernow, and J. Wahren

Obesity and Diabetes

Clinical and Metabolic Aspects on Obesity in Childhood ... 521
   G. Sterky

Metabolism During Exercise in Patients with Diabetes 529
   J. Östman, E. Cerasi, L.-G. Ekelund, R. Luft, and
   S. Nordlander

Discussion 537

Index 553

MUSCLE METABOLISM DURING EXERCISE IN MAN

A historical survey

      Erling Asmussen

      University of Copenhagen

     The animal body - including man's - is a machine that can convert chemical energy into mechanical energy: force and work. This ability is especially developed in the muscles, and the theme of our symposium, muscle metabolism during exercise, deals specificly with this process.

     It has probably been known since the oldest times that for animals and men to be able to perform work, extra food is needed.

     How the foodstuffs were utilized and turned into useful work was, however, hardly known. This question had to await the birth of a new science: chemistry, and the consequent development of methods for measuring the amounts of substances taken up and given off by the living organisms. This happened in the late eighteenth century: the great period of Enlightenment and of the Encyclopedists. Oxygen was found and isolated by the Englishman <u>Priestly</u> and by the Swede <u>Scheele</u>, and some few years later, in the years just before the French Revolution, <u>Lavoisier</u> in Paris performed his experiments on the metabolism of plants and animals. <u>Lavoisier</u> was also the first experimenting work physiologist. With his colleague, <u>Séguin</u>, he performed experiments in which the subject lifted weights by stamping on a pedal. He measured the work output over a 15 min period in foot-pounds and the concomitant consumption of "air vital" (i.e. oxygen) in cubic inches. He compared results on a fasting subject with those obtained on the same subject in the absorptive state after a meal, and he took notice of the temperature in the room and apparently also had a helper taking the pulse rate of the subject. Converted to present day terms, his subject worked at a rate of approximately 100 kpm/min with an oxygen uptake of about 1.0 l O_2/min - a not unreasonable result considering the work conditions. At rest his subject consumed about 0.4 l O_2/min.

<u>Lavoisier</u> thus was the first to show definitely that work of the human organism increased the combustion in the body, and his experimental procedures could be accepted even to-day. He was a most brilliant scientist, far ahead of his times, and his meaningless death in the hands of political fanatics probably delayed the development of work physiology very considerably.

Concomitantly with the research on the exchange of substances by the living and working organism a development of the ideas concerning the heat output took place. <u>Crawford</u>, a Scotchman working in Glasgow, had already in 1778 measured the heat output of small animals by means of a water calorimeter. His ideas about combustion of inflammable substances such as wax, oils and tallow were, however, dominated by his adherence to the phlogiston theory which held that fire is a distinct substance. Although he showed that there was a close correlation between heat output and CO_2 production he still believed that the heat stemmed from the "elementary fire" contained in the air and set free by its combination with the combustible substances.

<u>Lavoisier</u>, working with <u>de la Place</u>, at about the same time, did corresponding experiments on guinea pigs and concluded that the heat was produced by a combination of the same kind as that taking place when coal is burning, only it progressed very slowly. He believed that this combustion took place in the lungs and that the substance burned was an "air fixe" brought to the lungs by the blood. It should take a long time before <u>H.G. Magnus</u>, working in Leipzig, in 1837 came to the conclusion that the combustion took place not only in the lungs but in the capillaries of the entire body.

During the unruly first half of the 19th century which saw the end of the Napoleonic wars and their aftermath of political upheavals rather few and scattered observations on muscle metabolism were made. Most of the investigators found it enough to register the changes in the carbon dioxide percentage of the expired air and only a few were able to measure quantitatively the CO_2-output. The German, Vierordt, is probably the best known name from this epoch. He was able to show that after a brisk walk the CO_2-output was increased by 80 ml per minute (1845).

But then around the middle of the century some very important steps forward were taken, greatly contributing to the understanding of what metabolism is. One of these was the formulation by the German, J.R. von Mayer in 1845, of the principle later to be known as "the law of the conservation of energy". This law was an important weapon in the polemic against the adherents of the idea of a "vital-force" that could step in and suspend or again set free the processes in the living body. The renown nutritionist Justus von Liebig, among others, was a supporter of this idea, which von Mayer and a little later von Helmholz (1861) vigorously attacked. It was, however, not until 1893 when Max Rubner made his classical experiments on dogs that the law of the conservation of energy was finally accepted as valid for the animal body. In the following years the American, W.O. Atwater, working at Wesleyan University in Connecticut with the expiration calorimeter verified and extended Rubner's results on human subjects.

During the last half of the 19th century it became generally accepted, that the metabolic processes in rest and work to a great extent were combustions. New methods for measuring accurately both the uptake of oxygen and the output of CO_2 were developed, often combined with calorimetric measurements of the heat produced. Also more reliable ways of determining the external work were designed, and various ergometers, including both the bicycle ergometer and the motor driven treadmill were taken into use. For this period such names as Pettenkofer and Voit, Zuntz and Geppert of Germany, Chauveau and Tissot of France, Sondén and Tigerstedt of Sweden and Finland are synonymous with new apparatus and basic research in exercise physiology. The problems that were studied, and in some cases solved, were the quantitative relations of exercise to metabolism, including the efficiency of the human machine, but also the qualitative questions of what material was actually catabolized during exercise. Besides studying the effect of ordinary positive work, both static exercise and negative work were considered, the last named type by for instance Chauveau (1896) in experiments on ascending and descending stairs.

The question of what substances - carbohydrate alone, or fat and carbohydrate mixed - were the deliverers of the energy set free during work divided opinions. (Proteins had been eliminated as a possible source by the studies of urinary N-output by Pettenkofer and Voit in 1862). Zuntz in Germany and his adherers believed that work was produced on the cost of the same mixture of nutrients as during the preceding rest period, whereas Chauveau of France and his school maintained that carbohydrate was the exclusive source of muscular energy. Both schools based their opinions on measurements of the respiratory quotient, RQ, during work. The experience of physiologists was now so advanced, that the usual pitfalls in determining true values of RQ - i.e. lactate formation, and hyper- or hypoventilation - were known and could be avoided. Chauveau and his co-workers in work experiments that exhausted the subject in little more than an hour, found that at least at the end of the work period the RQ was near unity, whereas Zuntz and his students found - in work that was not too great and could be carried out for a long time without excessive fatigue - that the RQ was very close to the values during rest and even would decrease if the work period was very long. Chauveau further maintained that if fat was used as fuel it must be converted into carbohydrate first, and this conversion, if it followed the route suggested by Chauveau, would cause a loss in energy of about 30%. Such a decrease in efficiency was not observed by any of the two competing schools, and the question as to whether carbohydrate alone, or fat and carbohydrate combined delivered the energy for muscular work was thus left undecided.

This was the situation at the turn of the century. The following 30-40 years saw partly a continuation of the discussion of which substances were burned in the exercising human subjects, partly the beginning of a line of research in biochemistry that used isolated muscles and minced muscles ("Muskelbrei" in German) to solve the problems of muscle metabolism. These two lines of research, working towards the same final goal, for a long time seemed to run in parallel, and only towards the end of the period serious attempts were made of integrating the human physiology line with that of the muscle biochemists.

On the "human physiology line" there are 3 great series of experiments which helped clear away the discrepancies between the earlier results of the Chauveau-school and the Zuntz-school. These are Benedict and Cathcart (USA): Muscular work (1913), Krogh and Lindhard (Denmark): The relative value of fat and carbohydrate as sources of muscular energy (1920) and Christensen and coworkers (Denmark): Investigations on heavy muscular work (1936).

Benedict and his coworkers at the Carnegie Institute of Washington studied bicycling on the bicycle ergometer constructed by Benedict (a forerunner of Krogh's better known ergometer) at different rates of exercise and on diets that were either rich or poor in carbohydrates. They found that the RQ during exercise varied with the diet, being lower when the body's carbohydrate reserves presumably were low; and further that RQ increased over that at rest with increasing severity of exercise. This latter observation was in 1930 elaborated by Dill and coworkers at the Harvard Fatigue Laboratory in Boston into the equation:
$RQ = 0.04 \times (metabolic\ rate) + 0.817$.

For Krogh and Lindhard the main concern was to reconcile the new theories of Fletcher, Hopkins, A.V. Hill a.o. that lactic acid formation from carbohydrate or carbohydrate-like substances were the immediate energy sources of muscular contraction, with the findings and ideas of Zuntz, that muscles can burn both fat and carbohydrate. If fat was used they, therefore, supposed that it must somehow be transformed into carbohydrate - as suggested by Chauveau - and this transformation must cost some energy, although not necessarily as much as the 30% predicted by Chauveau.

Krogh and Lindhard's experiments were performed with the utmost care, and especially their gas analytical methods had an accuracy that later hardly has been equalled, not to speak of surpassed. From their results they concluded, that both at rest and during exercise fat and carbohydrate can be utilized as fuel,

the ratio fat/carbohydrate being a function of the
available supplies of these substances. They further
found that when fat was catabolized, supposedly first
after being transformed into carbohydrate, about 11%
of the caloric value of the fat was lost.

This latter loss in efficiency when fat is used
as fuel for muscular exercise has later been confirmed
by several investigators, and it has cost biochemists
many speculations to find out how it could be reconciled with the various possible roads of transformation.

World War I brought a temporary deceleration to
the international exchange of ideas, but the end of
the hostilities and the ensueing attempts at preventing
future wars by international cooperation such as the
shaping of the League of Nations, gave new impetus to
the study of work physiology. Under the auspices of
the League of Nations, Christensen and his coworkers,
students of Krogh and Lindhard, continued and expanded
the research program of their teachers with greater
understanding of the importance of standard conditions
and steady state in their various work experiments.
The publication of their results in 1939 solved the
old Zuntz-Chauveau controversy. Their main conclusions
were: Both fat and carbohydrate can be used as fuel
for muscular work. The ratio fat/carbohydrate depends
on several factors: In light to moderate exercise of
short duration, mainly on the diet. In long continued
moderate exercise the relative fat utilization increases gradually, due to the reduction of the carbohydrate stores. In exercise of increasing intensity
the relative carbohydrate utilization begins to increase from a certain intensity of work, approaching
100% at maximal work.

The work intensity at which the shift to an increasing carbohydrate utilization took place depended
on the working capacity of the subject and on the
working capacity of the muscle group that performed
the work. With the same subject it happened at a
lower O_2-uptake in arm-work than in leg-work. This
break-point coincided with the intensity at which
lactate acid concentration began to increase in the
blood. On a fat diet they found a greatly diminished
capacity for prolonged exercise of moderate intensity,
and further that this was due to hypo-glycaemia
apparently caused by the depletion of the glycogen
stores in the liver. The liver glycogen, therefore,

was assumed to be the main source of carbohydrate for muscular exercise in fully aerobic conditions. The role of muscle glycogen was tacitly assumed to be that of a depot for anaerobic work only, leading to formation of lactic acid.

These crucial observations were not published in detail until 1939 just before the outbreak of World War II - and they were published in German. When after the war research on these problems was resumed, English had become the scientific language, and probably for these reasons they did not obtain the wide observation they deserved.

It is now necessary to go back again to the beginning of this century, when a new line of research in muscle metabolism took its beginning. In 1907 Fletcher and Hopkins published an article in J.Physiol. (London) entitled: Lactic acid in amphibian muscle.

Lactic acid had been known for a long time as a metabolite in muscles but its origin was uncertain. One hypothesis (Hermann 1867) had it that a "giant" molecule, called inogen by Hermann, by an explosive combination with inbuilt oxygen gave rise to a formation of both lactic acid and carbon dioxide when the muscle contracted. In the recovery period oxygen was taken up and again built into the inogen. Fletcher and Hopkins, using an improved technique for extracting substances from isolated muscles, showed that the lactic acid is formed from glycogen stored in the muscle and that oxygen is used in recovery, solely for combustion of carbohydrate, leaving nothing to be stored in an intra-molecular form.

Fletcher and Hopkin's observations were followed by numerous others, not least by those of Otto Meyerhof and his school in Germany, and formed the basis for a hypothesis that, by combining the chemical, the thermodynamical, and the mechanical events known at that time, came to dominate muscle physiology for a long time. A.V. Hill, who integrated his own thermodynamic studies with those of the biochemists, summarized in 1924 the events in an anaerobic muscle contraction in the following recognisable steps:

A. glycogen disappears,
B. lactic acid appears in equivalent amounts,
C. preformed CO_2 is driven off,
D. heat is produced proportional to the lactic acid

formed,
E. The hydrogen ion concentration rises.

It was believed that the explosive formation of lactic acid by direct effect on some "Verkürzungsorten" (Meyerhof) evoked the mechanical response. Under aerobic conditions, oxygen was taken up in the recovery period and used to restore glycogen by burning about one fifth of the lactic acid to CO_2 and water.

The caloric equivalents for the various steps in the contraction and relaxation processes were measured and with some auxiliary assumptions brought to fit the direct measurements of heat production and mechanical energy. Only some few observations did not readily fit into the Hill-Meyerhof theory: Embden and coworkers demonstrated in 1921 that phosphate was liberated when lactate was produced and, therefore, assumed that hexose diphosphate was the precursor of lactic acid. And the American, W.O. Fenn, had shown that more heat was produced if the muscle performed work than if it worked isometrically or was lengthened during contraction. As Hill wrote in 1924: "Fenn has provided us with some very difficult problems".

In the mid-twenties of this century the Hill-Meyerhof theory seemed to explain practically all the main changes in metabolism during exercise: The sudden formation of lactic acid from glycogen, as the direct cause of the mechanical events; and the aerobic resynthesis of the muscle glycogen by combustion of part of the lactic acid. A "steady state" could be attained as a dynamic equilibrium between these two processes, and the "oxygen debt" paid after the cessation of work was naturally explained as the slow return to resting conditions from this steady state of equilibrium. Meyerhof and Hill shared the Nobel prize in physiology in 1922.

Meanwhile, in other laboratories, but not least in Meyerhof's institute, work was progressing on lines that soon would bring the beautiful theory tumbling down: Embden in Germany had already pointed out that phosphate liberation took place in connection with muscle contractions. Now Eggleton and Eggleton in Britain and Fiske and Subbarow in USA almost simultaneously (1927-28) discovered creatin-phosphate (CrP) as a substance in muscles, and showed that its concentration decreased during contractions and again increased in the recovery phase. Meyerhof and Lohmann in 1928 measured the energy liberated by the hydrolysis of one Mole of CrP to 12 kcal. A stunning blow to the lactic acid theory was the discovery made by Lundsgaard, working in Meyerhof's laboratory in 1930, that muscles poisoned by monoiodo-acetic acid could perform a series of anaerobic contractions without formation of lactic acid. The development of mechanical tension or of work was found to be proportional to the breakdown of CrP to Cr and P. This process, therefore, was assumed to be the direct precursor of the mechanical events, while the formation of lactic acid from carbohydrates took place after a delay and in order to rebuild CrP from the split products.

In nearly the same years another energy rich phosphate-compound was discovered: adenosin-triphosphate (ATP) by Lohmann in Germany and by Fiske and Subbarow in USA.

Meyerhof and Lohmann measured the energy set free by hydrolysis of ATP to ADP and P to 12 kcal per Mole. And most important: Lohmann found that CrP would not split into its two components unless ADP was present to take up the P. In muscle it means, that CrP can not deliver energy to the contraction mechanism, until ATP has been split into ADP and P. Everything thus seemed to point to ATP as the immediate energy source for the mechanical events.

Finally, just before the second World War in 1939, the step that linked the biochemical discoveries of the importance of the phosphate-compounds to the mechanical events in the muscle fibres was taken: Engelhardt and Ljubimowa could show that myosin, known as a constituent of the muscle fibrils, also had enzymatic powers. It acts as an ATPase, thus enabling the further energy transfer from CrP to ADP (the Lohmann-reaction) to take place consecutively to muscle contraction.

The gross metabolic processes, leading to muscle contractions and work, thus could be outlined: ATP delivers energy directly to the contractile mechanism by splitting into ADP and P. It is resynthetized either anaerobically from CrP - which again may be resynthetized from glycolysis, leading to lactic acid formation, - or aerobically from the energy liberated in the oxygenation of the foodstuffs in the muscle fibre.

An attempt from these years at bringing the new ideas in line with observations on humans in exercise is Margaria, Edwards and Dill's discussion of the oxygen debt in 1933. These authors could show that the original explanation of Hill that the oxygen debt was an expression of the slow elimination of the lactic acid, could not be maintained. There was after work of all intensities an oxygen debt, that was unrelated to the appearance of lactate in the blood, and only after strenuous exercise was there a further increase in the oxygen debt that was correlated with lactic acid formation. The "alactacid" oxygen debt was suggested to be related to the resynthesis of "Phosphagen", the name adopted at that time for CrP and ATP combined.

The years after the second World War saw immense progress in the understanding of muscle metabolism. This was to a degree due to the development of methods that enabled scientists to analyse infinitesimal small amounts of substances, even when in loco in the cells. The electron-microscope brought new evidence of the intra-cellular structures - the actin-myosin filaments, the mitochondria, the membranes etc. and gradually our present ideas of muscle metabolism evolved. The central role of the "energy rich phosphate-bonds", as they had been christened by Lipmann in 1941, was realized and accepted.

In parallel with the growing of our understanding of the specific muscle metabolism a tremendous development taking place in the biochemistry of the normal foodstuffs, the carbohydrates and the fats had taken place. Glycolysis, lactic acid production, the processes in the Krebs cycle and in the respiratory chain had been elucidated, and also here it was found that ATP was the key substrate that could preserve and transfer the energy set free by the enzymatic processes.

The problem for human work physiologists now was to unite the new biochemical and histochemical knowledge with the known features of metabolism of exercising man, with the concepts of oxygen deficit, oxygen debt, steady state, anaerobic and aerobic condition, RQ, lactate and pyruvate levels, blood glucose, and free fatty acids. But it was no longer enough to study the respiratory exchange of oxygen and carbon dioxide and the blood concentration of certain metabolites. It was necessary to know <u>where</u> the various steps in the catabolism of the foodstuffs took place, <u>where</u> the actual stores of fat and carbohydrates were located, how and in what form the metabolites were moved to or from the sites where the actual chemical processes took place. For the study of these problems new tools had to be developed and taken into use. A most important method was the use of labeled compounds, notably C^{14}-compounds. Further, new techniques were used in exercising man, for instance the methods for catheterization of arteries and veins to individual muscles and organs, and the technique of making muscle biopsies in resting and exercising human subjects. These two techniques, combined with the refined analytical methods for enzymes and substrates, have done much to close the gap between those working on minced muscles or muscle extracts and those working with the whole human organism.

The catheterization technique which has been extensively used in USA by for example <u>Rowell</u> and others, in Germany by <u>Keul</u>, <u>Doll</u> a.o. and in Sweden by <u>Pernow</u> a.o. has opened our eyes for the relative importance of substrates in the muscle cells, and substrates brought to the muscle cells by the blood stream. Measurements of the a-v differences of such substances as glucose, free fatty acids, lactate and pyruvate in resting and exercising muscles, in liver and other organs, have done much in the way of elucidating the fate of such metabolites.

The muscle biopsy method as developed and extensively used by <u>Bergström and Hultman</u> and other Swedish investigators has conclusively shown the importance of muscle glycogen and of CrP and ATP stores in the muscle fibres for both anaerobic and aerobic exercise.

I shall not go into details with respect to all the new evidence that has been published within the last few years. Much of this has to be checked and rechecked before it can fit into the great mosaic of muscle metabolism in exercise. Many problems are still unsolved, or little understood, some of them old, dating back to the very first days of exercise physiology. The general outline of this mosaic was already laid out at the time of <u>Zuntz</u>. But this picture was only a rough draft. As in all true mosaics the single stones have to be selected and formed and put into place before the beauty of the whole work becomes apparent and this is what is going on today. To use a more materialistic metaphor: Earlier, in order to find out what was boiling, physiologists could only watch the smoke that rose from the chimney. These days, they can look directly into the pot. May this symposium help us to recognize and identify some of the ingredients of the stew!

This historical outline is not based on intensive studies of original literature - time did not allow that. As sources I have rather extensively used older and newer text books, monographs and proceedings. For this reason no literary references are supplied, and only authors that I have deemed of special importance have been mentioned by name and year of their contribution to the present topic.

HISTOCHEMICAL EFFECTS OF CONTRACTION ON RED AND WHITE RAT MUSCLE FIBRES. THEIR ORGANIZATION IN THE MOTOR UNIT

Eric Kugelberg and Lars Edström
Department of Neurology
Karolinska sjukhuset, Stockholm 60, Sweden

A cross-section of the anterior tibial muscle of the albino rat presents a mosaic pattern of fibres of different enzymatic activity. On the basis of the activity of mitochondrial enzymes such as succinic dehydrogenase, three different types of fibres have been distinguished (1). The A fibres show very weak activity, the B fibres intermediate and the C fibres strong activity. The A fibres are the classical white muscle fibres and the B and C two types of red fibres. The C fibres stain more strongly for myoglobin than do the B fibres and thus show a higher degree of redness. As regards glycogen no consistent difference between fibre types occur. The results of histochemical and biochemical studies indicate that the white muscle fibres depend mainly on glycolysis for their energy supply, whereas the red fibres to a high degree depend on oxidation of carbohydrates and lipids.

If now the nerve to the anterior tibial muscle is stimulated with repetitive supramaximal shocks it will be found that the A fibres are depleted of glycogen at surprisingly low stimulation frequencies, as judged from the PAS reaction (2). The fibres become PAS negative already at a frequency of 5 c/sec. after some 2000 contractions. At this stage the muscle is fatigued in proportion to the area occupied by these fibres, that is twitch tension has declined about 50 %. At a stimulus frequency of 10/sec. almost all B fibres will be PAS negative within 10 minutes. The C fibres, however, show rather an increase in the PAS reaction. This overshoot may disappear after longer periods of stimulation, but the PAS reaction remains as strong as before stimulation, even after two hours of stimulation. Stimulation with higher frequencies does not deplete the C fibres, as tested up to an hour. High frequencies will, however, rapidly cause a neuromuscular block which will reduce the metabolic strain on the fibres. Thus in the A and B fibres the break-down of glycogen occurs rapidly and is not compensated by resynthesis even at low frequency stimulation. The C fibre glycogen content is increased or unaffected within a wide range of stimulation frequencies. Break-down and resynthesis are in balance even during long periods of stimulation. The different effects of contraction on the glycogen content of various types of fibres have been observed in the pigeon pectoralis (3), but little seems to have been known of its effect on mammalian muscle. This is of some interest for those studying glycogen metabolism in whole muscle or muscle biopsies with biochemical methods, since at least in man striated muscles are composed of both white and red fibres. The change in glycogen content in such muscles may well under certain circumstances be confined to only one part of the population of muscle fibres.

The changes in the histochemical phosphorylase reaction are as impressive as the changes in PAS. The strongly active A fibres become negative at about the same time as the PAS reaction becomes negative. The less strongly active B fibres when stimulated at a frequency of 10/sec. become first strongly active and then negative together with PAS negativity. The weakly active C fibres rapidly become almost as strongly active as unstimulated A fibres and remain so for some time. After 1-2 hours stimulation the phosphorylase reaction in the different types of fibres has returned to about the same level as before stimulation.

The striking changes in phosphorylase activity do not tally with the common view that the histochemical method of Takeuchi and Kuriaki (4) and its modifications demonstrate the total phosphorylase content of the fibres. Biochemical studies have shown that total phosphorylase does not change during contractions. The changes observed would be less surprising if the histochemical method demonstrated active rather than total phosphorylase. The active form is known to increase during contractions and disappear on fatigue (5). Accordingly, activation of phosphorylase in the C fibres in the early stage of contractions would induce an increased rate of glycogen break-down which, if so, is compensated by an increased rate of glycogen resynthesis, since glycogen increases rather than decreases. It has been assumed, however, that the phosphorylase system is interlocked with the main resynthetic enzyme, the glycogen synthetase system in such a way that activation of phosphorylase is coupled with inactivation of synthetase (6, 7). The obvious differences between different types of fibres observed with histochemical methods have, however, so far been little explored by biochemists.

The clear difference in the reactions of different types of fibres is more or less abolished if the muscle contracts during ischemia. The break-down of glycogen in B and C fibres is speeded up. All fibres become PAS negative at about the same time in connection with profound muscular fatigue. Furthermore, there is little change in phosphorylase activity. This situation probably occurs during strong isometric contraction, which impedes the blood flow as a result of a rise in intramuscular pressure. The rapid disappearance of glycogen in the C fibres shows the latent glycogenolytic capability of these fibres.

Whatever the underlying biochemical mechanism may be, the histochemical changes serve as a powerful tool in the study of the motor unit. Stimulation of a single motor nerve fibre of the anterior root supplying the muscle displays a histological map of the unit (2, 8, 9, 10, 11). The fibres lie scattered singly or a few together. Perhaps more important is the fact that the motor unit is composed of one type of fibre, as shown both in the rat (8, 10) and in the cat (11). Homogeneity of the motor unit is a prerequisite for the organization of the different types of fibres in various types of movements according to their functional capacity.

One physiological property which is closely associated with redness is fatiguability in the anterior tibial muscle of the rat. The A fibre motor units fatigue after 1500-2000 contractions at 5-10/sec., that is, a frequency of contraction encountered in tonic activity responsible for posture. C fibre units show no fatigue at these frequencies and the B units a spectrum of intermediate reactions. It is thus clear that the C units are designed for tonic activity and the A units for phasic activity. The metabolic differences in the different types of units must therefore also be reflected in the over-all metabolic changes induced by different types of exercise.

REFERENCES

1) Stein, J.M. and Padykula, H.A. (1962). Histochemical classification of individual skeletal muscle fibers of the rat. Amer. J. Anat., 110, 103-123.

2) Kugelberg, E. and Edström, L. (1968). Differential histochemical effects of muscle contractions on phosphorylase and glycogen in various types of fibres: relation to fatigue. J. Neurol., Neurosurg., Psychiat., 31, 415-423.

3) George, J.C. and Nene, R.V. (1965). The effect of exercise on the glycogen content of the red and white fibres of the pigeon pectoralis muscle. J. Anim. Morphol. Physiol., 12, No. 2, 246-248.

4) Takeuchi, T. and Kuriaki, H. (1955). Histochemical detection of phosphorylase in animal tissues. J. Histochem. Cytochem., 3, 153-160.

5) Cori, G.T. (1945). The effect of stimulation and recovery on the phosphorylase a content of muscle. J. biol. Chem., 158, 333-339.

6) Larner, J. (1966). Hormonal and nonhormonal control of glycogen metabolism. Trans. N.Y. Acad. Sci., 29, 192-209.

7) Staneloni, R. and Piras, R. (1969). Changes in glycogen synthetase and phosphorylase during muscular contraction. Biochem. Biophys. Res. Commun., 36, No. 6, 1032-1038.

8) Edström, L. and Kugelberg, E. (1968). Histochemical composition, distribution of fibres and fatiguability of single motor units. Anterior tibial muscle of the rat. J. Neurol., Neurosurg., Psychiat., 31, 424-433.

9) Brandstater, M.E. and Lambert, E.H. (1969). A histological study of the spatial arrangement of muscle fibers in single motor units within rat tibialis anterior muscle. Bulletin. American Association of Electromyography and Electrodiagnosis. Vol. 15-16, 82.

REVERSAL OF ENZYMATIC PROFILES AND CAPILLARY SUPPLY OF MUSCLE FIBERS

IN FAST AND SLOW MUSCLES AFTER CROSS INNERVATION

Flaviu C.A. Romanul

Neurological Unit, Boston City Hospital, and Department

of Neurology, Harvard Medical School, Boston, U.S.A.

During the past decade histochemical studies have contributed much to our understanding of skeletal muscle. As a result of these investigations it has become possible to bring together in a meaningful way some very basic observations from anatomy, physiology, and biochemistry. It is well known by now that the mammalian skeletal muscles are composed of a number of fiber types which differ in enzymatic activity. In every type of fiber the activities of the enzymes of glycolysis are inversely proportional to the activities of the enzymes of oxidative and lipid metabolism[1,2,3]. These differences indicate that some muscle fibers derive their energy of contraction mostly by anaerobic glycolysis while others obtain it chiefly by oxidative breakdown of lipids and other compounds.

There are several histochemical aspects of muscle which should be reviewed in this meeting on muscle metabolism during exercise. With regard to the number and nomenclature of the different histochemical types of fibers existing in skeletal muscle, some workers distinguished two, naming them I and II[4]. Other investigators described three, designating them as A, B, C[5]. We found a larger number of fiber types but did not label them[2]. For ease of reference we prefer to divide them into two groups and call them "glycolytic" or "oxidative" depending upon their predominant enzymatic activity.

Histochemical studies are most advantageiously carried out on serial cross sections of "sandwich blocks" composed of red and white skeletal muscles and a fragment of heart (Figure 1). After incubating the sections for different enzymes one can study the enzymatic profiles of individual fibers, as well as the population of fibers in each muscle. A red muscle such as the soleus, known to contract

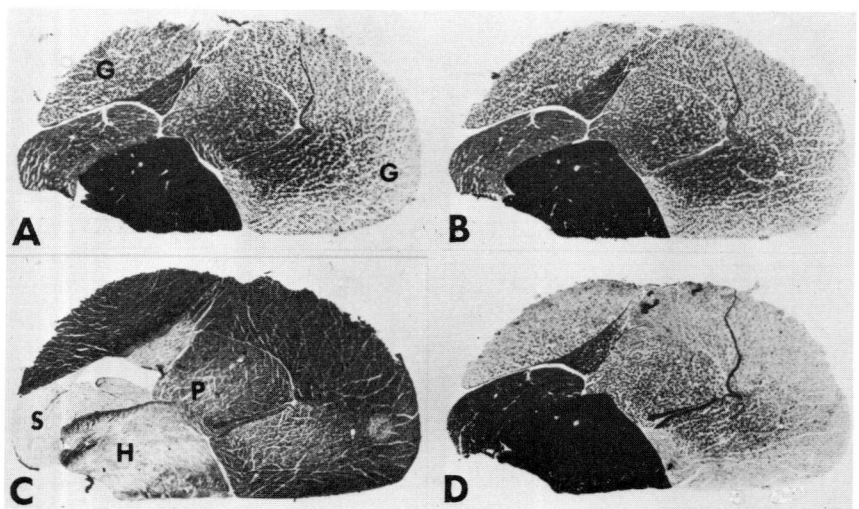

Fig. 1 Serial cross sections of a "sandwich block" consisting of gastrocnemius (G), plantaris (P), soleus (S), and heart (H) muscles incubated for cytochrome oxidase (A), succinic dehydrogenase (B), mitochondrial alpha-glycerophosphate dehydrogenase (C), and beta-hydroxybutyric dehydrogenase (D).

slowly, is composed almost entirely of oxidative fibers. The white muscles like the plantaris and gastrocnemius have fast contraction characteristics and consist of a majority of glycolytic fibers. It is important to note that in the white muscles the oxidative fibers are not evenly scattered but are more concentrated in the axial portions of the muscle heads near the tendons of insertion[2]. The existence of such an arrangement indicates a special physiological role of the oxidative fibers in white muscles. From the practical standpoint one can envisage the serious errors which may be made when taking muscle biopsies for chemical analysis. Different values may be obtained by sampling various parts of the same muscle, and even if the biopsies are confined to a single muscle head, the results will depend upon the depth of sampling[3].

Concerning the blood supply of the muscle fibers it has been known since the time of Ranvier[6] that the white muscles have fewer capillaries than the red muscles, but it has always been stated or assumed that every fiber comes in contact with an equal number of capillaries. Recent investigations have shown that the number of capillaries around each muscle fibers is directly proportional to the oxidative metabolism of the fiber[7].

From the above considerations it is apparent that the white, glycolytic muscle fibers are ideally suited for contractions which are necessary at all cost, as in fight or flight, since these fibers are almost self sufficient during contraction. They store their substrate in the form of glycogen which is degraded without consumption of oxygen. The meager capillary supply is needed only for the removal of the lactic acid formed and for the replenishment of the glycogen used. By contrast, the red, oxidative fibers have an energy metabolism similar to that of the cardiac muscle which is designed for repeated or continuous contractions for prolonged periods of time. With such requirements the fibers can not store enough substrate and they can not afford to break it down incompletely leaving the other tissues in the body to continue the biochemical work. Consequently these fibers must have constant supply of substrate and oxygen from the circulating blood, a need which is reflected in the large number of capillaries surrounding them.

There are several other lines of evidence indicating that the red, oxidative muscle fibers are used for prolonged work. 1) The red muscles or red portions of white muscles are important in reflexes concerned with the maintenance of posture[8]. 2) In the white muscles the oxidative fibers are concentrated in the axial portion of the muscle or muscle heads[2]. This is the location expected if the oxidative fibers are to contract effectively by themselves. 3) The arrangement of the three calf muscles which insert on the Achilles tendon, with the soleus situated deepest and the gastrocnemius most superficially, probably has the same physiological meaning[2]. 4) The red fibers are capable of maintaining their contraction for prolonged periods of time while the white fibers fatigue very easily[9,10]. 5) In the calf muscles of unanesthetized cats there is a decrease in blood flow confined to the red soleus during sleep and an increase in blood flow limited to the white gastrocnemius during the defense reaction[11]. 6) The patients with McArdle's syndrome, who lack muscle phosphorylase, develop painful cramps and contractures of muscles after rapid repetitive movements but are entirely asymptomatic if sitting or standing all day. Thus it appears that the continuous contraction of the muscles necessary for the maintenance of posture is carried out by the oxidative fibers.

The available histochemical and physiological data indicate that there is good correlation between the metabolism of the muscle fibers, their speed of contraction, and the physiological characteristics of the nerve fibers responsible for their innervation. Soleus, which consists almost entirely of oxidative fibers, receives only slowly conducting axons[11] and the speed of contraction of its individual motor units is slow having a unimodal distribution[12]. Gastrocnemius which is composed mostly of glycolytic but also of some oxidative fibers is innervated by fast axons as well as a few slowly conducting ones[11]. The motor units of the gastrocnemius have a bimodal distribution of contraction times with a large group of rapid units and a smaller group of slow ones[13]. As stated earlier white muscles have several histochemical fiber types[8]. Correspondingly, the motor nerves to such muscles contain fibers with a spectrum of conduction velocities which correlate with the speeds of contraction of their motor units[14]. Recently the motor units were shown to consist of single histochemical type of muscle fibers[10]. This complex organization of the nerves and muscles must have important meaning in terms of the activation of fibers with different energy metabolism during various types of muscular work.

If a muscle is denervated, all the fibers show a progressive diminution in the activities of the enzymes of energy metabolism. However, the glycolytic fibers lose more rapidly glycolytic enzymes and the oxidative fibers oxidative enzymes so that the metabolic differences between the fibers gradually decrease and tend to disappear[15,16]. Thus it becomes evident that after denervation the muscle fibers undergo a metabolic dedifferentiation.

Such findings inevitably raise the question whether the preferential energy metabolism of the muscle fibers is determined by their nerve supply. This problem was investigated by reinnervation and cross-innervation experiments[17,18] similar to those performed by Buller et al[19]. Young and adult rats and cats were placed under pentothal anesthesia and the nerves to soleus (representing a slow muscle) and flexor digitorum longus (FDL) or flexor hallucis longus (FHL) (representing fast muscles) were exposed and sectioned in one hind limb. In some animals the nerves were surgically reunited while in others they were cross-united. Four to ten months later the muscles were investigated physiologically recording their contraction characteristics in response to electrical stimulation of the nerves proximal to the site of suture. The corresponding muscles in the contralateral hind limb of the operated and non-operated animals served as controls. After the physiological studies were completed, the muscles from both hind limbs were removed and apposed to form a "sandwich block". The block was frozen and serial cross sections cut in a cryostat were incubated for the histochemical demonstration of a number of enzymes.

Compared to the normal counterparts the reinnervated muscles showed no change in the speed of contraction and histochemically they were composed of the same proportion of fibers of the appropriate histochemical types (Fig.2). By contrast, the cross-innervated muscles reversed their speed of contraction and the enzymatic characteristics of their fibers (Fig. 3). Thus, the high oxidative and low glycolytic enzymatic profile of the soleus muscle fibers was changed to the low oxidative and high glycolytic pattern of the normal FDL or FHL. Converse changes occurred in the FDL or FHL. The results proved that the energy metabolism of the muscle fibers is determined by the nerve supply.

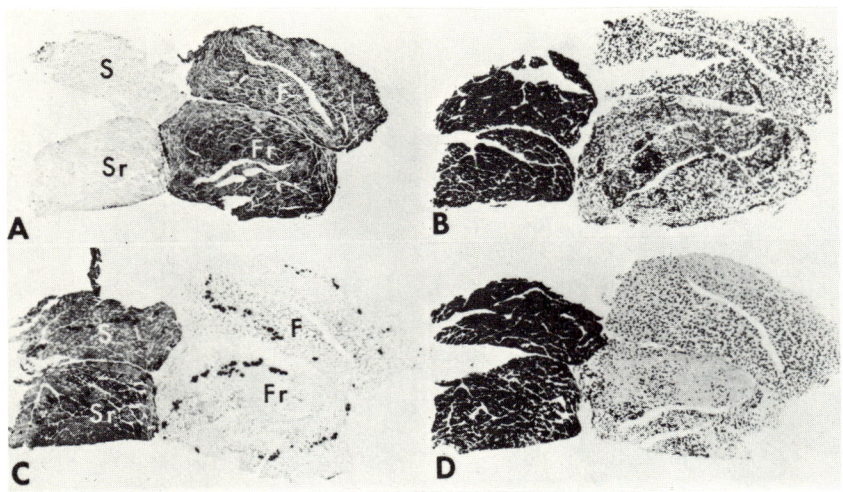

Fig. 2 Serial cross sections of a "sandwich block" of normal and reinnervated rat muscles: normal soleus (S), reinnervated soleus (Sr), normal FDL (F), and reinnervated FDL (Fr). The sections are incubated for mitochondrial alpha-glycerophosphate dehydrogenase (A), succinic dehydrogenase (B), esterase (C), and beta-hydroxybutyric dehydrogenase (D).

Several cross-innervated muscles were found to have some reinnervation by the parent nerve in spite of the precautions taken at the time of surgery. In such doubly innervated muscles stimulation of the soleus nerve caused the muscle to contract slowly, while stimulation of the FDL nerve elicited a fast contraction from the same muscle. Histochemically the muscle showed some areas with a population of fibers characteristic of soleus and other areas typical of FDL.

In normal muscles the fibers of different histochemical types were scattered among each other. In the cross-innervated and reinnervated muscles the fibers of the same histochemical type were arranged in small groups. These findings revealed that the innervation pattern of muscle fibers after regeneration of the nerve is different from that which occurs during development. A hypothesis was formulated to explain the difference. During normal development the arrival of the axons at the muscle can be expected to occur in a close time relationship. The muscle fibers nearest the tips of the incoming axons are the first to be innervated. As collaterals develop, they are unable to innervate adjacent fibers which have already been innervated by neighboring axons. Therefore, they grow until they reach fibers which have not yet been innervated. In

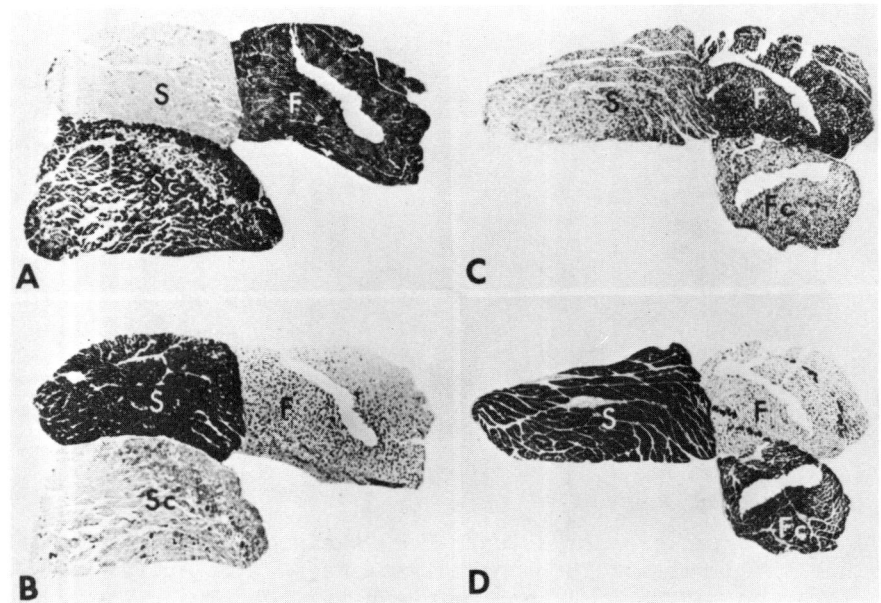

Fig. 3 Serial cross sections of a "sandwich block" of normal and cross-innervated rat muscles: normal soleus (S), cross-innervated soleus (Sc), normal FDL (F), and cross-innervated FDL (Fc). The sections are incubated for mitochondrial alpha-glycerophosphate dehydrogenase (A and C), beta-hydroxybutyric dehydrogenase (B), and esterase (D).

this manner a wide scatter and overlap of the motor units occurs. By contrast, the regenerating axons have to follow new and sometimes tortuous pathways in the sutured nerve and reach the muscle at different times. The first arriving axon can supply an entire group of adjacent muscle fibers which have not yet been innervated. Each subsequent axon can act in similar manner.

The distribution of capillaries in the normal, reinnervated, and cross-innervated muscles was also investigated[21]. The capillaries were rendered visible by the alkaline phosphatase reaction of their endothelium. In order to study the relationship between the oxidative metabolism and the capillary supply of the muscle fibers, serial cross sections of the "sandwich blocks" of muscles were incubated for succinic dehydrogenase and alkaline phosphatase. In the normal soleus, which consisted of oxidative fibers, each fiber was surrounded by many capillaries. In the normal FDL a dense network of capillaries was present only around the scattered oxidative fibers (Fig. 4). The reinnervated muscles showed the same direct

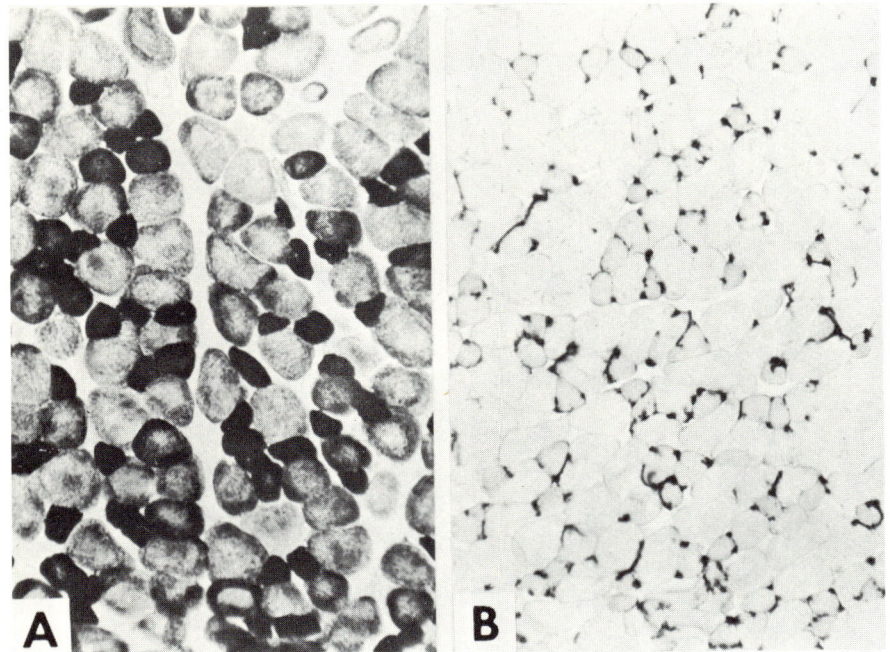

Fig. 4 Cross sections of normal FDL muscle of cat incubated for succinic dehydrogenase (A) and alkaline phosphatase (B).

relationship between the oxidative metabolism and the capillary density of the muscle fibers (Fig. 5). However, a reinnervated muscle differed from its normal counterpart in that the fibers of the same histochemical type occurred in small groups. The cross-innervated muscles showed a marked change in the oxidative enzymatic activity of the fibers. A soleus muscle cross-innervated by the FDL nerve was enzymatically indistinguishable from a reinnervated FDL muscle and had a corresponding distribution of capillaries. Conversely, a FDL muscle cross-innervated by the soleus nerve appeared enzymatically like a normal or reinnervated soleus muscle and had a large number of capillaries around each fiber. In a muscle which was partially cross-innervated and partially reinnervated by its parent nerve, large areas composed of oxidative fibers with dense capillary networks were juxtaposed to areas consisting of glycolytic fibers with poor capillary supply (Fig. 6).

The results of this study show that the changes in the energy metabolism of the muscle fibers induced by cross-innervation are accompanied by a corresponding change in the density of the capillary

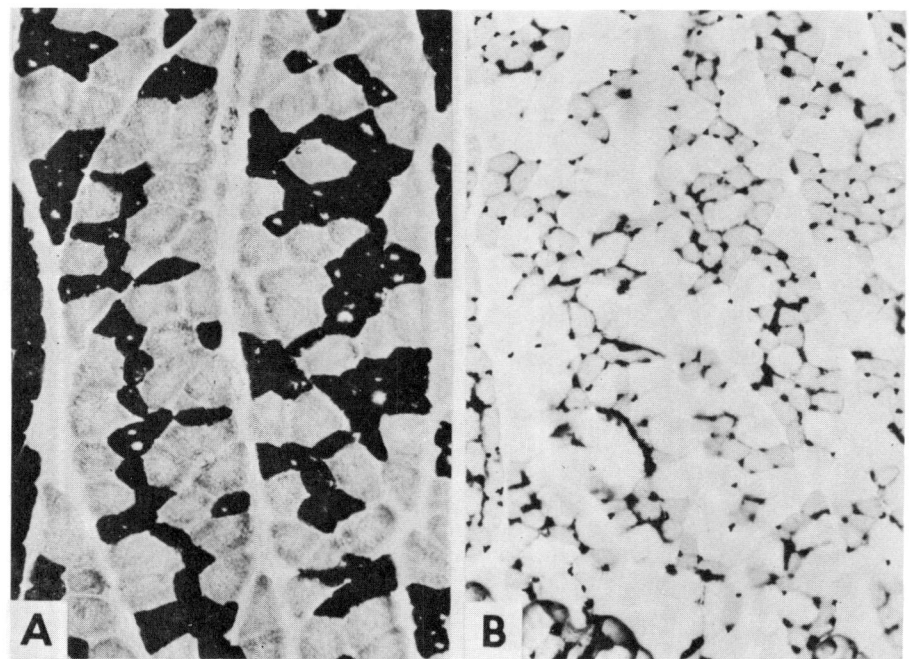

Fig. 5 Cross sections of reinnervated FDL muscle of cat incubated for succinic dehydrogenase (A) and alkaline phosphatase (B).

networks around the fibers. The number of capillaries surrounding each fiber changes in such a way as to parallel the modifications in the oxidative metabolism of the fiber. Thus, the capillary system is a very plastic and dynamic one adjusting continuously to the local metabolic needs. In order to test whether biochemical changes induced by others means in the muscle fibers are also accompanied by such a change in capillaries, other studies were carried out. Calf muscles of rats rendered thyrotoxic by administration of triiodothyronine were compared histochemically with the corresponding muscles of control animals[21]. On serial sections of "sandwich blocks" containing these muscles the muscle fibers of the thyrotoxic rats were found to have a pronounced increase in oxidative enzymatic activity and a corresponding increase in the density of the capillaries.

The means whereby the motor nerves determine the physiological properties and the energy metabolism of the muscle fibers is unknown. According to one theory the motor nerves supply hypothetical "specific trophic substances", different for fast and slow muscles[22].

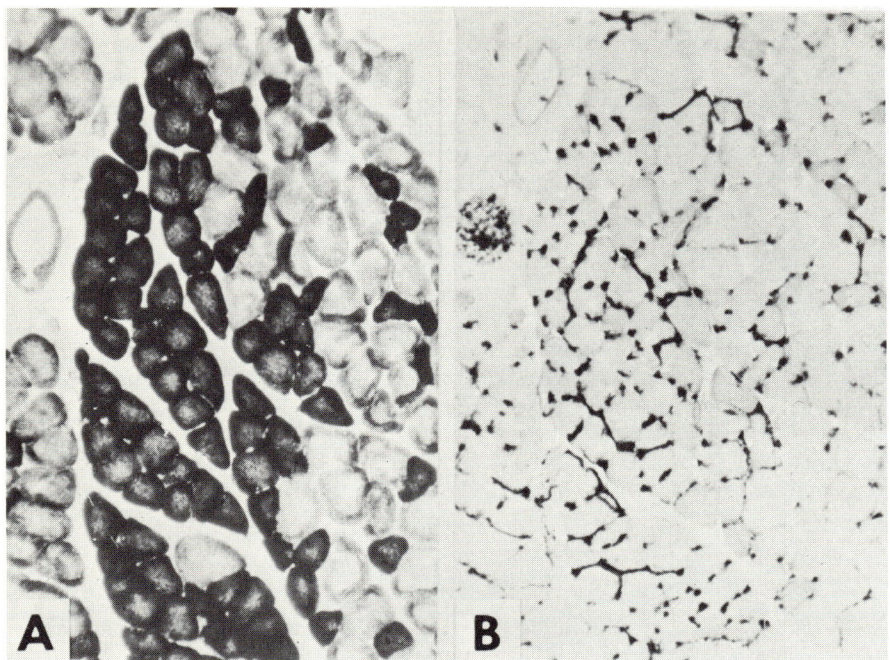

Fig. 6 Cross sections of cat FDL muscle partially cross-innervated by the soleus nerve and partially reinnervated by its original nerve. The sections are incubated for succinic dehydrogenase (A) and alkaline phosphatase (B).

An alternate proposal is that the neural influences on both types of muscle are mediated only through acetylcholine, which is the single known neuromuscular transmitter. This possibility received considerable support from a histochemical study in which the effects of Clostridium botulinum toxin on muscle were compared with those of denervation[23]. Calf muscles of rats were paralyzed by repeated intramuscular injections of C. botulinum toxin. Other rats had the calf muscles denervated for the same length of time by sciatic nerve excision. Histochemically the enzymes of energy metabolism showed identical changes in the C. botulinum-treated and in the denervated muscles. Since the C. botulinum toxin is well established to block the release of acetylcholine from the nerve terminals but not to have a direct effect on muscle, the findings prove that cholinergic transmission is essential for the maintenance of the enzymatic differences between red and white muscle fibers. It is very tempting to conclude that the energy metabolism of the muscle fibers is determined by the axons only through the release of acethylcholine, different metabolic pathways being induced in muscle by different rates of axonal discharge. However, one can not exclude the possibility that the findings in this experiment may have been due to the blocking by the C. botulinum toxin of as yet undiscovered misterious substances in the nerve terminals. Consequently, a crucial experiment remains to be done in order to determine whether the nerve's trophic influences on muscle are mediated through acetylcholine alone or through "mysterines".

Aknowledgements

The author's studies were supported by a Research Grant (NB-02603) from the National Institute of Neurological Disease and Blindness, U.S. Public Health Service.

The illustrations are reproduced by the kind permission of the American Medical Association and Little, Brown and Company, Boston.

References

1. Dubowitz, V., and Pearse, A.G.: Comparative Histochemical Study of Oxidative Enzymes and Phosphorylase Activity in Skeletal Muscle, Histochem. 2:105-117, 1960.

2. Romanul, F.C.A.: Enzymes in Muscle: I. Histochemical Studies of Enzymes in Individual Muscle Fibers, Arch. Neurol. 11:355-368, 1964.

3. Dawson, D.M., and Romanul, F.C.A.: Enzymes in Muscle: II Histochemical and Quantitive Studies, Arch. Neurol. 11:369-378, 1964.

4. Engel, W.K.: Essentiality of Histo- and Cytochemical Studies of Skeletal Muscle in Investigations of Neuromuscular Disease, Neurology (Minneap.) 12:778-784, 1962.

5. Stein, J.M., and Padykula, H.A.: Histochemical Classification of Individual Skeletal Muscle Fibers of Rat, Amer. J. Anat. 110:103-123, 1962.

6. Ranvier, L.: Note sur les vaisseaux sanguins et la circulation dans les muscles rouges, Arch. Physiol. Norm. Pathol. 1:446, 1874.

7. Romanul, F.C.A.: Capillary Supply and Metabolism of Muscle Fibers, Arch. Neurol. 12:497-509, 1965.

8. Denny-Brown, D.: On the Nature of Postural Reflexes, Proc. Roy. Soc. 104B:252-301, 1929.

9. Kugelberg, E., and Edström, L.: Differential Histochemical Effects of Muscle Contractions on Phosphorylase and Glycogen in Various Types of Fibers: Relation to Fatigue, *J. Neurol. Neurosurg. Psychiat.* 31:415-423, 1968.

10. Edström, L., and Kugelberg, E.: Histochemical Composition, Distribution of Fibers and Fatiguability of Single Motor Units, *J. Neurol. Neurosurg. Psychiat.* 31:424-433, 1968.

11. Eccles, J.C., Eccles, R.M., and Lundberg, A.: The Action Potentials of the Alpha Motoneurones Supplying Fast and Slow Muscles, *J. Physiol.* 142:275-291, 1958.

12. McPhedran, A.M., Wuerker, R.B., and Henneman, E.: Properties of Motor Units in a Homogeneous Red Muscle (Soleus) of the Cat, *J. Neurophysiol.* 28:71-84, 1965.

13. Wuerker, R.B., McPhedran, A.M., and Henneman, E.: Properties of Motor Units in a Heterogeneous Pale Muscle (M. Gastrocnemius) of the Cat, *J. Neurophysiol.* 28:85-99, 1965.

14. Bessou, P., Emonet-Denand, F., and Laporte, Y.: Relation entre la vitesse de conduction des fibres nerveuses motrices et le temps de contraction de leurs unités motrices, *C.R. Acad. Sci.* 256:5625-5627, 1963.

15. Romanul, F.C.A., and Hogan, E.L.: Enzymatic Changes in Denervated Muscle: I. Histochemical Studies, *Arch. Neurol.* 13:263-273, 1964.

16. Hogan, E.L., Dawson, D.M., and Romanul, F.C.A.: Enzymatic Changes in Denervated Muscle: II. Biochemical Studies, *Arch. Neurol.* 13:274-282, 1964.

17. Romanul, F.C.A., and Van Der Meulen, J.P.: Reversal of the Enzyme Profiles of Muscle Fibres in Fast and Slow Muscles by Cross-Innervation, *Nature* 212:1369-1370, 1966.

18. Romanul, F.C.A., and Van Der Meulen, P.: Slow and Fast Muscles After Cross-Innervation. Enzymatic and Physiological Changes, *Arch. Neurol.* 17:387-402, 1967.

19. Buller, A.J., Eccles, J.C., and Eccles, R.M.: Interactions Between Motoneurones and Muscles in Respect of the Characteristic Speeds of Their Responses, *J. Physiol.* 150:417-439, 1960.

20. Close, R.: Effects of Cross-Union of Motor Nerves to Fast and Slow Skeletal Muscles, *Nature* 206:831-832, 1965.

21. **Romanul**, F.C.A., and Pollock, M.: "The Parallelism of Changes in Oxidative Metabolism and Capillary Supply of Skeletal Muscle Fibers". In Locke, S (ed): <u>Modern Neurology</u>, <u>Papers in Tribute to Derek Denny-Brown</u>, Boston: <u>Little,Brown & Co.</u>, 1969 pp 203-213.

22. Eccles, J.C.: The Effects of Nerve Cross-Union on Muscle Contraction, In Milhorat AT (ed): <u>Exploratory Concepts in Muscular Dystrophy</u>, <u>Excerpta Medica Foundation</u>, 1967, pp. 151-160.

23. Drachman, D.B., and Romanul, F.C.A.: Effects of Neuromuscular Blockade on Enzymatic Activities of Muscles, <u>Arch. Neurol.</u> 23:85-89, 1970.

METABOLIC DIFFERENTIATION OF DISTINCT MUSCLE TYPES AT THE LEVEL OF ENZYMATIC ORGANIZATION

Dirk Pette

Fachbereich Biologie, Universität Konstanz

Metabolic differentiation of muscle tissue may be understood from the viewpoints of quantitative and qualitative adaptation. Quantitatively, it is the consequence of a balance of input of chemical energy and output of mechanical work in the myofibrillar apparatus. Qualitatively, it is the expression of an adjustment to the functional characteristics of the muscle, such as the quality and temporal pattern of energy expenditure (e.g. steady and continuous, steady and discontinuous or dicontinuous performance of work. The kinds of metabolic differentiation possible are based on the fact that muscle cells may draw energy from the oxidation of various fuels. The caloric values of the various fuels determine their different theoretical maximum energy-yields. The nature of the fuel also determines the type of its metabolism, especially with regard to an obligatory or nonobligatory aerobic catabolism. Energy-output in cell metabolism depends finally on the catabolic rate. Nature of the fuel, type of metabolism and catabolic rate thus represent fundamental elements in metabolic differentiation.

Energy-supplying metabolism in muscles of different type or species is based on the function of homologous enzymes. However, main principles of metabolic differentiation emerge from comparative analysis of enzyme activity patterns [1-4]. The activity of an enzyme determined under standardized conditions may be considered as a measure of enzyme concentration. The following limitations must, however, be taken into account: minor species or organ dependent variations in the molecular activity of an individual enzyme as well as possible variations in its isoenzyme pattern are reflected at the level of enzyme activity. Identical concentrations of an individual enzyme in different tissues therefore do not necessarily correspond to strictly identical activities or vice versa. By no means, enzyme activities as measured at optimum conditions in vitro, can be set equal to physiological flux rates in vivo. They may, however, be regarded as relative measures of metabolic capacities in comparative analysis.

An example is given in Table 1 which compares activity levels of selected key enzymes of energy-supplying metabolism in flight muscles of honey bee and silk moth. Each of the enzymes listed represents a definite metabolic segment or system. A main similarity of the two muscles consists in the deficiency of lactate dehydrogenase, a common characteristic of insect flight muscles, as was shown first by ZEBE and McSHAN [5]. The absence of a capacity of anaerobic energy-supply in insect flight muscles may be understood as a consequence of the rich and direct oxygen supply of these muscles by tracheolar respiration. Regardless of the fuel used, energy-supplying metabolism of insect flight muscles is thus strictly based on aerobic catabolism. Aerobic catabolism is certainly a necessity for maintained high energy expenditure.

Table 1

ACTIVITY LEVELS OF KEY ENZYMES OF ENERGY-SUPPLYING METABOLISM IN INSECT FLIGHT MUSCLES

		APIS MELLIFERA (U/g w.w.)	BOMBYX MORI (U/g w.w.)
GLYCOGEN PHOSPHORYLASE	(EC 2.4.1.1)	8.3	0.6
HEXOKINASE	(EC 2.7.1.1)	43	1.8
TRIOSE-P DEHYDROGENASE	(EC 1.2.1.12)	685	90
LACTATE DEHYDROGENASE	(EC 1.1.1.27)	1	1.2
c-GLYCEROL-P DEHYDROGENASE	(EC 1.1.1.8)	220	13
m-GLYCEROL-P DEHYDROGENASE	(EC 1.1.99.5)	110	11
CITRATE SYNTHASE	(EC 4.1.3.7)	215	130
3-HYDROXYACYL-CoA DEHYDROGENASE	(EC 1.1.1.35)	< 1	280
ADENYLATE KINASE	(EC 2.7.4.3)	82	590
ARGININE KINASE	(EC 2.7.3.3)	35	160

This is especially emphasized in these muscles, since prior to the functional development of the flight muscles, an anaerobic energy-supplying capacity exists [6].

Obviously, the main difference between the two flight muscles reflects the nature of the fuel for energy-supply. Considering the activity levels of 3-hydroxyacyl-CoA dehydrogenase, the capacity of bee flight muscle for oxidizing fatty acids is negligible or even zero, while in the flight muscle of the silk moth fatty acid oxidation appears to be the main pathway of energy-supplying metabolism. Since flight muscles of other insects are known which are capable of oxidizing fat as well as carbohydrate [7,8], the two muscles compared in Table 1 represent extreme specializations, for example due to differing conditions of life and nourishment. The strictly aerobic catabolism of carbohydrate in the flight muscle of honey bee is based not only on high activities of enzymes of glycogenolysis, glycolysis and citric acid cycle, it is also linked to a high capacity of the glycerolphosphate cycle [9]. Thus, the two glycerolphosphate dehydrogenases involved in the extra-intramitochondrial hydrogen transfer are found at very high activities in flight muscle of the bee and are present only at low concentrations in the flight muscle of the silk moth.

With regard to the preponderance of fatty acid oxidation in the flight muscle of the silk moth, it is interesting that the activity of adenylate kinase is much higher in this muscle than in the flight muscle of the bee. A possible explanation of this finding may be seen in the necessity of re-phosphorylating the adenylic acid which accumulates in connection with the activation of fatty acids. It is clear, however, that this explanation is incomplete, since the role of adenylate kinase in cell metabolism is not linked only to this function. As is evident from Table 1, the activity of arginine kinase is also higher in the flight muscle of the silk moth than in the flight muscle of the honey bee. The different activity levels of these two phosphotransferases involved in the balance of the energy-stores of phosphagen and the ATP/ADP/AMP system are probably related to a different role of these stores of energy-rich phosphates in the initiation of contraction in the two flight muscles.

As has been outlined elsewhere [3,4], metabolic differentiation at the level of enzymatic organization

Table 2

ENZYME ACTIVITY RATIOS IN FLIGHT MUSCLES OF HONEY BEE AND SILK MOTH

	APIS	BOMBYX
$\dfrac{\text{Triose-P dehydrogenase}}{\text{3-Hydroxyacyl-CoA dehydrogenase}}$	> 685	0.3
$\dfrac{\text{3-Hydroxyacyl-CoA dehydrogenase}}{\text{Citrate synthase}}$	< 0.005	2.15
$\dfrac{\text{Hexokinase} \times 10}{\text{Citrate synthase}}$	2.0	0.14
$\dfrac{\text{Glycogen phosphorylase} \times 10^2}{\text{Triose-P dehydrogenase}}$	1.2	0.7
$\dfrac{\text{c-Glycerol-P dehydrogenase}}{\text{m-Glycerol-P dehydrogenase}}$	2.0	1.2
$\dfrac{\text{m-Glycerol-P dehydrogenase}}{\text{Triose-P dehydrogenase}}$	0.16	0.12

is reflected not only by differences in the absolute levels of enzyme activities but is even more apparent upon comparing enzyme activity ratios. Activity ratios of some of the enzymes compared in Table 1 are presented in Table 2. The analysis of these values refers actually not only to activity ratios but reveals in some cases ratios of different metabolic systems. It has been shown in earlier investigations that essential segments of energy-supplying metabolism [e.g. glycolysis, citric acid cycle, fatty acid oxidation, repiratory chain) are characterized by enzyme groups of constant proportions [1,2]. Each of these groups represents a more or less invariable unit of organization and may therefore be envisaged as representative of the respective metabolic system. At the level of enzymatic organization, the ratios of these groups reflect thus relations of metabolic systems. Numerical values of these ratios do of course not represent ratios of flux rates *in vivo*. They may, however, be taken as relative measures in a comparative study. Thus, constancy of these ratios indicates very probably a constant order of the respective metabolic systems at the level of molecular organization. Variations, on the other hand, may be regarded as the expression of metabolic differentiation. The activity ratios of triosephosphate dehydrogenase, citrate synthase and 3-hydroxyacyl-CoA dehydrogenase which are listed in Table 2, represent "system relations" of glycolysis, citric acid cycle and ß-oxidation of fatty acids. The numerical values of these system relations vary largely in the two muscles, and emphasize thus differences in the metabolic type already mentioned.

Table 3

ACTIVITY LEVELS OF KEY ENZYMES OF ENERGY-SUPPLYING METABOLISM IN RED AND WHITE RABBIT SKELETAL MUSCLE

		M. SOLEUS	M. PSOAS	ACTIVITY RATIO
		ENZYME ACTIVITY		M. PSOAS / M. SOLEUS
		(U/g w.w.)	(U/g w.w.)	
Glycogen phosphorylase	(EC 2.4.1.1)	2.70	45	16.7
Hexokinase	(EC 2.7.1.1)	1.50	0.33	0.22
Triose-P dehydrogenase	(EC 1.2.1.12)	199	1730	8.7
Lactate dehydrogenase	(EC 1.1.1.27)	113	1320	11.7
Hexosediphosphatase	(EC 3.1.3.11)	0.06	1.89	31.5
m-Glycerol-P dehydrogenase	(EC 1.1.99.5)	0.22	1.42	6.5
Citrate synthase	(EC 4.1.3.7)	7.1	1.2	0.17
3-Hydroxyacyl-CoA dehydrogenase	(EC 1.1.1.35)	7.4	1.3	0.18
Cytochrome a	(EC 1.9.3.1)	6.5*	1.3*	0.20
Adenylate kinase	(EC 2.7.4.3)	60	790	13.2
Creatine kinase	(EC 2.7.3.2)	228	350	1.5

* 10^{-9} M/g w.w.

On the other hand, it is obvious that similarities also exist. Thus, the activity ratios phosphorylase/triosephosphate dehydrogenase, extramitochondrial glycerolphosphate dehydrogenase/mitochondrial glycerolphosphate dehydrogenase and mitochondrial glycerolphosphate dehydrogenase/triosephosphate dehydrogenase are very similar in both muscles.

So far, two muscles of similar functional characteristics have been compared. Maintained high energy expenditure of these muscles was shown to be based on a strictly aerobic substrate catabolism due either to the oxidation of carbohydrate or fat. In the following, the study of metabolic differentiation will be extended to a comparison of two muscles of different functional type. These are "white" or fast and "red" or slow mammalian skeletal muscle.

Activities of key enzymes in energy-supplying metabolism in two typical representatives of red and white muscle are listed in Table 3. The values illustrate the well known differences in the activity levels of enzymes involved in anaerobic and aerobic metabolic systems (e.g. [10,11]). Thus, white muscle is characterized by high activities of glycogenolytic and glycolytic enzymes and relatively low activities of the enzymes of citric acid cycle, fatty acid oxidation and respiratory chain. Red muscle, on the contrary, reveals relatively low levels of glycogenolytic and glycolytic enzymes and relatively high activities of the enzymes of citric acid cycle, fatty acid oxidation and respiratory chain. It appears thus that energy-supplying metabolism in white muscle is based mainly on anaerobic carbohydrate catabolism, whereas aerobic substrate oxidation is predominant in red muscle. With regard to this suggestion, the question arises, whether the comparable energy expenditure of white and red skeletal muscle can be related to such pronounced differences in the type of energy-supplying metabolism.

Under the simplifying assumption that carbohydrate is the main fuel in both muscle types, the answer to this question depends on two points: 1. the different quantitative energy-yields of anaerobic and aerobic carbohydrate catabolism, and 2. the different catabolic rates in carbohydrate metabolism of white and red muscle. With regard to the theoretical maximum energy-yields, complete oxidation of glucose (38 moles ATP/mole of glucose), gives a 19-fold higher ATP yield than does lactic fermentation (2 moles ATP/mole of glucose). In the case that glucose is derived from glycogen, its aerobic catabolism (39 moles ATP/mole of glucose)leads to a 13-fold higher ATP yield than its anaerobic break-down (3 moles ATP/mole of glucose). These differences may be compensated by reciprocal differences in the glycolytic rate. At the level of enzymatic organization, corresponding differences should then exist in the glycolytic capacity or in the activity levels of the glycolytic enzymes.

An inspection of the data in Table 3 reveals that neither white nor red muscle can be envisaged as exclusively "anaerobic" or "aerobic" tissue. Both muscles contain activities of enzymes involved in anaerobic and aerobic carbohydrate catabolism. It is evident also that carbohydrate is not the only fuel for energy-supply, since the enzyme activity pattern of both muscles shows a capacity for fatty acid oxidation. However, the aerobic capacities of the two muscles compared vary greatly and the differences existing in the activity levels of key glycogenolytic and glycolytic enzymes are therefore significant with respect to the points discussed before: The activity of glycogen phosphorylase in white muscle is higher than that in red muscle by a factor of about 17. Triosephosphate dehydrogenase activity is higher in white than in red muscle by a factor of about 9. These differences indi-

Table 4

VARIABLE OR DISCRIMINATIVE ENZYME ACTIVITY RATIOS IN RED AND WHITE RABBIT SKELETAL MUSCLE

	M. SOLEUS	M. PSOAS
$\dfrac{\text{Glycogen phosphorylase}}{\text{Hexokinase}}$	1.8	136
$\dfrac{\text{Hexokinase} \times 10^3}{\text{Triose-P dehydrogenase}}$	7.5	0.2
$\dfrac{\text{Hexosediphosphatase} \times 10^2}{\text{Hexokinase}}$	4.0	572
$\dfrac{\text{Triose-P dehydrogenase}}{\text{3-Hydroxyacyl-CoA dehydrogenase}}$	27.0	1330
$\dfrac{\text{Triose-P dehydrogenase}}{\text{Citrate synthase}}$	28.0	1440
$\dfrac{\text{m-Glycerol-P dehydrogenase} \times 10^2}{\text{Cytochrome a}}$	6.8*	218*
$\dfrac{\text{Adenylate kinase}}{\text{Creatine kinase}}$	0.3	2.3

* ($10^3 \times \text{min}^{-1}$)

cate the different capacities of glycogenolysis and glycolysis in these two muscles. As a matter of fact, the higher activity of triosephosphate dehydrogenase and the other glycolytic enzymes [10] in white muscle corresponds almost exactly to its higher glycolytic throughput when compared with red muscle. Thus, measurements in tetanically stimulated white and red rabbit skeletal muscle have shown that the rate of lactate production in white muscle exceeds that in red muscle by a factor of 7.7 [12]. Calculations of the metabolic throughput at the enolase step as performed by BÜCHER and SIES [13], gave a 8.6 fold higher glycolytic flux in maximally stimulated white than red rabbit skeletal muscle.

It is remarkable that the activity of hexokinase shows a different distribution than glycogen phosphorylase. Its activity in white muscle amounts to only one fifth of its activity in red muscle. Carbohydrate catabolism appears thus to be based mainly on glycogenolysis in white muscle. A further characteristic of this muscle is its relatively high activity of hexosediphosphatase, which is present in red muscle in only negligible quantities. The distribution of hexosediphosphatase is the inverse of that of hexokinase. On the other hand, this distribution resembles that of mitochondrial glycerolphosphate dehydrogenase and a coordinated function of these two enzymes has therefore been suggested [4,14,15]. Especially in white muscle, this might serve to reform glucose from triosephosphate which is generated by oxidation of glycerolphosphate accumulating during contraction.

Differences between energy-supplying metabolism of white and red muscle consist obviously not only in different "anaerobic" and "aerobic" metabolic capacities, but apply also to other characteristics. The full extent of these differences is illustrated by the values of enzyme activity ratios which are given in Table 4, and have been calculated from the data in Table 3. The extremely varied relations of the metabolic systems of glycolysis, citric acid cycle and fatty acid oxidation are reflected by the activity ratios triosephosphate dehydrogenase/3-hydroxyacyl-CoA dehydrogenase and triosephosphate dehydrogenase/citrate synthase. Similar differences exist also in the values which indicate the differentiation of the two muscles with regard to special characteristics of carbohydrate metabolism (e.g. glycogen phosphorylase/hexokinase, hexokinase/triosephosphate dehydrogenase, hexosediphosphatase/hexokinase).

Finally, white and red muscle vary with regard to the ratios adenylate kinase/creatine kinase and mitochondrial glycerolphosphate dehydrogenase/cytochrome a [10]. This latter ratio is of special interest, since it points to a main difference of the mitochondria in white and red muscle: Although white muscle has a lower cellular content of mitochondria, as illustrated by its lower cellular concentration of enzymes involved in "standard" mitochondrial functions (e.g. citric acid cycle, fatty acid oxidation, oxidative amino acid metabolism, respiratory chain), its mitochondria contain a specifically higher activity of glycerolphosphate dehydrogenase. With regard to the function of this enzyme in the glycerolphosphate cycle for transfer of extramitochondrial ("glycolytic") hydrogen to the respiratory chain, this higher glycerolphosphate dehydrogenase activity in the glycolytically more active white muscle was interpreted in terms of a "compensation phenomenon" [2,10].

It is evident from Table 4 that each of the activity ratios listed represents a "discriminative index". As was shown elsewhere [3,4], these variable ratios make it possible to elucidate metabolic differentiation not only for white and red muscle but also in the case of intermediate forms where there are varied proportions of white and red fibres in "mixed" muscles.

Fig.1

Microscope-photometric activity test for succinate dehydrogenase. Bipositional recording in a red (r) and intermediate (i) fibre within the same 5μ cross-section of soleus muscle of the rat.

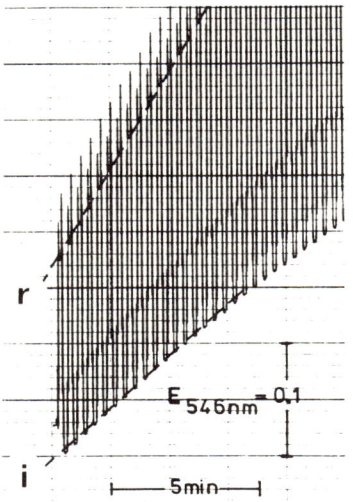

It may be stated, however, that the variations of these ratios are due to individual characteristics of the respective fibre types. White and red fibres represent thus only extreme examples of possible differentiation. Especially in mixed skeletal muscle, the overall metabolic type of the muscle may not only result from the proportion of white and red but also from the share of the "intermediate" fibres.

The intermediate fibre type has been characterized enzymologically mainly by qualitative histochemical techniques (e.g. [16-20]). Fig.1 illustrates preliminary results [21] which confirm the qualitative histochemical data and suggest that a quantitative characterization of the enzyme activity pattern of this fibre type is possible. Fig. 1 presents reproductions of original test charts as obtained by comparative activity determinations of succinate dehydrogenase in a cross-section of red and intermediate fibres in rat soleus muscle. The measurements were performed with a microscope-photometric technique [22] which was developed from the previously described gel-film method [23,24]. As may be seen, there exist significant differences in the activity levels of succinate dehydrogenase, and the activity ratio intermediate: red fibre is 1:1.7 as calculated from the recorded reaction rates.

In spite of the fundamental variations discussed so far, energy-supplying metabolism in muscle is linked to common characteristics. At the level of enzymatic

Table 5

CONSTANT OR NON-DISCRIMINATIVE ENZYME ACTIVITY RATIOS IN RED AND WHITE RABBIT SKELETAL MUSCLE

	M. SOLEUS	M. PSOAS
$\dfrac{\text{GLYCOGEN PHOSPHORYLASE} \times 10^2}{\text{TRIOSE-P DEHYDROGENASE}}$	1.36	2.60
$\dfrac{\text{M-GLYCEROL-P DEHYDROGENASE} \times 10^3}{\text{TRIOSE-P DEHYDROGENASE}}$	1.10	0.82
$\dfrac{\text{LACTATE DEHYDROGENASE}}{\text{TRIOSE-P DEHYDROGENASE}}$	0.57	0.76
$\dfrac{\text{3-HYDROXYACYL-CoA DEHYDROGENASE}}{\text{CITRATE SYNTHASE}}$	1.04	1.08
$\dfrac{\text{CITRATE SYNTHASE}}{\text{CYTOCHROME A}}$	2.18*	1.85*
$\dfrac{\text{HEXOKINASE} \times 10}{\text{CITRATE SYNTHASE}}$	2.12	2.74

* ($10^3 \times \text{min}^{-1}$)

organization, these common characteristics have been shown to be reflected by constant or non-variable ratios of definite enzymes [3,4]. Examples concerning white and red rabbit skeletal muscle are given in Table 5. The similarity of the listed activity ratios is obvious. It is emphasized by the marked differences in the absolute levels of the respective enzymes (Table 3) and is contrasted by the extreme variations of those activity ratios presented in Table 4. Glycogenolytic and glycolytic capacities are obviously correlated quantatively in muscle metabolism. This correlation is in agreement with the accepted role of glycogen as the primary substrate of carbohydrate catabolism in muscle.

A corresponding relation may also be deduced for glycolysis and the glycerolphosphate cycle (glycerolphosphate dehydrogenase/triosephosphate dehydrogenase). Quantitatively, this ratio is lower by two orders of magnitude in white and red muscle than in the two insect flight muscles compared in Table 2. This difference is certainly due to the fact that deficiency of lactate dehydrogenase is compensated by the high activity of the glycerolphosphate cycle in insect flight muscle. In vertebrate muscle, on the contrary, carbohydrate catabolism is facultatively anaerobic and glycolytic substrate oxidation at the triosephosphate dehydrogenase step is coupled not only to the glycerolphosphate cycle but also to the function of lactate dehydrogenase. Hydrogen transfer occurs thus not only

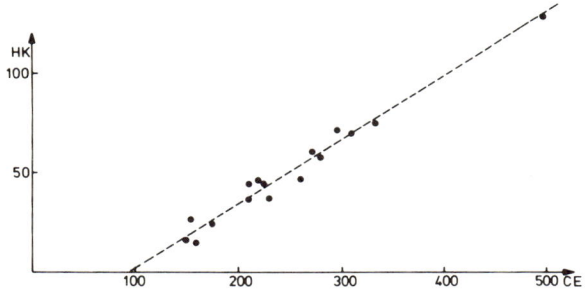

Fig. 2 Correlation of hexokinase (HK) and citrate synthase (CE) activities in various muscles of the human. Enzyme activities were plotted as [μM/h x g w.w.].

into the mitochondrial but also into the extracellular space. The role of lactate dehydrogenase in red skeletal muscle may, however, not be coupled solely to lactic fermentation. The predominance of the H-type isoenzyme in red muscle (e.g. [11]) suggests that this enzyme functions also in oxidizing lactate. Probably this also applies to the oxidation of exogenous lactate. The contiguity of white and red fibres in mixed skeletal muscle might be related to an oxidation by red muscle of lactate produced in white muscle.

A further common characteristic may be seen in the constant activity ratios of key enzymes of the citric acid cycle, fatty acid oxidation and the respiratory chain. These constant system relations reflect identical features of metabolic organization in mitochondria of white and red muscle. As has been shown elsewhere [3,4], this constancy is of almost general validity and exists in muscles of very different types. Thus, the numerical value of the activity ratio 3-hydroxyacyl-CoA dehydrogenase/citrate synthase in white and red muscle is very similar to that found in the flight muscle of the silk moth (Table 2) or of the desertlocust [4]. In the flight muscle of the honey bee with its metabolism specialized for carbohydrate oxidation, this ratio is of course quite different. On the other hand, the activity ratio glycogen phosphorylase/triosephosphate dehydrogenase in flight muscle of the bee (Table 2) agrees well with the value found in vertebrate skeletal muscle (Table 5). This holds also for the activity ratio hexokinase/citrate synthase which was shown to be comparable or identical in a large variety of different muscles [3,4,25]. A comparison of the activity levels of hexokinase and citrate synthase in various human skeletal muscles in Fig. 2 [26] provides further evidence for the constancy of this system relation.

It is of special interest that the activity level of the enzyme responsible for glucose phosphorylation is found in constant relation with the activity level of citrate synthase or other enzymes of the citric acid cycle. As is obvious from Table 5, this constancy holds also for the investigated key enzymes of the respiratory chain. The phenomenon of constant proportions of hexokinase and key enzymes of substrate end-oxidation reflects thus, at the level of enzymatic organization, a quantitative correlation between the capacity of fuel input into energy-supplying metabolism with the capacity of "energy-output" of this system. Changes in this relation which were observed under the influence of thyroid hormones, apparently justify this interpretation [27].

It appears therefore that metabolic differentiation is limited to variations of only certain magnitudes which are reflected at the level of enzymatic organization by the variable or discriminative enzyme activity ratios. With the exception of certain cases of metabolic specialization, constant or non-variable enzyme activity ratios or system relations are not subject to metabolic differentiation. This conclusion

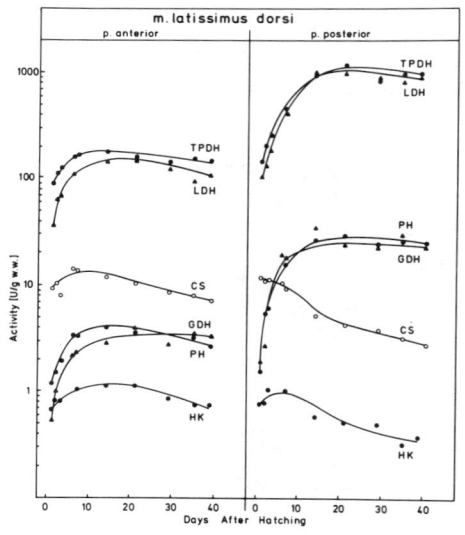

Fig.3
Enzyme activities in the anterior and posterior part of the latissimus dorsi muscle during postnatal development of the chicken [28].
Abbreviations: CS = citrate synthase, GDH = α-glycerolphosphate dehydrogenase, HK = hexokinase, LDH = lactate dehydrogenase, PH = glycogen phosphorylase, TPHD = triosephosphate dehydrogenase.

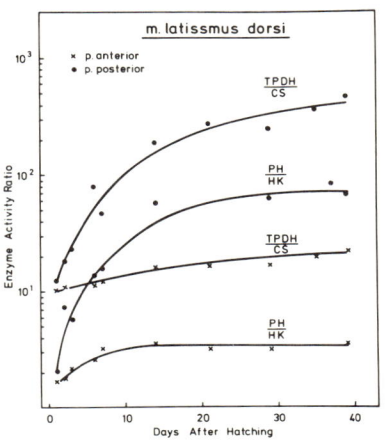

Fig. 4
Discriminative enzyme activity ratios in the anterior and posterior part of the latissimus dorsi muscle of the chicken during postnatal development. For abbreviations see legend of Fig.3.

is fully supported by the following findings.

Fig.3 illustrates the differentiation of enzyme activity patterns in two distinct muscle types of the chicken during postnatal development [28]. In the adult animal, the anterior part of m.latissimus dorsi represents a typical "slow" muscle, whereas the posterior part is a "fast" muscle [29]. As may be seen from Fig.3, the postnatal differentiation of the enzyme activity patterns of these two muscles implies the same characteristics which have been also shown by comparison of different muscles to be representative of metabolic differentiation. Constant proportion enzyme activities increase in parallel, e.g. glycogen phosphorylase, triosephosphate dehydrogenase, extramitochondrial glycerolphosphate dehydrogenase and lactate dehydrogenase. A similar parallel exist in the levels of hexokinase and citrate synthase. Consequently, postnatal differentiation of the enzyme activity patterns of these two muscles leads to a change in enzyme activity ratios which have been classified as discriminative. Changes of the discriminative ratios glycogen phosphorylase/triosephosphate dehydrogenase and triosephosphate dehydrogenase/hexokinase are illustrated in Fig.4.

Concomitant changes in the enzyme activity pattern of energy-supplying metabolism may also be observed in the case of induced changes of the functional type of muscle. An example is demonstrated in Fig.5 which shows typical changes in the enzyme activity pattern of

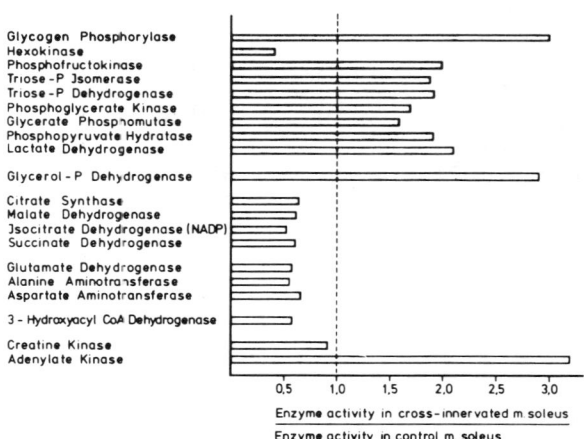

Fig.5 Effects of cross-innervation on the enzyme activity pattern of energy-supplying metabolism in the soleus muscle of the rabbit. Enzyme activities in the cross-innervated muscle are refered to that of the normal soleus muscle in the non-operated contralateral leg.

a red muscle as induced by foreign innervation [30,31]. In this experiment, the soleus muscle of the rabbit was innervated with the peroneal nerve. As was shown also by others (e.g.[32,33]),foreign innervation of the "slow" soleus muscle, with "fast" motor neurons results in a transformation of its enzyme activity pattern. Changes of this pattern, however, are restricted only to "discriminative" magnitudes. Cross-innervation induces an increase in enzyme activities of glycogenolysis,glycolysis and mitochondrial glycerolphosphate oxidation.The activity ratios of these enzymes remain more or less constant.On the other hand,a decrease is observed in the activities of hexokinase as well as the enzymes of the citric acid cycle,amino acid metabolism and fatty acid oxidation. The enzyme activity pattern of the cross-innervated slow muscle thus reveals characteristics of a fast muscle.This holds also for the induced change in the activity ratio adenylate kinase/creatine kinase.The increase in mitochondrial glycerolphosphate dehydrogenase appears to be of special interest,since the cellular content of mitochondria obviously decreases under the influence of foreign innervation.It may thus be suggested that a mitochondrial population with a specifically higher activity of glycerolphosphate dehydrogenase is induced in the soleus muscle under the influence of "fast" motor neurons.

These findings indicate a specific influence of innervation on the metabolic type of skeletal muscle. This influence is obviously related to the control of enzyme synthesis. At the present state of our knowledge it is only possible to speculate about the exact nature of this mechanism which may act at any stage between the neuromuscular junction and the protein synthetizing apparatus of the post-synaptic cell. The phenomenon itself may be described as an altered expression of the genome.

Differentiation of energy-supplying metabolism in muscle is thus the consequence of various factors. Ultimately, it may be understood as an optimum adaptation of energy-supply to the requirements of muscle function. Therefore, different metabolic types may be ascribed to different functional types. Differences in the metabolic type are probably reflected best at the level of enzymatic organization. Variable and non-variable characteristics of this organization may be recognized from the comparative analysis of enzyme activity patterns which therefore may be envisaged as the expression of the metabolic program of a tissue.

References

1. Bücher,Th., in "Erbliche Stoffwechselkrankheiten" (edited by F.Linneweh), Urban und Schwarzenberg, München-Berlin 1962, p.125.
2. Pette,D., Naturwissenschaften, 52 (1965) 597.
3. Pette D., in "Symposion über progressive Muskeldystrophie" (edited by E.Kuhn). Springer-Verlag, Berlin,Heidelberg,New York, 1966, p. 492.
4. Bass, A., Brdiczka,D.,Eyer,P.,Hofer,S.,and Pette,D., Europ.J.Biochem.10 (1969) 198.
5. Zebe,E., and McShan,W.H., J.gen.Physiol.40 (1957) 779.
6. Brosemer,R.W.,Vogell,W.,and Bücher,Th., Biochem.Z. 338 (1963) 854.
7. Zebe,E., Biochem.Z. 332 (1960) 328.
8. Beenakkers,A.M.Th.,J.Insect Physiol.15 (1969) 353.
9. Zebe,E.Delbrück,A.,and Bücher,Th., Biochem.Z. 331 (1959) 254.
10. Pette,D.,and Bücher, Hoppe-Seyler's Z.Physiol.Chem. 331 (1963) 180.

11. Dawson,D.M., and Romanul,F.C.A.,
    Arch.Neurol. 11 (1964) 369.
12. Krieg,U., Dissertation,Faculty of Medicine,
    University of Munich 1967.
13. Bücher,Th., and Sies,H.,
    Europ.J.Biochem. 8 (1969) 273.
14. Krebs,H.A., and Woodford,M.,
    Biochem.J. 94 (1965) 436.
15. Opie,L.H., and Newsholme,E.A.,
    Biochem.J. 103 (1967) 391.
16. Stein,J.M.,and Padykula,H.A.,
    Amer.J.Anat.110 (1962) 103.
17. Romanul,F.C.A., Arch.Neurol.,11 (1964) 355.
18. Henneman,E.,and Olson,C.B.,
    J.Neurophysiol. 28 (1965) 581.
19. Nishiyama,A.,Acta Med.Okayama 19(1965) 177.
20. Edgerton,V.R.,and Simpson,D.R.,
    J.Histochem.Cytochem. 17 (1969) 828.
21. Nolte,J.,and Pette,D.,in preparation.
22. Nolte,J.,and Pette,D.,in "Recent Advances
    in Quantitative Histochemistry",Verlag
    Hans Huber,Bern,Stuttgart,in press.
23. Pette,D., and Brandau,H.,
    Enzym.Biol.Clin. (Basel) 6 (1966) 79.
24. Sigel,P.,and Pette,D.,
    J.Histochem.Cytochem. 17 (1969) 225.
25. Burleigh,I.G.,and Schimke,R.T.,
    Biochem.J.113 (1969) 157.
26. Hofer,S.,Dissertation,Faculty of Medicine,
    University of Munich 1970.
27. Kubišta,V.,Kubištová,J.,and Pette,D.,
    Europ.J.Biochem.,in press.
28. Bass,A.,Lusch,G.,and Pette,D.,
    Eur.J.Biochem.13 (1970) 289.
29. Gutmann,E.,and Syrový,I.,
    Physiol.bohemoslov. 16 (1967) 232.
30. Golisch,G.,Pette,D., and Pichlmaier,H.,
    Z.Klin.Chem. 3 (1965) 202.
31. Golisch,G.,Pette,D.,and Pichlmaier,H.,
    Europ.J.Biochem. 16 (1970) 110.

32. Prewitt,M.A., and Salafsky,B.,
    Am.J.Physiol. 213 (1967) 295.
33. Guth,L.,Watson,P.K., and Brown,W.C.,
    Exptl.Neurol. 20 (1968) 52.

BIOCHEMICAL ADAPTATIONS TO ENDURANCE EXERCISE IN SKELETAL MUSCLE

J.O. Holloszy, L.B. Oscai, P.A. Molé, and I.J. Don
Department of Preventive Medicine, Washington University
School of Medicine, St. Louis, Missouri, U.S.A. 63110

The adaptations that occur in response to a vigorous program of prolonged exercise, such as long distance running or swimming, manifest themselves in functional terms as an increase in endurance. In the present context, an increase in endurance can be defined simply as an increase in the period of time for which an individual can maintain a given submaximal level of exercise. A logical first question in an investigation of the physiological basis for this increase in endurance is - What are the factors that force an individual to stop exercising, and how do the adaptations to physical training modify these factors?

Severe fatigue is much easier to experience than it is to explain. Nevertheless, although our knowledge in this area is still quite limited, it does seem clear that the fatigue or discomfort that forces one to stop exercising is not a single physiological state, but has a variety of etiologies and forms. The present discussion will only deal with certain aspects of two of these.

When exercise is of an intensity that can be maintained for long periods (i.e. about 60 to 180 min.), the discomfort that forces cessation of exercise often takes the form of a "heavy" or "dead" feeling in the exercising muscles. This forces the exercising individual to slow his pace or stop despite the absence of respiratory or other distress. This form of fatigue appears to correlate well with the depletion of muscle glycogen [1,2]. Why this should be so, when large amounts of substrate, in the form of fatty acids, are still available, remains a mystery. However, despite our current inability to explain this phenomenon in biochemical terms, it seems firmly established empirically that depletion of muscle glycogen stores results in fatigue (1-3). When muscle glycogen stores are increased by feeding a high carbohydrate diet following glycogen depletion, endurance is increased (3). Thus, it would appear that any adaptation that has a sparing effect on glycogen stores during exercise should result in an increase in endurance. It is interesting in this context that during submaximal exercise physically trained men and animals derive a greater percentage of their energy from fatty acid oxidation than do untrained individuals at comparable work levels (4-6).

When exercise is sufficiently strenuous to force an individual to stop after approximately 10 to 20 minutes, blood lactate levels increase progressively, as do respiratory distress and muscular discomfort. The rise in blood lactate levels may reflect a 2 to 3 fold greater increase in muscle lactate levels (7,8).

Over 40 years ago Hill and Kupalov (9) provided convincing evidence that accumulation of lactate in muscle causes fatigue. It is relevant in this context that physical training of the endurance type markedly modifies the increase in lactate that occurs during submaximal exercise. It is well established that physically trained individuals maintain lower levels of blood lactate during submaximal exercise than do untrained (10-13). Thus, a highly trained individual may maintain a plateau blood lactate level only twice the resting value while exercising for hours at a work load that produced a progressive rise in blood lactate and forced him to stop after 10 to 20 minutes when he was untrained.

Until recently, the development of muscle hypoxia during exercise was thought to be responsible for the large amounts of lactate produced by untrained individuals during exercise. It is well established that major cardiovascular adaptations occur in response to endurance exercise, and it has been customary to attribute the much lower rates of lactate production, during submaximal exercise in trained individuals to improved delivery of blood and O_2 to the working muscles.

Information from a number of recent studies now makes this explanation untenable. In the first place, oxygen consumption is the same in the untrained and in the trained state at a given work level of submaximal intensity (13-16). In the second place, cardiac output is either unchanged or actually decreased when previously sedentary individuals are retested at a given submaximal exercise load following a program of endurance exercise (14,16,17). Finally, blood flow to the working muscles is lower in trained than in sedentary individuals during submaximal exercise (16). The muscles of the trained individual compensate for this decreased blood flow by increased extraction of O_2, as evidenced by a greater arterio-venous O_2 difference (14).

In view of the foregoing, it seems clear that the lower rates of lactate production by the trained, as compared to the untrained, individual during submaximal exercise cannot be explained on the basis of cardiovascular adaptations. A new explanation for this difference is, therefore, needed.

Biochemical Adaptations in Skeletal Muscle. In recent years, it has become evident that, in addition to the changes that occur in the cardiovascular system, major biochemical adaptations occur in skeletal muscle in response to regularly performed endurance exercise. These adaptations can result in as much as a two-fold increase in the capacity of skeletal muscle for aerobic metabolism.

Our studies in this area are conducted on male rats of a Wistar strain (specific pathogen-free CFN rats from Carworth Farms). In the exercise program we developed empirically and now use routinely, animals weighing approximately 120 g at the beginning of the study, run on a motor-driven treadmill consisting of a wide endless belt riding on metal rollers. A lucite box partitioned into individual compartments 30 cm long by 10 cm wide is suspended over the belt providing a limited area for each animal to run in. Motivation is provided by a shock grid at the rear of the compartments. The treadmill is set at an $8°$ incline. Animals are exercised five days per week. The speed and duration of the run is gradually increased from 10 min at 22 m per min to 2 hrs at 31 m per min with 12 intervals at 42 m per min lasting 30 to 60 sec interspersed at 10 min intervals through the workout. "Exercising control" rats are trained to run on the treadmill and then maintained on a program of 10 min of running per day at 31 m per min. The purpose of this program is to maintain running skill and familiarity with the procedure while keeping the training stimulus minimal.

We routinely use two groups of sedentary controls; a freely eating group is provided with food ad libitum, and a paired-weight group has its food intake restricted so as to maintain rate of weight gain the same as that of the exercisers.

The running program results in a remarkable increase in exercise capacity. In recent studies, "all-out" run time to the point of severe fatigue at 31 m per min up an $8°$ incline has varied from 4-8 hours in the trained rats. In contrast, the exercising controls become exhausted after approximately 30 min at 31 m per min. When the trained animals are tested at 42 m per min immediately following their two hour training run, they can continue for an additional 50 to 60 min.

Effects of the Running Program on Muscle Mitochondria. The running program described above results in an approximately two-fold increase in the capacity of the mitochondrial fraction from skeletal muscle (gastrocnemius and quadriceps) to oxidize pyruvate (18,19). This difference is not secondary to differences in the percentage yield of mitochondria, as a similar increase in oxidative capacity is evident when whole homogenates of muscle from trained animals are studied (18,19). Underlying this increase in oxidative capacity is a two-fold increase in the levels of activity of the respiratory enzymes involved in the oxidation of DPNH and succinate (18) and of some of the enzymes of the citric acid cycle (20). This rise in enzyme activity appears to be due to an increase in enzyme protein as evidenced by a doubling of the concentration of cytochrome c and a 50% to 60% increase in mitochondrial protein (18,21). The responses of cytochrome c and of succinate dehydrogenase, which serve as markers for the mitochondrial respiratory chain, and of the citric acid cycle enzyme citrate synthase, are illustrated in Table 1. A number of investigators have provided electronmicroscopic evidence that this increase in mitochondrial enzymes is secondary to an increase in the size and number of mitochondria (22,23). However, the finding that the protein content of the mitochondrial fraction from skeletal muscle is increased only about 55% (cf. refs. 18,21,Table 1) suggests that there is also a change in mitochondrial composition. This is borne out by the finding that the levels of activity, expressed per g of muscle, of some mitochondrial enzymes such as mitochondrial α-glycerophosphate dehydrogenase (21), mitochondrial adenylate kinase, and

Table 1. Effect of the Running Program on the Levels of Some Mitochondrial Constituents in Gastrocnemius Muscle

Animals were not exercised for 66 hours prior to sacrifice. Cytochrome c was determined by the method of Williams and Thorp (38). Succinate dehydrogenase activity was measured as described by King (39). Citrate synthase activity was assayed by the method of Srere (40). Values are expressed as means $\pm$ S.E. Number of animals is given in parentheses. Enzymatic activity is reported as μ moles of substrate utilized per min per g wet weight of muscle.

Group	Cytochrome c	Succinate dehydrogenase	Citrate synthase	Mitochondrial protein
	n moles/g	μ moles/min/g		mg/g
Sedentary	8.8 $\pm$0.7(14)	4.2 $\pm$0.4(6)	25.6 $\pm$1.1(12)	5.74 $\pm$0.29(10)
Exercising	17.8[a] $\pm$0.9(14)	8.4[a] $\pm$0.4(6)	51.5[a] $\pm$2.2(12)	3.77[a] $\pm$0.18(10)

[a] Exercising versus sedentary, $p < 0.001$

mitochondrial creatine kinase (24) do not increase in response to training. Some other enzyme activities such as glutamate dehydrogenase, malate dehydrogenase, and α-ketoglutarate dehydrogenase increase only 30% to 50% (20). During the course of studies on the oxidation of pyruvate (18) and of palmityl carnitine (25), it was found that mitochondria obtained from muscles of the exercised animals exhibited tightly coupled oxidative phosphorylation. This finding indicates that the two-fold increase in mitochondrial electron transport capacity is associated with a parallel rise in the capacity to generate ATP aerobically, and suggests that the coupling factors involved in coupling electron transport to oxidative phosphorylation are also involved in the adaptive response to exercise. This interpretation is borne out by the finding that mitochondrial ATPase activity is increased two-fold in the muscles of the trained animals (24). Mitochondrial ATPase activity, assayed under appropriate conditions, is a measure of the amount of coupling factor 1 (F_1) present in disrupted mitochondria (26-28). In intact mitochondria, F_1 catalyses the phosphorylation of ADP to ATP (26,28,29). However, the activity of this enzyme is assayed more readily in the reverse direction, hence the name ATPase.

Oxidation of fatty acids provides a major portion of the energy for muscle metabolism during prolonged submaximal exercise (30-32). A progressively greater proportion of the needed energy is derived from carbohydrate, as the intensity of the exercise increases (4,33). However, the relative amounts of carbohydrate and fat utilized at different work loads depend on the level of physical fitness, with fatty acid oxidation serving as a more important source of energy in physically trained, than in untrained, men and animals (4-6,33). It is of interest in this context that the capacity of gastrocnemius muscle to oxidize palmitate is significantly increased in rats subjected to the above described running program (34). Associated with, and probably underlying, this increase in the capacity to oxidize palmitate is a rise in the levels of activity of the three enzymes which have been postulated to be rate limiting for the oxidation of palmitate (25). These are ATP dependent palmityl-CoA synthetase, palmityl coenzyme A - carnitine palmityltransferase, and palmityl-CoA dehydrogenase.

In addition to these biochemical adaptations in muscle, regularly performed exercise appears to result in adaptations in the lipolytic system in adipose tissue. These are reflected in a greater rate of fatty acid release from adipose tissue in physically trained than in untrained individuals during exercise (5,6). It appears probable that this increased rate of fatty acid mobilization acts in synergy with the increase in the capacity of skeletal muscle for fatty acid oxidation, to enable physically trained individuals to obtain a greater proportion of the energy required for exercise from fat.
Effects of the Running Program on Muscle Myoglobin. It has

been postulated on the basis of direct measurements on contracting muscles and indirect measurements obtained on humans that myoglobin acts as an O_2 store and helps to support aerobic metabolism by releasing O_2 to cytochrome oxidase when O_2 becomes limiting during muscle contraction. It seems unlikely that this role of myoglobin is of much importance in endurance exercise, as this form of exercise does not involve prolonged, forceful muscle contractions.

Myoglobin has been shown to increase O_2 transport across a fluid layer. This suggests that myoglobin facilitates O_2 utilization by enhancing O_2 transport through the cytoplasm from the cell membrane to the mitochondrial cristae. It is therefore of interest that the myoglobin content of the exercising rats' leg muscles increases to roughly the same extent as do muscle mitochondrial cristae (35).

Effect of the Running Program on Glycolytic and Glycogenolytic Capacity of Muscle. Since the rate of lactate production during exercise may play a role in the development of one type of muscle fatigue, and since the rate of glycogen depletion may play a role in the development of a second type of fatigue, it seemed of interest to study the effect of training on the glycolytic and glycogenolytic capacity of muscle. A marked decrease in the glycolytic capacity of skeletal muscle as a result of a decrease in the level of activity of one or more rate limiting enzymes, could, conceivably, contribute to decreased rates of glycogen utilization and lactate production.

As shown in Table 2, the level of activity of hexokinase is

Table 2. Effect of Exercise Program on the Levels of Activity of Certain Enzymes of Carbohydrate Metabolism

Enzyme activities were measured as described previously (41). Values are expressed as means $\pm$ S.E. Number of animals is given in parentheses. Enzymatic activity is reported as μ moles of substrate utilized per minute per g wet weight of gastrocnemius.

Group	Phosphorylase	Hexokinase	PFK[a]	Pyruvate kinase	Lactate DH[b]
		μ moles/min/g muscle			
Sedentary	98.8 $\pm$16.4(3)	1.05 $\pm$0.10(7)	45.2 $\pm$4.3(12)	726 $\pm$39(9)	676 $\pm$42(11)
Exercising	96.8 $\pm$19.5(3)	2.19[c] $\pm$0.20(7)	37.5 $\pm$3.4(10)	779 $\pm$43(9)	608 $\pm$36(11)

[a] Phosphofructokinase.
[b] Lactate dehydrogenase.
[c] Exercising versus sedentary, $p < 0.001$.

approximately twice as high in the leg muscles of the treadmill runners as in those of the sedentary controls. This finding confirms the observations of Peter, Jeffress and Lamb (36) and of Lamb et al. (37) on exercised guinea pigs. This adaptation to exercise, which may play a role in the super-compensation of muscle glycogen levels following depletion by exercise, does not appear to be a "training" effect in the same sense as the increase in muscle mitochondria. The increase in hexokinase levels, in contrast to the increase in muscle mitochondria, can apparently be induced in untrained animals by means of a few relatively light bouts of exercise; even a single bout of exercise appears to be followed by an increase in hexokinase activity.

The levels of activity of phosphorylase, phosphofructokinase, pyruvate kinase, and lactate dehydrogenase in gastrocnemius muscles from trained and sedentary animals are shown in Table 2. Phosphorylase, pyruvate kinase, and lactate dehydrogenase activities were not significantly different in the two groups. Phosphofructokinase activity tended to be somewhat lower in the muscles of the exercised animals. However, this difference was inconsistent, and the average values for the two groups are not significantly different. The values given for phosphofructokinase activity were obtained on muscle that had been frozen for six or more weeks and are somewhat lower than the maximal values that have been reported for rat skeletal muscle; they should, therefore, be considered preliminary.

The effect of training on glycogenolytic capacity was also studied by measuring lactate production by soleus and extensor digitorum longus muscles stimulated to contract by means of supramaximal direct current shocks, in vitro, under anaerobic conditions. As shown in Table 3, there was no significant difference between the exercising and sedentary groups in the amount of lactate produced. The isometric tension developed by comparable muscles from animals in the two groups was similar.

Even if statistically significant changes in the levels of activity of the glycogenolytic enzymes had occurred in response to training, it is questionable whether they would be of much physiologic significance during submaximal exercise. The maximal catalytic capacities of the glycogenolytic and glycolytic enzymes are representative of the potential of a muscle rather than of its actual performance. Under conditions of submaximal exercise the rate of conversion of glucose and glycogen to pyruvate and lactate is determined not so much by the concentrations of the rate limiting enzymes as by the operation of a number of control mechanisms.

<u>A Hypothesis Regarding the Control Mechanisms Responsible for Maintaining Lactate Production by Trained Skeletal Muscle at Lower Levels During Submaximal Exercise</u>. As a result of the tight

Table 3. Lactate Production by Muscles Stimulated to Contract Under Anaerobic Conditions

Extensor digitorum longus and soleus muscles were stimulated to contract isometrically with supramaximal, rectangular, direct current shocks, in vitro, at a rate of 120 shocks per min for 20 min. Muscles were suspended in Ringer's bicarbonate solution containing 10 mM glucose gassed with 95% N_2 and 5% CO_2 at $37^\circ C$. The stimulation system has been described previously (42). Values are means $\pm$ S.E. The number of muscles is given in parentheses.

Group	Muscle	Lactate production
		μ moles/g muscle/20 min
Sedentary	EDL[a]	56.8 $\pm$ 7.9(5)
	Soleus	34.5 $\pm$ 4.3(4)
Exercising	EDL[a]	50.4 $\pm$ 4.1(5)
	Soleus	33.8 $\pm$ 3.2(4)

[a] Extensor digitorum longus

coupling of oxidative phosphorylation to electron transport in intact mitochondria, the rate of O_2 consumption is determined by the availability of P_i and ADP; the rate of respiration appears to be an inverse function of the ratio $ATP/(ADP+P_i)$ (43-48). At rest, the levels of P_i, ADP and O_2 consumption are low. With the onset of exercise, ATP and creatine phosphate are split, and the levels of P_i and ADP rise progressively until the rates of mitochondrial electron transport and oxidative phosphorylation increase sufficiently to balance the rate of ATP hydrolysis. At this point, a functional "steady state" is attained. Oxygen consumption increases and plateaus in parallel with the rate of electron transport. As a result, O_2 uptake rises gradually, rather than immediately, to a steady state level during adjustment to a submaximal work load (49-51). This lag in the rise in O_2 consumption is roughly equivalent to the "O_2 debt" contracted by muscle, i.e. the O_2 required following exercise for the oxidative regeneration of the ~P that was split - but not regenerated - during the initial adjustment to exercise (51).

Skeletal muscle that has adapted to strenuous endurance exercise contains approximately twice as many mitochondrial cristae per gram as untrained muscle. Therefore, the functional "steady state" in which O_2 consumption balances ATP hydrolysis during submaximal exercise must be attained at lower concentrations of P_i and ADP in the working muscle cells of trained, as compared to sedentary, individuals. This must be so because O_2 consumption is the same in the untrained as in the trained state for a given submaximal work load.

In addition to controlling the rate of respiration, the intracellular levels of P_i, ADP, and ATP also, to a large extent, control the rates of glycolysis and glycogenolysis (52-55, and cf. refs 56-58). Glycolysis and glycogenolysis should, therefore, because of lower "steady state" levels of intracellular P_i and ADP, occur at a slower rate in the muscles of trained individuals during submaximal exercise. As a result, pyruvate and DPNH should be formed at a slower rate at a given submaximal work load, accounting for a lower lactate production in trained as compared to sedentary muscles.

In addition to the role of the Pasteur effect postulated in the foregoing working hypothesis, another factor which probably contributes importantly to, and acts synergistically with, the decreased rate of glycolysis to account for the lower lactate production in the trained individual, is the relatively greater utilization of fatty acid oxidation to fulfill the energy requirements of submaximal exercise (4-6, 33).

Acknowledgements. This work was supported by Research Grant HD01613, and Training Grant AM05341 from the U.S. Public Health Service. J. O. Holloszy is the recipient of U.S.P.H.S. Research Career Development Award K4-HD-19573.

We wish to express our appreciation to Mrs. May Chen for skillful technical assistance, and to Mrs. Celeste Amitin for assistance in the preparation of this manuscript.

REFERENCES

1. Ahlborg, B., Bergström, J., Ekelund, L.-G., and Hultman, E., Acta physiol. scand., 70, 129 (1967).
2. Hermansen, L., Hultman, E., and Saltin, B., Acta physiol. scand., 71, 129 (1967).
3. Bergstrom, J., Hermansen, L., Hultman, E., and Saltin, B., Acta physiol. scand., 71, 140 (1967).
4. Christensen, E. H., and Hansen, O., Skand. Arch. Physiol., 81, 160 (1939).
5. Havel, R. J., Carlson, L. A., Ekelund, L.-G., and Holmgren, A., J. Appl. Physiol., 19, 613 (1964).
6. Issekutz, B., Miller, H. I., Paul, P., and Rodahl, K., J. Appl. Physiol., 20, 293 (1965).

7. Diamant, B., Karlsson, J., and Saltin, B., Acta physiol. scand., 72, 383 (1968).
8. Carlsten, A., In Physical Activity and Aging (D. Brunner and E. Jokl, eds.), p. 37, University Park Press, Baltimore (1970).
9. Hill, A. V., and Kupalov, P., Proc. Roy. Soc. (London) Ser. B., 105, 313 (1929).
10. Edwards, H. T., Brouha, L., and Johnson, R. E., Le Travail Humain, 8, 1 (1939).
11. Robinson, S., and Harmon, P. M., Am. J. Physiol., 132, 757 (1941).
12. Holmgren, A., and Strom, G., Acta Medica Scand., 163, 185 (1959).
13. Cobb, L. A., and Johnson, W. P., J. Clin. Invest., 42, 800 (1963).
14. Varnauskas, E., Bergman, H., Houk, P., and Bjorntorp, P., Lancet, 2, 8 (1966).
15. Frick, M. H., and Katila, M., Circulation, 37, 192 (1968).
16. Clausen, J. P., Larsen, O. A., and Trap-Jensen, J., Circulation, 40, 143 (1969).
17. Frick, M. H., Konttinen, A., and Sarajas, H. S. S., Am. J. Cardiol., 12, 142 (1963).
18. Holloszy, J. O., J. Biol. Chem., 242, 2278 (1967).
19. Fuge, K. W., Crews, E. L., Pattengale, P. K., Holloszy, J. O., and Shank, R. E., Am. J. Physiol., 215, 660 (1968).
20. Holloszy, J. O., Oscai, L. B., Don, I., and Molé, P. A., Biochem. Biophys. Res. Commun., In Press.
21. Holloszy, J. O., and Oscai, L. B., Arch. Biochem. Biophys., 130, 653 (1969).
22. Gollnick, P. D., and King, D. W., Am. J. Physiol., 216, 1502 (1969).
23. Kraus, H., Kirsten, R., and Wolff, J. R., Pflugers Arch., 308, 57 (1969).
24. Oscai, L. B., and Holloszy, J. O., (In Preparation).
25. Molé, P. A., Oscai, L. B., and Holloszy, J. O., (In Preparation).
26. Penefsky, H. S., Pullman, M. E., Datta, A., and Racker, E., J. Biol. Chem., 235, 3330 (1960).
27. Fessenden, J. M., and Racker, E., J. Biol. Chem., 241, 2483 (1966).
28. Pullman, M. E., and Monroy, G. C., J. Biol. Chem., 238, 3762 (1963).
29. Racker, E., and Horstman, L. L., J. Biol. Chem., 242, 2547 (1967).
30. Havel, R. J., Naimark, A., and Borchgrevink, C. F., J. Clin. Invest., 42, 1054 (1963).
31. Paul, P., and Issekutz, B., J. Appl. Physiol., 22, 615 (1967).
32. Zierler, K. L., Maseri, A., Klassen, G., Rabinowitz, D., and Burgess, J., Trans. Ass. Amer. Physicians, 8, 266 (1968).
33. Issekutz, B., Miller, H. I., and Rodahl, K., Fed. Proc., Fed. Amer. Soc. Exp. Biol., 25, 1415 (1966).

34. Molé, P. A., and Holloszy, J. O., *Proc. Soc. Exp. Biol. Med.*, 134, 789 (1970).
35. Pattengale, P. K., and Holloszy, J. O., *Am. J. Physiol.*, 213, 783 (1967).
36. Peter, J. B., Jeffress, R. N., and Lamb, D. R., *Science*, 160, 200 (1968).
37. Lamb, D. R., Peter, J. B., Jeffress, R. N., and Wallace, H. A., *Am. J. Physiol.*, 217, 1628 (1969).
38. Williams, J. N., and Thorp, S. L., *Biochem. Biophys. Acta*, 189, 25 (1969).
39. King, T. E., *Methods Enzymol.*, 10, 322 (1967).
40. Srere, P. A., *Methods Enzymol.*, 13, 3 (1969).
41. Dart, C. H., and Holloszy, J. O., *Circulation Res.*, 25, 245 (1969).
42. Holloszy, J. O., and Narahara, H. T., *J. Biol. Chem.*, 240, 3493 (1965).
43. Lardy, H. A., and Wellman, H., *J. Biol. Chem.*, 201, 357 (1953).
44. Chance, B., and Williams, G. R., *Advan. Enzymol.*, 17, 65 (1956).
45. Chance, B., *Ann. N. Y. Acad. Sci.*, 81, 477 (1959).
46. Chance, B., and Maitra, P. K., In *Control Mechanisms in Respiration and Fermentation* (Wright, B., ed.), p. 307, Ronald Press, New York (1963).
47. Klingenberg, M., and Schollmeyer, P., *Biochem. Z.*, 335, 231 (1961).
48. Klingenberg, M., and von Hafen, H., *Biochem. Z.*, 337, 120 (1963).
49. Henry, F. M., and DeMoor, J., *J. Appl. Physiol.*, 8, 608 (1956).
50. Margaria, R., Mangili, F., Cuttica, F., and Cerretelli, P., *Ergonomics*, 8, 49 (1965).
51. Piiper, J., Di Prampero, P. E., and Cerretelli, P., *Am. J. Physiol.*, 215, 523 (1968).
52. Wu, R., and Racker, E., *J. Biol. Chem.*, 234, 1029 (1959).
53. Chance, B., Garfinkel, D., Higgens, J., and Hess, B., *J. Biol. Chem.*, 235, 2426 (1960).
54. Lowry, O. H., Passonneau, J. V., Hasselberger, F. X., and Schulz, D. W., *J. Biol. Chem.*, 239, 18 (1964).
55. Uyeda, K., and Racker, E., *J. Biol. Chem.*, 240, 4689 (1965).
56. Racker, E., In *Mechanisms in Bioenergetics*, p. 252, Academic Press, New York (1965).
57. Lehninger, A. L., In *The Mitochondrion*, p. 152, W. A. Benjamin, Inc., New York (1965).
58. Mahler, H. R., and Cordes, E. H., *Biological Chemistry*, p. 422, Harper and Row, Publishers, New York (1966).

INFLUENCE OF EXERCISE ON ELECTRON TRANSPORT CAPACITY OF HEART MITOCHONDRIA

Hans Kraus

University Pediatric Department, 3400 Göttingen

Germany

The ability of mammalian organisms to maintain homeostasis depends on their capacity to alter the rate of the metabolic events which underlie the physiological processes at molecular level. Since metabolic reactions are mediated by enzyme systems, an elucidation of the factors which regulate cellular enzyme activity is essential in understanding the ability of the organism to adapt and maintain a dynamic equilibrium (19).

Regarding cardiac metabolism, different laboratories have shown that the energy required by the heart of normal resting individuals is supplied by oxidative pathways. The major fuels are in turn fatty acids, glucose and lactate (2). Keul et al. (12) reported that with increasing work loads the heart of trained individuals utilizes increasing amounts of lactate. At maximal work load lactate can account for 80 per cent of the total metabolic rate. These data were obtained by measuring differences in the composition of blood collected from an artery and from the coronary sinus. But, because of the possible storage of substrates and their interchange in heart muscle, the coronary arteriovenous difference reflects only a balance and permits no conclusions as to the pathways of intermediary metabolism in the heart muscle cell. It is therefore essential to combine metabolic balance studies with investigations of cellular metabolism. The mitochondrial barrier is impermeable for NADH (15, 17) and - as shown by Isaacs et al. (10) - a limitation on lactate oxidation is present in normal hearts. Thus it is not clear by which mechanism the increased amount of hydrogen released in the cytoplasma of the exercising heart can be transferred to the respiratory chain. Of the several metabolic shuttles that may be involved in the transfer of hydrogen from extramitochondrial NADH through the mitochondrial membrane, physiological significance can be attributed only to the α-glycerophosphate cycle (3). This cycle, originally discovered in insect flight muscle (4, 6), appears to be operative in mammalian skeletal muscle and probably as well in brain (11). In myocardial tissue, however, levels of mitochondrial α-glycerophosphate dehydrogenase have been found to be low (18). The effects of this low activity may be reflected in an apparent limitation on carbohydrate oxidation in the heart. Thus the isolated perfused heart has been reported to consume glucose or lactate at a rate that can account for only about 50 per cent of the total oxygen uptake (20). Pyruvate, however, is taken up at a rate sufficient to account for approximately all of the oxygen consumed (21). This indicates a limiting step in the conversion of lactate to pyruvate. Since lactate dehydrogenase is present in high activity in the heart (7), it is possible that the oxidation of lactate is limited by the availability of NAD^+, as has been suggested by Williamson (21) and Butow and Racker (5). In conclusion these data indicate that the rise in lactate utilization in the trained heart ought to be parallelled by a significant increase in the specific activity of mitochondrial glycerol-P dehydrogenase, if the α-glycerophosphate cycle is operating in cardiac tissue.

It was the aim of the series of experiments presented to clarify this interrelation, i.e. to study whether strenuous physical exercise may cause an increase in mitochondrial glycerol-P dehydrogenase activity of the rat heart. In addition, succinic dehydrogenase activity is determined as a "marker" enzyme (16) of the intramitochondrial electron transport. The results demonstrate that glycerol-P dehydrogenase activity, expressed per g of mitochondrial protein, doubles in response to the training. Succinic dehydrogenase activity increases approximately 40 per cent. The rise of both enzymes' activities apparently is due to an increase in enzyme protein. This is suggested by the finding of a general increase in mitochondrial protein.

RESULTS

The data for enzymatic activities and mitochondrial protein content of the heart ventricles are presented in Fig. 1. Expressed per gram of mitochondrial protein, the activities of glycerol-P dehydrogenase and succinate dehydrogenase in untrained animals are 9.5 ± 1.7 μmoles/min and 935.0 ± 17.0 μmoles/min, respectively. Glycerol-P dehydrogenase activity (GP-OX) increases continuously during strenuous swimming exercise, having approximately doubled at the end of the experimental period. The increase in succinate dehydrogenase activity (SDH) is less pronounced, running up to about 40 per cent in relation to the controls. Total mitochondrial protein content of the heart also increases significantly,

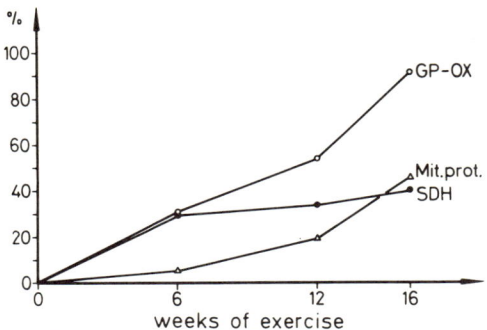

Fig. 1. Glycerol-P dehydrogenase (GP-OX) and succinate dehydrogenase (SDH) activities in rat heart mitochondria, and mitochondrial protein content of cardiac tissue in response to exercise. Values are expressed in per cent of those from untrained controls. The training consisted in swimming to exhaustion twice a day. Increasing weights were fixed to the tail of the animals.

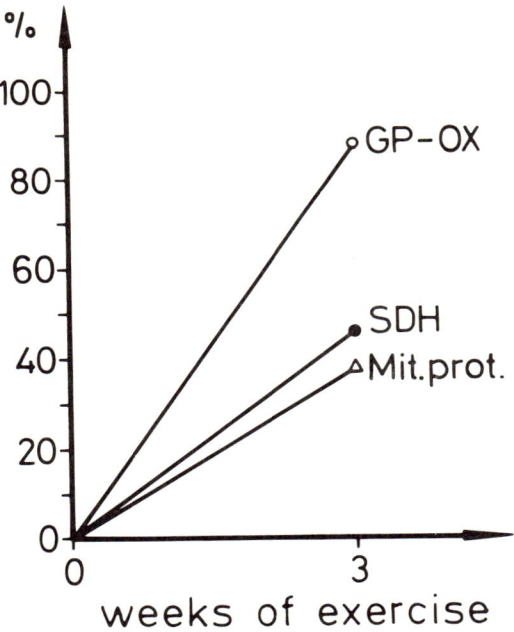

Fig. 2. Enzyme levels in rat heart mitochondria, and mitochondrial protein content of cardiac tissue in response to exercise. Values are expressed in per cent of those from untrained controls. The exercise program consisted in running voluntarily in a treadmill connected with the cage.

suggesting that the rise in oxidative enzyme activities is due to de novo-synthesis of enzyme protein. Almost the same results are found after running exercise (Fig. 2). The exercising group was trained by running voluntarily in a treadmill connected with the cage. This kind of exercise provides a physiological work load and avoids stressing the animal. After 3 weeks, glycerol-P dehydrogenase activity and succinate dehydrogenase activity increase approximately 90 per cent and 40 per cent, respectively. The rise in mitochondrial protein content (about 40 per cent) is also statistically significant.

DISCUSSION

The most important result of the present investigation is the twofold increase in glycerol-P dehydrogenase of heart mitochondria in response to the training. In view of the finding that, in situ, lactate is utilized in large amounts by the athlete heart (12), this data provides evidence for the operation of the α-glycerophosphate cycle in cardiac tissue. The rise in the flux of this system in connexion with the increased acitivity of lactic dehydrogenase during exercise (7) constitutes an adaptation which can associate the metabolic pathways of heart and skeletal muscle. Thus lactate produced by the skeletal muscle in proportion to the work load is taken up by the heart cell and transferred to the respiratory chain.

In comparison with the enhancement of glycerol-P dehydrogenase, the increase in succinate dehydrogenase is only half as great. Probably this enzyme, a representative of the intramitochondrial electron transport (16), is not rate-limiting to the same degree as glycerol-P dehydrogenase. The present literature reveals contradictory reports concerning the effect of exercise on succinc dehydrogenase activity (8, 22, 23). This disagreement may be caused by differences in the exercise programs or the assay methods employed. The results described above indicate that succinic dehydrogenase increases in cardiac tissue in the same way as in skeletal muscle (9, 13) during rigorous exercise. The rise of both enzymes' activities apparently is due to a net increase in enzyme protein. This is suggested by the finding of a general increase in mitochondrial protein and it is also in line with morphological observations that physical exercise can produce an increase in the mitochondrial mass of the heart muscle cell (1, 14).

REFERENCES

1. ALDINGER, E.E. Effect of thiamine deficiency on potential myocardial contractility. <u>Circulat. Res.</u> 16: 238, 1965.

2. BING, J.R. Cardiac metabolism. Physiol. Rev. 45: 171, 1965.

3. BORST, P. Hydrogen transport and transport metabolites. Funktionelle und morphologische Organisation der Zelle. Rottach-Egern: Springer-Verlag, 1963.

4. BÜCHER, TH., AND M. KLINGENBERG. Wege des Wasserstoffs in der lebendigen Organisation. Angew. Chemie. 70: 552, 1958.

5. BUTOW, R.A., AND E. RACKER. On the mechanism of respiratory control. J. Gen. Physiol. 49: 149, 1965.

6. ESTABROOK, R.W., AND B. SACKTOR. Alphaglycerophosphate oxidase of flight muscle mitochondria. J. Biol. Chem. 223: 1014, 1958.

7. GOLLNICK, P.D., AND G.R. HEARN. Lactic dehydrogenase activities of heart and skeletal muscle of exercised rats. Amer. J. Physiol. 201: 694, 1961.

8. HEARN, G.R., AND W. WAINIO. Succinic dehydrogenase activity of heart and skeletal muscle of exercised rats. Amer. J. Physiol. 185: 348, 1956.

9. HOLLOSZY, J.O. Biochemical adaptions in muscle. Effects of exercise on mitochondrial oxygen uptake and respiratory enzyme activity in skeletal muscle. J. Biol. Chem. 242: 2278, 1967.

10. ISAACS, H.G., B. SACKTOR, AND T.A. MURPHY. The role of the α-glycerophosphate cycle in the control of carbohydrate oxidation in heart and in the mechanism of action of thyroid hormone. Biochim. Biophys. Acta 177: 196, 1969.

11. KADENBACH, B. Der Einfluss von Thyreoidhormonen in vivo auf die oxydative Phosphorylierung und Enzymaktivitäten in Mitochondrien. Biochem. Z. 344: 49, 1966.

12. KEUL, J., E. DOLL, H. STEIM, U. FLEER, UND H. REINDELL. Über den Stoffwechsel des Herzen bei Hochleistungssportlern. Z. Kreisl.-Forsch. 55: 477, 1966.

13. KRAUS, H., R. KIRSTEN, UND J.R. WOLFF. Die Wirkung von Schwimm- und Lauftraining auf die celluläre Funktion und Struktur des Muskels. Pflüg. Arch. ges. Physiol. 308: 57, 1969.

14. LAGUENS, R.P., B.B. LAZADA, AND C.L. GOMEZ-DUMM. Effect of acute and exhaustive exercise upon the fine structure of heart mitochondria. Experientia. 22: 224, 1966.

15. LEHNINGER, A.L. Phosphorylation coupled to oxidation of diphosphopyridine nucleotide. J. Biol. Chem. 190: 345, 1951.

16. ROODYN, D.B. The mitochondrion. In: *Enzyme Cytology*. New York: Academic Press, 1967, p. 103.

17. SACKTOR, B. The role of mitochondria in respiratory metabolism of flight muscle. *Ann. Rev. Entomol.* 6: 103, 1961.

18. SACKTOR, B., AND A.R. DICK. Oxidation of extramitochondrial diphosphopyridine nucleotide by various tissues of the mouse. *Science* 145: 606, 1964.

19. WEBER, G., S.K. SRIVASTAVA, AND R.L. SINGHAL. Role of enzymes in homeostasis. *J. Biol. Chem.* 240: 750, 1965.

20. WILLIAMSON, J.R. Effects of insulin and diet on the metabolism of L (+) lactáte and glucose in the perfused rat heart. *Biochem. J.* 83: 377, 1962.

21. WILLIAMSON, J.R. Effects of insulin and starvation on the metabolism of acetate and pyruvate by the perfused rat heart. *Biochem. J.* 93: 97, 1964.

22. YAMPOLSKAYA, L.J. Biochemical changes in the muscle of trained and untrained animals under the influence of small loads. *Sechenow J. Physiol. U.S.S.R.* 39: 91, 1952.

23. YAMPOLSKAYA, L.J., AND N.N. YAKOVLEV. The influence of muscle activity on muscle proteins. *Sechenow J. Physiol. U.S.S.R.* 37: 110, 1951.

ULTRASTRUCTURAL AND ENZYME CHANGES IN MUSCLES WITH EXERCISE

Philip D. Gollnick, C. David Ianuzzo, and Douglas W. King

Exercise Physiology Laboratory, Department of Physical

Education for Men, and Electron Microscope Center

Washington State University. Pullman, Washington 99163

It is generally accepted that the immediate energy source for muscular contraction comes from hydrolysis of the terminal phosphate bond of adenosine triphosphate (ATP). ATP is supplied to the contractile apparatus primarily from the oxidation of the carbohydrates and fats stored in the muscle or brought to it by the circulation. Most of the ATP is produced by the aerobic pathways of the mitochondria. Thus, the functional capacity of a muscle may be related to the ability of the mitochondria to provide a continuing, adequate supply of ATP. The purpose of this paper is to consider some of the changes that occur in the ultrastructure and enzyme activity, particularly those of the mitochondria, of skeletal and cardiac muscle as a result of acute or chronic exercise.

MATERIALS AND METHODS

Male Sprague-Dawley rats with initial body weights between 180 and 200 grams were used in the animal experiments. Hypophysectomized and thyroidectomized rats were obtained from Hormone Assay, Chicago, Illinois. The animals were housed individually in 7 x 10 x 7-inch cages in an air-conditioned room with the ambient temperature maintained at 24 ± 1 C. They were fed and watered ad libitum. The day was artificially divided into 12 hr of light and 12 hr of darkness. The trained animals were exercised in motor-driven wheels. Further details of the training programs will be given in the Results and Discussion.

In two of the experiments (13,27) rats were sacrificed at the point of exhaustion or 2 to 24 hr later. Trained animals were exhausted by running at speeds between 26.8 and 53.6 m/min. Animals not trained to run in the work wheels will not run long enough or fast enough to produce a state of fatigue comparable to that of the trained animals; therefore, the untrained rats swam in groups of 4 in water at 35 C until exhausted. Swimming was done in 46 cm diameter tanks filled to a depth of 46 cm. Exhaustion was defined as the point where the rats could not run at a speed of 26.8 m/min or regain the surface of the water after submersion during swimming. All of the rats were sacrificed by decapitation and tissue samples removed as rapidly as possible.

Two investigations were also conducted with muscle samples obtained from humans. These samples were taken from the vastus lateralis using the needle biopsy technique described by Hultman (25). In the first study (14), 3 male volunteers in good physical condition and accustomed to riding the bicycle ergometer served as subjects. Muscle samples were taken before and immediately after the subjects had worked to exhaustion on a bicycle ergometer (Monark). The work was performed at a pedal frequency of 60 rpm with a work load requiring an oxygen consumption of 75 to 80% of their maximal aerobic capacity (bicycle test). The subjects were considered exhausted when they failed to maintain the 60 rpm pedal frequency. In the second study, muscle samples from 3 humans were obtained from Dr. Bengt Saltin of the Gymanstik-och idrottshogskolan in Stockholm. They were taken at rest, after a maximal exercise test, following a 4 hr period of submaximal work, and, finally, after a maximal work test that had been preceded by a 4 hr submaximal work bout.

Tissue samples were prepared for electron microscopy as described elsewhere (13,14,27). Briefly, we have used osmium fixation and epon embedment. Thin sections (gold or silver) were cut with either glass or diamond knives with a Porter-Blum or Reichert OMU-2 microtome and stained with uranyl acetate and lead citrate. Sections were examined with a Philips 100B or Hitachi HU-125E electron microscope.

Succinic dehydrogenase (SDH) activities were determined spectrophotometrically as described by Cooperstein et al. (9). The mitochondrial fractions of the heart and skeletal muscle were isolated from the homogenates by differential centrifugation as described by Holloszy (24). The protein content of the mitochondrial fraction was determined either by the biuret (15) or Folin-Ciocalteau technique described by Lowry et al. (33). Tissue glycogen concentrations were determined using the anthrone method as modified for small samples (14).

RESULTS AND DISCUSSION

Effect of training on enzyme activity and mitochondrial concentration. Rats that had completed a 10-week training program were used in the first experiments (13,27). During the final 6 weeks of this program each rat ran 40 min at 26.8 m/min followed by 20 min at 38.5 m/min each day for 7 days/week. The trained rats were run to exhaustion on the final day of the training program and sacrificed along with the controls 24 hr later. Samples of the left myocardium and central portion of the gastrocnemius were rapidly removed at the time of sacrifice and prepared for examination with the electron microscope.

The basic ultrastructure of skeletal muscle from trained rats sacrificed 24 hr after an exhaustive exercise bout was similar to that of sedentary animals (Fig. 1,2). There was, however, clear evidence of a proliferation and hypertrophy of the mitochondria in the muscles from the trained rats (Fig. 2). These enlarged mitochondria were densely packed with cristae. Mitochondrial density in muscle of control rats sacrificed at rest and after an exhaustive swim, and in trained rats killed 24 hr after an exhaustive run and at the point of exhaustion is presented in Fig. 3. As can be seen, training produced a significant increase in the mitochondrial concentration.

The observation of an increased mitochondrial density after training is consistent with the biochemical data that has demonstrated an increased oxidative capacity and mitochondrial protein per gram of tissue of rat (24,29) and guinea pig (2) skeletal muscle after training. Kraus and co-workers (29) have also presented

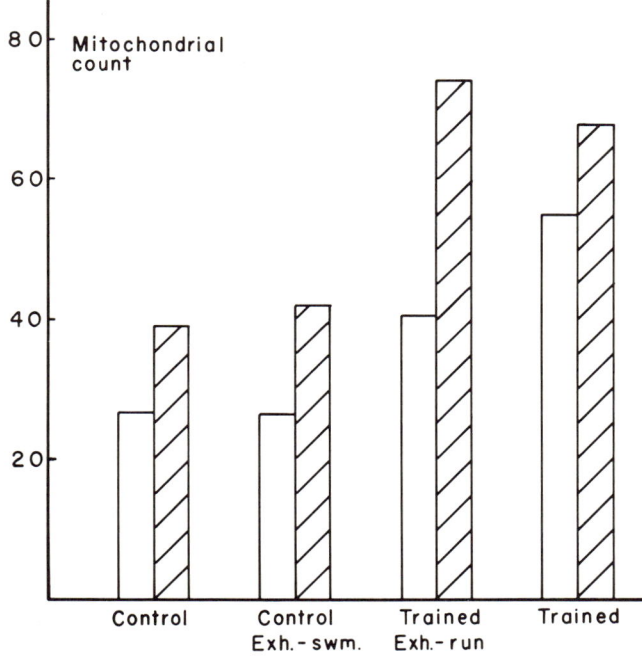

Fig. 3. Mitochondria in rat gastrocnemius muscle after training and/or exhaustive (Exh) exercise (swimming or running). Open bars are mitochondria/100 sq microns and cross hatched bars are mitochondria/100 sarcomeres of muscle.

electron microscopic evidence for an increased concentration and size of mitochondria in rat skeletal muscle following training. Morgan and associates (35) reported that the lipid composition of human skeletal muscle was altered by training in a manner consistent with an increase in either mitochondrial size or number.

The duration and intensity of the exercise needed to induce the enhanced oxidative capacity and mitochondrial concentration is unknown. Hearn and Wainio (19), Gould and Rawlinson (16), and Holloszy (24) failed to find any change in the activity of some of the oxidative enzymes in the skeletal muscle of rats trained by swimming 30 min per day for periods ranging from 5 to 8 weeks. Holloszy (24) did find increases in the oxygen uptake, activity of some oxidative enzymes, and protein content of the mitochondrial fraction of rat muscle after a 12-week program of strenuous running. We (Gollnick and Ianuzzo, unpublished data) have found increases in the SDH activity and mitochondrial protein per gram of fresh muscle of rats trained 9 to 10 weeks by running. Barnard and associates (2) recently found 18 weeks of treadmill running were needed to produce similar changes in guinea pigs. On the other hand, Kraus and co-workers (29) found significant increases in the oxidative capacity, the activities of some enzymes, and mitochondrial protein in rat skeletal muscle after only 3 weeks of running or 2 weeks of swimming (120 to 180 min of swimming per day). In a recently completed series of experiments, we (Gollnick and Ianuzzo) did not find any change in the SDH activity or mitochondrial protein in the muscle of rats after 3 weeks of endurance (running 26.8 m/min for 60 min/day during the final week) or sprint (77.0 m/min for 30 sec alternated with 30 sec rest for 15 sprints during the final week) training. This exercise program was then continued for six weeks, with the endurance speed increased to 37 m/min and the sprint speed to 80.4 m/min (up to 20 sprints). At the end of this 6-week period there was a significant increase in the SDH activity and mitochondrial protein concentration in only the sprint group. These data suggest that intensity may be as important as duration in producing this change.

Cardiac muscle, as compared to skeletal muscle, possesses high oxidative capacity and therefore contains a high concentration of large mitochondria that are densely packed with cristae (Fig. 4). Although the mitochondria of the heart were slightly swollen 24 hr after an exhaustive run (Fig. 5), there was no evidence of an increase in number. Furthermore, in recent experiments we (Gollnick and Ianuzzo) have not found any change in the SDH activity or mitochondrial concentration per gram of myocardium. In contrast to our findings, however, Arcos et al. (1) have reported that the protein content of the mitochondrial fraction of rat myocardium is significantly greater in rats after endurance training (group swimming up to 6 hr/day for 6 days/week). This occurred in rats that had completed from 140 to 180 hr of swimming. They also reported that the oxygen uptake of the mitochondria (O_2 uptake/mg mitochondrial dry weight/hr) in the presence of various substrates was not altered by training. When recalculated as O_2 uptake/g of fresh muscle, these data do suggest that a significant increase in oxidative capacity of the myocardium had occurred. Laugens and Gomez Dumm (31) have also reported that a single bout of swimming (30 to 180 min) produces mitochondrial enlargement in rat myocardium resulting from a replication of the internal components of the mitochondria. They (32) also reported that an increased synthesis of mitochondrial DNA occurs during exercise.

Present evidence indicates that training can increase the oxidative capacity and mitochondrial protein concentration per gram of skeletal muscle in both the rat and guinea pig. Although direct experimental evidence is lacking at this time, it has been postulated (28,35) that this also occurs in human muscle. Two questions concerning this adaptation are presently unanswered. First, what is the physiological significance of this change to the muscle and to the whole organism during exercise? Secondly, how are these adaptations induced?

As for the first question, it is obvious that the maximal oxygen uptake of man does not increase in proportion to the maximal change in oxidative capacity as reported for the rat, that is, 2 to 3 fold (24,29). (Our results and those of Barnard et al. (2), however, indicate only a 30 to 50% change). Also, we know of no data that has related these changes in muscle to the maximal oxygen consumption of the experimental animal in which they were demonstrated. Furthermore, nearly all of the change in maximal oxygen uptake that occurs in man can be accounted for by an increased cardiac output and a wider A-V oxygen difference (10,41). The wider A-V oxygen difference after training may be the result of an improved capacity to shunt blood to the working muscle (38).

It has been hypothesized that the increase in mitochondrial concentration that occurs in skeletal muscle during training is important during submaximal exercise (8,24,28). The lower cardiac output and muscle blood flow, and wider A-V oxygen difference during submaximal work after training and in athletes support this possibility (6,8,17,28). These responses to training have been interpreted as resulting from an increased ability of the muscle to utilize oxygen (8,28), or that more of the work is done by red fibers (6). It does not appear, however, that the aerobic capacity of the muscle is limiting since a reduction in cardiac output and widening of the A-V oxygen difference can also be induced in man during submaximal work by beta adrenergic blockade (11).

One possibility is that the metabolic character of the muscle is changed by training to enable a greater portion of the muscle to participate in the oxidative work required for endurance activity.

This hypothesis is supported by the recent observation of Barnard et al. (2) that training increases the concentration of red fibers in both the white and red regions of the muscle. This could result in a spreading of the work load throughout a greater portion of the muscle.

Concerning the second question, it is also obvious that the increased mitochondrial and enzyme content of skeletal muscle results from an increased protein synthesis. Protein synthesis is normally controlled by hormone or substrate induction. In skeletal muscle this may be regulated to some extent by the motor nerves as the functional and biochemical properties of muscle can be modified by its motor nerves (7,37,40). In an attempt to investigate some of the aspects of this problem we have studied the possible involvement of some hormones in the adaptation to training. In these studies we have used normal, thyroidectomized, and hypophysectomized rats that had been trained for 9 to 10 weeks. The intensity of the training program was adjusted to the endurance capacity of each group. During the last weeks the thyroidectomized, normal, and hypophysectomized groups were running 80 min at 35.4 m/min, 60 min at 35.4 m/min, and 60 min at 26.8 m/min each day, respectively. One group of thyroidectomized rats was also given triiodothyronine (T_3) replacement during the last 3 weeks of the experiment (2.0 ug/100 g body weight/day). As can be seen (Fig. 6), removal of the pituitary or thyroid gland did alter the SDH activity and protein content of the mitochondrial fraction of skeletal muscle. The large reduction in enzyme activity with only a small change in mitochondrial protein content in the thyroidectomized group is consistent with the findings of Gustafsson et al. (18). SDH activity returned to normal after 3 weeks of T_3 replacement even though mitochondrial protein remained constant. In every group, training increased both the SDH activity and mitochondrial protein concentration of skeletal muscle but not of the myocardium. The increases in enzyme to mitochondrial protein ratio remained relatively constant, indicating that the change was the result of a synthesis of additional mitochondrial components having enzymatic properties similar to those of the untrained controls. These data indicate the adaptation in skeletal muscle that occurs during training is probably induced by one or more substrates or by a modification in the motor nerve.

Effect of acute exercise on muscle ultrastructure. The ultrastructure of the gastrocnemius and heart muscle of rats was significantly altered by exhaustive exercise (Fig. 7,8). This occurred in skeletal muscle only after exhaustive running whereas it existed in the myocardium of rats exhausted either by swimming or running. The most prominent change was a swelling of the mitochondria. Many of the cristae within these swollen mitochondria appear to have degenerated and those that remained were disorientated. The amplitude of the swelling was greatest in the myocardial mitochondria of rats exhausted by running. In some instances all of the cristae of these

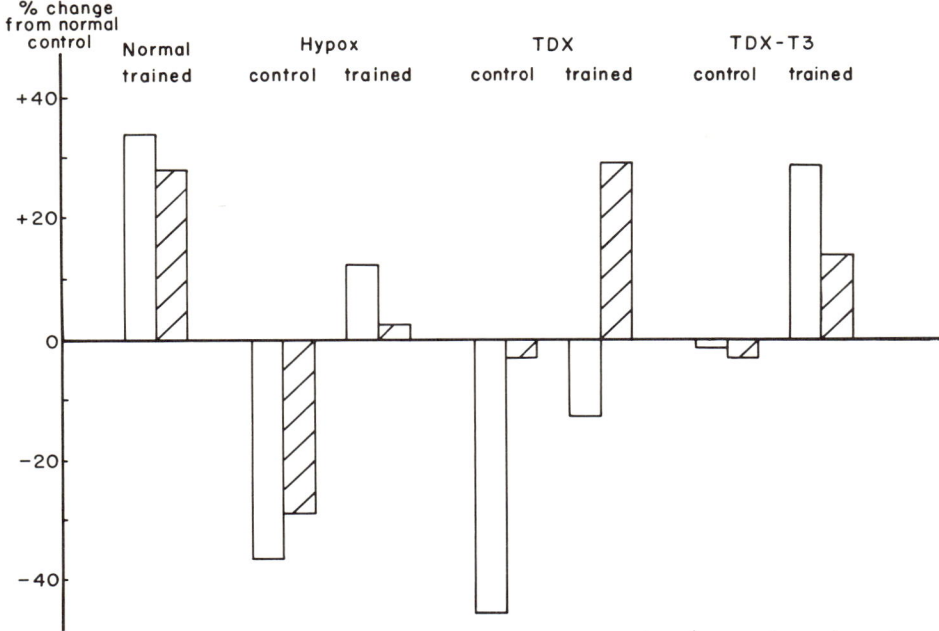

Fig. 6. Effect of training on the SDH activity (open bars) and mitochondrial protein per g of muscle (cross hatched bars) of normal and hormone deficient rats. Mean values from untrained normal rats were 4.4 umole/g/min and 4.26 mg/g for SDH and mitochondrial protein, respectively. Abbreviations are: Hypox=hypophysectomy, TDX=thyroidectomy, and TDX-T3=throidectomy plus daily injections of triiodothyronine for last 3 weeks of experiment.

grossly swollen mitochondria had degenerated, leaving a vacuole-like structure in the tissue (Fig. 9). It should be pointed out that these changes were not present in all mitochondria and that various stages of swelling and disruption existed in the tissue with some mitochondria appearing normal. Other morphological changes after exhaustive exercise included an expansion of the sarcoplasmic tubules (most prominent in the heart) and some loss of the normal banding that is typical of striated muscle. There was also a distinct widening of the interfibrillar space in the gastrocnemius muscle from rats exhausted by running. This was suggestive of a generalized edema in the tissue.

Laugens and co-workers (30) were perhaps the first to show that a single bout of exercise could alter mitochondrial structure. They found that the mitochondria in dog myocardium became enlarged and swollen after a single bout of exhaustive swimming. In a subsequent study with rats they found that a single bout of swimming (30 to 120 min) produced mitochondrial enlargement in the myocardium but did not alter their basic morphology. They assumed this occurred by a replication of the internal mitochondrial components. Pelosi and Agliati (36) also found the mitochondria of rat myocardium to be enlarged and somewhat swollen after a single 8 hr swim. Vallyathan et al. (43) recently reported that after 5 hr of electrical stimulation, the mitochondria in the pectoralis muscle of pigeons became swollen and that the cristae of these mitochondria had a tubular-like appearance. The mitochondrial changes we observed after exercise, however, were far more extensive than those reported from other laboratories. Recently, Banister and associates (personal communication) found changes in the mitochondria of rat myocardium similar to those which we have reported. The disparity between our results and those from other laboratories is probably due to differences in the severity of the exercise test. Thus, although longer work periods have been used, they have not been as severe or exhaustive. In this regard it is known that individual, free-swimming rats can remain afloat for extended periods of time when the water temperature is maintained slightly below body temperature. This, however, cannot be considered strenuous exercise. When rats are forced to swim in groups in a relatively small area, as we have done, they work much harder and become exhausted in a short time. The running program used to exhaust the rats was also arduous as evidenced by the fact that at exhaustion colonic temperature and heart rates exceeded 41 C and 500 beats/min, respectively (12).

The changes in the fine structure of heart and skeletal muscle following exhaustive exercise were transient. Thus, although some mitochondrial swelling was evident in the myocardium of rats sacrificed 2 hr after exhaustive swimming (Fig. 10), it was far less than that present in the tissue of rats sacrificed at exhaustion (Fig. 7). Furthermore, the skeletal muscle of rats sacrificed 24 hr after exhaustive running was normal (Fig. 2) and the myocardial mitochondria only slightly enlarged (Fig. 5). The mitochondria in the myocardium of rats sacrificed 24 hr after exhaustion resembled those described by Arcos et al. (1) in the myocardium of rats sacrificed 15 to 18 hr after completion of a training program consisting of group swimming up to 6 hr daily, 6 days/week. Some swelling was evident in the myocardium of rats that had completed 361-490 hr of swimming. They did not sacrifice any rats at the completion of the last exercise bout. From our results it might be postulated that a large amplitude swelling and cristae disruption had existed in the myocardium of these animals and that these changes had been largely reversed by the 15 to 18-hr recovery period.

No ultrastructural changes were evident in the vastus lateralis muscle of the 3 men following 90 to 138 min of pedaling a bicycle ergometer (Fig. 11,12). The only difference between the pre- and post-exercise samples was a depletion of the glycogen stores. This was evident in the micrographs and was confirmed by direct chemical analysis. This is consistent with previous findings of Hermansen et al. (21).

In the samples obtained from Dr. Saltin, mitochondrial swelling existed in the muscle of one of the three subjects (Fig. 13-15). This was present only in the samples taken after completion of a maximal work test following a 4 hr submaximal work bout. These changes were similar to those seen in rat skeletal and heart muscle. They were found throughout the entire muscle cell, that is, deep in the fiber (Fig. 13), immediately under the sarcolemma (Fig. 14), and in the perinuclear area (Fig. 15). It was also interesting to note that various degrees of mitochondrial swelling could be found in a given area. Thus, intact mitochondria were found immediately adjacent to those moderately swollen and to those grossly swollen with badly disrupted and degenerated cristae.

The mechanism producing the mitochondrial damage in heart and skeletal muscle during exhaustive exercise is unknown. Similar changes in myocardial mitochondria, however, have been reported in several species following ischemia (4,5,20,22), hypoxia (34,42), and chronic exposure to altitude (3). Changes produced by hypoxia can be reversed by returning the animals to an atmosphere containing normal oxygen tension (42). The similarity of changes following hypoxia and exercise is striking, and it is tempting to speculate that the mechanisms producing the mitochondrial swelling is the same for each. Some hypoxia may indeed occur in working muscle during severe exercise as it is known that the blood leaving these tissues is almost completely devoid of oxygen. In addition, both exercise and hypoxia reduce the ATP and phosphocreatine concentrations of tissue (23,26,34,39). This reduction in high energy phosphate could upset normal ionic balance in the mitochondria and produce an inward movement of water with a subsequent disruption of the cristae. This hypothesis, however, is tempered by the finding that the ATP and phosphocreatine concentrations in the muscle samples obtained from Dr. Saltin were all similar and that only one of these contained swollen and disrupted mitochondria.

The significance of the mitochondrial changes that occur during exhaustive exercise is unknown. It is known that the metabolic capacity of the mitochondria is adversely altered when the basic morphology is lost (44). If impairment of the metabolic capacity does occur, it could be a factor in hastening the onset of fatigue during prolonged exercise. In our studies we have not found any difference in the SDH activity of skeletal muscle after exhaustive running. This is a major problem requiring further investigation.

This work supported by U.S.P.H.S., National Heart Institute, Grant-in-AID HE-08262 and Washington State University Medical and Biological research fund project 0404.

REFERENCES

1. ARCOS, J. C., R. S. SOHOL, S. SUN, M. F. ARGUS, and G. S. BURCH. Changes in ultrastructure and respiratory control in mitochondria of rat heart hypertrophied by exercise. Exptl. Mol. Pathol. 8: 49-65, 1968.
2. BARNARD, R. J., V. R. EDGERTON, and J. B. PETER. Effect of exercise on skeletal muscle. I. Biochemical and histochemical properties. J. Appl. Physiol. 28: 762-766, 1970.
3. BISCHOFF, M. W., W. D. DEAN, T. J. BUCCI, and L. A. FRIES. Ultrastructural changes in myocardium of animals after five months residence at 14,110 feet. Federation Proc. 28: 1268-1273, 1969.
4. BRYANT, R. W., W. A. THOMAS, and R. M. O'NEIL. An electron microscopic study of myocardial ischemia in the rat. Circulation Res. 6: 699-709, 1958.
5. CAULFIELD, J., and F. KLIONSKY. Myocardial ischemia and early infraction. An electron microscopic study. Am. J. Pathol. 35: 489-523, 1959.
6. CLAUSEN, J. P., and J. TRAP-JENSEN. Effect of training on muscular blood flow during exercise. Acta Physiol. Scand. 74: 23A, 1968.
7. CLOSE, R. Dynamic properties of fast and slow skeletal muscles of the rat after nerve cross-union. J. Physiol., London 204: 331-346, 1969.
8. COBB, L. A., P. H. SMITH, S. IWAI, and F. A. SHORT. External iliac vein flow: its response to exercise and relation to lactate production. J. Appl. Physiol. 26: 606-610, 1969.
9. COOPERSTEIN, S. J., A. LAZOROW, and N. J. KURFESS. A microspectrophotometric method for the determination of succinic dehydrogenase. J. Biol. Chem. 186: 129-139, 1950.
10. EKBLOM, B., P. O. ASTRAND, B. SALTIN, J. STENBERG, and B. WALLSTROM. Effect of training on circulatory response to exercise. J. Appl. Physiol. 24: 518-528, 1968.
11. FURBERG, C., and G. V. SCHMALENSEE. Beta-adrenergic blockade and central circulation during exercise in sitting position in health subjects. Acta Physiol. Scand. 73: 435-446, 1968.
12. GOLLNICK, P. D., and C. D. IANUZZO. Colonic temperature response of rats during exercise. J. Appl. Physiol. 24: 747-750, 1968.
13. GOLLNICK, P. D., and D. W. KING. Effect of exercise and training on mitochondria of rat skeletal muscle. Am. J. Physiol. 216: 1502-1509, 1969.
14. GOLLNICK, P. D., C. D. IANUZZO, C. WILLIAMS, and T. R. HILL. Effect of prolonged, severe exercise on the ultrastructure of human skeletal muscle. Intern. Z. angew. Physiol. 27: 257-265, 1969.
15. GORNALL, A. G., C. J. BARDWILL, and N. A. DAVID. Determination of serum proteins by means of the biuret reaction. J. Biol. Chem. 177: 751-766, 1949.
16. GOULD, M. K., and W. A. RAWLINSON. Biochemical adaptation as a response to exercise. I. Effect of swimming on the levels of

lactic dehydrogenase, malic dehydrogenase and phosphorylase in muscles of 8-, 11-, and 15-week-old rats. Biochem. J. 73: 41-44, 1959.
17. GRIMBY, G., E. HÄGGENDAL, and B. SALTIN. Local xenon 133 clearance from the quadriceps muscle during exercise in man. J. Appl. Physiol. 22: 305-310, 1967.
18. GUSTAFSSON, R., J. R. TATA, D. LINDBERG, and L. ERNSTER. The relationship between the structure and activity of rat skeletal muscle mitochondria after thyroidectomy and thyroid hormone treatment. J. Cell. Biol. 26: 555-578, 1965.
19. HEARN, G. R., and W. W. WAINIO. Succinic dehydrogenase activity of heart and skeletal muscle of exercised rats. Am. J. Physiol. 185: 348-350, 1956.
20. HERDSON, P. B., H. M. SOMMERS, and R. B. JENNINGS. A comparative study of the fine structure of normal and ischemic dog myocardium with special reference to early changes following temporary occlusion of a coronary artery. Am. J. Pathol. 46: 367-386, 1965.
21. HERMANSEN, L., E. HULTMAN, and B. SALTIN. Muscle glycogen during prolonged severe exercise. Acta Physiol. Scand. 71: 129-139, 1967.
22. HOELSCHER, B., O. H. JUST, and H. J. MERKER. Studies by electron microscope on various forms of induced cardiac arrest in dog and rabbit. Surgery 49: 492-499, 1961.
23. HOHORST, H. J., M. REIM, and H. BARTELS. Studies on the creatine kinase equilibrium in muscle and the significance of the ATP and ADP levels. Biochem. Biophys. Res. Commun. 7: 142-146, 1962.
24. HOLLOSZY, J. O. Effects of exercise on mitochondrial oxygen uptake and respiratory enzyme activity in skeletal muscle. J. Biol. Chem. 242: 2278-2282, 1967.
25. HULTMAN, E. Muscle glycogen in man determined in needle biopsy specimens. Method and normal values. Scand. J. Clin. Lab. Invest. 19: Suppl. 94, 1-63, 1967.
26. HULTMAN, E., J. BERGSTROM, and N. M. ANDERSON. Breakdown and resynthesis of phosphorylcreatine and adenosine triphosphate in connection with muscular work in man. Scand. J. Clin. Lab. Invest. 19: 56-66, 1967.
27. KING, D. W., and P. D. GOLLNICK. Ultrastructure of rat heart and liver after exhaustive exercise. Am. J. Physiol. 218: 1150-1155, 1970.
28. KLASSEN, G. A., G. M. ANDREW, and M. R. BECKLAKE. Effect of training on total and regional blood flow and metabolism in paddlers. J. Appl. Physiol. 28: 397-406, 1970.
29. KRAUS, H., R. KIRSTEN, and J. R. WOLFF. Die Wirkung von Schwimm- und Lauftraining auf die cellulare Funktion und Struktur des Muskels. Pflügers Arch. 308: 57-79, 1969.
30. LAUGENS, R. P., B. B. LOZADA, C. L. GOMEZ DUMM, and A. R. BERAMENDI. Effect of acute and exhaustive exercise upon the fine structure of heart mitochondria. Experientia 22: 244-246, 1966.
31. LAUGENS, R. P., and C. L. A. GOMEZ DUMM. Fine structure of myocardial mitochondria in rats after exercise for one-half to

two hours. Circulation Res. 21: 271-279, 1967.
32. LAUGENS, R. P., and C. L. GOMEZ DUMM. Deoxyribonucleic acid synthesis in the heart mitochondria after acute and exhaustive exercise. Experientia 24: 163-164, 1968.
33. LOWRY, O. H., N. J. ROSEBROUGH, A. L. FARR, and R. J. RANDLE. Protein measurement with the Folin phenol reagent. J. Biol. Chem. 193: 265-275, 1951.
34. MOLBERT, E. Die Herzmuskelzelle nach akuter Oxydationshemmung im elecktronenmikroskopishen Bild. Beirt. Pathol. Anat. Allgem. Pathol. 118: 421-435, 1958.
35. MORGAN, T. E., F. A. SHORT, and L. A. COBB. Effect of long-term exercise on skeletal muscle lipid composition. Am. J. Physiol. 216: 82-86, 1969.
36. PELOSI, G., and G. AGLIATI. The heart muscle in functional overload and hypoxia. Lab. Invest. 18: 86-93, 1968.
37. ROMANUL, F. C. A., and J. P. VAN DER NEULEN. Slow and fast muscles after cross innervation. Arch. Neurol. 17: 387-402, 1967.
38. ROWELL, L. B. "Fatigue and 'Disorders' of Normal Cardiovascular Regulation." In Physiology and Pathology of Fatigue. E. Simonson, ed. Springfield: C. C. Thomas, In Press.
39. SACKTOR, B., and E. C. HURLBUT. Regulation of metabolism in working muscle in vivo. II. Concentration of adenosine nucleotides, arginine phosphate, and inorganic phosphate in insect flight muscle during flight. J. Biol. Chem. 241: 632-634, 1966.
40. SALMONS, S., and G. VRBOVA. The influence of activity on some contractile characteristics of mammalian fast and slow muscles. J. Physiol., London 201: 535-549, 1969.
41. SALTIN, B., G. BLOMQVIST, J. H. MITCHELL, R. L. JOHNSON, JR., K. WILDENTHAL, and C. B. CHAPMAN. Response to exercise after bed rest and after training. Circulation 38: Suppl. 7, 1-78, 1968.
42. SULKIN, N. M., and D. F. SULKIN. An electron microscopic study of the effects of chronic hypoxia on cardiac muscle, hepatic, and autonomic ganglion cells. Lab. Invest. 14: 1523-1546, 1965.
43. VALLYATHAN, N. V., I. GRINYER, and J. C. GEORGE. Effect of fasting and exercise on lipid levels in muscle. A cytological and biochemical study. Can. J. Zool. 48: 377-383, 1970.
44. WEINBACK, E. C., J. GARBUS, and H. G. SHEFFIELD. Morphology of mitochondria in the coupled, uncoupled, and re-coupled stages. Exptl. Cell Res. 46: 129-143, 1967.

LEGENDS FOR MICROGRAPHS (LONGITUDINAL SECTIONS)

Fig. 1. Skeletal muscle from a control animal sacrificed at rest. X 10,500
Fig. 2. Skeletal muscle from a trained rat killed 24 hr after an exhaustive run showing an increase in the size and concentration of the mitochondria. X 10,500
Fig. 4. Left myocardium of a control rat. Mitochondria are numerous

and large with densely packed cristae. X 12,600

Fig. 5. Heart muscle of a trained rat sacrificed 24 hr after completion of a run to exhaustion. Mitochondria are noticeably swollen. X 12,600

Fig. 7. Skeletal muscle of a rat sacrificed at the point of becoming exhausted while running. Mitochondria are grossly swollen. Mitochondrial cristae are disoriented and degenerated. X 10,500

Fig. 8. Myocardium of a rat sacrificed at the point of exhaustion while swimming. Mitochondria are swollen and the sarcoplasmic tubules are expanded. X 12,600

Fig. 9. Heart muscle from an exhausted rat (running) showing extensive degeneration of the cristae, in some instances leaving a vacuole-like structure in the tissue. X 12,600

Fig. 10. Cardiac muscle of a rat killed 2 hr after becoming exhausted while swimming. Some mitochondrial swelling is evident but is much less than in rats sacrificed at the point of exhaustion either after swimming or running. X 12,600

Fig. 11. Skeletal muscle sample from subject CW at rest. Note glycogen particles (small black dots) in the interfibrillar space and between the myofilaments in the I band region. X 16,200

Fig. 12. Skeletal muscle sample from subject CW after completion of 105 min of bicycle exercise requiring 76% of his aerobic capacity. Tissue appears normal except for depletion of glycogen. X 16,200

Fig. 13-15. Skeletal muscle from subject TP obtained after maximal work test that had been preceded by a 4 hr submaximal work bout. Figures show mitochondria in various stages of swelling deep within the fiber (13), immediately beneath the sarcolemma and around a nucleus (14), and in the subsarcolemmic regions of 2 adjacent fibers (15). X 14,800

Fig. 4 and 7-10 are from King and Gollnick (27). Reproduced by permission of the American Journal of Physiology.

Fig. 11 and 12 are from Gollnick et al. (14). Reproduced by permission of the Intern. Z. angew. Physiol.

Micrographs to follow.

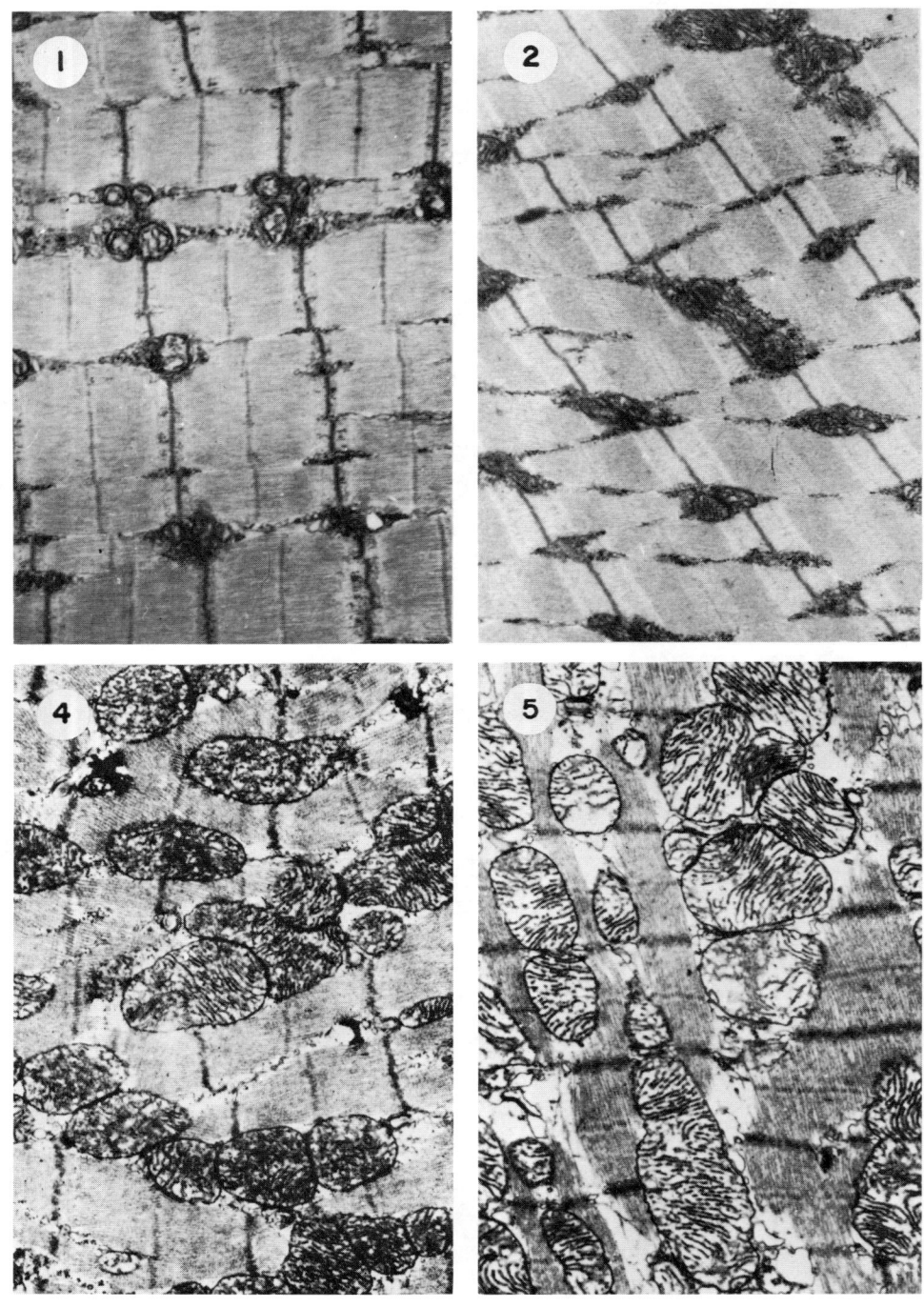

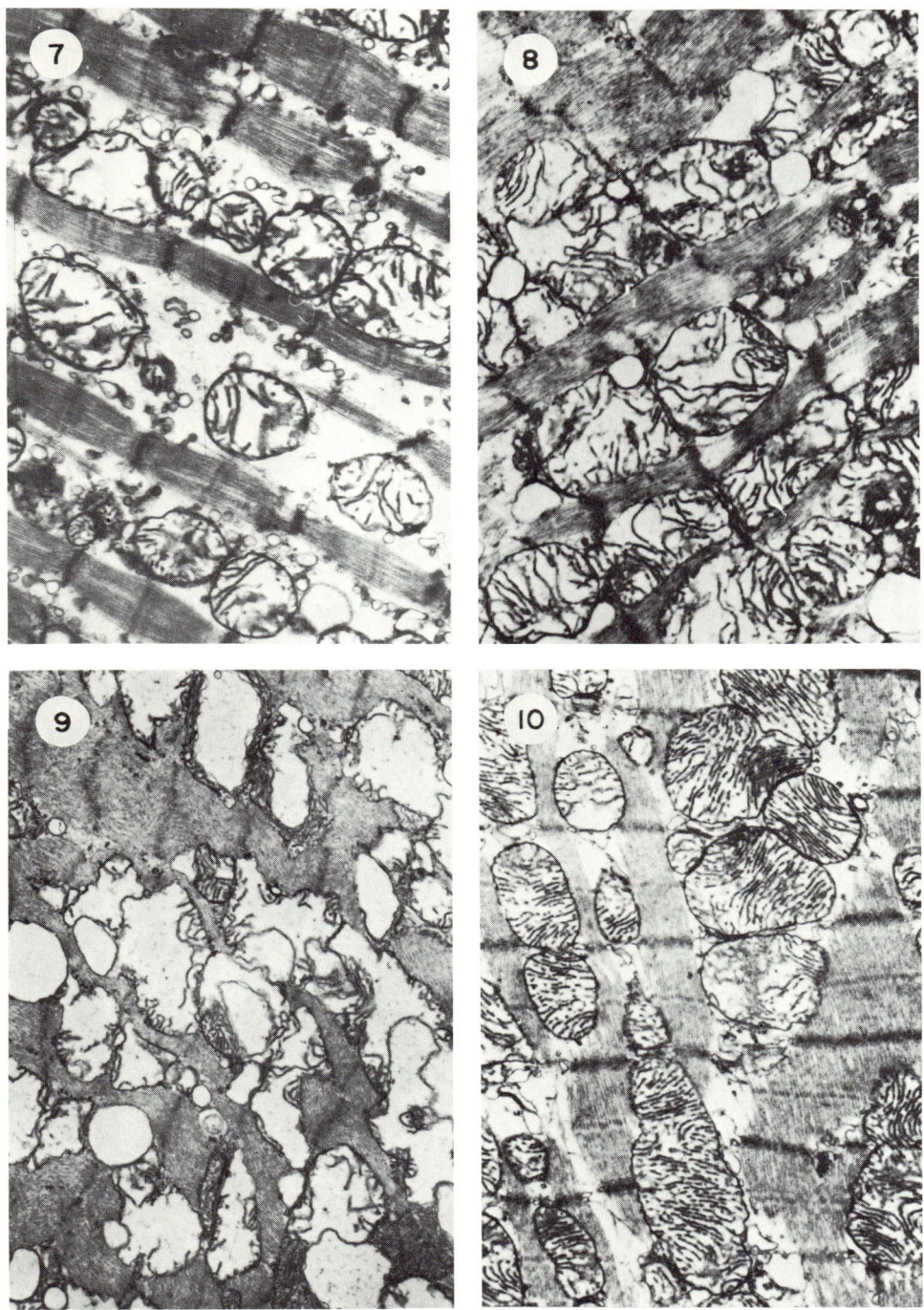

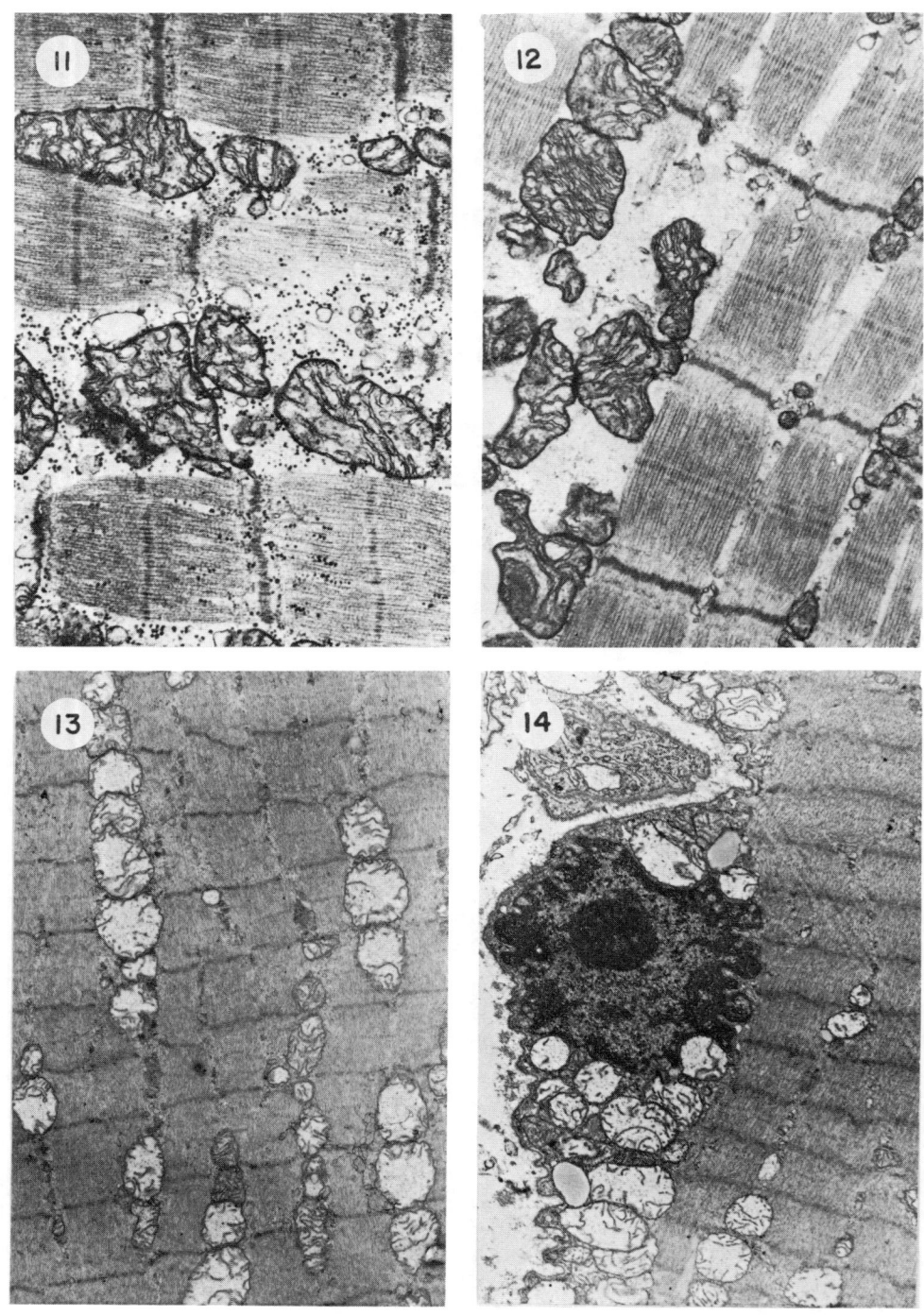

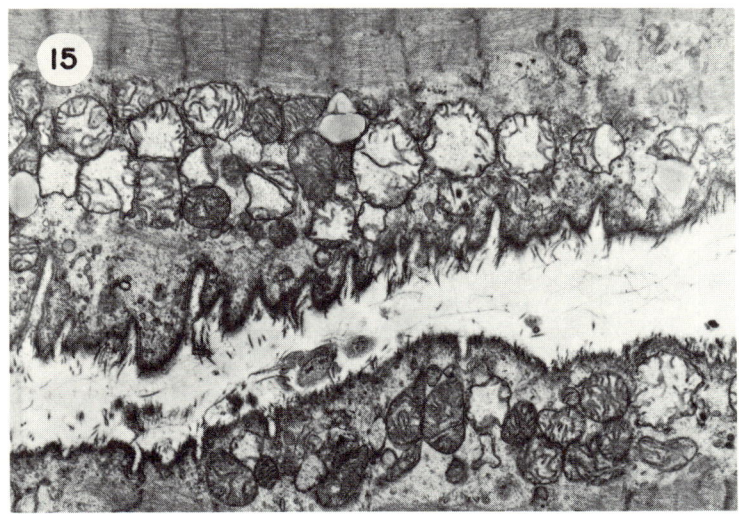

EFFECTS OF LONG-TERM EXERCISE ON HUMAN MUSCLE MITOCHONDRIA

T.E.Morgan, L.A.Cobb, F.A.Short, R.Ross and D.R.Gunn

Departments of Medicine, Pathology and Orthopedics

University of Washington School of Medicine, Seattle

In this study we have examined muscle adaptations to long term exercise in man. Studies of Holloszy (1), Kraus (2), Gollnick (3) and their coworkers leave little doubt that exercise training stimulates muscle mitochondrial changes. We will present further evidence that exercise training in man stimulates mitochondrial growth, oxidative capacity, and capacity for syntheses of glycogen and lipid.

Ten men exercised one leg 2 hours daily for one month by pedalling a bicycle ergometer. The exercise load was increased progressively from 300 to 900 kg-m/min. Each subject pedalled with one leg while his other leg remained at rest. Thirty-six to 48 hours after the last exercise period, the quadriceps muscles of both the trained and untrained legs of each of the ten subjects were biopsied. Small portions of the biopsy specimen were prepared for histochemical and electron microscopic examination; the bulk of the biopsy was used for chemical analysis, enzyme assay, in vitro incubation, and separation of mitochondria. This experimental model of exercise training enabled us to compare trained muscle and untrained muscle from the same subject, minimizing the problem of wide differences in muscle metabolism among individuals. Statistical analysis of the data was done by t-test of paired observations on control and trained muscle of each subject. The areas of muscle metabolism studied were glycogen metabolism and glycolysis; fatty acid and lipid metabolism, and mitochondrial metabolism and morphology.

We first focused on glycogen metabolism and glycolysis. We have shown in previous studies that glycogen content is increased in trained human muscle at rest (4). We measured glycogen concen-

Table I
Glycolytic Enzymes in Human Quadriceps Muscle

	Control	Trained	P
Hexokinase	0.75 ± 0.19^a	1.15 ± 0.24^a	<0.01
Glycogen synthetase D	$0.0058 \pm .0035$	0.0077 ± 0028	<0.03
Phosphorylase	28.4 ± 12.3	31.3 ± 11.9	n.s.
αGlycerophosphate dehydrogenase	15.6 ± 4.0	11.4 ± 2.7	n.s.
Phosphofructokinase	24.5 ± 3.4	17.9 ± 2.7	n.s.
Lactic dehydrogenase	236 ± 41	183 ± 19	n.s.
Aldolase	46.5 ± 6.3	43.9 ± 5.4	n.s.
Pyruvate kinase	247 ± 28	240 ± 22	n.s.

^{a}Micromoles/min/g. wet muscle, means ± standard error of the mean.
n.s. = not significant differences

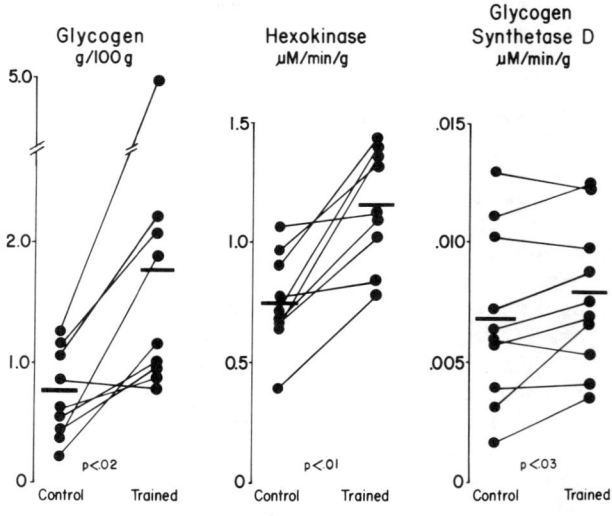

Figure 1. Results of paired duplicate determinations. Horizontal bars indicate means of control or trained (exercised) values.

-tration and certain enzymes involved in glycogen synthesis and glycolysis in homogenates of biopsied muscle. The enzymes involved in glycogenolysis and glucose metabolism, including phosphorylase, phosphofructokinase, aldolase, alpha-glycerophosphate dehydrogenase, pyruvate kinase, and lactate dehydrogenase were not stimulated by exercise training (Table I). Hexokinase, glycogen synthetase D and glycogen were affected, however, as shown in Figure 1 and Table 1. Exercise training stimulated an increase of glycogen concentration in all but one subject and hexokinase activity in each subject; glycogen synthetase D increased in 7 of 10 subjects.

We next examined the influence of exercise training on lipid metabolism. The possibility that intracellular muscle lipid might serve as a source of fatty acid has been debated for several years (5,6). Intracellular triglyceride values, Table II, were similar to those reported by Froberg (7). Triglyceride was not significantly increased in response to training in contrast to studies reported earlier (6); however, intracellular stores are large and complete oxidation of these lipids could provide 2000 to 2400 kcal. of energy in a 70 kg. man.

Table II
Lipid values in Human Quadriceps Muscle

	Control	Trained	P
Triglyceride[a]	8.78 ± 1.35	10.56 ± 1.22	n.s.
Carnitine[a]	3.49 ± 0.23	4.07 ± 0.33	n.s.
Acylcarnitine transferase[b]	7.12 ± 1.26	6.45 ± 1.26	n.s.

[a] μM/g wet muscle.   [b] μM acylcarnitine/g wet wt./hr.
means ± s.e.   n.s. = not significant differences

Figure 2 diagrams the metabolism of muscle incubated _in vitro_ with fatty acid. Entry of fatty acid into the intracellular fatty acid pool and incorporation into triglyceride of trained muscle were increased. Oxidation of palmitate to CO_2 was not stimulated by exercise training. Incorporation into phosphatidyl choline, which is a predominantly structural lipid, was likewise not affected. The work of Wittels and Spann (8) on failing guinea pig heart suggested that the control point of fatty acid oxidation by muscle may be the carnitine mediated transfer of fatty acid at the mitochondrial membrane. Fatty acid movement into mitochondria is effected by coupling of fatty acid with carnitine to form an acylcarnitine, catalyzed by the enzyme acylcarnitine transferase. Our data, however, indicate that this transfer is not a likely control point in human muscle. First, carnitine and acylcarnitine transferase levels were not affected by exercise training (Table II)

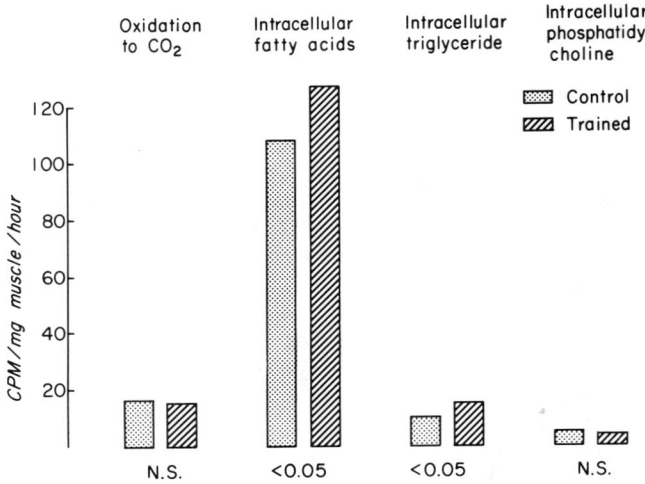

Figure 2. 100 mg teased fibers were incubated 2 hrs. in 2.0 ml Krebs-Ringer Bicarbonate containing 200,000 cpm palmitate-2-C^{14} on albumin carrier. Gas phase 5%CO_2-95%O_2, 37. Fibers were extracted, chromatographed on thin layer, eluted and counted. CO_2 was trapped in Hyamine.

and second, added carnitine did not alter palmitate oxidation by muscle fibers *in vitro*. A partial explanation of this lack of effect may be related to the observation that the muscle of man contains ten times as much carnitine as rodent muscle (Table II). Rodent values in our laboratory were comparable to those reported by others (9).

Holloszy has shown that muscle mitochondrial metabolism in rats is stimulated by training (1). We have confirmed in our human subjects that muscle mitochondrial protein, phospholipid, respiratory enzymes, and oxygen consumption were stimulated by training, and we have correlated the biochemical and morphologic evidence for mitochondrial growth. Mitochondria isolated from both trained and control muscle exhibited tightly coupled oxidative phosphorylation and normal respiratory control when malate-pyruvate oxidation was stimulated by added ADP. The data of Figure 3 show that mitochondrial protein and phospholipid, components of mitochondria membrane, were significantly increased in the trained muscle. We inferred from this data that the number of mitochondria, or their size, or both, might be increased by training.

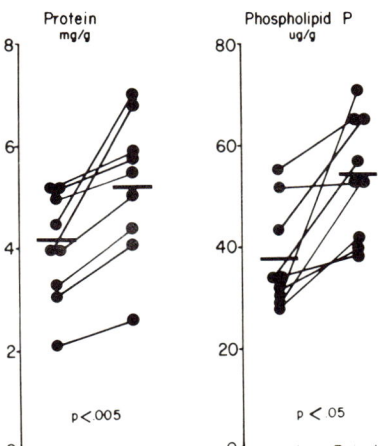

Figure 3.

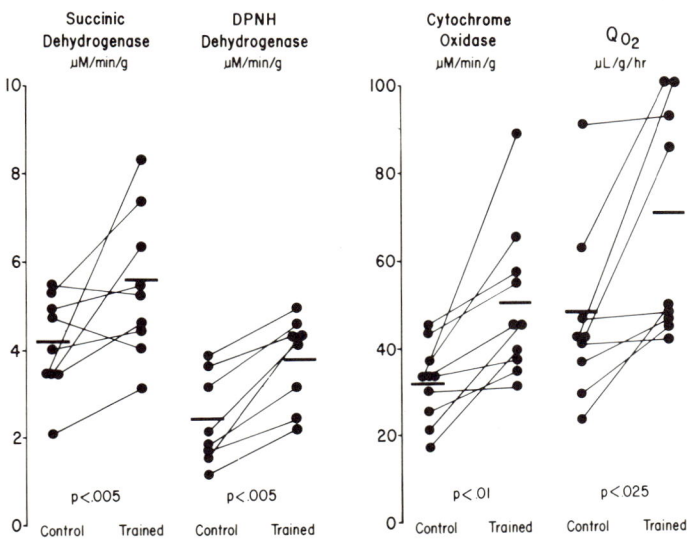

Figure 4.

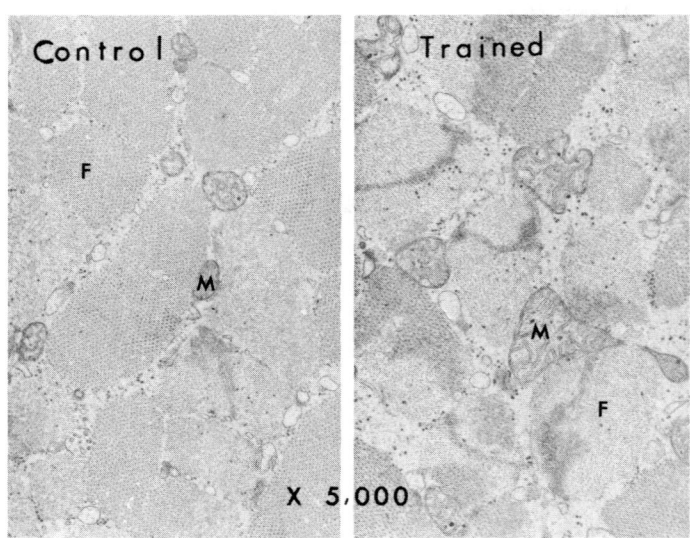

Figure 5. Representative electron micrographs of control and trained (exercised) muscle. Morphometric analysis was performed on similar photographs taken at random after the method of Weibel. Biopsies were taken in clamps, fixed in O_sO_4 and dehydrated. The clamps were removed before embedding in Epon. M - mitochondria, F - myofibrils.

Table III
Morphometric Analysis of Mitochondria

	Control	Trained	P
Volume	3.42 ± 0.29	5.31 ± 0.37	<0.01
Surface: Volume	3.23 ± 0.17	2.61 ± 0.12	<0.01
Number	21.0 ± 1.8	25.6 ± 2.1	<0.10

Arbitrary units; means ± s.e. Micrographs from 8 subjects were studied by the method of Weibel (see Fig. 5 and text).

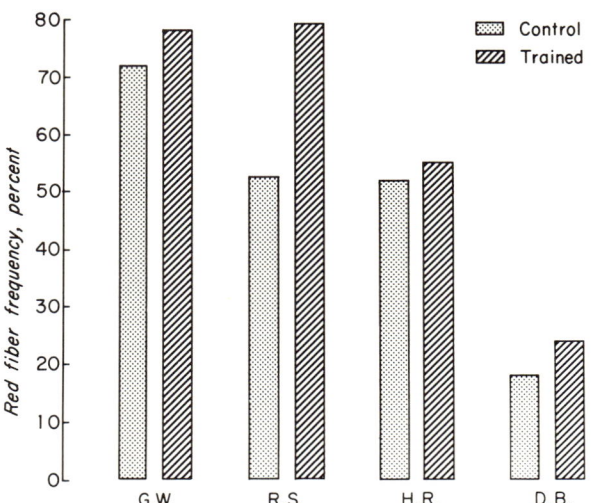

Figure 6. Optical densities of fibers determined photometrically on random frozen cross-sections stained for succinic dehydrogenase activity. Red fibers were those having an O.D. less than 60%.

Figure 4 depicts the marked stimulation in trained muscle mitochondria of succinic dehydrogenase, DPNH dehydrogenase, cytochrome oxidase, and oxygen consumption. All of these changes are statistically significant.

Finally, we have related biochemical to microscopic changes. Figure 5 illustrates the remarkable contrast in morphology of mitochondria in electron micrographs of cross sections from control and trained muscle of a representative subject. The mitochondria of trained muscle are obviously larger than those of the control but to quantify the striking morphological change, we calculated the mitochondrial surface and volume from electron micrographs, using the stereological technique described by Weibel (10). Table III tabulates the effect of training on mitochondria of 8 subjects. Mitochondrial volume increased in the trained muscle to a highly significant degree, but the number of mitochondria was not affected by training. The ratio of mitochondrial surface area to volume decreased, as would be expected with increased mitochondrial cross-sectional area. This change, too, is highly significant and simply adds support to the data demonstrating the increase in volume. The question whether the increase in mitochondrial mass is associated with an increase in proportion of red (mitochondria-rich fibers (11) was evaluated in 4 subjects. In these subjects frozen cross-sections were stained for succinic dehydrogenase. The stained sections were photographed at random. Fiber optical densities measured photometrically might be expected to increase as mitochondria increase and in Figure 6 a tendency toward more red fibers was found in the four subjects studied.

SUMMARY

Exercise training in man stimulated the activities of muscle hexokinase and glycogen synthetase, and increased the concentration of muscle glycogen. Palmitic acid incorporation into intracellular fatty acid and triglyceride were increased in trained muscle incubated *in vitro*. Mitochondrial membrane components, respiratory enzymes, and oxygen consumption were all stimulated by training. Mitochondrial size were increased remarkably.

It is evident that this form of endurance training is accompanied by increased oxidative capacity of trained muscle, and by enhanced capacity to synthesize two potential intracellular energy stores, glycogen and triglyceride.

ACKNOWLEDGEMENTS

This work was supported in part by grants from the U.S.P.H.S. (ROI-22354 and HE-07478) and from the Washington State Heart Association. This study was performed in part at the University of Washington Clinical Research Center at Harborview Medical Center which is supported by N.I.H. Grant. No. RR-133.

Drs. Cobb and Morgan are recipients of U.S.P.H.S. Research Career Awards (K3-HE-4570 and K3-HE-07268).

REFERENCES

1. Holloszy, J.O., Biochemical adaptations in muscle. Effects of exercise on mitochondrial oxygen uptake and respiratory enzyme activity in skeletal muscle. J. biol. Chem. 242: 2278 (1967)

2. Kraus, H., R. Kirsten and J.R. Wolff, Die Wirkung von Schwimm- und Lauf-training auf die celluläre Funktion und Struktur des Muskels. Pflügers Arch. ges. Physiol. 308: 57 (1969)

3. Gollnick, P.D. and W.D. King, The immediate and chronic effect of exercise on number and structures of skeletal muscle mitochondria. In Biochemistry of exercise. Basel/New York, S. Karger 1969, 239.

4. Short, F.A., L.A. Cobb and T.E. Morgan, Influence of exercise training on in vitro metabolism of glucose and fatty acid by human skeletal muscle. In Biochemistry of exercise. Basel/New York, S. Karger 1969, 122.

5. Masoro, E.J., L.B. Rowell, R.M. McDonald and B. Steiert, Skeletal muscle lipids. II. Nonutilization of intracellular lipid esters as an energy source for contractile activity. J. biol. Chem. 241: 2626 (1966)

6. Morgan, T.E., F.A. Short and L.A. Cobb, Effect of long-term exercise on skeletal muscle lipid composition. Amer. J. Physiol. 216: 82 (1969)

7. Fröberg, S., L.A. Carlson and L.G. Ekelund, Local lipid stores and exercise. This volume.

8. Wittels, B. and J.F. Spann, Defective lipid metabolism in the failing heart. J. clin. Invest. 47: 1787 (1968)

9. Brockhuysen, J., C. Rosenblum, M. Ghislain and G. Deltour, Distribution of carnitine in the rat. In Recent research on carnitine. Cambridge, M.I.T. Press, 1965, 23.

10. Weibel, E.R., G.S. Kistler and W.F. Scherle, Practical stereological methods for morphometric cytology. J. Cell. Biol. 30: 23 (1966)

11. Padykula, H.A. and G.F. Gauthier, Morphological and cytochemical characteristics of fiber types in normal mammalian skeletal muscle. In Exploratory concepts in muscular dystrophy and related disorders. Excerta med. Int. Congress, ed. A.T. Milhorat, Series No. 147: 117 (1967)

EFFECT OF PHYSICAL TRAINING ON ULTRASTRUCTURAL
FEATURES IN HUMAN SKELETAL MUSCLE

K.-H. Kiessling, K. Piehl and C.-G. Lundquist

Institute of Zoophysiology, University of Uppsala

Box 560, Uppsala, Sweden

It has been demonstrated that prolonged physical training of rats is followed by an increase in enzyme activity and in number and size of the skeletal muscle mitochondria (1-4). Furthermore, running to exhaustion produced marked swelling of the muscle mitochondria in rat (3) whereas acute exercise to exhaustion produced no morphological changes in the skeletal muscle mitochondria in human subjects (5).

The present study was undertaken in order to determine whether mitochondrial changes occur in human skeletal muscle after endurance exercise. Fourteen volunteers, all men between the ages of 18 to 25 took part in a regular exercise program during 28 weeks. The program consisted of the ordinary exercise involved in their military service and, in addition, running a distance of 4 miles 2 to 3 times a week. Muscle biopsies were taken from Vastus lateralis on three occasions, at the beginning of the period, after 14 weeks and after 28 weeks. Maximal oxygen uptake (determined with the Douglas bag method) at the three occasions is given in Table 1. Three subjects had a higher value after 14 than after 28 weeks which partly explains why there is only an insignificant rise in mean value between 14 and 28 weeks.

The muscle pieces were fixed in 1 % osmiumtetroxid and embedded in Epon. Photographs of longitudinal sections and of perinuclear zones were taken and the electron micrographs were evaluated quantitatively by means of the double lattice system described by Weibel, Kistler and Scherle (6). The number of mitochondria per 100 μ^2 of tissue and the mean area of the mitochondria were estimated in the interfibrillar space

Time of exercise (weeks)	Body length (cm)	Body weight (kg)	Oxygen uptake (ml/kg x min)	P
0	178	70.0	47.8 ± 1.0	
14	179	71.5	54.6 ± 2.1	0.01 - 0.001
28	178	70.9	55.2 ± 1.9	

Table 1. Maximal oxygen uptake. The figures are obtained from a separate investigation on the same material as used by B. Saltin (unpublished data).

Time of exercise (weeks)	Interfibrillar space		Perinuclear space	
0	19.4 ± 1.8		81.5 ± 10.5	
		p 0.01 - 0.001		p 0.2 - 0.1
14	29.0 ± 2.7		107.3 ± 14.8	
		$p < 0.001$		p 0.02 - 0.01
28	42.6 ± 2.7		154.1 ± 9.7	(0-28: $p < 0.001$)
		p 0.3		p 0.2 - 0.1
Athletes	47.8 ± 4.3		173.1 ± 8.0	

Table 2. Number of mitochondria/100 μ^2. Fifteen photographs from three different muscle pieces have been evaluated from each subject. The figures are mean values with their standard errors. On account of the typical localization of the mitochondria in the interfibrillar spaces longitudinal sections allow a more reliable estimation of their frequency. In transverse sections the small mitochondria are easily overlooked.

Time of exercise (weeks)	Interfibrillar space		Perinuclear space	
0	11.1 ± 0.6		11.1 ± 0.8	
				p 0.6 - 0.5
14	12.0 ± 0.9		11.9 ± 1.1	
				p 0.7
28	11.1 ± 0.9		12.4 ± 0.8	
		$p < 0.001$		p 0.1
Athletes	15.7 ± 0.9		15.0 ± 1.4	(0-athl.: p 0.05 - 0.02)

Table 3. Mitochondrial size ($\mu^2 \times 10^{-2}$). Fifteen photographs from three different muscle pieces have been evaluated from each subject. The figures are mean values with their standard errors.

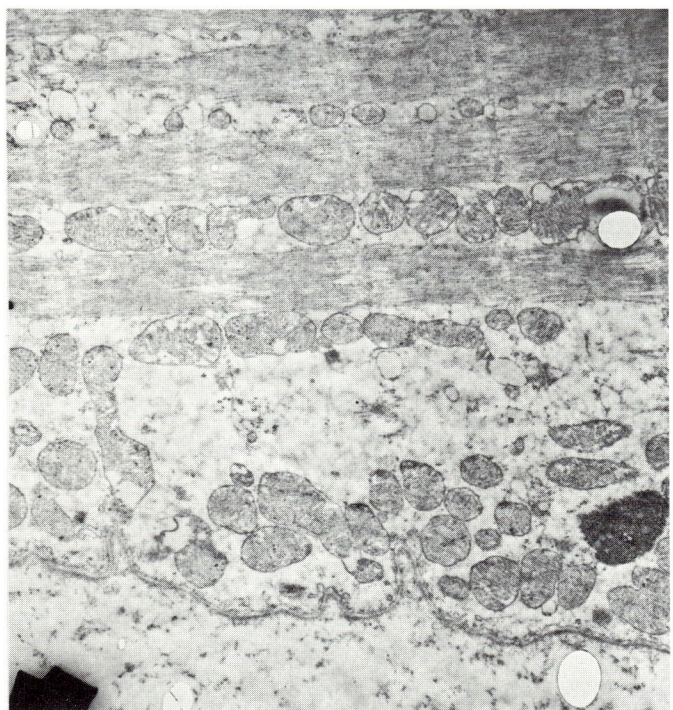

Fig. 1. Electronmicrograph of longitudinal section of skeletal muscle from man (vastus lateralis). Mitochondria of various size and shape are seen in the interfibrillar and the perinuclear spaces. 12000x.

and in the perinuclear zone.

Table 2 shows the number of mitochondria after 14 and 28 weeks of training. For comparison results from seven young men belonging to the elite of Swedish athletes in endurance events are also presented. Their speciality is either cross-country running or cross-country skiing.

Interfibrillary, a remarkable increase in number is seen in the fourteen volunteers already after 14 weeks. After 28 weeks there are twice as many mitochondria as in the beginning of the training period. The number of mitochondria in the athletes is however only insignificantly higher than in the volunteers after 28 weeks of training.

Also in the perinuclear space a corresponding increase in the number of mitochondria can be observed as a consequence of prolonged training.

In contrast the size of the mitochondria is mainly unaltered. Although mitochondria of various sizes can be found between the fibrils as well as around the nuclei (Fig.1) the mean size is strikingly constant throughout the whole training period (Table 3). Only in the interfibrillar space of the athletes a significant increase in size is found. This increase also exists in the perinuclear zone, although not to the same degree.

The results indicate that prolonged but moderate training causes an increase of the number of mitochondria without concomitant alteration of their size. A heavy training program causes only a slight further rise of the number but a pronounced increase of the mitochondrial size.

On the whole our results from human volunteers agree with those from rats observed by Holloszy (1) by Kraus and coworkers (2) and by Gollnick and King (3). Holloszy reports an increase of mitochondrial protein whereas the electron microscopical studies by Kraus et al and by Gollnick and King reveal that both the size and number of mitochondria had increased after training. However, to our knowledge, it has not been observed in animals that an increase in the number of mitochondria precedes an increase in size during training. Our observations that this occurs in man may either be art specific or a consequence of differences in the training programme.

The advantage of an increase in mitochondrial activity is at least 2-fold. An increased capacity to form ATP is certainly most important. The balance between mitochondrial function and lactate level may also be of importance. Hill and Kupalov showed many years ago that a rise in the lactate level in muscle was important in the development of muscle fatigue during exercise (7). Asmussen and coworkers (8) suggested that maximal performance might be limited by critical values of lactate concentration in the working muscles. Results by Karlsson et al (personal communication) support this suggestion. Metabolic changes counteracting this rise would therefore favour an increased capacity for prolonged submaximal physical activity, as has also been pointed out by Holloszy (1). An increased capacity for fatty acid oxidation would slow down glycolysis with decreased formation rate of pyruvate and extramitochondrial NADH as a consequence. This effect together with the increased mitochondrial oxidation of pyruvate and probably also of extramitochondrial NADH by means of shuttle systems would contribute to a reduced lactate level.

REFERENCES

1. Holloszy,J.O. Biochemical adaptions in muscle. Effects of exercise on mitochondrial oxygen uptake and respiratory enzyme activity in skeletal muscle. J.Biol.Chem. 242 (1967) 2278.
2. Kraus,H.,Kirsten,R. and Wolff,J.R. Die Wirkung von Schwimm- und Lauftraining auf die celluläre Funktion und Struktur des Muskels. Pflügers Arch. 308 (1969) 57.
3. Gollnick,P.D. and King,D.W. Effect of exercise and training on mitochondria of rat skeletal muscle. Am.J.Physiol. 216 (1969) 1502.
4. Kowalski,K.,Gordon,E.E., Martinez,A. and Adamek,J. Changes in enzyme activities of various muscle fiber types in rat induced by different exercises. J. Histochem.Cytochem. 17 (1969) 601.
5. Gollnick,P.D.,Ianuzzo,D.C. and D.W. King. Ultrastructural and enzyme changes in muscles with exercise. Karolinska Institutet Symposia: Muscle Metabolism During Exercise, Stockholm Sept. 1970.
6. Weibel,E.R. Kistler,G.S. and Scherle,W.F. Practical stereological methods for morphometric cytology. J.Cell.Biol. 30 (1966) 23.

METABOLIC ROLE OF MUSCLE

George F. Cahill, Jr.

Joslin Research Laboratory

Harvard Medical School, Boston, Massachusetts

Muscle's primary function is contraction, but thanks to its relatively great mass and protein content, it serves several vital metabolic roles. It is a most important buffer to acid-base alterations and even changes its own metabolic reactions as a response to increases or decreases in hydrogen ion concentration. It also serves as a buffer to other ion changes, such as potassium or magnesium. Less emphasized has been its role as the body's principle nitrogen reservoir, or more accurately, as the body's depot of substrate for gluconeogenesis. This brief note addresses itself to this last point.

Adipose tissue is man's predominant energy store thanks to the high caloric content and anhydrous nature of triglyceride. Normal man carries 10-20 kilograms of triglyceride, adequate calories for survival for 2-3 months. In contrast, available carbohydrate reserves as liver and muscle glycogen and glucose in body fluids account for only 1000-1500 calories. Body protein in normal man approximates 10-11 kilograms of which 1/2 to 2/3 is in muscle, but unlike adipose triglyceride and glycogen in liver or muscle, every molecule of protein in both muscle and other tissues is serving a primary role for other than caloric storage. For example, the protein may be an enzyme or a hormone, or may be providing an oncotic purpose, such as albumin, or a structural role like collagen. In muscle, the bulk of protein is as actin and myosin, serving thereby its principle role in the contractile process. As far as we know, no protein in man is stored for storage alone, it is either integrated into the body machinery or, if the amino acid intake is in excess, it is metabolized, the nitrogen excreted and the caloric equivalents consumed or stored as adipose tissue triglyceride.

The earliest isotopic data of Rittenberg and Schoenheimer gave the impression of a definite but slow turnover of muscle protein. More recent data, particularly those of Waterlow and associates (1,2) with ^{14}C-lysine in rats and man and ^{14}C-glycine in rats (3) neither of which transaminate in muscle to any great degree, have shown turnovers in young men of 3 grams protein/kg body weight/day. Confirmatory data have been obtained studying net amino acid exchange across muscle beds in the post-absorptive state, in which a significant release of amino nitrogen has been observed (4-6). That this was not simply release of stored amino acids was confirmed by its magnitude and its persistence for weeks of total starvation (6). Thus muscle protein is not only continuously turning over, but also expanding and diminishing in accord with dietary intake.

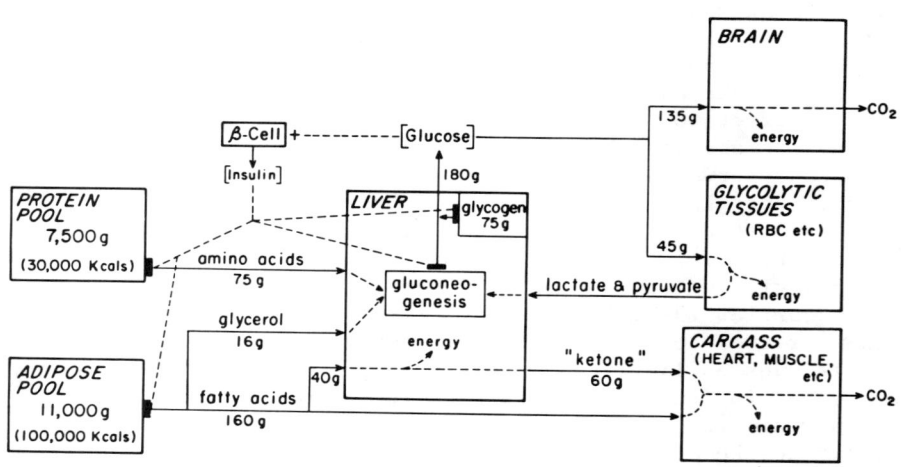

Fig. I

Idealized interrelations of various fuels, their sites of origin and sites of utilization in a fasting man. Insulin levels, as determined by Beta cell release of insulin which, in turn, is determined by ambient glucose and amino acid levels, appear to control the rate of amino acids from muscle protein and free fatty acids from adipose tissue. Of note is glucose exclusion from all tissues other than nerve and the obligatory glycolyzers such as the red cells.

Since activation of amino acids, their incorporation into protein and their hydrolytic cleavage back to free amino acids are energy-wasting reactions, the continuous expansion and contraction of muscle protein appears nonsensical unless it serves an important role when integrated into the fuel homeostasis of the entire body.

Human fuel interrelationships: Combining data from indirect calorimetry and values obtained form arterio-venous differences of substrates, my associates and I have tried to assemble schemes of fuel transfer from depots to consumers (7-9). Fasting man, near basal at 1800 calories/day, mobilizes 75 g of muscle protein and 160 g of adipose triglyceride. Splanchnic glucose production approximates 180 g of which 144 g are metabolized by the central nervous system, and the difference is utilized by obligatory glycolyzers such as erythrocytes, renal medulla, peripheral nerve and probably others. The remainder of the carcass consumes free fatty acids or the products of hepatic oxidation of free fatty acids, β-hydroxybutyrate and acetoacetate. These idealized figures are shown in Fig 1.

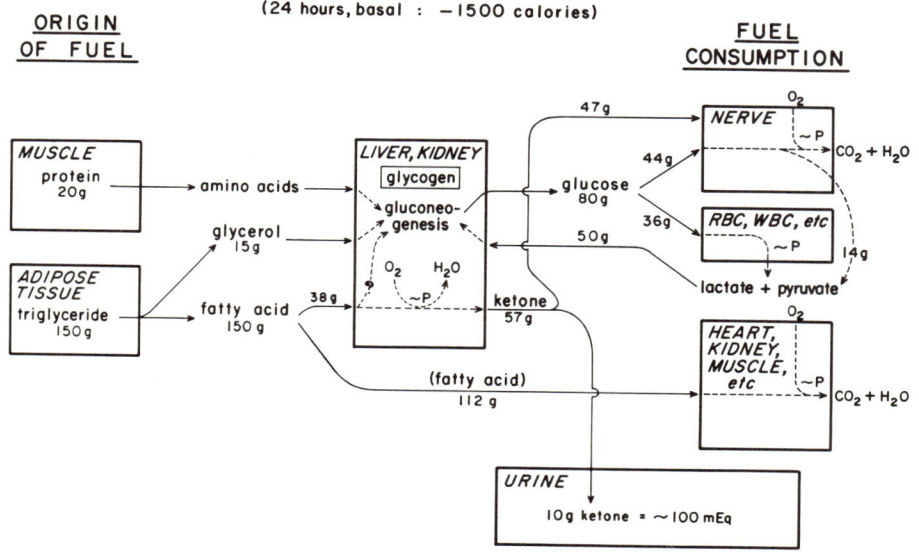

Fig. 2

Idealized interrelations of various fuels, their sites of origin and sites of utilization in a man fasting for several weeks. Brain has adapted to ketoacid utilization. Muscle has diminished its own ketoacid utilization to help provision the brain. The resultant of each of these is a sparing of daily nitrogen catabolism, thereby prolonging survival.

As fasting continues, brain apparently adapts to utilization of β-hydroxybutyrate and acetoacetate, sparing thereby, the major proportion of muscle protein catabolism (10). These data are shown in Fig. 2.

The important point of showing these schemes is to illustrate how muscle is integrated into the needs of the body. Firstly, it excludes glucose as fuel, and instead consumes free fatty acids and ketoacids. Secondly, as brain adapts to ketoacids, muscle undergoes another alteration and consumes free fatty acids in preference to the ketoacids. This latter adaptation has been recently shown conclusively by Owen and Reichard (11). From a nitrogen viewpoint, muscle proteolysis in both short term and prolonged fasting provides precisely the glucogenic precursor for hepatic (and renal) glucose production. Likewise exclusion of glucose, and later, of ketoacids spares nitrogen. The next question is how does the body integrate muscle metabolism into its overall needs.

Muscle amino acid metabolism: Forearm muscle of fasting man shows a pattern of amino acid release distinct from muscle protein composition (6), in contrast to its uptake during states of rapid protein synthesis such as after insulin (12). Table I lists the arteriovenous differences in plasma amino acids in both the postabsorptive state and after prolonged starvation. Also listed is an approximation of the rate of release as compared to muscle protein content derived not from direct analysis of human muscle amino acid composition but from rough calculations from known compositions of muscle constituents in the literature. The predominance of alanine and glutamine are apparent. Thus muscle in the anabolic state, as shown by Lotspeich (12), takes up amino acids in a pattern conforming to its composition, and during catabolism, metabolizes a proportion of the amino acids to alanine and glutamine, oxidizes others such as leucine isoleucine and valine to CO_2 and releases others such as glycine and lysine intact. The predominance of alanine and glutamine are logical, since alanine is the principle glucogenic amino acid extracted by liver, and glutamine, a principle amino acid for hepatic and renal extraction.

Fig. 3 schematically summarizes muscle protein synthetic and catabolic reactions. Much interest has been directed to whether exercise, androgen, growth hormone and other hormonal or physical factors, such as innervation, effect muscle protein amino acid trapping, protein synthesis or protein catabolism. Many effects have been observed and will not be discussed here, but in view of the daily lability of muscle protein, it appears that the crucial mechanism is not related to altered synthesis or catabolism, but rather to the homeostatic processes within the muscle cell itself which signals it to contain a certain number of molecules of contractile protein, the optimal "setting" which each of us possess for our age, sex and physical status.

MUSCLE PROTEIN METABOLISM

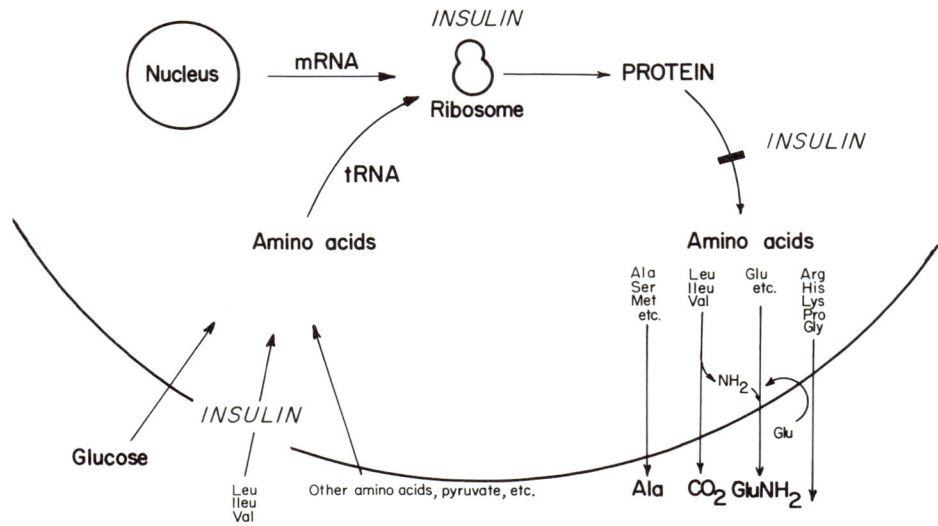

Fig. 3
Scheme for muscle protein synthesis and catabolism. Insulin augments uptake of amino acids, glucose and other anabolic factors such as phosphate and potassium. Insulin also exerts an effect in the ribosomal aggregate, and data also suggest that insulin may directly decrease proteolysis. Amino acid release patterns are discussed in the text.

In common language, increasing dietary protein intake to levels far above that in an average satisfactory intake with adequate essential amino acids and non-essential nitrogen does not increase muscle mass, in spite of training table mythology! Also, subjects with acromegaly or islet cell tumors do not progressively expand their muscle mass.

In summary, muscle is man's principle nitrogen depot. Its metabolism is thereby integrated into that of the remainder of the body, providing glucogenic substrate to liver to feed brain in short-term starvation and adapting to utilization of fatty acids and ketoacids. In prolonged starvation, it diminishes its catabolic rate and also adapts to using selectively fatty acids in preference to ketoacids. Muscle protein content in each of us varies about an "optimal norm" which is a function of age, sex, exercise, and probably also heredity or cellularity. About this steady-state "norm" amino acids move in and out depending on dietary habits and needs for glucogenic precursor. This "norm" has separate chemical pathways for synthesis and catabolism, each with its own control, analogous to the separate pathways for synthesis and catabolism of other essential molecules such as glycogen, triglyceride, phospholipids, etc.

Table I

Amino acid	Arteriovenous differences		Approximate ratio of release to muscle protein content
	Post-absorptive	Prolonged-starvation	
Taurine	-3 ± 1	$+3 \pm 3$	1
Threonine	-28 ± 6	-11 ± 5	1
Serine	-7 ± 6	$+1 \pm 6$	1/4
Proline	-29 ± 10	-6 ± 7	1
Citrulline	$+3 \pm 1$	0 ± 1	1
Glycine	-36 ± 7	-19 ± 10	1
Alanine	-111 ± 20	-29 ± 9	2 to 3
α-aminobutyrate	-2 ± 1	0 ± 2	1
Valine	-16 ± 5	-2 ± 4	2/3
Cystine	$+9 \pm 3$	$+3 \pm 2$	—
Methionine	-6 ± 1	-3 ± 2	1/2
Isoleucine	-11 ± 4	-3 ± 1	1/2
Leucine	-14 ± 3	-6 ± 2	1/3
Tyrosine	-9 ± 2	-3 ± 1	1
Phenylalanine	-8 ± 2	-3 ± 1	1
Ornithine	-2 ± 2	0 ± 1	1
Lysine	-37 ± 9	-12 ± 2	1
Histidine	-14 ± 3	-4 ± 2	1
Tryptophan	-4 ± 2	—	1
Arginine	-23 ± 5	-2 ± 2	1
Glutamate	$+55 \pm 5$	$+27 \pm 4$	-1
Glutamine	-175 ± 28	-66 ± 14	2

Values in µM in plasma. Arteriovenous differences for all amino acids except glutamate and glutamine are from reference (6). Values for glutamate and glutamine have been collected by Drs. T. T. Aoki and E. B. Marliss using enzymatic procedures. The overall reduction in all amino acids released with prolonged starvation complies with the reduction in daily nitrogen excretion to 1/3rd of its original rate early in starvation. Nevertheless, the predominance of alanine and glutamine persist. Also, the fact that muscle in man can be such a prominent glutamine producer is surprising, and recent data collected by Drs. Aoki and Marliss have shown splanchic bed in man to exhibit a net removal and not a production of glutamine (in preparation). Net uptakes of glutamate and cystine are shown.

REFERENCES

1) Waterlow, J. C. and Stephen, J. M. L. The measurement of total lysine turnover in the rat by intravenous infusion of L-(U-^{14}C) lysine. Clin. Sci. 33: 489, 1967.
2) Waterlow, J. C. Lysine turnover in man measured by intravenous infusion of L-(U-^{14}C)lysine. Clin. Sci. 33: 507, 1967.
3) Garlick, P. J. Measurement of muscle protein turnover by constant intravenous infusion of (^{14}C)glycine. Biochem. J. 113: 7p 1970.
4) Carlsten, A., Hallgren, B., Jagenburg, R., Svanborg, A. and Werko, L. Arterio-hepatic venous differences of free fatty acids and amino acids. Acta Med. Scand. 181: 199, 1967.
5) Pozefsky, T., Felig, P., Tobin, J. D. Soeldner, J. S. and Cahill, G. F. Jr. Amino acid balance across tissues of the forearm in postabsorptive man. Effects of insulin at two dose levels. J. Clin. Invest. 48: 2273, 1969.
6) Felig, P., Pozefsky, T., Marliss, E. and Cahill, G. F. Jr. Alanine-key role in gluconeogenesis. Science 167: 1003, 1970.
7) Cahill, G. F. Jr. and Owen, O. E. Some observations on carbohydrate metabolism in man in Carbohydrate Metabolism and its Disorders: Chapt 16. Dickens, F., Randle, P. J. and Whelan, W. J. eds 1968 Academic Press, London pp 497-522
8) Owen, O. E. Felig, P., Morgan, A. P., Wahren, J. and Cahill, G. F. Jr. Liver and Kidney metabolism during prolonged starvation. J. Clin. Invest. 48: 574, 1969.
9) Felig, P., Owen, O. E., Wahren, J. and Cahill, G. F. Jr. Amino acid metabolism during prolonged starvation. J. Clin. Invest. 48: 584, 1969.
10) Owen, O. E. Morgan, A. P., Kemp, H. G., Sullivan, J. M., Herrera, M. G. and Cahill, G. F. Jr. Brain metabolism during fasting. J. Clin. Invest. 46: 1589, 1967.
11) Owen, O. E. and Reichard, G. A. Jr. Substrate extraction and/or production by forearm muscle during progressive starvation. Clin. Research 18: 461, 1970
12) Lotspeich, W. The role of insulin in the metabolism of amino acids. J. Biol. Chem. 179: 175, 1949.

ADRENERGIC NEURO-HUMORAL CONTROL OF LIPOLYSIS IN ADIPOSE TISSUE

Sune Rosell and Kathryn Ballard

Department of Pharmacology, Karolinska Institutet

104 01 Stockholm 60, Sweden

During exercise there is an elevated plasma level of catecholamines, especially noradrenaline (1,2). Since catecholamines are potent lipolytic agents it has been suggested that the adrenergic neurohumoral system is to a significant degree responsible for the mobilization of fat during exercise. In fact, infusion of catecholamines in animals and man elevates the free fatty acid (FFA) level in plasma (3).

However, the question arises whether elevated adrenergic neuro-humoral activity - within physiological limits - can induce lipolysis in adipose tissue from different localities and furthermore, whether both links, i.e. the sympathetic innervation of adipose tissue and circulating catecholamines from sources outside the adipose tissue are of equal quantitative importance.

To study these problems experiments were performed on subcutaneous, mesenteric and omental tissue in anaesthetized dogs. Adipose tissue blood flow and release of glycerol and FFA were measured during electrical stimulation of the appropriate sympathetic nerves or during intra-arterial infusion of catecholamines (4, 5).

In the inguinal region subcutaneous adipose tissue can easily be prepared to allow circulatory and metabolic studies (4). Maximal release rate (1.5 µEq/100g/min) was obtained at a stimulation frequency of about 3/s. At higher frequencies the release rate tended to decline, presumably due to increasing vasoconstriction (Fig. 1).

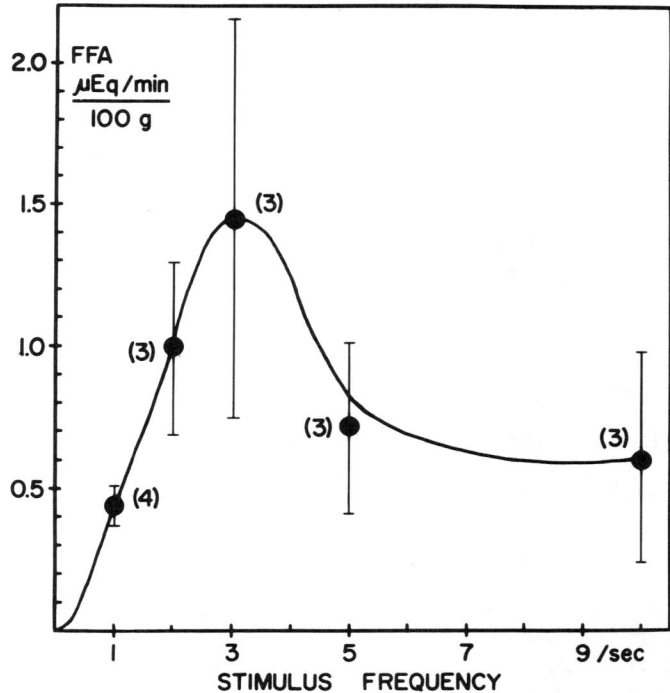

Fig. 1. Relation between stimulation frequency and the rate of net release of FFA. Canine subcutaneous tissue. From ref. (4).

These experiments show that the sympathetic nerve activity is effective in producing lipolysis at activity levels supposed to occur under physiological conditions. To judge from the work of Folkow (6), the impulse rate in the postganglionic nerves in the sympathetic nervous system, at least to skeletal muscle, is the order of 1-2/s and rarely exceeds 6-8/s. It is reasonable to assume that the sympathetic nerve fibres to adipose tissue have the same characteristics.

There is a rapid onset of the lipolytic response as shown in Fig. 2. A supramaximal stimulation frequency was chosen and the stimulation period was varied. Before the nerve stimulation there was a net uptake of FFA. Stimulation for 1 min. resulted in a small net release which was more pronounced with a stimulation period of 2 min. However, no significant response was noticed if the stimulation lasted for only half a minute.

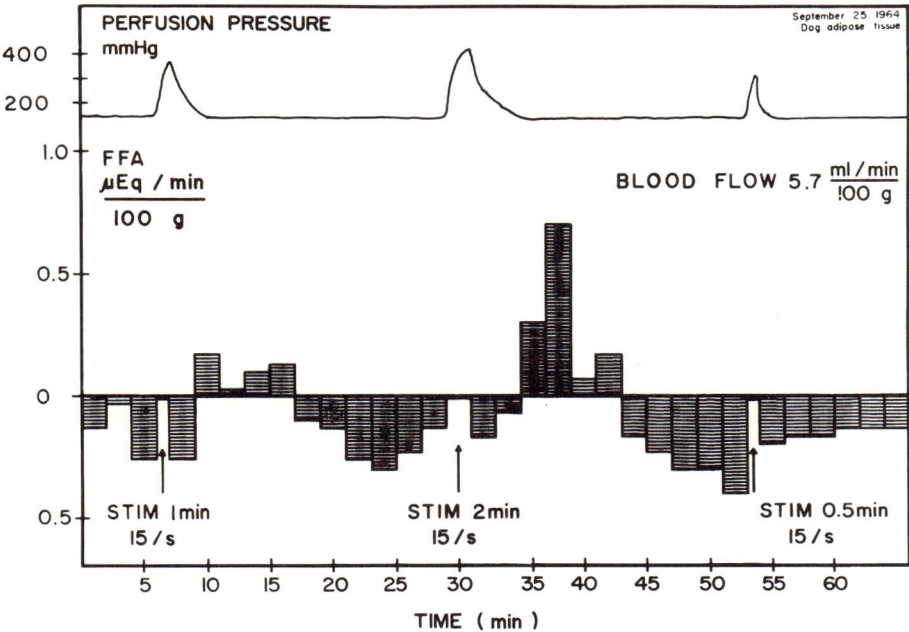

Fig. 2. Electrical stimulation of the sympathetic nerve to canine subcutaneous adipose tissue. Release of FFA following supramaximal stimulation (15/s). From ref. (4).

Infusion of adrenaline or noradrenaline also produced lipolysis as expected. However, to increase lipolysis the adrenaline or noradrenaline levels in arterial blood had to exceed concentrations reported to be present in resting dogs. As a matter of fact, significant augmentation of lipolysis was only seen at concentrations above 0.01 µg/ml. Such levels have been reported in humans during supramaximal work (2).

Comparatively speaking, activity in sympathetic nerves to subcutaneous adipose tissue appears to be more effective in mobilization of fat than circulating catecholamines. As shown in Fig. 1. stimulation of the nervous supply with 2/s released about 1 µmole/min/100 g FFA. To produce a comparable release rate by infusion of noradrenaline to the subcutaneous adipose tissue, the plasma concentration had to be raised to between 0.01 and 0.05 µg/ml. Since the frequency of 2/s is supposed to occur during resting conditions and the plasma concentrations mentioned during severe stress it seems obvious that the nervous supply plays a greater role for lipolysis than circulating catecholamines.

In the mesenteric adipose tissue we did not observe any change in the venous outflow of FFA or glycerol following sympathetic nerve stimulation. This consistant failure to induce fat mobilization constitutes an important difference in comparison with the responses in subcutaneous adipose tissue. Interestingly enough, it was also necessary to use very high doses of noradrenaline to induce lipolysis. The threshold dose for enhanced lipolysis in subcutaneous adipose tissue was about 0.1 µg i.a. whereas a similar degree of fat mobilization in the mesentery required injections of 20 µg, or more. Thus the influence of either released noradrenaline from sympathetic nerves in the mesentery or circulating noradrenaline is probably of no physiologic significance for the lipolysis. Instead other regulatory mechanisms may be of greater importance.

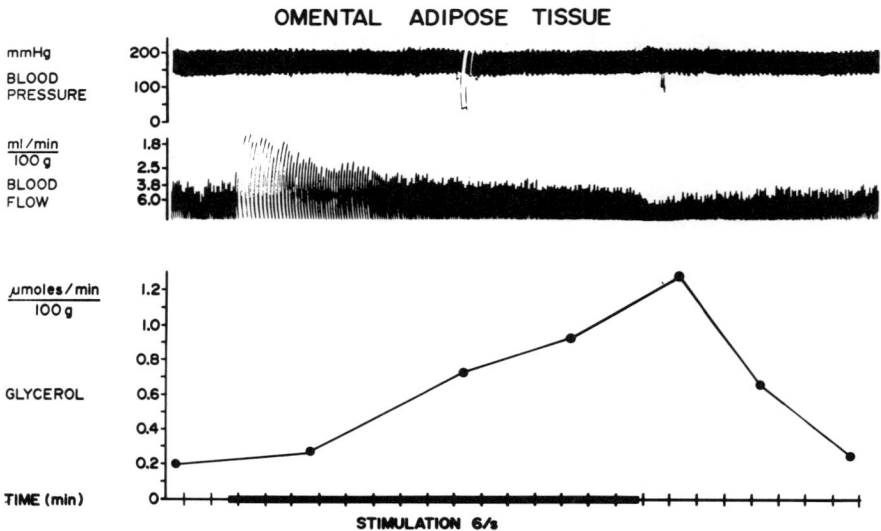

Fig. 3. Systemic blood pressure, blood flow and release of glycerol in canine omental adipose tissue. The nervous supply was stimulated electrically for 15 min.

In the omentum, on the other hand, both links in the adrenergic neuro-humoral system seem to be effective for fat mobilization. As can be seen in Fig. 3. stimulation of the sympathetic nerves caused an initial vasoconstriction and an elevated venous outflow of glycerol. Likewise, infusion of noradrenaline induced similar effects. However, as in the subcutaneous adipose tissue it seems as if concentrations above the physiologic level have to be approached to cause a significant lipolysis.

Our experiments thus indicate that there are regional differences of physiologic importance in the ability of the adrenergic neuro-humoral system to elevate the lipolytic rate. Furthermore, the sympathetic nerves to adipose tissue seem to play a greater role than circulating catecholamines in fat mobilization.

Once established that sympathetic nerve activity, at least to some regional fat depots, induces lipolysis, the question arises concerning the quantitative importance of the resulting inflow of fat for the energy requirement of the body during exercise. To elucidate this problem, we have taken the liberty of comparing our data with those by P. Paul published recently (9). Dr. Paul let dogs run for 4 hrs on a treadmill at three different levels of energy expenditure and on the basis of their oxygen consumption during exercise they were classified into three groups, A, B, C. The rate of oxidation of plasma FFA was calculated and in Table 1 the oxidation rate of plasma FFA during the first 2 hrs of work is depicted. To compare those data on plasma FFA oxidation with ours, we have assumed that a 10 kg dog has about 1 kg of adipose tissue. It is furthermore assumed that the release rate seen in Fig. 1. is characteristic not only for subcutaneous tissue but also for adipose tissue in other locations where sympathetic nerves exert a regulatory function. It is evident that the release rates found by us are not enough to account for the energy requirements during exercise. Moreover, not even the maximal release rates, produced by several means including combination of high frequency nerve stimulation with blockade of phosphodiestrerase with theophylline (10) are enough to account for more than a fraction of the FFA required for fat oxidation during exercise. Of course these calculations are open to critizism, but nontheless they point out that the adrenergic neuro-humoral system is not the only factor of importance in the promotion of lipolysis during exercise. Perhaps the sympathetic nerves are of prime importance for turning on the inflow of fuel and that other mechanisms including several hormones are required to maintain the supply of fatty acids for muscular work of long duration.

Table 1.   FFA METABOLISM DURING EXERCISE
            Data from P. Paul (9).

Work load	O_2 cons. ml/kg/min	Plasma FFA OXIDIZED during the first 2 hrs. µEq/kg b. weight/min
Rest	5	3
A	18	24
B	27	27
C	43	42

Data from S. Rosell (4).
Net FFA inflow following symp. nerve stim.
(assumption: Dog, 10 kg b. weight, 1 kg adipose tissue).

Stim freq.	Net inflow µEq/kg b. weight/min
1/s	0.5
2/s	1.0
3/s	1.5
Max. value	10

Acknowledgements Investigations reported here were supported by grants from the Swedish Medical Research Council (No: 14X-731), Svenska Läkaresällskapet and from Karolinska Institutet.

REFERENCES

1. Vendsalu, A., Studies on adrenaline and noradrenaline in human plasma. Acta physiol.scand. suppl. 173: 1 (1960)

2. Häggendal, J., Role of circulating noradrenaline and adrenaline. This volume

3. Havel, R.J., Autonomic nervous system and adipose tissue In: Handbook of Physiology, sect. 5: 575 (1965)

4. Rosell, S., Release of free fatty acids from subcutaneous adipose tissue in dog following sympathetic nerve stimulation. Acta physiol.scand. 67: 343 (1966)

5. Ballard, K., C.A. Cobb and S. Rosell, The relationship between lipolytic and vasoactive effects of adrenaline and noradrenaline on canine subcutaneous adipose tissue. Acta physiol. scand. (1971)

6. Folkow, B., Impulse frequency in sympathetic vasomotor fibre correlated to the release and elimination of the transmitter. Acta physiol. scand. 25: 49 (1952)

7. Watts, D.T., Adrenergic mechanisms in hypovolemic shock. Grune and Stratton 385 (1965)

8. Ballard, K. and S. Rosell, The unresponsivness of lipid metabolism in canine mesenteric adipose tissue to biogenic amines and to sympathetic nerve stimulation. Acta physiol. scand. 77: 442 (1969)

9. Paul, P., FFA metabolism of normal dogs during steady-state exercise at different work loads. J. Appl.Physiol. 28: 127 (1970)

10. Fredholm, B.B., Studies on the sympathetic regulation of circulation and metabolism in isolated canine subcutaneous adipose tissue. Acta physiol. scand. suppl. 354. (1970)

ROLE OF CIRCULATING NORADRENALINE AND ADRENALINE

Jan Häggendal

Department of Pharmacology, University of Göteborg

Göteborg, Sweden

The catecholamines (CA) adrenaline (A) and noradrenaline (NA) have evident cardiovascular and metabolic effects in man when given as an i.v. infusion within the dose range of 0.1 to 0.5 µg/kg/min, which is often used. The direct effect of CA on the heart is mainly ascribed to activation of β-receptors. Many of the metabolic effects appear also to be β-receptor mediated. The picture, however, is complicated by the observations that different types of β-receptors can be distinguished (11,18,19). Furthermore, the mechanisms of e.g. the hyperglycemic action of CA seem to be very complex. Of great interest for the basic knowledge of CA action is the effect of CA on the adenyl cyclase-cyclic AMP system (23) since cyclic AMP is discussed to mediate CA action (for rev. e.g. 21).

Plasma Levels of CA at Different Conditions

During CA infusion. During i.v. infusion of CA at a constant infusion rate, the concentration of circulating CA will increase and reach a plateau (for ref. 1,8). When NA was infused at a low rate, 0.08 µg/kg/min, a steady state venous plasma conc. of about 1.5 µg/litre was reached after about 10 min (24). The levels of the steady state were directly related to the rate of the NA infusion. With a higher infusion rate, about 0.25 µg/kg/min, the NA plateau has been found between 2 and 3 µg/l plasma (6).

At rest and during exercise. The levels of CA in blood are normally low at rest. Most authors have found about 0.2 to 0.4 µg NA/l blood plasma, with slight differences between arterial and venous blood. The values for A are lower, 0.1 to 0.3 µg/l for arterial plasma and 0.0 to 0.2 µg/l for venous plasma (for rev. 1).

According to my own experience, it is rare that even traces of A are detected in arterial plasma (with the method used (13)).

The levels of circulating CA may increase due to many factors, such as drugs (e.g. insulin) or exercise. Due to the aim of this Symposium, I will restrict myself to discuss the situation at muscle exercise.

The blood levels of CA appear to be reflected in the urinary excretion of CA. During mild to moderate exercise, the increase of the urinary excretion of NA is rather low when compared to the excretion at rest. The urinary NA excretion seems to be marked only at strenuous work where also the A excretion increases (c.f.7).

The levels of circulating NA in venous blood have been shown to increase at increasing work loads (e.g.24). From about 0.3 µg/l plasma at rest the NA levels increased to about 1 µg/l plasma at a work load of 900 kpm/min in healthy subjects. The low levels of A found at rest, were only slightly increased during the exercise. Vendsalu also performed experiments involving catheterization of the left renal vein (24). The A and NA levels in samples from this vein increased only to a minor degree during mild to moderate muscle exercise.

The CA levels in arterial blood plasma during muscle exercise have recently been correlated to the relative work load (15) Fig. 1.

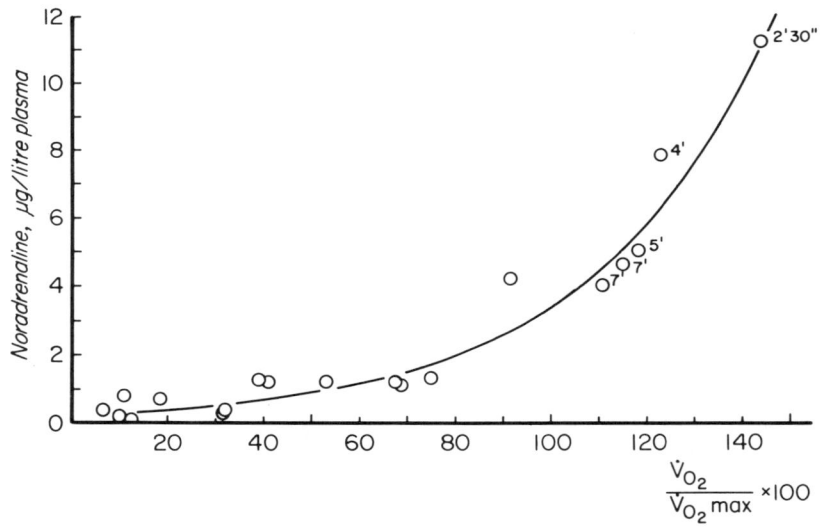

Fig. 1. The arterial noradrenaline levels (µg/l) in 5 males, in relation to the relative work load (estimated or calculated O_2 uptake in per cent of the individual max. O_2 uptake, c.f. Dr. Saltin this Symposium). The min for the supramax. work loads are indicated (from 15).

The NA levels increased slowly up to a work load corresponding to an oxygen consumption of about 75 per cent of the maximal oxygen consumption reaching levels of 1 to 1.5 µg/l plasma. Thereafter the levels increased rapidly and were about 10 µg/l plasma at supramaximal work loads. No detectable amounts of A were present.

Sources of Circulation CA

When discussing the role of circulating CA at muscular work it seems to be of importance to consider the source of these circulating CA:

The adrenals. The pattern of urinary CA excretion, the lack of significant amounts of A in blood (which can be found with the same estimation techniques at other situations, e.g. after insuline) and the results obtained after catheterization of the left renal vein, all together indicate that the adrenals contribute very little to the CA levels during muscular work, at least as long as the work load is mild or moderate.

The adrenergic nerve terminals. An overflow of NA from the adrenergic nerve terminals has been shown to occur during stimulation of e.g. the sympathetic vasoconstrictor fibres in skeletal muscles provided that the blood flow is not markedly reduced during the stimulation (e.g. 2, 22). Except for effects due to certain factors, e.g. variations in blood flow, inactivation of circulating NA by e.g. the liver and the kidneys, as well as uptake in other tissues, the blood NA levels may be considered to reflect the nerve activity in the sympathetic nervous system. NA released from the nerve terminals into the synaptic gaps can locally be inactivated in at least three different ways: 1) by enzymatic destruction, 2) by active reuptake into the nerve terminals by the "membrane pump", and 3) by diffusion into the blood stream (for ref. and disc. 10).

It may thus be asked: does the circulating NA represent only a waste product, released from the adrenergic nerve terminals, or does it have an important direct effect on tissue receptors? This seems to be a complex question. I am not going to discuss problems such as a possible effect of circulating NA on the CNS, but restrict myself to some aspects on the cardiovascular system, on the metabolism of adipose tissue, and on the metabolism of carbohydrates in striated muscles in man.

CA Released by Nerve Activity

It seems reasonable to suggest that in tissues which are well innervated by adrenergic nerve terminals the NA released by nervous activity would be the dominating factor for the activation of receptors, as compared to effects of circulating CA. During nerve activity the concentrations of NA created in the vicinity of the nerve terminals are probably so high that the addition of NA from the blood stream would have minor significance for the physiological response.

With respect to the cardiovascular system, results obtained from animal studies with a physiological approach (5,9) indicate that the control of the cardiovascular functions occurs predominantly via the adrenergic nerves which directly innervate heart and blood vessels. Blood-borne NA, on the other hand, seems to be of minor importance. Morphological studies have shown that the blood vessels to e.g. the skeletal muscles are innervated by adrenergic fibres, (12). However, it must be observed that in the blood vessels far from all smooth muscle cells are directly innervated, and also that the blood vessels in different regions differ markedly in density of innervation. Thus, in some blood vessels diffusion of transmitter may be necessary for the activation of those smooth muscle cells, which are not directly innervated by adrenergic fibres. However, a probably more common mechanism, myogenic conduction of nerve induced activity, appears to be most important for the activation of these not directly innervated cells (c.f. 20).

There seems to be no direct adrenergic innervation of the adipose tissue in man (25), but a rather rapid response at sympathetic stimulation with respect to e.g. release of FFA. This problem has been discussed by Dr. Rosell at this Symposium, who pointed out that circulating CA appears to be of small importance when compared to the nerve mediated control. Some observations after ganglionic blocking drugs indicate that the sympathetic system is of less importance than other FFA-mobilizing mechanisms during moderate exercise (4).

The individual skeletal muscle cells appear to lack direct adrenergic innervation (12). However, as mentioned above, the blood vessels of skeletal muscle contain adrenergic nerves (12). CA released from these vasoconstrictor nerve fibres may by diffusion reach the striated muscle cells in concentrations which at least locally may be high. From these local areas, CA induced effects may possibly be initiated. Particularly in the rythmically working muscle the possibilities for diffusion may be favourable, due to tentative fast movements of the extracellular fluid.

Possible Role of Circulating CA

With respect to the points discussed above, the circulating NA may thus be of less importance than the NA directly released from the nerve terminals, which can affect receptors in the vicinity of the nerve terminals or possibly also reach more distant receptors by diffusion in a concentration which exceeds the concentration created via the blood stream. However, for such a discussion, more knowledge seems to be necessary with respect to such factors as CA concentration at different distances from the nerve terminals, localization of different receptors, and threshold values of CA concentration for receptor response in different tissues.

After having discussed the possible direct effects of the circulating NA, to the major extent originating from the nerve terminals (see above), we may now close the circle. Circulating NA can be reuptaken into the sympathetic nerve terminals by the efficient membrane pump and at least in part be released again at nerve activity. Thus, at least some of the circulating NA may be looked upon as a "temperary extraneuronal transmitter store".

A non-neuronal uptake of CA into tissues has also been discussed (16). At present the physiological significance of such an uptake seems to be questionable. However, CA present in the tissues may modify the amount of NA released per nerve impulse from the nerve terminal. This may be due to interference with local feedback mechanisms regulating the quantity of transmitter released per nerve impulse.

When A is released from the adrenals into the blood stream high levels of circulating A can be created at least for short periods of time. The receptors of the tissues are likely to be reached by the A via the blood stream. However, interesting enough also in this case the situation can be complicated due to the properties of the adrenergic nerve terminals. The membrane pump for uptake of circulating amines into the nerve terminals is efficient also for circulating A. At nerve activity A, taken up into the nerve terminals, may be released. At present, Dr. Folkow and myself are investigating the possibility of such a "shift" of transmitter from NA to A in sympathetic nerves. Preliminary results indicate that such a partial shift may occur in some situations. However, more data are needed before the meaning of such a tentative shift can be discussed from a physiological point of view.

Conclusions

The levels of circulating CA are, at mild and moderate exercise, so low that direct effect of circulating amines is probably of less importance for most circulatory and metabolic effects than amines released directly from the nerves. However, the morphological basis for such a nervous control is in many cases somewhat obscure. Furthermore, in some tissues receptors may be present which are very sensitive to small fluctuations in the circulating CA levels.

The interpretation of the rather high levels of circulating CA found during hard muscular exercise seems at present to be rather difficult. As outlined briefly above, the adrenals, the blood CA, and the adrenergic nervous system may from a functional point of view be looked upon as a "unit" where the different components may co-operate. Before their relative importance can be evaluated in relation to other metabolic stimulating mechanisms at exercise, it seems necessary to obtain more experimental data.

Thus, more information is needed regarding the arterial contra venous plasma CA levels at rest and exercise, with or without CA infusion, (preferably using the same CA estimation method, since e.g. the normal CA levels at rest differ somewhat when different methods are used). The results have to be correlated to the onset and development of e.g. metabolic changes. Also the part played by the nervous system, particularly the adrenergic nervous system, must be studied. In these studies, pharmacological tools ought to be useful, such as drugs known to act upon the adrenergic nervous system at different levels (e.g. reserpine - blocking the NA storage of the amine granules; bretylium - interfering with the release of amines at nerve activity; membrane pump blockers - inhibiting the uptake of CA into the nerve terminals). Of particular importance may be the use of different α- and β-receptor blocking substances (e.g. 3, 17, 21), not to mention stimulation of the sympathetic nervous system. It probably also is necessary to obtain more detailed knowledge ot the basic mechanisms giving metabolic changes during exercise ; mechanisms on which CA can excert its activating or modulating effects.

References

1) Callingham, B.A. The catecholamines. Adrenaline; noradrenaline. In: Hormones in Blood. Ed.: C.H.Gray and A.L.Bacharach. Acad. Press. 2 ed. Vol. 2. 519, 1968.

2) Carlsson, A., B. Folkow and J. Häggendal. Some factors influencing the release of noradrenaline into the blood following sympathetic stimulation. Life Sci. 3, 1335, 1964.

3) Carlsson, C., S.J.Dencker, G. Grimby and J. Häggendal. Circulatory studies during physical exercise in mentally disordered patients. I. Effects of large doses of chlorpromazine. Acta med. scand. 184, 499, 1968.

4) Carlsten, A., J. Häggendal. B. Hallgren, R. Jagenburg, A. Svanborg and L. Werkö. Effects of ganglionic blocking drugs on blood glucose, amino acids, free fatty acids and catecholamines at exercise in man. Acta physiol. scand. 64, 439, 1965.

5) Celander, O. The range of control exercised by the 'sympathoadrenal system'. Acta physiol. scand. 32, Suppl. 116, 1954.

6) Cohen, G. and M. Goldenberg. The simultaneous fluorimetric determination of adrenaline and noradrenaline in plasma - II Peripheral venous plasma concentrations in normal subjects and in patients with pheochromocytoma. J. Neurochem. 2, 71, 1957.

7) Euler, U.S.v. Noradrenaline, Charles C. Thomas Publ. Springfield. Ill. U.S.A. 1956.

8) Euler, U.S.v. The catecholamines. Adrenaline; noradrenaline. In: Hormones in Blood. Acad. Press. 515, 1961.

9) Folkow, B., B. Löfving and S. Mellander. Quantitative aspects of the sympathetic neuro-hormonal control of the heart rate. Acta physiol. scand. 36, 363, 1956.

10) Folkow,B., J. Häggendal and B. Lisander. Extent of release and elimination of noradrenaline at peripheral adrenergic nerve terminals. Acta physiol. scand. Suppl. 307, 1967.

11) Furchgott, R.F. The pharmacological differentiation of adrenergic receptors. Ann. N.Y. Acad. Sci. 139, 553, 1967.

12) Fuxe, K. and G. Sedvall. The distribution of adrenergic nerve fibres to the blood vessels in skeletal muscle. Acta physiol. scand. 64, 75, 1965.

13) Häggendal, J. An improved method for fluorimetric determination of small amounts of adrenaline and noradrenaline in plasma and tissue. Acta physiol. scand. 59, 242, 1963.
14) Häggendal, J. Some further aspects on the release of the adrenergic transmitter. In: New Aspects of Storage and Release Mechanisms of Catecholamines. Internat. Symp. Oct. 1969, 1970, in press.
15) Häggendal, J., L.H. Hartley and B. Saltin. Arterial noradrenaline concentration during exercise in relation to the relative work levels. Scand. J. clin. Lab. Invest. 1970, in press.
16) Iversen, L.L. The uptake and storage of noradrenaline in sympathetic nerves. Cambridge Univ. Press. 1967.
17) Kral, J.G., B. Åblad, P. Björntorp, L. Ek and G. Johnsson. Metabolic and cardiovascular effects of adrenaline and alprenolol (Aptin®) in human subjects. Pharmacol. Clin. 2, 40, 1969.
18) Lands, A.M., A. Arnold, J.P. McAuliff, F.P. Luduena and T.G. Brown Jr. Differentiation of receptor systems activated by sympathomimetic amines. Nature 214, 597, 1967.
19) Lands, A.M., F.P. Luduena and H.J. Buzzo. Differentiation of receptor response to isoprenaline. Life Sci. 6, 2241, 1967.
20) Ljung, B. Nervous and myogenic mechanisms in the control of a vascular neuroeffector system. Acta physiol. scand. Suppl. 349, 33, 1970.
21) Lundholm, L., E. Mohme-Lundholm and N. Svedmyr. Metabolic effects of catecholamines. In: Biol. Basis of Medicine. Ed. Bittar and Bittar. Acad. Press. London. Vol. II. 101, 1969.
22) Rosell, S., I.J. Kopin and J. Axelrod. Fate of H^3-norepinephrine in skeletal muscle before and following sympathetic stimulation. Amer. J. Physiol. 205, 317, 1963.
23) Sutherland, E.W. and T.W. Rall. The relation of adenosine-3´,5´-phosphate and phosphorylase to the actions of catecholamines. Pharmac. Rev. 12, 265, 1960.
24) Vendsalu, A. Studies on adrenaline and noradrenaline in human plasma. Acta physiol. scand. 49, Suppl. 173, 1960.
25) Wirsén, C. Studies in lipid mobilization. With special reference to morphological and histochemical aspects. Acta physiol. scand. 65, Suppl. 252, 1965.

THE LIVER AS AN ENERGY SOURCE IN MAN DURING EXERCISE[1]

Loring B. Rowell[2]

Departments of Physiology and Biophysics and of Medicine

University of Washington School of Medicine, Seattle
Washington 98105

INTRODUCTION

Apparently only 25-50 % of the available metabolic substrate for prolonged work in fasted man can be accounted for by oxidation of free fatty acids (FFA) transported to muscle via plasma (1-2). A key question is, "What is the source and nature of the unaccounted for substrate?"

Because hepatic blood flow (HBF) declines inversely with the severity of exercise (3), the liver has been discounted as a potentially significant energy source during moderate to heavy exercise. However, diminished HBF does not preclude release of considerable highly concentrated substrate by the liver into its venous blood. For example, patients with congestive heart failure who have low HBF at rest, still have normal hepatic glucose production (4). In addition, hepatic-splanchnic oxygen consumption can be maintained at normal resting values by increased extraction of oxygen despite an up to 80 % reduction in blood flow (3,5,6).

The question raised in this discussion is whether hepatic production of glucose and various lipid compounds can be augmented during exercise to a level where a significant fraction of the total

---

[1] This work was supported by U.S.P.H.S., National Heart Institute, Grant-in-Aid HE-09773 and by the Clinical Research Center Facility of the University of Washington supported by the National Institutes of Health (FR-37) and the Washington State Heart Association.
[2] Established investigator of the American Heart Association.

metabolic substrate is supplied. More specifically, can this augmentation occur when HBF is markedly reduced? The specific metabolic pathways which might be utilized can be dealt with only indirectly.

METHODOLOGY

All methods and procedures, including measurement of blood gases, indocyanine green (ICG), plasma lipids, lactate, etc., have been described elsewhere (3,7,9). Only those features having major bearing upon interpretation of the results are summarized briefly below.

The subjects were 23 normal, healthy young men ranging in age from 21-28 years. Some were sedentary, some physically active, and several were well-trained athletes. Physical status appeared to affect only the extent to which HBF was reduced at a given level of oxygen uptake (3). Consequently, data for oxygen uptake are normalized where appropriate for the percentage of maximal oxygen uptake required - i.e., relative oxygen uptake.

With one exception (7), in which HBF was measured via a single injection technique (3), the constant infusion method with indocyanine green dye originally described by Bradley, et al (10) was used. In one subject we combined both techniques. During 9 to 20 min of exercise, HBF was 487 ml/min according to the single injection technique and 426 ml/min during 43-61 min by the constant infusion technique. Thus, the two methods checked rather well (8).

Selection of an appropriate hepatic venous sampling site is important. The major disadvantage of current techniques for measuring HBF and hepatic metabolism is the lack of a sampling site into which pooled venous blood from various regions of the liver is drained and mixed. The largest of five hepatic veins drains the right lobe and provides the greatest freedom from catheter-induce sampling errors (9-12). This vein also drains a major fraction of the liver, reducing somewhat the effects of variation in venous contents from various portions of the organ. Available evidence indicates that regional differences for BSP (10,12) or ICG extraction (12) are relatively small. Thus, when possible, this large right hepatic vein was used (Fig. 1). All samples were taken slowly with little or no resistance to withdrawal. In this way the sampling problems discussed by Brauer (13) and by Sapirstein and Reininger (14) were largely avoided.

The greatest sampling difficulties occurred when one of the smaller left hepatic veins was used; sampling from a partially wedged catheter was difficult to avoid. In one very large subject we succeeded in simultaneously measuring ICG and concentrations of various substrates from an upper and lower right hepatic vein.

THE LIVER AS AN ENERGY SOURCE

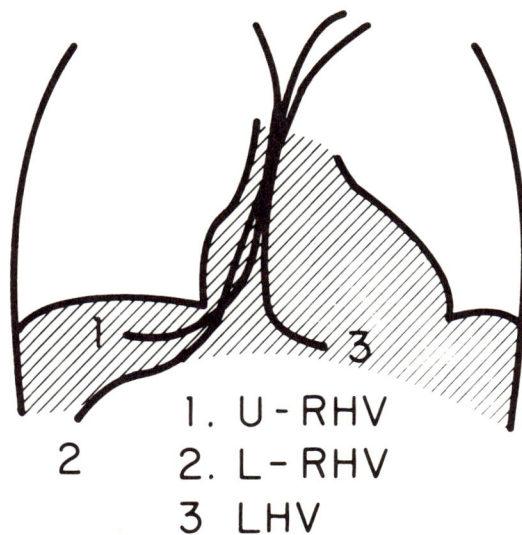

1. U-RHV
2. L-RHV
3. LHV

Fig. 1. Simulated fluoroscopic view of hepatic-venous catheter positions in various subjects. Site 2 is that of the major right hepatic vein – the preferred and most frequently used site. Data in Fig. 2 were derived from sites 1 and 2 – the upper and lower right hepatic veins. In 2 subjects data were from the left hepatic vein (LHV), site 3. All data obtained show relatively consistent responses from the 3 sites (site 1 alone was used several times) (7,8). Compare data from sites 1 and 3 in Fig. 6.

Data from this individual (Fig. 2) support the concept that some functional homogeneity exists within various regions of the liver. This is further supported by the general similarity of responses obtained when the more difficult left hepatic venous sampling sites were used. In general, the findings reported in this paper were not influenced in any major way by the sites used for hepatic venous sampling.

Basically, two procedures were followed. Initially the subjects were studied during prolonged (about 1 hour) exercise which required 2.1-2.5 LO_2/min. In these experiments the absolute workload was constant but the relative workload varied individually, requiring 48-70 % of maximal oxygen consumption. These studies were conducted at cool (16-21°C) ambient temperatures. Exercise sometimes was terminated with the subject at or near exhaustion.

Our second approach was to produce the greatest possible re-

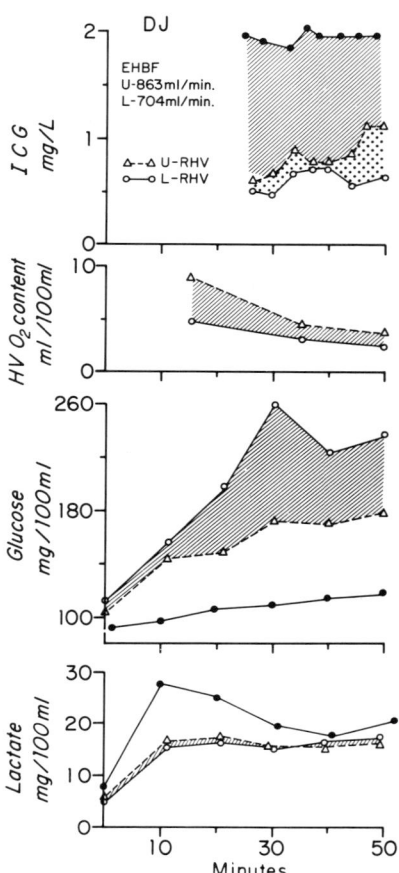

Fig. 2. Data from one subject with simultaneous sampling from an upper right (U-RHV, △ --- △) and a lower right hepatic vein (L-RHV, o——o) during upright exercise at 2.3 LO_2/min in a 48.9°C environment. Arterial values are shown as solid (●——●). Upper portion shows arterial and HV dye concentrations vs. time of exercise. Small descrepancy in dye extractions from two HV sites shown as stipled region with difference in calculated flow being 160 ml/min (Estimated hepatic blood flow (EHBF) for two sites is listed). U-RHV contents of oxygen, glucose and lactate reveal differences at the two sites. Differences in FFA concentration at these sites were such that calculated rates of uptake were the same. Calculated oxygen consumption rose with time as follows: U-RHV, from 103 to 139 ml/min; L-RHV, from 106 to 124. Glucose production: U-RHV, from 406 to 535 mg/min; L-RHV, from 416 to 840.

duction in HBF for as long as possible. To achieve this, the relative workload was set at 42-55 % of maximal oxygen uptake and extended to an end point of subjective exhaustion of the subject. Environmental temperature was 48.9°C dry bulb and 24-26.6°C wet bulb. Adding thermal stress to that of exercise was previously shown to procedure greater than normal decrements in HBF at any given level of oxygen uptake (15).

Hepatic Lipid Metabolism

During heavy exercise requiring 10-12 kcal/min, hepatic blood flow (HBF) averaged 995 ml/min (range 630-1300 ml/min) (7). Lipids released by the liver could account for no more than 5 % of the total energy expenditure - even if all were oxidized.

The liver took up FFA, at an average rate of 64 µM/min (range 30-116 µM/min). Ketone body (KB) production in five men averaged 20 mg/min (range 1.8-93 mg/min). A sixth subject - who was a well-trained athlete - produced 154-188 mg/min. Although reproducibility of ketone body analysis was poor ($\pm$20 % for duplicate analyses), even with an analytical error of 100 % hepatic KB production could contribute on the average only 1-2 % of total metabolic substrate utilized. For example, to account for 1 % of total metabolism 23 mg/min of KB would be required.

Splanchnic arterial-venous (A-V) concentration differences for phospholipids (PL) and triglycerides (TG), unlike other potential substrates, varied randomly with time from positive to negative. An analysis of our analytical precision in the measurement of hepatic A-V concentration differences led us to conclude that we could see a 2-5 % contribution to total energy production by PL or TG had either occurred. The data strongly suggested no net production of these substrate moieties by the liver (Fig. 3).

Hepatic Glucose and Lactate Metabolism

Glucose was the only metabolic substrate released from the liver in quantitatively significant amounts (7). On the basis of hepatic A-V concentration differences in whole blood, average glucose production was 536 mg/min (range 195-1240 mg/min)[2]. This 6-fold

---

[1]Caloric values (kcal/g) of each substance were estimated to be as follows: phospholipids, 7; triglycerides and FFA, 9; ketone bodies, 4.7; and glucose, 4.
[2]Our originally published estimate was 295 mg/min (Fig. 3) (7) - based conservatively on plasma glucose concentrations. The above stated estimate assumes equilibration of glucose with red blood cells (9).

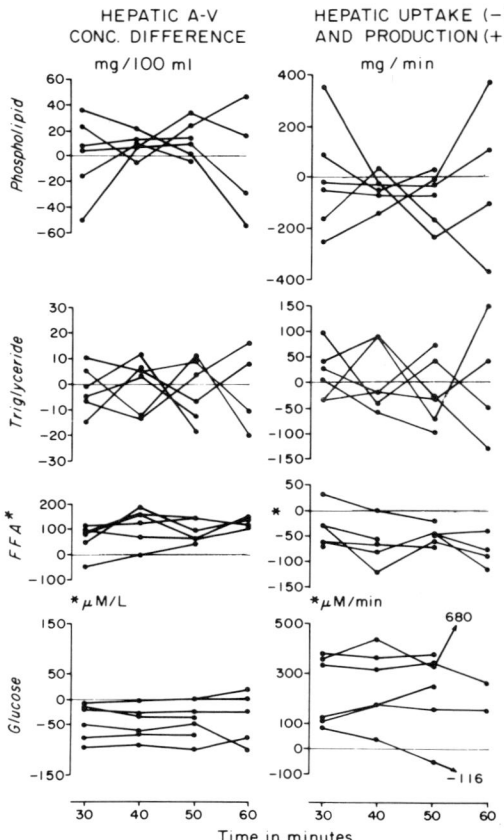

Fig. 3. Hepatic A-V concentration differences and rates of uptake or production of various substrates (8). Reproduced by permission of Journal of Applied Physiology.

increase in glucose production above resting values (fasting), reported by others (4,28), could contribute 8-42 % (average 19 %) to the total metabolic rate if completely oxidized. The high value of 1240 mg/min was a single value from one subject near the end of exercise.

The preceding calculations as well as the following on lactate uptake by the liver, rest on two basic assumptions: first, that extra-hepatic tissues in the splanchnic circuit make negligible contribution and second, that only minor intralobular variations in calculated HBF or hepatic glucose or lactate metabolism occur. As for the former assumption, no significant extra-hepatic contribution to either lactate or glucose metabolism is apparent in post-absorptive man (4,18). Evidence presented earlier in this paper indicates that intralobular variation does exist but that direction and relative magnitude of response in different hepatic regions is similar. Unfortunately intrahepatic variation in uptake or production of naturally occurring metabolites has not been examined as thoroughly as variation in flow indicators (dyes, etc.).

We conclude from these studies that the liver is not a significant source of <u>lipid</u> substrate for skeletal muscle metabolism. However, despite major reductions in HBF, glucose production by the liver can contribute substantially to the total pool of metabolic substrate during moderate to heavy exercise. Similar findings were reported by Bergström and Hultman (5).

Observations of hepatic lactate removal during exercise have suggested a possible contribution of lactate - via gluconeogenesis - to hepatic glucose release (7). It is well known that when blood lactate concentration is increased by exercise, in dogs, a fraction is removed by the liver (16). An attempt was undertaken to quantify the uptake of lactate from blood during prolonged mild and heavy exercise in man (8). HBF and splanchnic A-V difference for lactate were determined during prolonged (60-70 min) work at 30.5 ml O_2/kg /min - or 40-70 % of maximal oxygen uptake. As observed previously (17) blood lactate concentration peaked by the 10th min of exercise and decreased thereafter with a half-time of 22-36 min (average 28 min).

Hepatic lactate removal averaged approximately 0.8 ±0.25 %/min of estimated total body lactate. The actual rates of removal shown in fig. 4, vary with time and among different subjects. Removal rates were erratic during the first 20 min of exercise; thereafter, the rates declined progressively and in direct proportion to the fall in arterial concentration. Of the estimated total lactate removed during the 60-70 min of exercise 32-73 % (average 43 %) was removed by the liver.

The fate of lactate removed by the liver could not be determined directly. Total splanchnic oxygen uptake and CO_2 production were estimated from the product of HBF and serially determined splanchnic A-V O_2 and CO_2 differences. Splanchnic oxygen uptake rose from 74 to 93 ml/min or 26 % during exercise with a splanchnic RQ which varied widely around an average of 0.43. Calculation of the total <u>extra</u> O_2 consumed and CO_2 produced (i.e. above initial values at the start of exercise) suggest that only 20-40 % of the total lactate removed by the liver could have been oxidized. This calculation, of course, is based on the assumption that lactate would be oxidized in addition to rather than in substitution for normally metabolized substrate.

On the other hand if the 30 mM of lactate removed by the liver

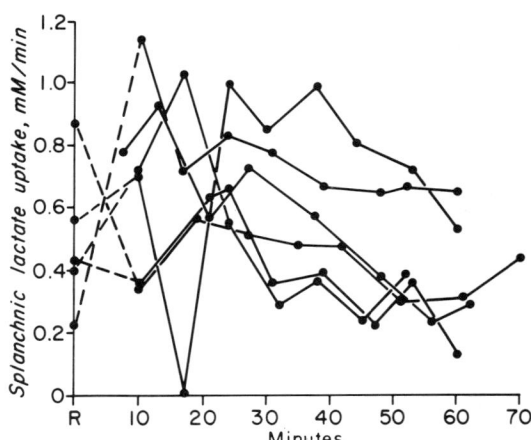

Fig. 4. Splanchnic (hepatic) removal rates for lactate during exercise in 6 men. Uptake declined with falling concentration in arterial blood whereas extraction remained nearly constant (average 45 %) (7). Reproduced by permission of Journal of Applied Physiology.

were converted to glucose, about 1 litre of O_2 would be required (i.e. assuming lactate to pyruvate requires 11.2 ml O_2/mM and pyruvate to glucose, 22.4 ml O_2/mM). Thus, the observed extra O_2 uptake of 810 ml by the liver could account for 80 % of the cost of this conversion.

Because glycolytic capacity of the liver is very low (19) and the presence of lactate is a potent stimulus to gluconeogenesis – the preferred pathway for hepatic lactate metabolism (19) – the most likely fate of this lactate is conversion to glucose. A less likely possibility in addition to oxidation is hepatic storage of lactate or its conversion to alanine.

From the above, two important points can be made. The first is that lactate produced by working skeletal muscle presumably can, via the Cori Cycle, be recycled by the liver during exercise and converted back to glucose, which in turn may be reutilized by muscle (16,20). This along with what is oxidized by other tissues during work in substitution for the usual substrate (21) constitutes a "repayment" of oxygen debt during exercise. Quantitatively the significance of this process depends upon the severity of work. If moderate to relatively heavy work is prolonged, total metabolism of liver and other non-working tissues is sufficient to consume all lactate produced while work continues - leaving behind an "oxygen debt" of up to 3 litres (7,17). Under these conditions lactate production usually does not continue past the early minutes of exercise (17,22,23). In contrast, when work intensity approaches maximal oxygen uptake, the fraction of lactate which can be removed by resting organs is very small. This is obvious since their metabolic rates have become very small in proportion to total metabolism. Also, lactate production continues or increases with time. Furthermore, work cannot be maintained long enough to observe any significant uptake of lactate.

The second point is that glucose release from the liver far exceeded that which could be generated via gluconeogenesis from lactate. Thus, another mechanism must be sought by which hepatic glucose release during exercise can be so significantly augmented.

Hepatic Metabolism in Hyperthermia

During heavy exercise in a hot ($48^\circ C$) environment, HBF was low, averaging 667 ml/min (range 469-870 ml/min) in 8 men. Five subjects showed little progressive reduction in HBF as work progressed despite falling arterial blood pressure (9).

Splanchnic oxygen consumption increased during the 38-65 min of exercise, averaging 82 ml/min at 10 min and 115 ml/min at exhaustion. Percentage extraction of oxygen increased markedly. Hepatic venous oxygen content fell as low as 0.6-0.7 ml/100 ml in 3 men at exhaustion and averaged 2.56 ml/100 ml at that time (Fig. 5).

Despite the low blood flow and oxygen tension of the liver, FFA were taken up at the rate of 133 mM/min (range 60-230 mM/min) - which is greater than observed in our previous study (7). Arterial lactate concentration was abnormally high for the percentage of maximal oxygen uptake required. Percentage extraction of lactate by hepatic-splanchnic tissues was subnormal but removal rates were similar to those seen previously, averaging 0.6 mM/min (range 0.07-1.9 mM/min) (Fig. 6).

The outstanding result of the hyperthermia study was the progressive rise in glucose output during exercise to extremely high values at exhaustion (Fig. 6). Net hepatic glucose release reached 1000-1200 mg/min. Average glucose production rates were 351 and 749 mg/min at 10 min of exercise and at exhaustion, respectively. Had this glucose been completely oxidized it would have provided 9-20 % (average 15 %) and 7-57 % (average 32 %) of total energy expenditure at the above stated times, respectively.

Again analysis of possible mechanisms of this glucose release is speculative. We know that lactate removal rates were far too low

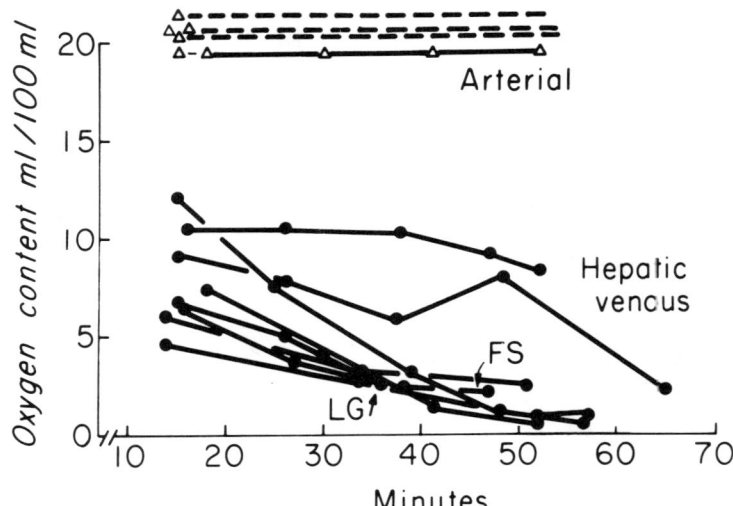

Fig. 5. Decline in hepatic venous O_2 content in 8 subjects during work to exhaustion at 48.9°C ambient temperature. Subjects reaching lowest values tended to show greatest hepatic glucose output. Modified from fig. 2 in Rowell et al (9).

(by a factor of 7) to account for glucose production via gluconeogenesis. Two schemes may be offered to explain the results: a) glucose was produced via gluconeogenesis from a variety of precursors and b) glucose was released via glycogenolysis from hepatic glycogen stores.

Several factors support (a), the first scheme. The high concentration of FFA and lactate perfusing the liver stimulates gluconeogenesis if adequate oxygen is available (24). Cortisol and epinephrine also stimulate gluconeogenesis (25); under these conditions, levels of both were probably elevated. If the risky extrapolation is made from maximal rates of gluconeogenesis by rat liver (26) to human liver based on mass difference, estimated human hepatic gluconeogenic capacity and the highest rates of hepatic glucose release we saw are of the same order of magnitude.

In our hyperthermia study, two major factors affecting gluconeogenesis - glucagon and insulin (25) - were measured at rest and at the end of exercise in two subjects. Both had very high hepatic glucose output at exhaustion but showed no change in either hormone.

THE LIVER AS AN ENERGY SOURCE

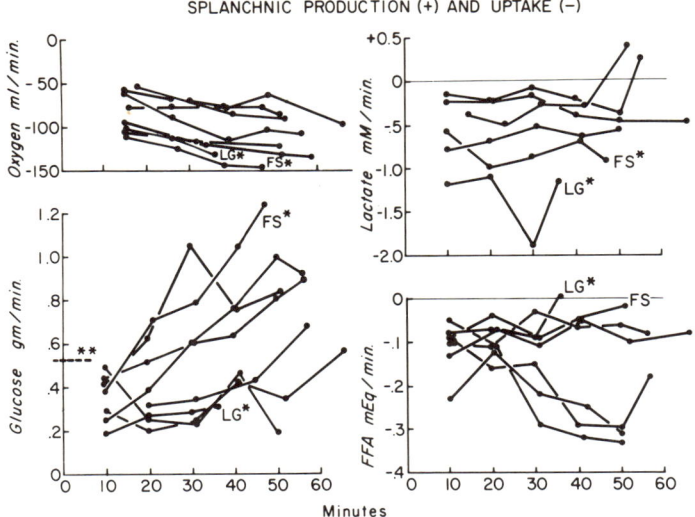

Fig. 6. Splanchnic (hepatic) production of glucose and uptakes (negative values) for oxygen, lactate and FFA during work to exhaustion in a 48.9°C environment. Initials FS^x and LG^x show data from left hepatic venous samples (site 3, Fig. 2). Dashed line (xx) crossing the axis of glucose production marks the average rate of production in an earlier study (8) in a cool environment. Modified from fig. 3 in Rowell et al (9).

The second scheme (b) - increased glycogenolysis - is also supported by several findings. 1) Estimation of hepatic glucose stores from liver biopsy in normal man fasted for 12 hours (27) suggest that these stores are normally adequate to release the quantities of glucose we observed. The magnitude of hepatic glucose output after glucagon and epinephrine stimulation also supports this contention (28). 2) Calculation of extra (above resting) glucose release by the liver over the average duration of exercise during which measurements were made - approximately 23 gm in 53 min - and extra O_2 consumed by splanchnic tissues over this interval - 1.06 litres - suggests that no more than one-third of the glucose could have been released via gluconeogenesis. This calculation requires the assumption that a minimum of 22.4 ml O_2 are required to convert 2 mmoles of pyruvate to 1 mmole of glucose. Actually gluconeogenesis is probably more costly. 3) Hepatic hypoxia appears to be a potent stimulus to local glycogenolysis (13,29,30). Two findings in our hyperthermic subjects suggest the development of hypoxia within the liver, namely, a marked fall in hepatic venous oxygen content and, in two men, net lactate release at exhaustion. However, hepatic venous serum potassium in three men showed no marked elevation suggestive of local hypoxia at the end of exercise. Rather, net potassium release closely matched Hultman's (6) predications of how much potassium would be relased from hepatic glycogen during glycogenolysis and also with what he observed during heavy exercise.

In conclusion, hepatic glucose release during exercise has not yet been fitted into an integrated scheme of overall metabolic regulation. Its occurrence is possibly no more than the consequence of various local factors, such as decreasing O_2 tension or changing pH, blood catechol levels, etc. which may become marked at exhaustion. Another factor of unknown consequence was the development of severe local hyperthermia at 48.9°C; hepatic venous temperature rose to nearly 42°C (9). The 1.4-fold rise in splanchnic oxygen uptake closely fits the 1.3-fold increment predicted from the Arrhenius equation (assuming 12-kcal activation energy). Whatever the cause, Bergström and Hultman (5) found that accelerated hepatic glucose release appears to coincide with near depletion of muscle glycogen stores. Postulation of feedback control of hepatic glycogen release via some stimulus from working muscle may be tempting but is also very premature.

We are left with the problem of resolving two lines of evidence about the missing metabolic substrate for exercise: the findings of Havel and colleagues (1,2) and Issekutz et al (31) suggest that endogenous muscle lipids supply the missing substrate. However, the fact that triglyceride concentration in human (32) and monkey (33, 34) skeletal muscle is normally very low argue against this concept. Furthermore, total lipid content does not decline during exercise in either species (11); turnover of the various muscle lipid moeities in the monkey is not increased (34); and finally release of glycerol into venous blood draining working muscle in man is insignificant (35).

On the other hand our data and Bergström's and Hultman's (5) show the liver to be a significant source of energy substrate - in the form of glucose only - during moderate to heavy exercise. This contribution can exceed, at times, 50 % of the total metabolic requirement. <u>Total</u> glucose release, including resting production rate, summed over the actual duration of exercise was 28 gm (22 to 33 gm). This could account for 20 % of the total energy expended during exercise.

REFERENCES

1. HAVEL, R.J., A. NAIMARK, AND C.F. BORCHGREVINK. Turnover rate and

oxidation of free fatty acids of blood plasma in man during exercise: Studies during continuous infusion of palmitate-I-C^{14}. J. clin. Invest. 42: 1054-1063, 1963.

2. HAVEL, R.J., L.A. CARLSON, L-G. EKELUND, AND A. HOLMGREN. Turnover rate and oxidation of different free fatty acids in man during exercise. J. appl. Physiol. 19: 613-618, 1964.

3. ROWELL, L.B., J.R. BLACKMON, AND R.A. BRUCE. Indocyanine green clearance and estimated hepatic blood flow during mild to maximal exercise in upright man. J. clin. Invest. 43: 1677-1690, 1964.

4. MYERS, J.D. Net splanchnic glucose production in normal man and in various disease state. J. clin. Invest. 29: 1421-1429, 1950.

5. BERGSTRÖM, J., AND E. HULTMAN. A study of glycogen metabolism during exercise in man. Scand. J. clin. Lab. Invest. 19: 218-228, 1967.

6. HULTMAN, E. Studies on muscle metabolism of glycogen and active phosphate in man with special reference to exercise and diet. Scand. J. clin. Lab. Invest. 19: Suppl. 94, 1-63, 1967.

7. ROWELL, L.B., K.K. KRANING II, T.O. EVANS, J.W. KENNEDY, J.R. BLACKMON, AND F. KUSUMI. Splanchnic removal of lactate and pyruvate during prolonged exercise in man. J. appl. Physiol. 21: 1773-1783, 1966.

8. ROWELL, L.B., E.J. MASORO, AND M.J. SPENCER. Splanchnic metabolism in exercising man. J. appl. Physiol. 20: 1032-1037, 1965.

9. ROWELL, L.B., G.L. BRENGELMANN, J.R. BLACKMON, R.D. TWISS, AND F. KUSUMI. Splanchnic blood flow and metabolism in heat-stressed man. J. appl. Physiol. 24: 475-484, 1968.

10. BRADLEY, S.E., F.J. INGELFINGER, G.P. BRADLEY, AND J.J. CURRY. The estimation of hepatic blood flow in man. J. clin. Invest. 24: 890-897, 1945.

11. HULTMAN, E. Blood circulation in the liver under physiological and pathological conditions. Scand. J. clin. Lab. Invest. 18: Suppl. 92, 27-41, 1966.

12. WINKLER, K., J.A. LARSEN, T. MUNKNER, AND N. TYGSTRUP. Determination of the hepatic blood flow in man by simultaneous use of five test substances measured in two parts of the liver. Scand. J. clin. Lab. Invest. 17: 423-432, 1965.

13. BRAUER, R.W. Liver circulation and function. Physiol. Rev. 43: 115-213, 1963.

14. SAPIRSTEIN, L.A., AND E.J. REININGER. Catheter induced error in hepatic venous sampling. Circulat. Res. 4: 493-498, 1956

15. ROWELL, L.B., J.R. BLACKMON, R.H. MARTIN, J.A. MAZZARELLA, AND R.A. BRUCE. Hepatic clearance of indocyanine green in man under thermal and exercise stresses. J. appl. Physiol. 20: 384-394, 1965.

16. HIMWICH, H.E., Y.D. KOSKOFF, AND L.H. NAHUM. Studies in carbohydrate metabolism. I. A glucose-lactic acid cycle involving muscle and liver. J. biol. Chem. 85: 571-584, 1930.

17. BANG, O. The lactate content of blood during and after muscular exercise in man. Scand. Arch. Physiol. 74: Suppl. 10, 51-82, 1936.

18. SHERLOCK, S., AND V. WALSHE. The use of a portal anastomotic vein for absorption studies in man. Clin. Sci. 6: 113-123, 1946.

19. KREBS, H. Gluconeogenesis. Proc. roy. Soc. B. 159: 545-564, 1964.

20. DEPOCAS, F., Y. MINAIRE, AND J. CHATONNET. Rates of formation and oxidation of lactic acid in dogs at rest and during moderate exercise. Canad. J. Physiol. and Pharm. 47: 603-610, 1969.

21. ISSEKUTZ, B. JR., H.I. MILLER, P. PAUL, AND K. RODAHL. Effect of lactic acids and glucose oxidation in dogs. Am. J. Physiol. 209: 1137-1144, 1965.

22. FLOCK, E.V., D.J. INGLE, AND J.L. BOLLMAN. Formation of lactic acid, an initial process in working muscle. J. biol. Chem. 129: 99-110, 1939.

23. SACKS, J., W.C. SACKS, AND J.R. SHAW. Carbohydrate and phosphorus changes in prolonged muscular contractions. Am. J. Physiol. 118: 232-240, 1937.

24. WEBER, G., H.J. HIRD-CONVERY, M.A. LEA, AND N.B. STAMM. Feedback inhibition of key glycolytic enzymes in liver: Action of free fatty acids. Science 154: 1357-1360, 1966.

25. EISENSTEIN, A.B. Current concepts of gluconeogenesis. Am. J. clin. Nutr. 20: 282-289, 1967.

26. EXTON, J.H., AND C.R. PARK. Control of gluconeogenesis in liver. I. General features of gluconeogenesis in the perfused livers of rats. J. biol. Chem. 242: 2622-2636, 1967.

27. HILDES, J.A., S. SHERLOCK, AND V. WALSHE. Liver and muscle glycogen in normal subjects, in diabetes mellitus and in acute hepatitis. I. Under basal conditions. Clin. Sci. 7: 287-295, 1949.

28. KIBLER, R.F., W.J. TAYLOR, AND J.D. MYERS. The effect of glucagon on net splanchnic balances of glucose, amino acid nitrogen, urea, ketones and oxygen in man. J. clin. Invest. 43: 904-915, 1964.

29. BALLINGER, W.F., H. VOLLENWEIDER, AND E.H. MONTGOMERY. The response of the canine liver to anaerobic metabolism induced by hemorrhagic shock. Surg. Gynec. Obstet. 112: 19-26, 1961.

30. BERNELLI-ZAZZERA, A., AND G. GAJA. Some aspects of glycogen metabolism following reversible or irreversible liver ischemia. Exptl. Molecular Pathol. 3: 351-368, 1964.

31. ISSEKUTZ, B. JR., H.I. MILLER, P. PAUL, AND K. RODAHL. Source of fat oxidation in exercising dogs. Am. J. Physiol. 207: 583-589, 1964.

32. MORGAN, T.E., F.A. SHORT, AND L.A. COBB. Effect of long-term exercise on skeletal muscle lipid composition. Am. J. Physiol. 216(1): 82-86, 1969.

33. MASORO, E.J., L.B. ROWELL, AND R.M. MC DONALD. Skeletal muscle lipids. I. Analytical method and composition of monkey gastrocnemius and soleus muscles. Biochim. biophys. Acta 84: 493-506, 1964.

34. MASORO, E.J., L.B. ROWELL, R.M. MC DONALD, AND B. STEIERT. Skeletal muscle lipids. II. Nonutilization of intracellular lipid esters as an energy source for contractile activity. J. biol. Chem. 241: 2626-2634, 1966.

35. HAVEL, R.J., B. PERNOW, AND N.L. JONES. Uptake and release of free fatty acids and other metabolites in the legs of exercising men. J. appl. Physiol. 23(1): 90-99, 1967.

LIVER GLYCOGEN IN MAN. EFFECT OF DIFFERENT DIETS AND MUSCULAR EXERCISE

E. HULTMAN and L. H. NILSSON

From the Department of Clinical Chemistry, S:t Eriks Sjukhus, Stockholm, Sweden

It is known that during heavy exercise the glucose production from the liver is increased (3, 4, 8, 9). The glucose production appears to increase successively during continued exercise, which also means that the glucose production from the liver increases concomitantly with decreasing glycogen stores in working skeletal muscle. Glucose production can increase from rest values of 100--150 mg/min to 900-1,100 mg/min at the end of heavy exercise (4) (Fig. 1). This increased output can be due to glycogenolysis of the glycogen store in liver or to an increased rate of gluconeogenesis. Gluconeogenic substrates are increased during exercise, both lactate and glycerol levels in blood being elevated. On the other hand, the splanchnic uptake of lactate is not increasing during the exercise period (4, 8). It was also shown many years ago that no increase in urea production occurred during exercise (6, 10). These facts are not consistent with a pronounced increase in gluconeogenesis during the exercise period.

In a series of experiments (1) the performance capacity of normal subjects was measured three times with 3 days interval between the exercise tests. These were all performed to complete exhaustion at 75 % of the subject's maximal oxygen uptake. During the 3 days' interval between the tests, the diet was changed from an ordinary mixed diet before the first exercise test to a normocaloric carbohydrate-free diet before the second test and thereafter to a carbohydrate-rich diet before the last test. The muscle glycogen was measured before and after each exercise period and it was found that the preformed glycogen was very much dependent on the diet, so that low glycogen levels were found after the carbohydrate-free diet, while the carbohydrate-rich diet gave a pronounced increase in muscle glycogen, giving values two to three

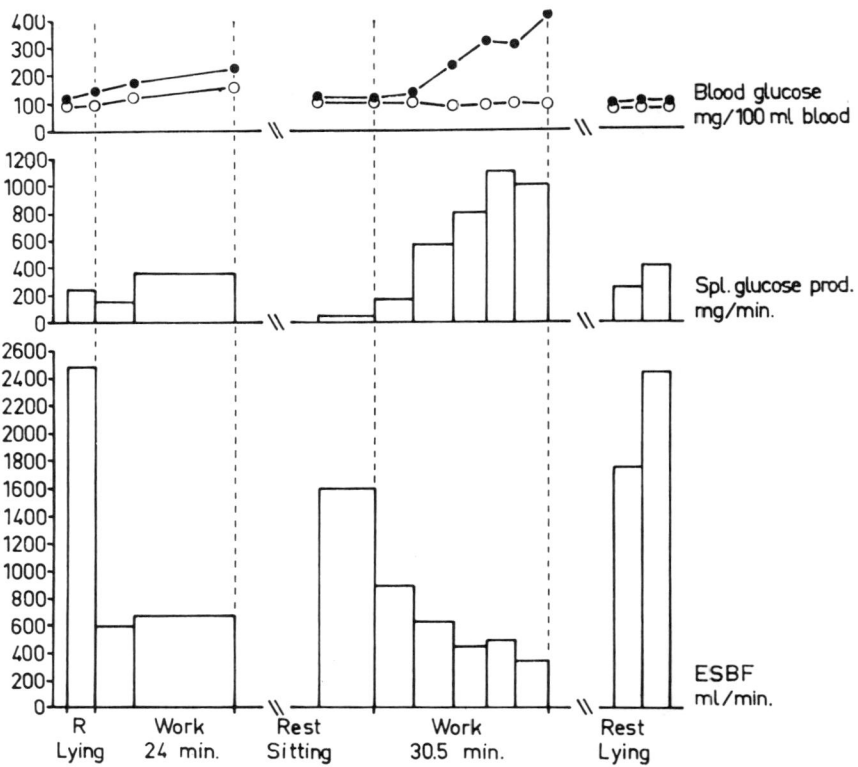

Fig. 1. Splanchnic blood flow and glucose output during heavy exercise (O_2 uptake corresponding to 71 % of the subject's max). o = hepatic vein blood; ● = arterial blood; ESBF = estimated splanchnic blood flow, ml/min.

times higher than normal. It was also found that the subject's performance capacity was well correlated with the muscle glycogen store in the muscles before exercise. Another observation was also made, i.e. that the blood glucose level decreased markedly during the exercise test after the carbohydrate-free diet (Fig. 2). In at least one of the subjects the decrease in blood glucose after 60 min of exercise was so pronounced that it could well explain his inability to continue the exercise. The blood glucose value at exhaustion was 32 mg/100 ml blood, and after 60 min of recovery it was still only 34 mg/100 ml. After the period on the carbohydrate--rich diet, the blood glucose level showed only small changes except for an initial decrease at the beginning of exercise.

LIVER GLYCOGEN: EFFECT OF DIETS AND EXERCISE

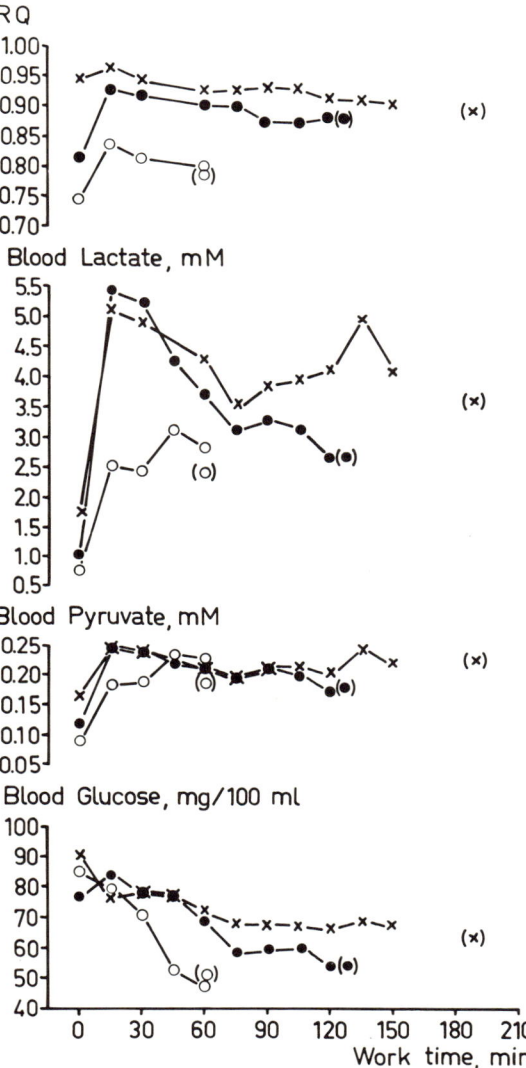

Fig. 2. Mean value of RQ, blood lactate, pyruvate, and glucose in connection with exercise after different diets in 6 subjects.

    x = carbohydrate diet
    ● = mixed diet
    o = fat diet
    + = protein diet
    ( ) = denotes the value at end of exercise

In order to correlate these findings with the liver glycogen store we measured the concentration of glycogen in needle biopsy material from normal subjects under different physiological conditions. A Menghini biopsy needle with a diameter of 1.2-1.4 mm was used. The tissue samples were weighed, homogenized in ice cold TCA-solution and glycogen determined as glucose after alcohol precipitation and acid hydrolysis. The first biopsy was obtained in the morning after an overnight fast and thereafter liver tissue was obtained at different times with total starvation or on a carbohydrate-free diet. The glycogen decrease during the first 4 hours of total starvation and rest is demonstrated in Fig. 3. The mean liver glycogen in the morning after an overnight fast was 50.9 g/kg wet liver tissue (range 14.4-69.2). Two hours later the mean value was 42.9 and after another 2 hours the glycogen content was 37.2 g/kg liver. The decrease in liver glycogen during complete rest and starvation was 50.8 mg glycogen per kg liver tissue per min (range 25.7-77.1) or 300 µmol/kg/min (Table 1).

The glucose production from the splanchnic area has been estimated by means of liver vein catheterization studies by several investigators and found to be in the order of 120-240 mg glucose per min. A reasonable estimate of the basal value at rest is 120--150 mg/min for a normal man not in a stress situation.

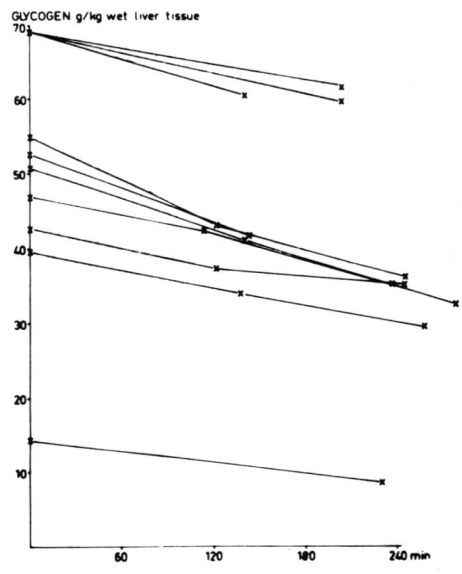

Fig. 3. Liver glycogen levels during short term starvation.

Table 1. Effect of short starvation on liver glycogen, g glycogen/kg wet liver tissue.

SUBJECT	FIRST BIOPSY AFTER AN OVERNIGHT FAST	BIOPSY AFTER 135-145 min	BIOPSY AFTER 210-270 min	GLYCOGEN DECREASE mg/kg WET LIVER/min
1	39.6	34.0	29.4	39.7
2	69.1	60.5		61.4
3	52.3	41.9		72.7
4	69.2		61.5	37.9
5	68.7		59.6	44.8
6	55.1	42.9	36.2	77.1
7	50.8	41.2	32.4	66.4
8	14.4		8.5	25.7
9	46.9	42.4	34.9	50.6
10	42.7	37.5	35.0	31.6
mean	50.9	42.9	37.2	50.8

Gluconeogenesis from amino acids, pyruvate, lactate, and glycerol after an overnight fast is estimated to be 40-50 mg glucose/min (2, 5, 7). The glycogenolysis measured in this series was equal to 55 mg glucose/kg liver tissue/min. This would in a normal man with a liver weight of 1.7 kg correspond to 90 mg glucose/min. Together with the gluconeogenesis (40-50 mg), this gives a total glucose production of 130-140 mg/min, a value which well corresponds to direct measurements of splanchnic glucose production. This would mean, on the other hand, that glycogenolysis in those circumstances, i.e. rest and complete starvation, empties the liver glycogen stores in less than 24 hours.

Fig. 4 shows an experiment in which the glycogen content in liver tissue was measured before and at different times during carbohydrate starvation for up to 10 days. In two of these cases (L.A. and T.W.) a total caloric starvation was employed. The caloric intake was 2,100 calories a day with less than 5 g carbohydrate. After this period the diet was changed to a normocaloric carbohydrate-rich diet containing 400 g carbohydrate. The liver glycogen contents increased rapidly up to 78-102 g/kg liver during the first 2 days (Table 2).

It was not possible, for technical reasons, to perform liver biopsy on normal subjects before and after muscular exercise, but on 14 occasions liver glycogen was determined after approximately one hour of heavy exercise on a bicycle ergometer. The mean

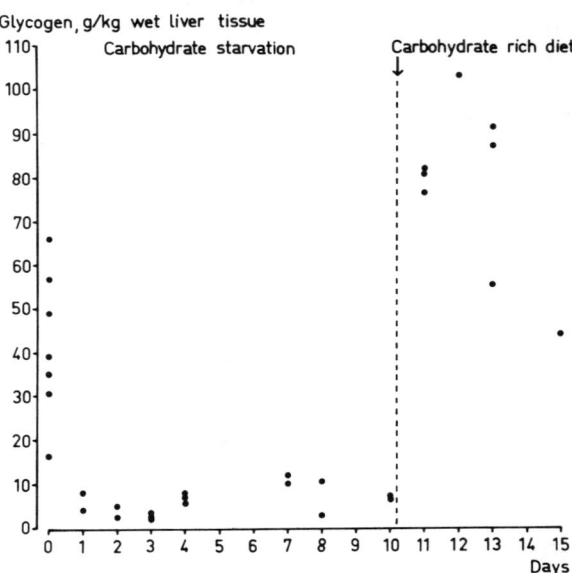

Fig. 4. Liver glycogen levels during long-term carbohydrate or total starvation followed by refeeding on a carbohydrate-rich diet.

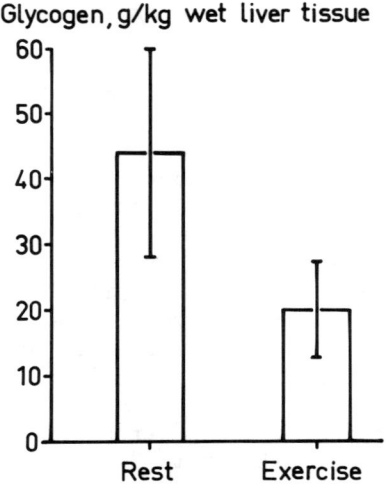

Fig. 5. Liver glycogen concentration determined in the morning after an overnight fast at rest ($n = 33$) and after one hour's heavy exercise ($n = 14$). Mean values and standard deviation are given. p value of difference <0.001.

Table 2. Effect of carbohydrate starvation and carbohydrate refeeding on liver glycogen, g glycogen/kg wet liver tissue.

SUBJ.	CARBOHYDRATE STARVATION, days											CARBOHYDRATE REFEEDING, days				
	0	1	2	3	4	5	6	7	8	9	10	1	2	3	4	5
L.A.	30.4	3.9	2.4	3.5								80.6				
T.W.	56.8	7.8														
J.S.	16.3				5.4			9.9			6.7	76.3		91.3		
K.E.	49.1				7.3			12.0			6.5	82.1		87.0		
M.A.	35.1			2.5					10.5				102.9			
C.F.	66.1			2.0					2.8							
E.J.	38.8		5.2		8.0									55.3		44.1

glycogen concentration was 20.0 g/kg wet liver (range 6.0-35.9) (Fig. 5). This value is significantly lower than the mean value of liver glycogen concentration determined at the same time of the day in another group of 33 normal resting subjects (mean 44.0; range 14.4-80.1). The mean value of liver carbohydrate used during the one hour's exercise would, thus, be 24 g/kg liver, or with a calculated liver weight of 1.7 kg, a total amount of 41 g.

CONCLUSIONS

The glycogen content depends on the composition of the food. After an ordinary mixed diet a considerable part of the glucose production from the liver is derived from glycogenolysis during the first 24 hours of rest and starvation. If carbohydrate is not supplied in the diet, the glycogen store in the liver will decrease rapidly to values less than 10 g/kg liver. During heavy exercise, and especially at the end of prolonged muscular work, appreciable amounts of glucose are produced by glycogenolysis in the liver. Values up to 1,100 mg/min have been observed, but still the blood sugar content tends to fall. After a carbohydrate-free diet the liver glycogen stores may be inadequate to meet this glucose need at the last period of heavy exercise. Excessive hypoglycaemia has been observed in this situation.

This investigation was supported by grants from the City of Stockholm.

REFERENCES

1. BERGSTRÖM, J., L. HERMANSEN, E. HULTMAN, and B. SALTIN. Diet, muscle glycogen and physical performance. Acta Physiol. Scand. 71: 140 - 150, 1967.
2. BORCHGREVINK, C.F. and R.J. HAVEL. Transport of glycerol in human blood. Proc. Soc. Exp. Biol. Med. 113: 946 - 949, 1963.
3. HULTMAN, E. Blood circulation in the liver under physiological and pathological conditions. Scand. J. Clin. Lab. Invest. 18: Suppl. 92, 27 - 41, 1966.
4. HULTMAN, E. Studies on muscle metabolism of glycogen and active phosphate in man with special reference to exercise and diet. Scand. J. Clin. Lab. Invest. 19: Suppl. 94, 1967.
5. KREBS, H.A. The metabolic fate of amino acids. In: Mammalian protein metabolism, edited by H.N. MUNRO and J.B. ALLISON. New York and London: Academic Press, 1964, I: 125 - 176.
6. MARGARIA, R. and P. FOA. Der Einfluss von Muskelarbeit auf den Stickstoffwechsel, die Kreatin- und Säureausscheidung. Arbeitsphysiol. 10: 553 - 560, 1939.

7. REICHARD, S.A.Jr., N.F. MOURY, N.F. HOCHELLA, A.L. PATTERSON, and S. WEINHOUSE. Quantitative estimation of the Cori cycle in the human. J. Biol. Chem. 238: 495 - 501, 1963.
8. ROWELL, L.B., K.K. KRANING, T.O. EVANS, J.W. KENNEDY, J.R. BLACKMON, and F. KUSUMI. Splanchnic removal of lactate and pyruvate during prolonged exercise in man. J. Appl. Physiol. 21: 1773 - 1783, 1966.
9. ROWELL, L.B., E.J. MASORO, and M.J. SPENCER. Splanchnic metabolism in exercising man. J. Appl. Physiol. 20: 1032 - 1037, 1965.
10. ZUNTZ, N. Ueber die Bedeutung der verschiedenen Nährstoffe als Erzeuger der Muskelkraft. Pfluegers Arch. Ges. Physiol. 83: 557 - 571, 1901.

METABOLISM OF FREE FATTY ACIDS AND KETONE BODIES IN SKELETAL MUSCLE

L. Hagenfeldt and J. Wahren

From the Departments of Clinical Chemistry and Clinical Physiology, Karolinska Institutet at Serafimerlasarettet Stockholm, Sweden

Evaluation of the free fatty acid (FFA) metabolism in skeletal muscle at rest and during exercise has been rendered difficult by the finding of simultaneous release and uptake of FFA from muscle tissue (6, 14). Analysis of local FFA uptake and oxidation in muscle has therefore been based on the use of isotopic techniques (7, 11). The human forearm is a useful model for such studies both at rest and during exercise since it posesses a dominant single artery blood supply into which a tracer substance may be infused and venous blood can be obtained almost exclusively from muscle tissue via the deep forearm veins (12).

METHODS AND PROCEDURE

Young, healthy male volunteers were studied in the postabsorptive state (12 - 14 hr fast). A group of four patients undergoing therapeutic prolonged starvation were studied after 3 weeks of total fasting. In all studies catheters were inserted percutaneously into a deep forearm vein in the distal direction and into the radial and brachial arteries. Exercise was performed on a hand ergometer with rhythmic, mainly isotonic, contractions (60 per min) at a work load of 10 kpm/min. Blood flow was determined with a dye dilution technique (15). Blood samples were collected at rest and repeatedly during exercise at timed intervals from the deep venous and the radial artery catheter.

Albumin-bound ^{14}C-FFA were infused into the brachial artery for periods of two minutes during exercise. A mixture of labeled palmitic, stearic, oleic and linoleic acid was used in the first studies and oleic acid only in subsequent investigations.

Individual FFA, 3-hydroxybutyrate and acetoacetate were determined by gas chromatography (3, 5) and specific activity of individual FFA by gas radiochromatography (7). $^{14}CO_2$ in blood was analyzed as discribed by Hagenfeldt (4). Radioactivity was also determined in a perchloric acid blood extract as described by Hagenfeldt and Wahren (7).

The uptake of individual FFA was determined from the fractional uptake of its radioactivity and its arterial concentration. A fractional uptake index was calculated as the ratio of the fractional uptake of an individual acid to that of total FFA. The fraction of the oxygen consumption used for fat oxidation was assumed to be $(1 - RQ)/0.293$ and the rate of fat oxidation was calculated on the basis of a mean oxygen equivalent of 24.7 moles per mole FFA.

RESULTS AND COMMENTS

Uptake of FFA

Labeled palmitic acid has been used most commonly as a tracer for the plasma FFA on the assumption that the differences between the individual acids in the FFA fraction can be disregarded (9). To test this contention a mixture of labeled palmitic, stearic, oleic and linoleic acid was infused into the brachial artery during exercise. The individual specific radioactivities of these acids were then determined in samples obtained from the radial artery and a deep vein during the infusion. The ratio of the fractional uptake of each individual acid to that of the four acids together are presented in Table I. Palmitic acid is taken up somewhat less effectively than the other three acids, which means that its use as a tracer for total FFA will lead to an underestimation of forearm muscle FFA uptake by about 10 per cent. Oleic acid has the same fractional uptake as the mixture of the four acids and since it is the main component of the FFA fraction it would seem to be the most suitable tracer substance for studies of muscle uptake of FFA.

Table I. Ratio of the fractional uptake of an individual FFA to that of total FFA.

Palmitic acid	0.91 ± 0.16	$p < 0.01$
Stearic acid	0.98 ± 0.44	N.S.
Oleic acid	1.03 ± 0.18	N.S.
Linoleic acid	1.07 ± 0.41	N.S.

Values are means $\pm$ SD of 34 observations in 15 subjects after 20, 40 and 60 min of forearm exercise at 10 kpm/min. The probability of the deviation from unity being caused by random factors is given.

METABOLISM OF FREE FATTY ACIDS AND KEYTONE BODIES

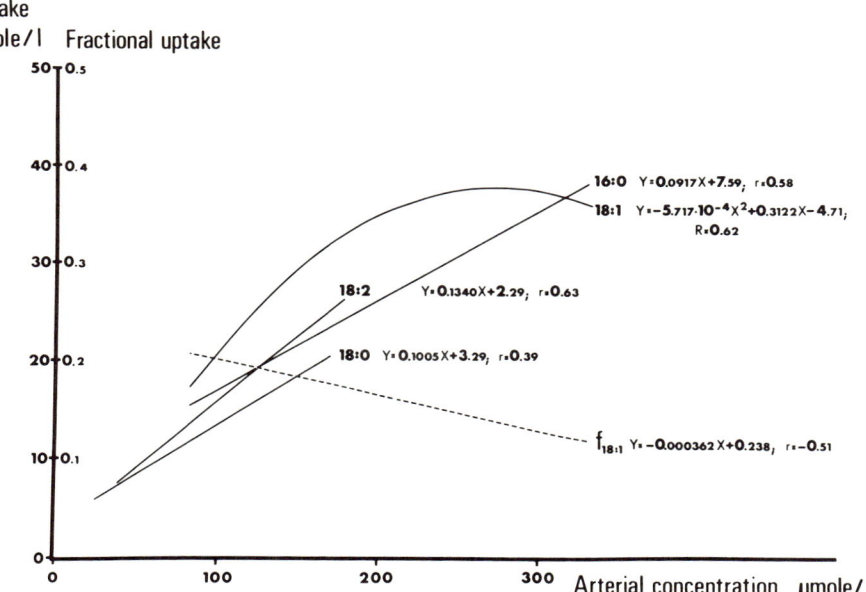

Fig. 1. Regressions of the uptake of individual FFA on their arterial concentrations and of the fractional uptake of oleic acid ($f_{18:1}$) on its arterial concentration. The regression equations are based on 34 observations in 15 subjects after 20, 40 and 60 min of forearm exercise at 10 kpm/min.

During exercise the muscle FFA uptake rises with the arterial concentration. However, it will be seen from the regressions for uptake of the individual acids during exercise at 10 kpm/min in Fig. 1 that, in these studies, the fractional uptake of oleic acid declined with rising arterial concentration, giving a significantly curved regression line for this acid. A negative correlation was seen when fractional uptake was correlated to plasma flow at different levels of work intensity (Fig. 2). These two sets of observations were combined with studies on four subjects during prolonged fasting (3 weeks) with very high levels of plasma FFA. It was then found that there is a linear regression of muscle FFA uptake (U_m) on FFA inflow (FFA_{inf}, calculated as the product of arterial concentration and muscle plasma flow) without any indication of curvilinearity at high FFA inflow values (Fig. 3). The relationship between these two variables can then be expressed by the linear equation:

$$U_m = b \cdot FFA_{inf} + a$$

Since U_m (in μmoles/min) is the product of the fractional uptake, the arterial FFA and the plasma flow, the regression equation of Fig. 3 can be divided by the factor (arterial FFA x plasma flow). This yields the following equation between the fractional uptake (f) and the FFA

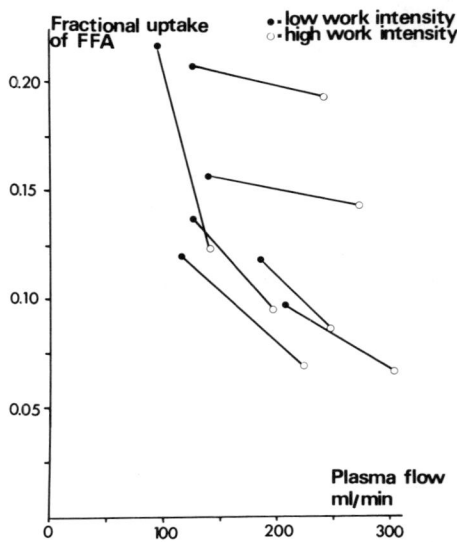

Fig. 2. Fractional uptake of FFA in relation to muscle plasma flow in 7 subjects exercising at two different work intensities.

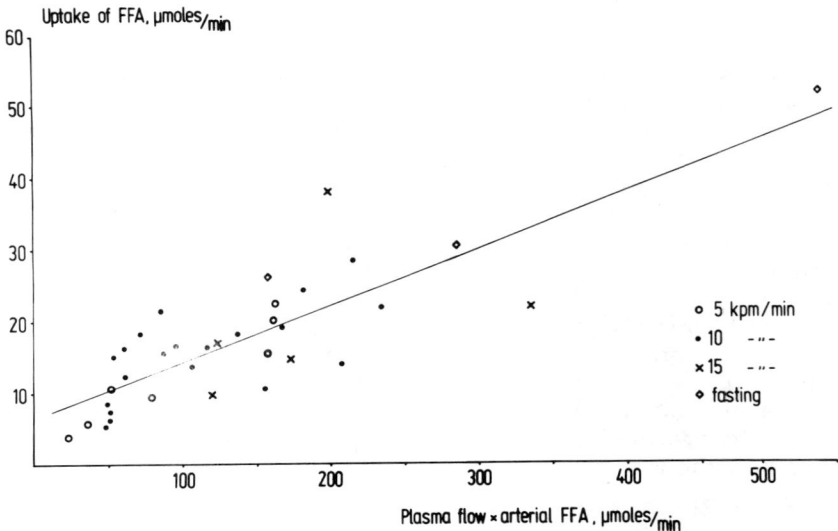

Fig. 3. Regression of rate of FFA uptake on FFA inflow (plasma flow x arterial FFA), $y = 0.079 x + 6.47$, $r = 0.83$, $p < 0.001$.

METABOLISM OF FREE FATTY ACIDS AND KEYTONE BODIES

inflow: $f = b + \dfrac{a}{FFA_{inf}}$; indicating that these variables should be inversely related unless the regression line in Fig. 3 passes through the zero point, in which case the fractional uptake would be constant. Since this is not the case ($p < 0.001$), the negative regressions mentioned earlier between the fractional uptake on the one hand and the arterial FFA and the plasma flow on the other should not be interpreted as indicating saturation of the transport mechanism, as suggested previously (7, 11). The muscle uptake of FFA thus rises linearly with the inflow and saturation conditions apparently do not arise even under the rather extreme circumstances in some of these studies. The results imply that the magnitude of FFA uptake is not regulated primarily by the muscle but is determined largely by outside factors such as the rate of lipolysis in adipose tissue.

The uptake of FFA during rhythmic forearm exercise of this type is not sufficient to cover the muscle's entire fat metabolism (Fig. 4). After 20 min exercise at 10 kpm/min in postabsorptive subjects the uptake of FFA is about 50 per cent of the fat oxidation as evaluated from blood RQ determinations. During prolonged fasting the uptake of FFA is higher (due to higher arterial FFA levels) while the fat oxidation is not much changed. In this situation the exercising

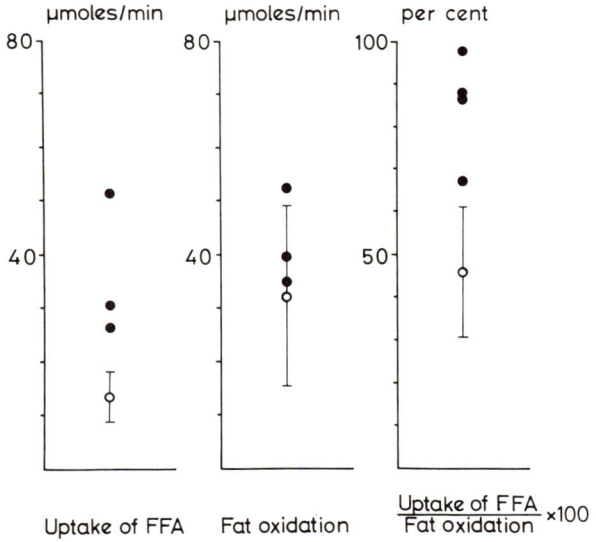

Fig. 4. Rate of FFA uptake, fat oxidation (calculated from RQ measurements) and percentage of the fat oxidation covered by FFA uptake in the forearm after 20 min exercise at 10 kpm/min in the postabsorptive state (open circles; means of 15 experiments, the bars corresponding to one SD) and during prolonged fasting (3 weeks, filled circles).

muscle thus seems to be capable of covering its fat metabolism more completely with plasma FFA.

Uptake of Ketone Bodies

At rest there are significant a - v differences across the forearm for both 3-hydroxybutyrate and acetoacetate. These a - v differences are correlated to the arterial concentrations for both substances (8). Both a - v differences decline during exercise and a net release of ketone bodies from the forearm has been observed during the latter half of a 60 min exercise period (8). The contribution of the ketone bodies to the exercising muscle's substrate supply is negligible at ordinary arterial concentrations; oxidation of the net uptake corresponds to 1 - 2 per cent of the oxygen consumption after 20 min exercise. Not even during prolonged fasting do the ketone bodies supply a significant amount of fuel for the exercising muscle, as indicated by the small variable a - v differences in Fig. 5. The arterial 3-hydroxybutyrate concentrations in these experiments were in the range 4 - 7 µmoles/ml. Equally small a - v differences, indicating non-utilization, were found for acetoacetate.

Oxidation of FFA

The oxidation of FFA in forearm muscle was studied by following

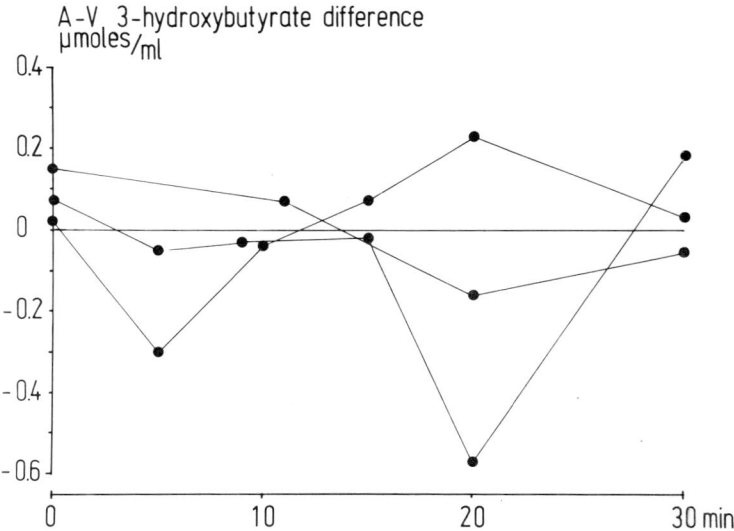

Fig. 5. The arterio-venous concentration differences of 3-hydroxybutyrate across the forearm at rest and during exercise at 10 kpm/min in three subjects after 3 weeks fasting.

the a - v difference of $^{14}CO_2$ after a two min intra-arterial infusion of labeled FFA during which the uptake of ^{14}C-FFA was determined. The release of $^{14}CO_2$ occurs rapidly, the maximum appearing less than two min after the end of the infusion. The $^{14}CO_2$ production curve then declines exponentially with an initial half-time of 2 - 3 min (7). The $^{14}CO_2$ elimination curve also shows a second phase with a half-time of 10 - 12 min. It is conceivable that this phase represents oxidation of FFA which has passed through an esterified pool. If so, this pool size can be estimated to approximately 0.6 µmoles per gram muscle, thus corresponding to the smaller pool of fatty acids in working muscle, calculated from whole body studies of FFA oxidation kinetics (10).

The total output of $^{14}CO_2$ during exercise at 10 kpm/min accounts for, on the average, 60 per cent of the ^{14}C-FFA uptake. At a lower work intensity the FFA oxidation is complete in most subjects (Table II). In subjects with less than 100 per cent oxidation it has been possible to recover the remaining radioactivity as water-soluble metabolites in a perchloric acid extract. Labeled 3-hydroxybutyrate has been identified among these metabolites but it has also been found in three subjects that more than 50 per cent of the v - a difference of water-soluble radioactivity is in the form of labeled acetate. This was shown by the addition of carrier acetate, precipitation with S-benzylthiouronium chloride and recrystallization to constant specific activity.

Table II. Percentage of FFA-^{14}C recovered as $^{14}CO_2$ and perchloric acid-soluble metabolites (PCA) at varying work intensities.

Exp.	Work intensity			
	low		high	
	$^{14}CO_2$	PCA	$^{14}CO_2$	PCA
1	101	-	52	-
2	100	-	101	0
3	100	-	76	24
4	107	-	95	5
5	73	-	40	59
6	95	-	40	-

- = not measured

The percentage recovery of $^{14}CO_2$ – the fractional oxidation – has been found to be dependent on the oxygen consumption ($\dot{Q}_{O_2}$, μmoles/min) and the lactate output ($\dot{Q}_{La}$, μmoles/min) according to the multiple regression equation:

% FFA oxidized = $0.048 \dot{Q}_{O_2} - 0.057 \dot{Q}_{La} - 1.0$

(12 observations after 20 min exercise at 10 kpm/min, R = 0.71, p < 0.05). A low oxygen consumption and a high lactate output are thus factors that in some way seem to be associated with a limited ability to oxidize acetyl-CoA. As 10 kpm/min is a rather moderate work load for the average forearm and the deep venous oxygen saturation during exercise is about 35 per cent (15), there is no reason to believe that the supply of oxygen to the exercising muscle limits the oxidative processes.

Since labeled acetate leaves the muscle, it seems likely that citrate synthesis is restricted. Oxidation of the total acetyl-CoA production would require less than 1.5 U per g muscle wet weight of citrate synthase activity (Fig. 6). Since human skeletal muscle probably contains several times this amount of the enzyme, citrate synthesis seems to be limited by inhibition. The lower part of Fig. 6

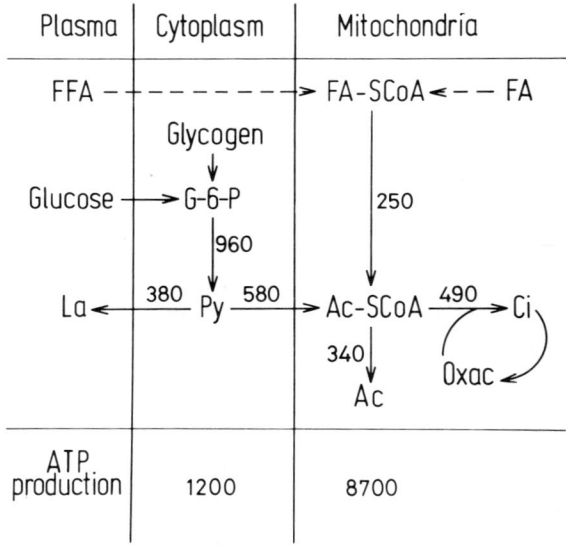

Fig. 6. A simplified scheme for the metabolic flow rates (expressed as μmoles/min of 2 or 3 carbon units) and ATP production rate (in μmoles/min) in the total forearm after 20 min exercise at 10 kpm/min. The values are means of 12 experiments. The following assumptions have been made: all acetyl-CoA that is not oxidized is deacylated to acetate; all substrates give maximal P:O ratios; 50 per cent of the glycolysis proceeds from glycogen, 50 per cent from plasma glucose.

shows the estimated ATP production in µmoles/min. Almost 90 per cent of the ATP is synthesized intramitochondrially in a situation when it is to be used extra-mitochondrially. The mitochondrial membrane is not freely permeable to adenine nucleotides, and the exchange between ATP produced in the mitochondria and cytoplasmic ADP requires an enzyme-like mechanism, the adenylate translocase (13). The present results are readily explained if this ATP-translocating mechanism is assumed to be limiting. Thus, the oxygen consumption would be restricted due to a limited supply of phosphate acceptors for the oxidative phosphorylation. Furthermore, the levels of ATP - or the energy charge of the adenine nucleotides - would be high intramitochondrially and this is known to inhibit citrate synthesis. The accumulation of acetyl-CoA could then restrict pyruvate oxidation and this together with factors favouring cytoplasmatic glycolysis, e.g. low ATP/AMP ratio and low levels of citrate (1), would account for the high rate of lactate output. Quantitative information on the adenylate translocase activity is available for rat liver (13) and beef heart mitochondria (2). These values both correspond to an estimated maximum rate of ATP translocation in a human forearm of less than 2000 µmoles/min or about 20 per cent of the intramitochondrial ATP production. Restriction of the capacity for ATP transfer between the mitochondria and the cytoplasm thus seems a possible explanation for the incomplete FFA oxidation under certain circumstances.

SUMMARY

The uptake and oxidation of FFA were studied in the human forearm muscle during exercise. The uptake of FFA rises linearly with the inflow to the muscle and covers in the postabsorptive state about 50 per cent of the muscle's fat oxidation as judged from blood RQ measurements. Oxidation of labeled FFA taken up by the muscle is complete at a low work intensity (5 kpm/min) but only about 60 per cent at a higher work intensity (10 - 15 kpm/min). The remaining radioactivity leaves the muscle as water-soluble metabolites, and labeled acetate seems to account for the major part of these. A hypothesis is presented according to which the ADP-ATP exchange between cytoplasm and mitochondria is the factor limiting oxidation of FFA in skeletal muscle during exercise.

ACKNOWLEDGEMENTS

This study was supported by grant 19X-722 from the Swedish Medical Research Council.

REFERENCES

1. ATKINSON, D.E. Citrate and the citrate cycle in regulation of energy metabolism. In: Metabolic roles of Citrate, ed. T.W. Goodwin. London: Acad Press, 1968, p. 23-40.
2. BYGRAVE, F.L., AND A.L. LEHNINGER. Properties of an oligomycin - sensitive adenosine diphosphate - adenosine triphosphate exchange reaction in intact beef heart mitochondria. J. Biol. Chem. 241:3894-3903, 1966.
3. HAGENFELDT, L. A gas chromatographic method for the determination of individual free fatty acids in plasma. Clin. Chim. Acta 13:266-268, 1966.
4. HAGENFELDT, L. A simplified procedure for the measurement of $^{14}CO_2$ in blood. Clin. Chim. Acta 18:320-321, 1967.
5. HAGENFELDT, L. Gas chromatographic determination of organic acids in blood. Arkiv Kemi. 29:63-73, 1968.
6. HAGENFELDT, L., AND J. WAHREN. Simultaneous uptake and release of individual free fatty acids in human forearm muscle during exercise. Life Sci. 5:357-364, 1966.
7. HAGENFELDT, L., AND J. WAHREN. Human forearm muscle metabolism during exercise. II. Uptake, release and oxidation of individual FFA and glycerol. Scand. J. Clin. Lab. Invest. 21:263-276, 1968.
8. HAGENFELDT, L., AND J. WAHREN. Human forearm muscle metabolism during exercise. III. Uptake, release and oxidation of B-hydroxybutyrate and observations on the B-hydroxybutyrate/acetoacetate ratio. Scand. J.Clin.Lab. Invest. 21:314-320, 1968.
9. HAVEL, R.J., L.A. CARLSON, L-G. EKELUND, AND A. HOLMGREN. Turnover rate and oxidation of different fatty acids in man during exercise. J. Appl. Physiol. 19:613-618, 1964.
10. HAVEL, R.J., L-G. EKELUND, AND A. HOLMGREN. Kinetic analysis of the oxidation of palmitate-1-^{14}C in man during prolonged heavy muscular exercise. J. Lipid. Res. 8:366-373, 1967.
11. HAVEL, R.J., B. PERNOW, AND N.L. JONES. Uptake and release of free fatty acids and other metabolites in the legs of exercising men. J. Appl. Physiol. 23:90-96, 1967.
12. IDBOHRN, H., AND J. WAHREN. The origin of blood withdrawn from deep forearm veins during rhythmic exercise. Acta Physiol. Scand. 61:301-313, 1964.
13. KLINGENBERG, M., AND E. PFAFF. Metabolic control in mitochondria by adenine nucleotide translocation. In: Metabolic Roles of Citrate, ed. T.W. Goodwin. London: Acad Press, 1968, p. 105-120.

14. RABINOWITZ, D., AND K.L. ZIERLER. Role of free fatty acids in forearm metabolism in man, quantitated by use of insulin. J. Clin. Invest. 41:2191-2197, 1962.
15. WAHREN, J. Quantitative aspects of blood flow and oxygen uptake in the human forearm during rhythmic exercise. Acta Physiol. Scand. 67: Suppl. 269, 1-93, 1966.

PLASMA INSULIN LEVELS DURING PROLONGED EXERCISE

E.D.R. Pruett

Institute of Work Physiology, Oslo, Norway

The clinical observation that diabetic patients who exercise require less insulin therapy than those who do not, led to interest in the behaviour of circulating insulin concentrations during exercise. An increase in insulin concentration should result in increased glucose transport, and thus a decrease in blood glucose concentration. Studies of diabetic patients have shown (Sanders et al 1964) that blood glucose concentration does, in fact, decrease as a result of exercise. However, since the same effect can be elicited in pancreatectomized animals (Ingle et al 1950, Goldstein et al 1953, Dulin and Clark 1961), it seems impossible that an increase in plasma insulin concentrations could be the only cause of the increased glucose transport. Thus a decrease in plasma insulin concentrations was also a possibility, and because of the necessity of preserving the blood and liver glucose for utilization by the central nervous system, this latter possibility seemed logical.
A few studies of the effect of various types of exercise on plasma immunoreactive insulin (IRI) concentrations have demonstrated a decrease (Cochrane et al 1966, Rasio et al 1966, Hunter and Sukkar 1968). A study of insulin secretion in both normal and diabetic patients showed no increase as a result of exercise (Nikkilä et al 1969). Studies of exercising animals have shown that insulin concentrations could be decreased during exercise even when glucose was infused (Issekutz et al 1967, Wright and Malaisse 1968).

As part of two rather comprehensive metabolic studies at this laboratory (in part in collaboration with University Hospital of Oslo), systematic investigations of changes in plasma immunoreactive insulin (IRI) levels with exercise have been carried out. In one of these investigations, healthy young men in the post-absorptive state exercised at work loads of from 20% to 90% of their previously measured maximal aerobic power (maximal oxygen uptake, max $\dot{V}_{O_2}$) (Pruett 1970a,b). In one part of this investigation the subjects lived on one of three carefully measured diets in which the fat and carbohydrate (but not the protein) content were varied. They exercised alternately on the bicycle ergometer and the treadmill in 45 minute work bouts interspaced with 15 minute rest intervals for six hours or to exhaustion, whichever came first, at work loads of 20%, 50% and 70% max $\dot{V}_{O_2}$. In another part of the investigation, the subjects lived on their own normal diet. Some exercised on the bicycle ergometer continuously, without pause, until exhausted at 70% max $\dot{V}_{O_2}$. Others exercised to exhaustion twice, with a 15 minute pause in between, at 85-90% max $\dot{V}_{O_2}$. A separate investigation was a study of pre-pubertal boys, exercising without pause on the treadmill for one hour at 70% max $\dot{V}_{O_2}$ (Oseid and Hermansen 1970). In all investigations blood glucose and plasma insulin levels before, during and after exercise were followed.

The purposes of these investigations were to ascertain: 1) whether or not plasma IRI concentrations did, in fact, decrease as a result of exercise; 2) whether or not a change in insulin concentration was dependent upon the duration and intensity, or the type of exercise; 3) whether or not diet had an appreciable effect upon a change in plasma IRI concentration with exercise; and 4) whether or not the normal interdependence of circulating insulin and glucose levels upon each other remained unaltered during exercise.

EFFECT OF INTENSITY, DURATION AND TYPE OF EXERCISE

<u>Interrupted exercise (20% to 70% max $\dot{V}_{O_2}$) in young men</u> (Pruett 1970a)

Plasma IRI concentrations fell consistently and significantly (Fig. 1) at work loads of approx 20%, 50% and 70% max $\dot{V}_{O_2}$ when subjects exercised alternately on the bicycle ergometer and the motor driven treadmill for 45 minute work bouts interrupted by 15 minute rest periods for six hours or until exhaustion, whichever came first. There was a small, but significant fall in 16 of 17 subjects during six hours of exercise at 20% max $\dot{V}_{O_2}$. In the seventeenth experiment, the subject exhibited a variable response with a net rise in plasma IRI after six hours. Individual variations from hour to hour were great at this work load. At work loads of approximately 50% and 70% max $\dot{V}_{O_2}$ the decrease in plasma IRI concentration was significant ($p < .001$) during the first 45 minute exercise period, and the IRI levels continued to decrease until, at exhaustion, they reached an average $60 \pm 7\%$ of the preexercise value (at 50% max $\dot{V}_{O_2}$) and $58 \pm 6\%$ of the preexercise value (at 70% max $\dot{V}_{O_2}$). Blood glucose concentrations during these same experiments decreased in approximately the same manner (Fig. 2). There was evidence at 50% max $\dot{V}_{O_2}$ that plasma IRI concentrations showed a greater decrease ($p < .05$) during the first hour of work than did blood glucose

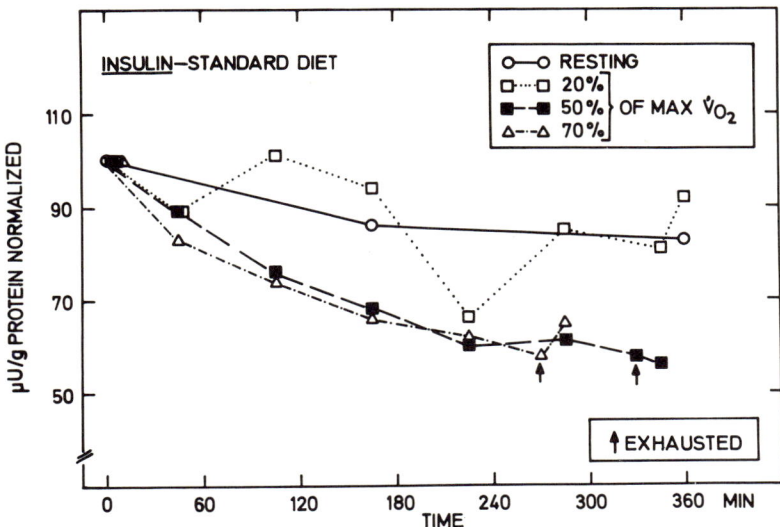

Fig. 1 The effect of rest and three levels of exercise on plasma immunoreactive insulin concentration in six subjects living on the standard diet. From: Pruett, E.D.R.: J. Appl Physiol. 1970a

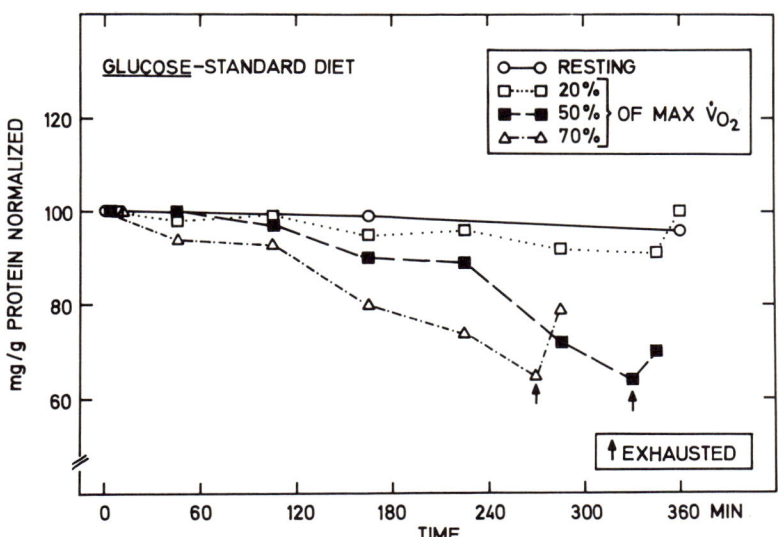

Fig. 2 The effect of rest and three levels of exercise on blood glucose concentration in seven subjects living on the standard diet. From: Pruett, E.D.R.: J.Appl.Physiol. 1970a

concentration. The decrease in blood glucose concentration during six hours of exercise at 20% max $\dot{V}_{O_2}$ was not significant. In subjects exercising to exhaustion at 50% and 70% max $\dot{V}_{O_2}$, it was to approx. 60% of the preexercise value (S.D.$\pm$ 15%) in all but 3 of 18 experiments, which will be discussed later under effects of diet. Thus, it is seen that at work loads from 20% to 70% max $\dot{V}_{O_2}$, plasma insulin concentrations decrease as a result of exercise, and that the normal interrelationship between circulating glucose and insulin concentrations generally holds.

Continuous exercise at 70% max $\dot{V}_{O_2}$ in young men

Two subjects exercised to exhaustion on the bicycle ergometer continuously, without pause at 70% max $\dot{V}_{O_2}$. In one of these subjects plasma IRI concentration decreased to 40% of the preexercise value while blood glucose concentration increased slightly. In the other subject, plasma IRI concentration fell to 50% of the preexercise value and blood glucose concentration remained essentially unchanged. Work time was considerably shorter than that during the interrupted exercise experiments at the same work load in the same subjects. However, the same general level of plasma IRI concentration was reached in both types of exercise. It is also seen that at 70% max $\dot{V}_{O_2}$ insulin levels can decrease even though glucose levels do not, indicating that glucose and insulin levels may not follow each other as exercise becomes more severe.

Interrupted exercise at near max $\dot{V}_{O_2}$ in young men (Pruett 1970b)

The results of the experiments described above indicated that the usual, well known relationship between circulating glucose and insulin levels might not hold during prolonged severe exercise. It was known from the experiments of Christensen (1931) that exercise at near max $\dot{V}_{O_2}$ resulted in an increase rather than a decrease in blood glucose concentration. Therefore, nine experiments at work loads of 85-90% max $\dot{V}_{O_2}$ were performed. Subjects exercised on the bicycle ergometer until they were exhausted. They then rested for 15 minutes and exercised to exhaustion again on the bicycle ergometer. As expected, plasma IRI concentrations invariably fell to an average 60% of the preexercise value, while glucose concentrations increased by the time exhaustion was reached during all exercise periods which were able to be maintained for longer than 10 minutes (Fig. 3). Because the necessity for prolonging the exercise for longer than 10 minutes appeared to be critical in obtaining the effect, a few of the same experiments were performed again, and this time blood samples (a total of 29) were drawn every two to five minutes during both the exercise and the rest periods. Fig.4 shows the results of one of these experiments. It is seen that during the first work period, blood glucose and plasma insulin

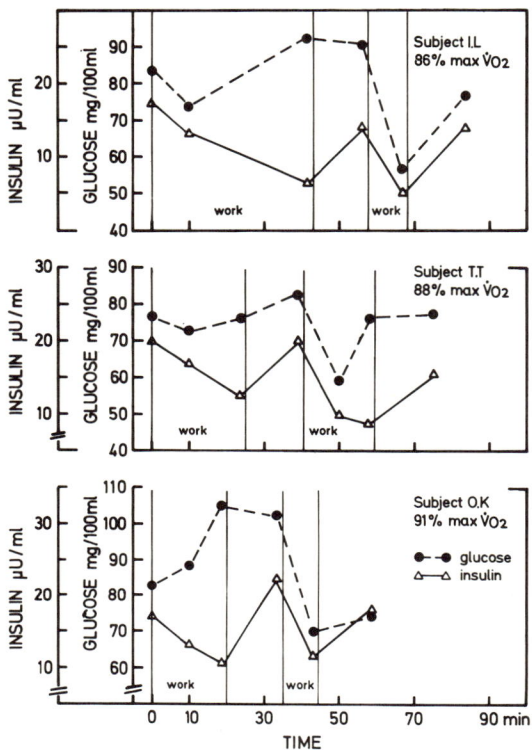

Fig. 3 Changes in blood glucose and plasma immunoreactive insulin concentrations due to severe exercise in three typical experiments in work loads at near max $\dot{V}_{O_2}$. From: Pruett, E.D.R.: J.Appl.Physiol. 1970b

concentrations remained relatively unchanged for 10 to 12 minutes. However, after that length of time, glucose concentration increased while insulin concentration fell. This effect was immediately reversed upon cessation of work. Both blood glucose and plasma insulin concentrations rose sharply within the first two minutes after work-stop, and continued to follow each other during the rest periods. When work was started again, plasma IRI and blood glucose concentrations fell during the first 10 minutes, after which IRI concentrations continued to fall and the glucose concentrations increased. Thus, it was shown that during very intense exercise, endogenous mechanisms can be activated by the exercise itself, which on the one hand produce hyperglycemia and on the other hand force circulating insulin concentrations to decrease. The usual relationship between circulating glucose and insulin levels is, thereby, altered.

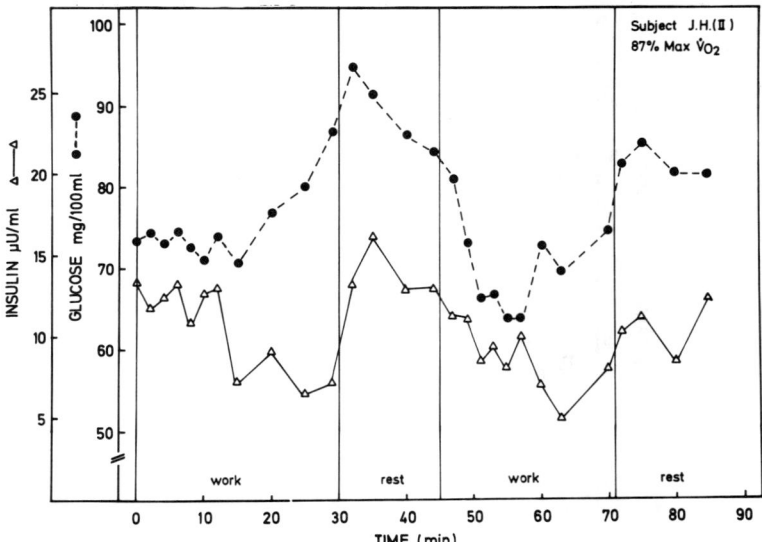

Fig. 4 Minute to minute changes in blood glucose and plasma IRI concentrations during exercise and recovery, during which a total of 29 blood samples were taken before, during and after the exercise. From: Pruett, E.D.R.: J.Appl.Physiol. 1970b

Continuous exercise at 70% max $\dot{V}_{O_2}$ in pre-pubertal boys
(Oseid and Hermansen 1970)

23 boys, twelve and one half years of age, in the post-absorptive state, ran on a motor driven treadmill at a work load of 70% max $\dot{V}_{O_2}$ for one hour. During this time plasma IRI concentrations were found to fall to approx. 50% of the preexercise value, and blood glucose concentrations were found to increase slightly (Fig. 5). Comparing these results with the results of the two continuous 70% max $\dot{V}_{O_2}$ bicycle ergometer experiments described above, it is seen that the effect of the exercise on plasma IRI and blood glucose concentrations at that work load was the same whether or not the work was performed on the bicycle or the treadmill, and whether or not the male subjects were adults or were of pre-pubertal age.

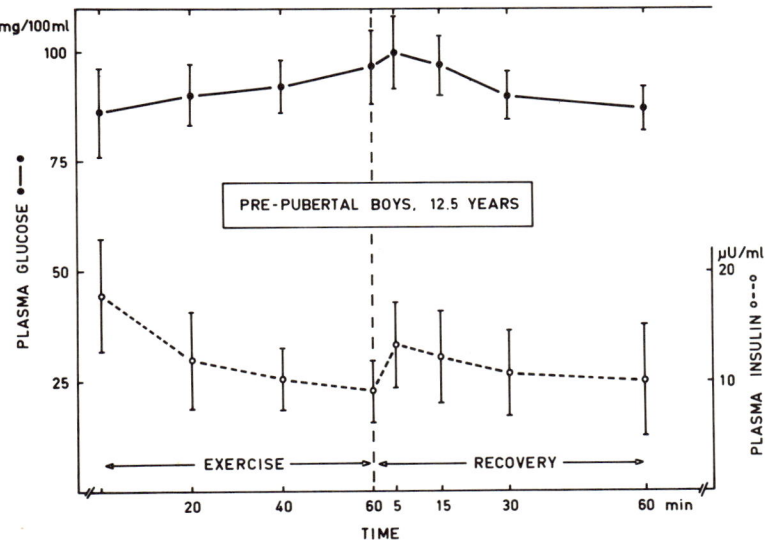

Fig. 5 Average values for blood glucose and plasma IRI concentrations during one hour of treadmill exercise at 70% max $\dot{V}_{O_2}$ in 23 pre-pubertal boys. From: Oseid, S. and L.Hermansen: Acta Paediat. Scand. 1970 (in press)

EFFECT OF DIET

The series of experiments described first in this article in which young men in the post-absorptive state performed bicycle and treadmill exercise, with pauses in between, for six hours or until exhausted, were carried out on three carefully weighed and measured diets:

	DAILY CONTENT			
DIET	Calories kcal	Protein g	Fat g	Carbohydrate g
STANDARD	3000	72	100	420
HIGH FAT	3000	72	200	180
HIGH CARBOHYDRATE	3000	72	30	620

Plasma IRI values were found to decrease with exercise in approximately the same manner as shown in Fig.1 no matter which diet was

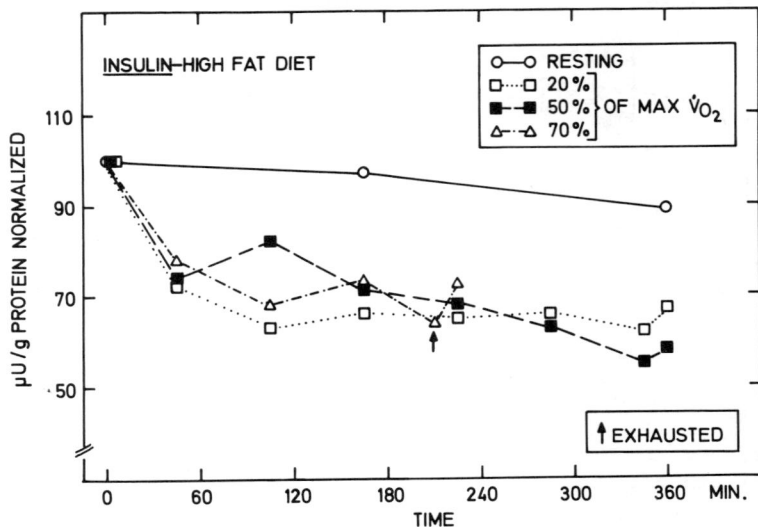

Fig. 6 The effect of rest and three levels of exercise on plasma IRI concentrations in five subjects living on a high fat diet.

used. Fig. 6 shows the decrease in plasma IRI at rest and during exercise at the three work loads. The decrease in IRI during the first hour of exercise at 20% max $\dot{V}_{O_2}$ was significant ($p < .001$) on the high fat diet, while this was not true on the other two diets. The decrease in plasma IRI concentration by the time exhaustion was reached at 50% and 70% max $\dot{V}_{O_2}$ was, however, the same for all diets, i.e. approximately 60% of the preexercise value. Blood glucose concentration decreased in all of the 17 cases of exercise at 50% max $\dot{V}_{O_2}$ irrespective of diet, and at 70% max $\dot{V}_{O_2}$ the same decrease was observed in 15 of 18 experiments. In the other three cases (two on the high fat and one on the high carbohydrate diet) blood glucose concentration either increased slightly or remained essentially unchanged. Thus it is seen that there was very little difference in the behaviour of plasma IRI concentrations during exercise which could be attributed to diet.

DISCUSSION

The results of the studies described above leave very little doubt that there is a decrease in plasma IRI concentrations during prolonged exercise, thus confirming earlier, more limited observations. In addition, it seems clear that if exercise is severe enough, it can, in itself, activate physiological mechanisms which alter the usual interrelationship between circulating glucose and insulin concentrations. This normal interrelationship is, however, restored almost immediately upon the cessation of exercise. The relative work load at which blood glucose and plasma IRI levels can be made to go in opposite directions appears to lie between 70% and 80% of the individual's maximal oxygen uptake. The exact position of this changeoverpoint may be somewhat dependent upon whether or not the exercise is performed with or without pauses. The age of the individual, his state of training and his diet appear from these investigations to play only a minor role.

The reduction in plasma IRI concentration during exercise might act as a partial barrier, hindering the transport of liver glucose from the blood to the cells of the exercising muscle cells. The effect of this barrier would be to preserve the available liver glycogen for use as the energy source for the central nervous system, which can only utilize glucose. This would leave the muscle glycogen as the primary carbohydrate energy source for muscular energy production. There is evidence that this is the case (Hultman 1967, Hermansen et al 1967, Karlsson et al 1969). However, it cannot be stated categorically from the observation that plasma IRI levels decrease with exercise, that the activity of insulin is also reduced. A reduced plasma concentration could mean either a decrease in insulin secretion or an increased peripheral utilization of insulin.

Other factors affect the permeability of the cell membrane to glucose. Extracellular factors produced during exercise and which increase the transport of glucose from the circulation have been suggested. One such factor is that demonstrated by Goldstein (Goldstein et al 1953). Intramuscular factors, such as the inhibition of the enzyme hexokinase by glucose-6-phosphate, a breakdown product of glycolysis, and many others must also be considered.

These experiments have shown that there is a physiological mechanism which induces simultaneously a hyperglycemia and a reduction in plasma IRI concentrations during extremely severe prolonged exercise. Many factors stimulate or inhibit the secretion of insulin. It is known that epinephrine is increased during exercise (von Euler and Hellner 1952, Vendsalu 1960) and that infusion of epinephrine in humans (Karam et al 1966, Porte et al 1966) results in a reduction of, or prevents in increase in circulating insulin even in cases of glucose induced hyperglycemia. Work by Porte (1967) suggested an alpha receptor mechanism for the epinephrine suppression of insulin release. Thus it is possible that an increased secretion of epinephrine during extremely severe exercise might be the mechanism which is responsible for the change during exercise from the usual resting interrelationship between circulating glucose and insulin concentrations.

SUMMARY

A significant fall in circulating insulin concentrations has been observed, almost without exception, during prolonged exercise at work loads of 20% of the subject's maximal aerobic power and higher. At higher work loads, the reduction in plasma IRI concentrations averages between 50% and 60% of the preexercise value. This is true in adult men and in pre-pubertal boys. As the work load becomes severe, blood glucose concentrations, instead of falling in approximately the same manner as the plasma IRI concentrations, increase as the exercise is prolonged for longer than ten minutes while IRI concentrations continue to fall. The work load level at which this division occurs appears to lie between 70% and 80% of the subject's max $\dot{V}_{O_2}$.

REFERENCES

Cochran, B., Jr., E.P.Marbach, R.Poucher, T.Steinberg and G.Gwinup: Effect of acute muscular exercise on serum immunoreactive insulin concentration. Diabetes, 15: 838-841, 1966

Dulin, W.E. and J.J.Clark: Studies concerning a possible humoral factor produced by working muscles. Its influence on glucose utilization. Diabetes, 10: 289-297, 1961

Goldstein, M.S., V.Mullick, B.Huddlestun and R.Levine: Action of muscular work on transfer of sugars across cell barriers: Comparison with action of insulin. Amer.J.Physiol., 173: 212-216, 1953

Hermansen, L., E.Hultman and B.Saltin: Muscle glycogen during prolonged severe exercise. Acta Physiol.Scand., 71: 129-139, 1967

Hultman, E.: Studies on muscle metabolism of glycogen and active phosphate in man with special reference to exercise and diet. Scand.J.clin. Lab. Invest., 19: (Suppl. 94): 1967

Hunter, W.M. and M.Y.Sukkar: Changes in plasma insulin levels during muscular exercise. Proc.Physiol.Soc., 110P-112P, 23-24 Feb. 1968

Ingle, D.J., J.E.Nezanis and K.L.Rice: Work output and blood glucose values in normal and diabetic rats subjected to the stimulation of muscle. Endocrinology, 46: 505-509, 1950

Issekutz, B., Jr., P.Paul and H.I.Miller: Metabolism in normal and pancreatectomized dogs during steady state exercise. Amer.J.Physiol., 213: 857-862, 1967

Karam, J.J., S.G.Grasso, L.C.Wegienka, G.M.Grodsky and P.M.Forsham: Effect of selected hexoses, of epinephrine, and of glucagon on insulin secretion in man. Diabetes, 15: 571-578, 1966

Karlsson, J., S.Rosell and B. Saltin: Blood-borne substrates. Acta Physiol.Scand., (Suppl. 330): 14, 1969

Nikkilä, E.A., M.-R.Taskinen, T.A.Miettinen, R.Pelkonen and H.Poppius: Effect of muscular exercise on insulin secretion. Diabetes, 17: 209-219, 1968

Oseid, S. and L. Hermansen: Hormonal and metabolic changes during and after prolonged muscular work in pre-pubertal boys. Acta Paediat.Scand., in press, 1970

Porte, D., Jr., A.L.Graber, T.Kuzuya and R.H.Williams: The effect of epinephrine on immunoreactive insulin levels in man. J.Clin.Invest., 45: 228-236, 1966

Porte, D.Jr.: A receptor mechanism for the inhibition of insulin release by epinephrine in man. J.Clin.Invest., 46: 86-94, 1967

Pruett, E.D.R.: Glucose and insulin during prolonged work stress in men living on different diets. J.Appl.Physiol., 28: 199-208, 1970a

Pruett, E.D.R.: Plasma insulin concentrations during prolonged work at near maximal oxygen uptake. J.Appl.Physiol. in press, August 1970b

Sanders, C.A., G.E.Levinson, W.H.Abelmann and N.Frienkel: Effect of exercise on the peripheral utilization of glucose in man. New.Engl.J.Med., 271: 220-225, 1964

Vendsalu, A.: Studies on adrenaline and noradrenaline in human plasma. Acta Physiol. Scand., 49 (Suppl. 173): 1-89, 1960

von Euler, U.S. and S. Hellner: Excretion of noradrenaline and adrenaline in muscular work. Acta Physiol.Scand.26: 183-191, 1952

Wright, P.H. and W.J.Malaisse: Effects of epinephrine, stress and exercise on insulin secretion in the rat. Amer. J.Physiol. 214: 1031-1034, 1968

GLUCOSE UPTAKE IN CONTRACTING, ISOLATED IN SITU DOG SKELETAL MUSCLE

N.S. Skinner, Jr., J.C. Costin, B. Saltin and G. Vastagh

Emory Univ., Atlanta, Ga.; Fysiologiska institutionen, Stockholm, Sweden; Southwestern Med. Sch., Dallas, Texas

Skeletal muscle constitutes the major mass of tissue in the human and other mammalian species, and this tissue has an enormous propensity for increasing both aerobic and anaerobic metabolism during contraction. For example, muscular contraction results in an increase in muscle blood flow from 5-10 to 60-70 ml/100 g/min and an increase in oxygen uptake by 30-40 times resting values. Vigorous physical activity can result in substantial reductions of glycogen stores in humans (Hermansen, Hultman and Saltin 1967).

While the above appear to be well established, the substrate or substrates that account for the large oxygen uptake in contracting skeletal muscle have not been defined clearly. Thus, a large number of candidates may serve to account for this increased oxygen uptake; e.g. fatty acids, ketones, triglycerides, glycerol, glucose, etc. The data to be reported are concerned only with the potential role of blood glucose.

It has been reported that some of the energy supply of inactive and contracting skeletal muscle is derived from blood glucose (Andres, Cader and Zierler 1956; Baltzan, Andres, Cader and Zierler 1962; Bass and Hudlická 1960; Bass and Hudlická 1964). Previous studies have also shown an increase in glucose uptake during periods of muscular exercise (Bass and Hudlická 1964; Christopher and Hayer 1958; Huyche and Kruhoffer 1955; Chapler and Stainsby 1968). The preponderence of evidence to support the conclusion that glucose uptake by muscle is increased by contraction has been obtained by either indirect methods or with use of heterogeneous tissue (e.g. hindlimb). Thus, the studies to be reported were done to investigate the uptake of glucose during skeletal muscle contraction in an isolated, in situ dog skeletal muscle where the experimental conditions were controlled carefully and in which major variables could be manipulated to evaluate in greater detail the effect of muscular contraction on uptake of glucose from blood. Portions of these data have been reported previously (Costin, Saltin, Skinner and Vastagh).

METHODS

The right gracilis muscle of mongrel dogs of both sexes, weighing 8-16 kg was used. Anesthesia was produced with pentobarbital sodium (25 mg/kg), and heparin was used for anticoagulation.

Autoperfused Gracilis Muscle (Free Flow Preparation)

The gracilis muscle was isolated from surrounding tissue, and the proximal and distal tendons were tied with a series of interrupted sutures. The nerve was cut 2-3 cm from the muscle and the muscle left in place. Except for the major artery and vein all blood vessels to and from the muscle were ligated. The femoral artery distal to the origin of the gracilis artery was cannulated with polyethylene tubing and through this cannula blood pressure was measured and recorded continuously. The gracilis vein was cannulated with polyethylene tubing after all branches leading into it were ligated so that the entire venous drainage from the muscle was directed through this cannula.

Constant Flow, Reservoir Perfused Gracilis Muscle

The muscle dissection was identical to the above with the exception that the artery to the muscle was cannulated and perfused with blood from reservoirs at constant flow with an oil displacement pump (Renkin and Rosell 1962). Venous drainage from the muscle was recorded continuously but was not returned to the animal. Two reservoirs were used and were filled simultaneously with arterial blood from a cannulated major artery. The blood was kept at 37.5 ± 1°C and stirred constantly with teflon coated magnetic stirrers.

Muscular contractions were produced with bipolar stimulation of the gracilis nerve (voltage, 15-30; impulse duration, 0.1 to 0.4 msec; frequency, 0.5 to 4.0/sec) with a Grass S4 generator. The contractions produced were single twitches and resulted in moderate shortening, since the leg of the animal was not bound tightly to the table. The force of contraction was not measured, but oxygen uptake was measured to estimate work level.

Procedure

The procedure used to study the autoperfused muscle was as follows. After isolation of the muscle and cannulation of the gracilis vein, 30 to 60 minutes were allowed to pass before blood samples were taken. Three samples of muscle were taken for glycogen analysis, following which 3 paired samples of arterial and venous blood were collected simultaneously at 10 minute intervals for glucose and lactate determinations. During this time paired samples of blood were also obtained for O_2 and CO_2 content. Stimulating electrodes were placed on the nerve and contractions begun. During the 60 minute contraction period paired samples of blood were obtained at 5, 10, 20, 30, 40, 50 and 60 minutes. Blood for CO_2 and O_2 contents were obtained at 10 and 40 minutes. Upon cessation of stimulation, 3 additional muscle samples were obtained for glycogen analysis. The entire muscle was removed and weighed (range 30 to 66 g).

Chemical Determinations

Plasma glucose was measured enzymatically with the glucose oxidase technique using glucostat^R (Worthington Biochemical Corp., Freehold, New Jersey). Even greater care was taken with glucose determinations in the autoperfusion studies. Paired samples of arterial and venous blood were collected and immediately a Somogyi filtrate was prepared. The glucose concentration of the filtrate was determined with the glucose oxidase technique. Each sample was examined in duplicate and read separately by two technicians. Five to 10 repeated analyses of one blood sample, a study done on 3 different days, showed the standard deviation did not exceed $\pm$ 1.0mg/100 ml when glucose concentrations were in the range of 80 to 100 mg/100 ml which was the range in which the majority of experiments were perfused.

Muscle glycogen content was determined with the method of Good, Kramer and Somogyi (1933). Three muscle samples (0.5-1.0 g each) from different sample sites were obtained immediately before and following muscle contraction for 60 min at 4 impulses/sec. Denaturation was accomplished in seconds by dropping these samples into tared tubes of hot 30% KOH.

Oxygen and CO_2 contents of arterial and venous blood were measured manometrically with the Van Syke method. Hemoglobin was determined by the ferrocyanide method and lactate by the method of Ström (1949).

RESULTS

Autoperfused Muscles

To evaluate the ability of this particular muscle preparation to take up glucose from blood, insulin was infused intra-arterially in the gracilis muscle of 4 animals at a rate of 0.015 units/min. The average control glucose uptake was 0.28 mg/100 g/min which increased to 1.08 mg/100 g/min following insulin injection. In each instance muscle blood flow remained relatively constant, and the arterio-venous glucose difference widened considerably.

Table I summarizes the data obtained from 6 muscles that were contracted at 2 impulses/sec for 60 minutes and from 6 additional muscles where stimulation was accomplished at 4 impulses/sec for one hour. The mean values ($\pm$ SEM) are shown. Some of these data are useful in assessing the physiological state of the preparation employed. These data indicate that a physiological hyperemia does occur and, simultaneously, O_2 extraction increases as indicated by widening of the arterio-venous O_2 difference. These changes were greater at the higher frequency of contraction. Lactate and CO_2 A-V difference also changed, and, again, the changes were greater at the higher contraction frequencies. Hemoglobin levels were essentially unchanged throughout the contraction period.

Figure 1 shows the relationship between O_2 uptake (ml/100 g/min) and glucose uptake (mg/100 g/min) in the gracilis muscles in the 12 animals studied. These values were obtained by integrating blood flow, O_2 A-V difference and glucose A-V difference for the 60 minute contraction period. The coefficient of correlation was 0.8423. The three clusters of points represent control, 2/sec contraction and contraction at 4 impulses/sec, respectively. Thus, there appears to be a reasonable relationship between glucose uptake and energy expenditure as reflected by oxygen uptake. Figure 2 shows blood flow (control, 2 and 4/sec contraction) along with glucose and oxygen uptake. This figure shows a tendency for glucose uptake to correlate in a rather linear fashion with the level of muscle activity as reflected by blood flow. In contrast, oxygen uptake tends to increase in a somewhat exponential fashion. Hence, the % of oxygen uptake that can be accounted for by oxidation of glucose taken up from the blood, assuming all of the glucose taken up was metabolized, was greater at the lower frequency of muscular contraction and less at the higher contraction frequency. Thus at 2/sec contraction the % VO_2 (glucose) was approximately 29% but only 15% at 4/sec contraction.

The biopsies of muscle obtained for glycogen analysis before and immediately following one hour of muscular contraction at 4/sec showed the mean glycogen content in resting muscle was 0.97 $\pm$ 0.18g/

TABLE I

	Glucose Uptake mg/100g/min	Glycogen g/100 g	Lactate A-V Diff mg %	O_2 A-V Diff vol %	O_2 uptake ml/100g/min	CO_2 A-V Diff vol %	Arterial Pressure mmHg	Blood Flow ml/100g/min
			MUSCLE CONTRACTION (2 impulses/sec) N=6					
Control	0.055 ±0.163	--	-0.52 ±0.30	4.70 ±1.77	0.28 ±0.08	-1.23 ±0.47	126 ±4	12.4 ±3.1
Contraction 60 min	1.533 ±0.477	--	-0.76 ±0.16	10.63 ±1.25	3.91 ±0.40	40.62 ±1.20	112 ±8	37.5 ±2.6
			MUSCLE CONTRACTION (4 impulses/sec) N=6					
Control	0.200 ±0.042	0.97 ±0.18	-0.70 ±1.82	3.83 ±0.65	0.27 ±0.03	-1.32 ±0.48	137 ±4	8.4 ±1.5
Contraction 60 min	2.368 ±0.234	0.26 ±0.08	-1.97 ±1.49	15.98 ±0.67	10.33 ±0.49	-14.87 ±0.35	124 ±5	61.9 ±3.7

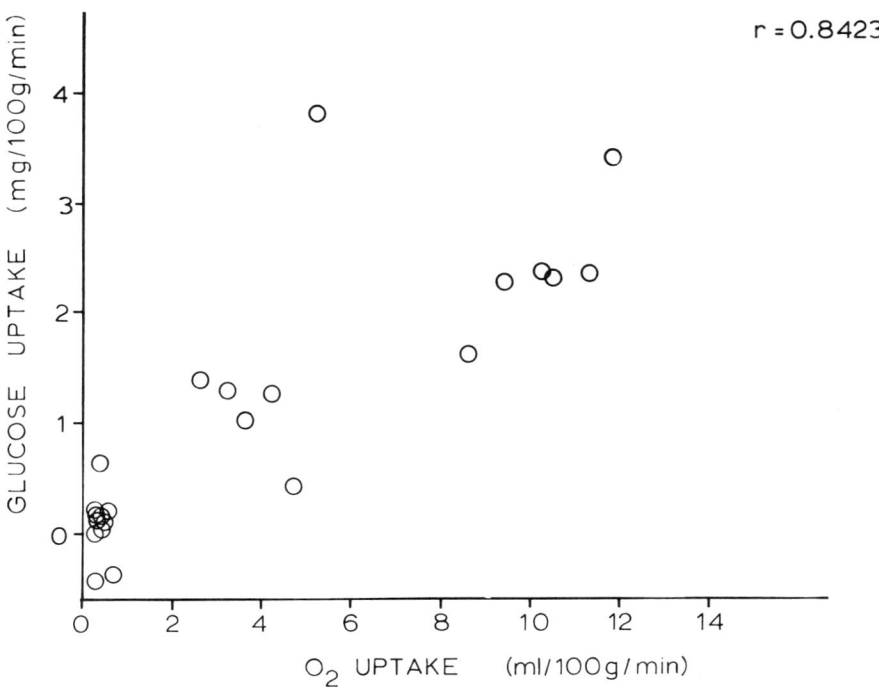

Figure 1: Plot of glucose uptake against O_2 uptake in autoperfused gracilis muscles. O_2 uptake less than 2 = resting values; 2 to 6 = 2/sec contraction; over 8 = 4/sec contraction. Coefficient of correlation = 0.8423.

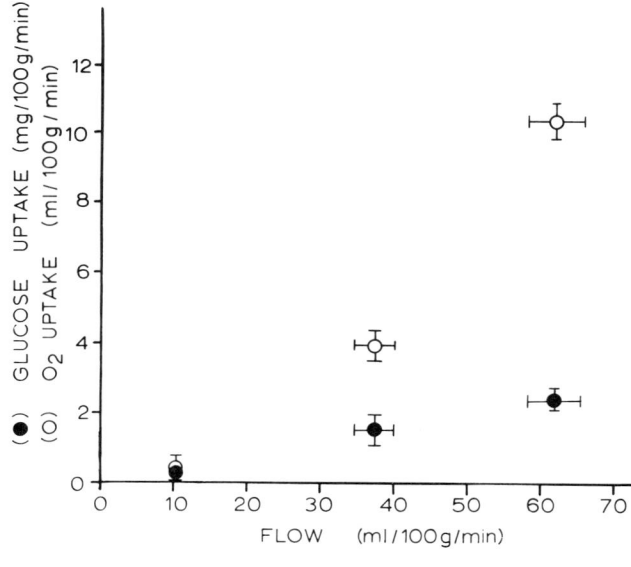

Fig. 2 Plot of mean ($\pm$ SEM) glucose uptake and O_2 uptake against blood flow in autoperfused gracilis muscle. Lowest blood flow represents resting conditions, intermediate flow = 2/sec contraction, highest flow = 4/sec contraction.

100 g (range 0.52 to 1.79). This decreased to 0.26 ± 0.98 g/100 g (range 0.05 to 0.64) at the termination of the contraction period, which represented a 73% reduction of muscle glycogen. No glycogen data were obtained for the 2/sec group.

In 5 additional autoperfused muscles, the effect of intravenously infused adrenalin on glucose A-V difference across the muscle was examined. Infusion lasted from 9 to 15 minutes; adrenalin delivery varied from 0.21 to 9.55 mg/min. This method of adrenalin administration increased glucose uptake from an average control of 0.07 mg/100 g/min (range -0.09 to +0.50) to an average of 0.78 mg/100 g/min (range 0.18 to 2.3). The greatest uptakes of glucose occurred with the highest infusion rates of adrenalin which, in turn, produced the greatest elevations in arterial glucose concentrations. In every instance glucose uptake was maximal during the first 5 minutes of adrenalin infusion.

Constant Flow, Reservoir Perfused Muscles

This type of preparation was used in order to exclude such factors as increased blood flow with contraction, and, hence, increased delivery of various substances that could influence glucose uptake in skeletal muscle. Also, in the autoperfused muscle the venous drainage from the contracting muscle entered the systemic circulation and was recirculated. Thus, the possibility existed that some substance was released, directly or indirectly, that could influence glucose uptake. Perfusion of muscles at constant blood flow from blood reservoirs allowed exclusion of these variables.

Ten experiments were performed using muscles from 8 animals. In all reservoir perfusion was employed. Five experiments were done in which blood flow was kept constant (range 5.3 to 7.1 ml/100 g/min) during muscular contraction at stimulation frequencies of 0.5 to 1.0 impulse/sec. The duration of contraction varied from 4 to 14 minutes. In every instance it was obvious from visual inspection that the venous blood draining the contracting muscle was markedly desaturated, but in no instance was there a widening of glucose A-V difference or an enhanced uptake of glucose. Instead, glucose A-V difference narrowed during contraction and, because blood flow was constant, glucose uptake decreased. In 4 additional experiments, blood flow from the reservoirs was increased during the contraction period (0.75 to 2.0 impulses/sec for 15 to 31 min). Again, glucose A-V difference failed to widen and in only one muscle was there an increase in glucose uptake. It is emphasized that like the autoperfused muscles, these constant flow (reservoir) perfused preparations responded to close arterial injection of insulin or addition of insulin to the blood reservoir with a marked widening of glucose A-V difference.

In 4 animals adrenalin was infused intra-arterially into the gracilis muscle in an amount sufficient to produce vasoconstriction. Adrenalin failed to produce an uptake of glucose greater than that present prior to infusion. In additional experiments, the combination of constant blood flow perfusion, close arterial adrenalin infusion and muscular contraction did not alter glucose uptake by the gracilis muscle.

In view of the correlation between increased arterial glucose concentrations induced by intravenous adrenalin infusions and glucose uptake by autoperfused gracilis muscles, the effect of abrupt alterations in arterial blood glucose concentration on uptake of glucose was studied in 6 experiments on 4 constant blood flow muscles. Two blood reservoirs were used and were filled simultaneously from an artery so that the blood in each was identical. Glucose was added to one of these. By changing perfusion from one reservoir to the other it was possible to increase abruptly the concentration of glucose in the blood used to perfuse the muscle and make frequent measurements of venous concentrations of glucose while blood flow remained constant. A substantial glucose uptake resulted with this procedure. Return of the arterial blood glucose concentration to normal (return to control reservoir) produced a transient net loss of glucose from the muscle tissue which would, of course, include interstitial fluid, etc. In two additional animals, repeating this same procedure in combination with muscular contraction produced essentially the same changes as when no contraction was employed.

DISCUSSION

The autoperfused (free flow) and constant flow, reservoir perfused dog gracilis muscle was studied. In both preparations intra-arterially injected insulin produced a pronounced widening of glucose arterio-venous difference and, hence, substantial increases in glucose uptake. In both preparations muscular contraction produced a marked widening of oxygen arterio-venous differences and resulted in lactate production. In six animals studied under free flow conditions, muscular contraction at 4 impulses/sec for one hour resulted in an average 73% reduction in glycogen content. Thus, these several criteria indicate that the preparations employed were capable of responding appropriately to various and common physiological stimuli.

The free flow experiments showed definite increases in glucose and oxygen uptakes and lactate production during the periods of contraction, and the changes were greater at the higher frequencies of contraction. In contrast, glucose uptake was not enhanced by muscular contraction in constant flow, reservoir perfused muscles. The constant flow preparation, by preventing the marked hyperemia associated with muscular contraction, facilitates the detection of glucose uptake by setting a stage for widening of glucose A-V differences. In the free flow preparation, in which blood flow increased 3 to 7 fold, even minor changes in glucose A-V differences can represent large changes in glucose uptake. Yet it was in these autoperfused muscles that glucose uptake during contraction was found.

The explanation for the different glucose uptake responses to contraction in the two preparations is not immediately clear but the finding does invite consideration for several possibilities. Changes in blood flow and, hence, glucose delivery may be a factor. The present data offer no help relative to this point. Part of the difference in glucose uptake response may relate to the difference in energy output. Obviously the high level of oxygen extraction and marked hyperemia in the free flow preparation indicates a substantial increase in energy output. In contrast, in the constant flow preparation, where blood flow during contraction is limited, the increase in energy output is also limited. Energy output, however, is never-the-less increased, since marked desaturation of the venous blood occurred as well as lactate production. Thus, while part of the difference in results between the two types of experiments may be the difference in energy output, other factors would also appear to be important. Because recirculation of blood from the contracting muscle did occur with the free flow preparation but was prevented completely in the constant flow technique, it is possible that contracting muscle results in release of substances that influence the uptake of glucose by skeletal muscle in a manner similar to that reported by Goldstein (1961). The present data do not offer any definitive information relative to this possibility.

There are several striking similarities and differences when the present data are compared with those from a similar study by Chapler and Stainsby (1968). Each employed *in situ* dog skeletal muscle; the gracilis was used in the present study and the gastrocnemius-plantaris in that by Chapler and Stainsby. In the gastrocnemius resting glucose uptake was 5 times that measured in the gracilis muscle. Resting oxygen uptake was 4 times greater in the gastrocnemius. During contraction at 5 impulses/sec, glucose uptake increased to an approximate average of 80 μg/g X min in the gastrocnemius but to only 25 μg/g X min in gracilis muscles contracted at 4 impulses/sec. Oxygen uptakes, however, were quite comparable in the two studies, being about 14 ml/100 g X min in the gastrocnemius (5 impulses/sec) and 11 ml/100 g X min in the gracilis (4 impulses/sec). Thus with comparable oxygen uptakes in the two studies at comparable frequencies of contraction, glucose uptake in the gracilis muscle was less than 30% of that seen in the gastrocnemius-plantaris muscle. This difference does not appear to be related to differences in fiber types, since both are mixed muscles as regards "red" and "white" fibers (Edström and Nyström 1969).

It is possible that glucose uptake during muscular contraction will be more or less depending upon the amount of glycogen breakdown. Chapler and Stainsby (1968) found less glycogen breakdown (36%) than measured in the present study (73%), although control glycogen contents were quite similar. As already emphasized, these authors also found greater glucose uptakes than reported in the present study. The number of animals in the present study is inadequate to allow any conclusion relative to the possible interrelationship between glycogen breakdown and glucose uptake. However, the data reported are sufficiently suggestive to warrant consideration for this possibility. Of course, many other factors could be involved (e.g. fatty acids, ketones, amino acids, etc) but were not evaluated in this study.

These data, if representative of the response in all skeletal muscle, would indicate that skeletal muscle with its enormous capacity for increasing metabolism does not depend to any major degree on exogenous glucose to meet metabolic demands. In view of the limited extramuscular pool of carbohydrates (liver and extra cellular fluid) which probably amount to 40-80 g in humans and the apparent inability of glucose-6-phosphate to leave the muscle cell, it seems logical that exogenous glucose is not a major fuel for skeletal muscle. Thus, the augmented glucose utilization measured during heavy work in humans or animal preparations - intact awake animals or isolated limbs - would appear to be more related to metabolism of nervous or adipose tissues, tissues that are dependent upon glucose as their energy source and have little or no stored glycogen. Hence, the relative lack of glucose uptake by contracting skeletal muscle would help preserve the limited extramuscular glucose pool, allow glucose to be better maintained at physiological levels to preserve the integrity of the organism by keeping this fuel available to those vital tissues that utilize only or primarily glucose.

BIBLIOGRAPHY

1. Andres, R., G. Cader and K. Zierler, The quantitatively minor role of carbohydrate in oxidative metabolism by skeletal muscle in intact man in the basal state. Measurements of oxygen and glucose uptake and carbon dioxide and lactate production in the forearm. J. Clin. Invest. 1956. 35. 671-682.

2. Baltzan, M., R. Andres, G. Cader and K. Zierler, Heterogeneity of forearm metabolism with special reference to free fatty acids. J. Clin. Invest. 1962. 42. 116-125.

3. Bass, A. and O. Hudlická, Utilization of oxygen, glucose, unesterified fatty acids, carbon dioxide and lactic acid in normal and denervated muscle in situ. Physiol. Bohemoslovenica. 1960. 9. 401-407.

4. Bass, A. and O. Hudlická, Interrelations between metabolism and blood flow in normal and denervated dog gastrocnemius muscle at rest and during stimulation. Physiol. Bohemoslov. 1964. 13. 48-61.

5. Chapler, C.K. and W.N. Stainsby, Carbohydrate metabolism in contracting dog skeletal muscle in situ. Amer. J. Physiol. 1968. 215. 995-1004.

6. Christophe, J. and J. Mayer, Effect of exercise on glucose uptake in rats and men. J. Appl. Physiol. 1958. 13. 269-272.

7. Costin, J.C., B. Saltin, N.S. Skinner, Jr. and G. Vastagh, Glucose uptake in contracting, isolated dog skeletal muscle. Acta Physiol. Scand. In Press.

8. Edström, L. and B. Nyström, Histochemical types and sizes of fibers in normal human muscles. Acta Neurol. Scand. 1969. 45. 257-269.

9. Goldstein, M.S., Humoral nature of hypoglycemia in muscular exercise. Amer. J. Physiol. 1961. 200. 67-70.

10. Good, C.A., H. Kramer and M. Somogyi, The determination of glycogen. J. Biol. Chem. 1933. 100. 485-491.

11. Hermansen, L., E. Hultman and B. Saltin, Muscle glycogen during prolonged severe exercise. Acta Physiol. Scand. 1967. 17. 129-139.

12. Huycke, E.J. and P. Kruhøffer, Effects of insulin and muscular exercise upon the uptake of hexoses by muscle cells. Acta Physiol. Scand. 1955. 34. 232-249.

13. Renkin, E.M. and S. Rosell, The influence of sympathetic adrenergic vasoconstrictor nerves on transport of diffusible solutes from blood to tissues in skeletal muscle. Acta Physiol. Scand. 1962. 54. 223-240.

14. Ström, G., The influence of anoxia on lactate utilization in man after prolonged muscular work. Acta Physiol. Scand. 1949. 17. 440-451.

GLUCOSE METABOLISM DURING EXERCISE IN MAN

J. Wahren, G. Ahlborg, P. Felig and L. Jorfeldt

From the Department of Clinical Physiology, Karolinska Institutet at Serafimerlasarettet, Stockholm, Sweden and the Department of Medicine, Yale University School of Medicine, New Haven, Conn., U.S.A.

It is generally recognized that in man, blood glucose is taken up to some extent by muscle during physical exercise, but the quantitative importance of its contribution to muscle substrate utilization has not been defined and is the subject of some controversy. Recent studies of glucose uptake by exercising muscle, based on determinations of a - v differences with and without simultaneous blood flow measurements, have indicated both considerable glucose uptake (15, 17, 18, 23) and a virtual non-utilization of this substrate (1, 2, 19). The a - v differences for glucose are often small and close to the error of the method for glucose analysis and few of the studies have been specifically directed towards defining muscle glucose metabolism. The present study was undertaken in an attempt to characterize the use of blood glucose as a substrate for small and large muscle groups during exercise of differing duration and intensity in postabsorptive man.

METHODS AND PROCEDURE

All subjects studied were healthy male volunteers (age range 21 - 39, mean age 25 years). None of the subjects was obese. The studies were performed in the morning, the subjects having fasted overnight. Two types of studies were made, involving forearm exercise and leg exercise respectively.

Forearm exercise series

Catheters were inserted percutaneously into the radial and brachial arteries and into a deep forearm vein. The subjects performed forearm exercise on a handergometer (30) at a moderately heavy work load (10 kpm/min) for periods up to 60 min and at heavier loads (15 - 20 kpm/min) for short periods until exhaustion. Blood samples from an artery and the deep forearm vein were collected at rest and at timed intervals during the exercise period for analyses of oxygen and carbon dioxide content, glucose and lactate concentrations. Forearm blood flow was determined using an indicator dilution technique (30) and uptake and production of oxygen, glucose and lactate were calculated per unit estimated forearm muscle mass as described earlier (17, 31).

Leg exercise series

Catheters were inserted percutaneously into a femoral vein in the distal direction, into an antecubital vein and into a brachial artery. A no 7 or 8 Goodale-Lubin catheter was positioned in the main right hepatic vein under fluoroscopic control after a venous cut down in either antecubital fossa. The catheter tip was placed 3 - 4 cm from the wedge position. Catheter position was checked repeatedly by fluoroscopy before and after the exercise period.

After the catheterization, a single intravenous injection of 18 - 20 mg indocyanine green dye was given and arterial blood samples were obtained at 2 min intervals during the following 8 min. The plasma volume was estimated from the zero time intercept of the exponential dye decay curve. The subjects were then studied at rest in the supine position and during upright exercise on a bicycle ergometer. The subjects exercised for 40 min at one of the work loads: 400, 800 or 1200 kpm/min. Blood samples were collected simultaneously from the femoral and hepatic veins and the brachial artery, twice at rest and repeatedly at timed intervals during exercise, for analysis of glucose, lactate, pyruvate, α-amino nitrogen, glycerol, hematocrit and oxygen content. Expired air was collected at rest and after 10 and 40 min of exercise for determination of oxygen uptake.

Hepatic blood flow was estimated at rest and during exercise using the continuous infusion technique (5) and indocyanine green dye (0.4 - 0.5 mg/min) (26). The infusion was started at rest and paired arterial and hepatic venous samples were collected at 5 min intervals throughout the study, starting after 15 min of infusion. After 30 min of infusion the subjects sat up on the bicycle and started the exercise. At the heavier work loads, a rise occurred in the arterial concentration of dye in several subjects. The rate of this rise was then determined graphically and subtracted from the rate of dye infusion in the calculation of blood flow. This correction factor did not exceed 0.04 mg $\cdot$ l^{-1} $\cdot$ min^{-1} in any subject and significant hepatic storage is thus unlikely to have occurred (4).

Analyses

Glucose was analyzed using the glucose oxidase reaction and a commercial analytical kit from Kabi Inc. (Stockholm, Sweden). Lactate was determined enzymatically as described by Wahren (30) and pyruvate mainly according to Segal, Blair and Wynngarden (28) and Bücher, Czok, Lamprecht and Latzko (6). Glycerol was analyzed enzymatically (32) and α-amino nitrogen by the colorimetric reaction of Moore, Spackman and Stein (21). Serum immunoreactive insulin was determined by a modification (29) of the double antibody technique of Morgan and Lazarow (22). Oxygen saturation was determined spectrophotometrically by a slightly modified form of the method of Drabkin (10). Hemoglobin concentration was analyzed with the cyanmethemoglobin technique (9). Indocyanine green was determined at the point of maximal absorption, using a Beckman B spectrophotometer. Hematocrit was determined using a microcapillary hematocrit centrifuge and a correction for trapped plasma (1.3 per cent, 12). Samples for RQ determination were analyzed by the van Slyke technique. Expired air was collected in Doublas bags and analyzed using the Scholander microtechnique.

Data in the text are given as mean $\pm$ SE.

RESULTS AND COMMENTS

Forearm exercise studies

The arterial concentration of glucose remained unchanged at rest and during forearm exercise for 60 min. At the onset of exercise there was first a reduction and a reversal of the a - v difference across the forearm, followed by a gradual increase during the next few minutes (Fig. 1). The glucose a - v differences observed during the initial phase of exercise correlated negatively to the simultaneous log lactate values ($r = 0.89$, $p < 0.001$, Fig. 2). The data given in Fig. 2 were obtained after 1 - 2 min of moderately heavy exercise (10 kpm/min) and at exhaustion after 2 - 4 min of quite strenuous forearm exercise (15 - 20 kpm/min). The correlation indicates that during heavy forearm exercise there is a gradually increasing net outflow of glucose from the muscle. Similar observations have been reported for stimulated dog skeletal muscle during tetanus (7). The mechanism behind this finding is not entirely clear since the enzyme glucose-6-phosphatase is not present in muscle tissue (20) but some clues as to its possible background are provided by the breakdown pattern of glycogen (Fig. 3).

The major part of glycogen is converted to glucose phosphate by the action of phosphorylase. However, in the debranching process, catalyzed by amylo-1,6-glucosidase and oligo-1,4-1,4-glucantrans-

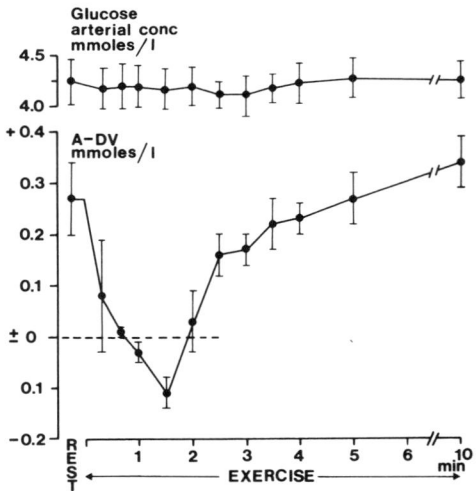

Fig. 1. Arterial concentration and arterial-deep venous concentration difference (A - DV) for glucose at rest and during forearm exercise at 10 kpm/min.

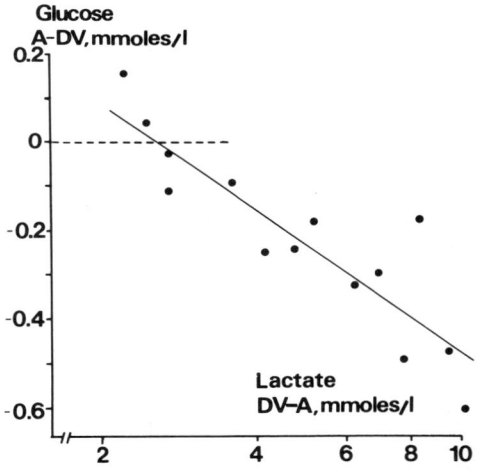

Fig. 2. Glucose arterial-deep venous difference in relation to log lactate deep venous-arterial difference measured during the initial phase of forearm exercise at 10 - 20 kpm/min.

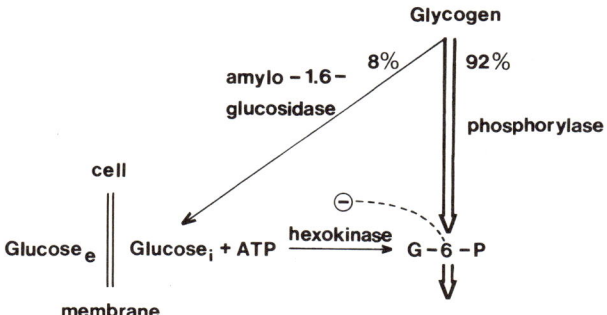

Fig. 3. Schematic representation of some aspects of glycogen degradation and glucose transport. See text for further explanation.

ferase, free glucose is formed corresponding to 8 - 10 % of the glycogen (11). During conditions of low to moderate rates of glycogenolysis this glucose is probably rapidly phosphorylated by the hexokinase reaction. However, at heavy exercise with very rapid glycogen degradation, increased amounts of glucose are formed at a time when the hexokinase reaction may be operating slowly or not at all, due to inhibition mainly by accumulating glucose-6-phosphate (7, 24) and possibly also by lowered levels of ATP. Under these circumstances it is thus conceivable that glucose transport across the cell membrane and into the cell is slowed down and actually reversed. The transient release of glucose to the blood noted in this study is quantitatively insignificant but the observation may possibly help to explain the divergent results reported for glucose uptake during exercise in man and it emphasizes the importance of relating such measurements both to work load and to duration of exercise.

After the first 10 min of forearm exercise, the glucose a - v difference gradually increases during the rest of a 60 min exercise period. The arterial glucose concentration remains unchanged. Blood flow to the forearm was measured with a view to quantifying the glucose metabolism (Fig. 4). A slight, continuous increase in blood flow during the exercise period was associated with a simultaneous fall in the a - v oxygen difference. The calculated forearm oxygen uptake remained constant between 10 and 60 min of the exercise. Calculated glucose uptake had increased 15 times after 10 min exercise and after 60 min had risen as much as ca. 35 times over the resting value, amounting at the latter time to approximately 40 µmoles · min^{-1} · 100 ml^{-1}.

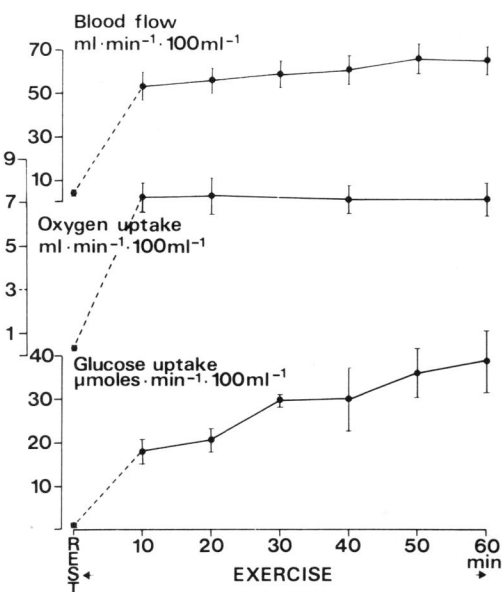

Fig. 4. Blood flow, oxygen and glucose uptake of the forearm at rest and during exercise for 60 min at 10 kpm/min.

The glucose uptake can be related to the estimated carbohydrate metabolism as evaluated from the local RQ (Fig. 5). At rest the RQ indicates a small carbohydrate oxidation, which is covered by the glucose uptake. At 10 min of exercise the RQ shows a larger fraction of carbohydrate oxidation in total oxidative metabolism, considerably more than can be accounted for by the glucose uptake, particularly as there is a large production of lactate from the forearm muscle at this time. The gap between glucose uptake minus lactate production and the estimated carbohydrate oxidation is probably filled by glycogen breakdown. These conditions change gradually during the course of the exercise period. After 60 min the RQ is again lower and indicates approximately 50 % carbohydrate oxidation. At this time the glucose uptake has increased and balances almost exactly the sum of the estimated carbohydrate oxidation and lactate production. Thus, it is concluded that during prolonged exercise of this type the glucose taken up from blood is an important substrate in muscle metabolism since it can sustain the entire carbohydrate oxidation and approximately half of the total oxidative metabolism of the forearm muscles.

The above calculations of estimated carbohydrate oxidation based on RQ should be treated with caution. The calculation assumes that local carbon dioxide stores are in equilibrium with the blood,

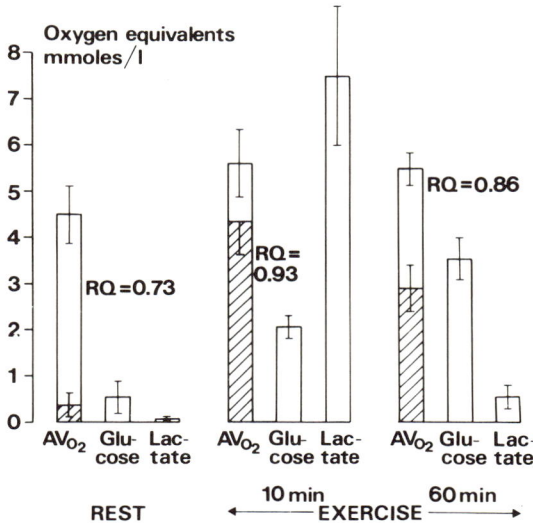

Fig. 5. Arterial-deep venous differences for oxygen and the oxygen equivalents of the a - v differences of glucose and lactate at rest and after 10 and 60 min of forearm exercise at 10 kpm/min. The hatched areas indicate the portion of the a - v oxygen difference which represents carbohydrate oxidation (evaluated from the RQ).

which may not be the case at the beginning of the exercise period. Moreover, recent findings concerning FFA oxidation indicate the FFA may undergo only partial degradation to acetate in muscle (13, 14). This is associated with oxygen consumption but no production of carbon dioxide, thus lowering the RQ value and resulting in a falsely low estimate for carbohydrate oxidation. However, these considerations do not affect the calculated contribution of blood glucose to the total oxidative metabolism in these studies.

Leg exercise studies

The results for glucose uptake to muscle during exercise with small muscle groups such as the forearm flexors may not necessarily be applicable to exercise with larger muscles. If, for example, the leg muscles were to take up glucose during exercise at a rate of approximately 40 μmoles $\cdot$ min^{-1} $\cdot$ 100 ml^{-1} - as found for forearm muscle - this would necessitate a sharply accelerated glucose production from the liver within 5 - 10 min. There is in fact evidence that the liver has the capacity to increase its glucose production drastically during physical exertion (3, 25). However, the available information on splanchnic glucose mobilization and hepatic gluconeogenesis in relation to intensity and duration of leg exercise is incomplete and the present study was undertaken in an attempt to provide such data.

Healthy subjects exercised on a bicycle ergometer at 400, 800 or 1200 kpm/min for 40 min. It was found that during exercise at 400 kpm/min all subjects showed mainly unchanged arterial levels of glucose. However, at both 800 kpm/min and 1200 kpm/min ($p < 0.01$) there was a significant rise in the arterial glucose level (Fig. 6). The mean increase in the 800 kpm/min group was 17 % and for those working at 1200 kpm/min 34 %. The rise in arterial glucose concentration during the exercise period correlated to both the work load ($r = 0.60$, $p < 0.05$) and the arterial lactate concentration (0.65, $p < 0.01$). These findings are in agreement with previous reports by Dill, Edwards and Mead (8) and recent results for heavy, intermittent tread-mill exercise (16). The a - v difference of glucose across the leg rose gradually during exercise in the 400 and 800 kpm/min groups, whereas at 1200 kpm/min there was a sharp increase at the beginning followed by only a moderate further rise during the second half of the exercise period (Fig. 6). Thus the increased arterial glucose concentration during heavy work is apparently not caused by a reduction in muscle glucose utilization, as has been discussed (8, 16).

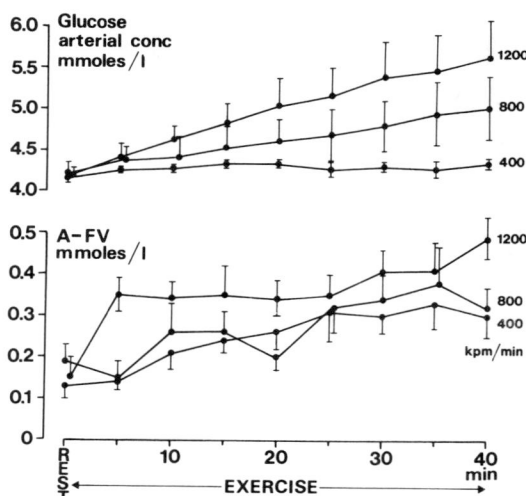

Fig. 6. Arterial concentration and arterial-femoral venous difference for glucose at rest and during leg exercise at 400, 800 and 1200 kpm/min.

The rise in glucose uptake was not accompanied by any significant change in arterial insulin levels. The insulin concentration at rest was 7 ± 1 µU/ml and after 20 and 40 min 8 ± 2 and 6 ± 1 µU/ml (n = 6). The finding supports the view, based on experimental studies and clinical evidence from insulin-dependent diabetic subjects (27), that the effect of exercise on glucose uptake does not depend on the ability to secrete increased quantities of insulin.

The glucose uptake may be viewed in relation to the total oxidative metabolism of the leg. In Fig. 7 the glucose a - v difference across the leg is shown expressed as oxygen equivalents. The values given are those observed after 40 min of exercise when a metabolic steady state had been reached and no significant production of lactate could be detected. At the lightest exercise the glucose uptake can sustain 80 % of the carbohydrate oxidation and approximately 25 % of the total metabolism. Similar results were obtained for exercise at 800 kpm/min. At the heaviest work load the RQ was higher and the glucose uptake may then account for 60 % of carbohydrate oxidation and as much as 40 % of the total oxidation. The results thus demonstrate that blood glucose is an important substrate in oxidative metabolism during prolonged leg exercise of this type.

Approximate calculations of leg muscle glucose utilization indicate that this form of exercise must be associated with rapid mobilization of glucose from carbohydrate stores or by gluconeogenesis. Direct estimates of splanchnic glucose production were

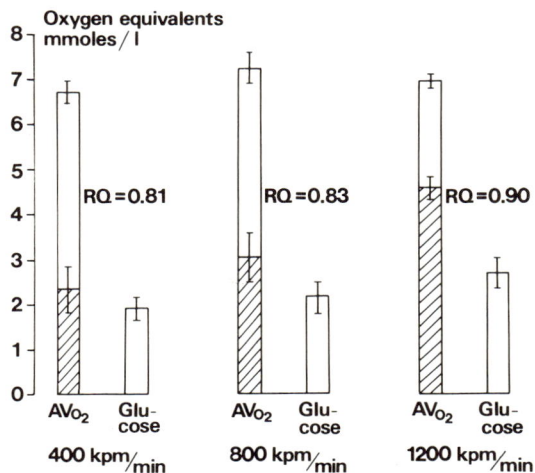

Fig. 7. Arterial-femoral venous differences for oxygen and glucose, expressed as oxygen equivalents, measured after 40 min exercise at 400, 800 and 1200 kpm/min. Hatched areas indicate the carbohydrate oxidation.

calculated as the product of the hepatic venous-arterial glucose difference and the estimated splanchnic blood flow. Splanchnic blood flow was 1150 ± 60 ml/min at rest and decreased markedly during exercise. The flow levels at the end of the 40 min exercise were 65 %, 47 % and 38 % of the resting value for work loads of 400, 800 and 1200 kpm/min, respectively. The splanchnic glucose production rose markedly during exercise (Fig. 8), the rise being most pronounced and approximately 5 1/2 times the resting value in the 1200 kpm/min group. The total splanchnic glucose production during the 40 min exercise period at the heaviest load was estimated to 100 mmoles or 18 grams.

While the interindividual dispersion was considerable in these determinations, the intra-individual variation was found to be relatively small. Fig. 9 shows results for estimated splanchnic glucose production in one subject who was studied on two different occasions. The first study involved catheterization of a right hepatic vein, and at the second study a left-sided hepatic vein was catheterized as well. There is fair agreement between the estimates of splanchnic glucose production obtained on the different occasions and on the basis of different sampling sites in the liver.

The question arises as to what portion of the splanchnic glucose production may derive from hepatic gluconeogenesis. The arterial-hepatic venous differences of lactate, pyruvate, glycerol and

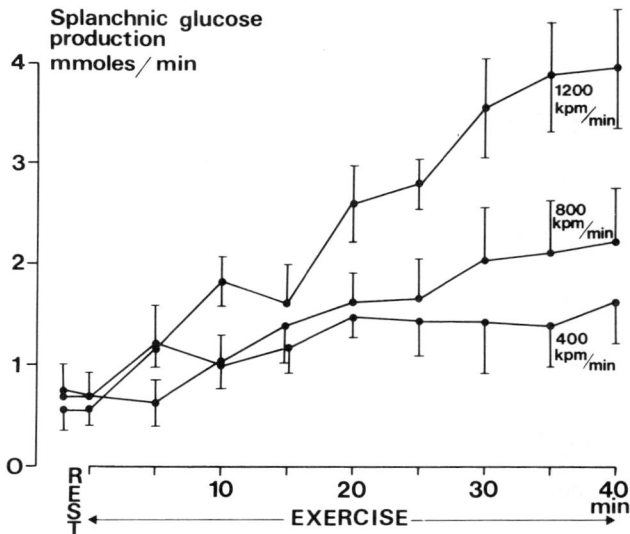

Fig. 8. Splanchnic glucose production at rest and during leg exercise at 400, 800 and 1200 kpm/min.

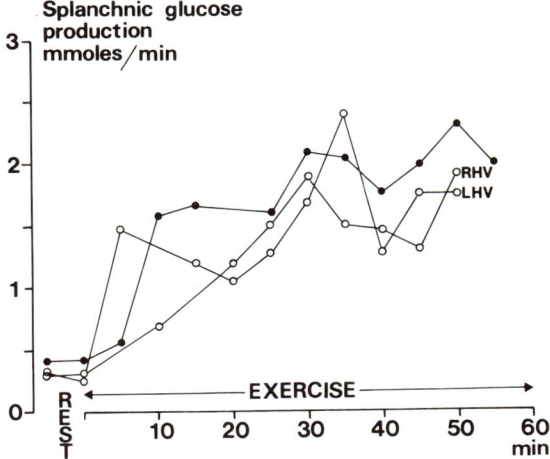

Fig. 9. Splanchnic glucose production at rest and during leg exercise at 1200 kpm/min on one occasion (filled circles) and three months later (open circles) when both a right (RHV) and a left (LHV) hepatic vein were catheterized.

α-amino nitrogen were measured and compared to the simultaneous glucose hepatic venous-arterial difference (Table 1). It is seen that, assuming maximum conversion efficiency, the measured glucose precursors may have provided the carbon skeleton for 28 % of the resting glucose output. This fraction is reduced during light exercise and even more so at the heavier work loads. Lactate is the dominant precursor but the contributions from both α-amino nitrogen and glycerol increase markedly during exercise. The hepatic gluconeogenesis theoretically possible from the measured precursors at rest is 0.18 ± 0.05 mmoles/min and at the work loads 400, 800 and 1200 kpm/min 0.31 ± 0.08, 0.27 ± 0.08 and 0.29 ± 0.06 mmoles/min respectively, the increase from rest to exercise being statistically significant ($p < 0.01$). It is concluded that although hepatic gluconeogenesis increases during exercise it can supply no more than 10 - 20 % of the total glucose production, the remainder presumably being derived from hepatic glycogenolysis.

SUMMARY AND CONCLUSIONS

Measurements of a - v glucose differences across the exercising forearm and leg have demonstrated that glucose uptake to exercising muscle is dependent on both the duration and the intensity of the

Table I. Hepatic venous-arterial differences (M ± SE, mmoles/l blood) for glucose and gluconeogenic precursors at rest and after 40 min exercise at the work loads 400, 800 and 1200 kpm/min. The sum of precursors 1 - 4 is given in glucose equivalents assuming complete conversion to glucose.

	REST	400 kpm/min	800 kpm/min	1200 kpm/min
Glucose	0.66 ± 0.12	2.15 ± 0.51	4.18 ± 1.19	6.46 ± 1.27
(1) Lactate	0.20 ± 0.06	0.50 ± 0.12	0.62 ± 0.16	0.56 ± 0.21
(2) Pyruvate	0.02 ± 0.01	0.03 ± 0.03	0.06 ± 0.05	0.05 ± 0.02
(3) Glycerol	0.03 ± 0.02	0.12 ± 0.06	0.12 ± 0.04	0.22 ± 0.06
(4) α-NH_3-nitrogen	0.12 ± 0.07	0.15 ± 0.05	0.24 ± 0.03	0.27 ± 0.07
Sum (1 - 4)	0.19	0.40	0.52	0.55
% of glucose output	28.0	18.6	12.4	8.5

work performed. During the initial 1 - 2 min of forearm exercise, measurements showed a transient net release of glucose; this reverted to a net uptake during the next few minutes and then continued to increase for the rest of the 60 min exercise period. The glucose being taken up at the end of the exercise could - if completely oxidized - account for the entire carbohydrate oxidation and half of the total oxidative metabolism of the forearm muscle.

In leg exercise, glucose a - v differences across the leg were found to increase gradually during 40 min exercise at work loads of 400, 800 and 1200 kpm/min respectively. At the end of exercise on these work loads glucose uptake could sustain 80, 70 and 60 % respectively of the estimated carbohydrate oxidation and 25, 25 and 40 % respectively of the total oxidative metabolism.

The increased glucose uptake during leg exercise was associated with a rapid and marked increase in splanchnic glucose production, related to the work load performed. The major part of the glucose output derived from hepatic glycogenolysis. Hepatic gluconeogenesis, estimated from splanchnic uptake of lactate, pyruvate, glycerol and α-amino nitrogen, could account for no more than 19, 12 and 9 % of the glucose production after 40 min at work loads of 400, 800 and 1200 kpm/min respectively.

It is concluded that as exercise proceeds beyond the first minutes, blood glucose assumes an increasingly important role as a substrate for muscle oxidation, coincident with the previously known decrease of local carbohydrate stores in muscle.

ACKNOWLEDGEMENTS

This study was supported by grants from the Swedish Medical Research Council (B71-19X-3108-01) and from Harald Jeanssons Stiftelse.

REFERENCES

1. BAKER, P.G., AND R.F. MOTTRAM. The metabolism of exercising human muscle. J. Physiol., London 194:64-65P, 1968.
2. BERGSTRÖM, J., AND E. HULTMAN. The effect of exercise on muscle glycogen and electrolytes in normals. Scand. J. Clin. Lab. Invest. 18:16-20, 1966.
3. BERGSTRÖM, J., AND E. HULTMAN. A study of the glycogen metabolism during exercise in man. Scand. J. Clin. Lab. Invest. 19:218-228, 1967.
4. BRADLEY, S.E. Liver function as studied by hepatic vein catheterization. In: Transactions of the Fifth Conference on Liver Injury. New York: Josiah Macy, Jr. Found., 1946, p. 38-43.
5. BRADLEY, S.E. Measurement of hepatic blood flow. In: Methods in Medical Research. Chicago: Year Book Publishers, 1948, vol. 1, p. 199-204.
6. BÜCHER, R., R. CZOK, W. LAMPRECHT, AND E. LATZKO. Pyruvat. In: Methoden der enzymatischen Analyse. Weinheim: Verlag Chemie, 1962, p. 253-259.
7. CORSI, A., M. MIDRIO, AND A.L. GRANATA. In situ utilization of glycogen and blood glucose by skeletal muscle during tetanus. Amer. J. Physiol. 216:1534-1541, 1969.
8. DILL, D.B., H.T. EDWARDS, AND S. MEAD. Blood sugar regulation in exercise. Amer. J. Physiol. 111:21-30, 1935.
9. DRABKIN, D.L., AND J.H. AUSTIN. Spectrophotometric studies. II. Preparations from washed blood cells; nitric oxide hemoglobin and sulfhemoglobin. J. Biol. Chem. 112:51-65, 1935.
10. DRABKIN, D.L. Measurement of O_2-saturation of blood by direct spectrophotometric determination. In: Methods in Medical Research. Chicago: Year Book Publishers, 1950, vol. 2, p. 159-161.
11. FIELD, R.A. Glycogen deposition diseases. In: The Metabolic Basis of Inherited Disease. New York: Mc Graw-Hill, 1960, p. 156-207.
12. GARBY, L., AND J.C. VUILLE. The amount of trapped plasma in a high speed microcapillary hematocrit centrifuge. Scand. J. Clin. Lab. Invest. 13: 642-645, 1961.

13. HAGENFELDT, L., AND J. WAHREN. Human forearm muscle metabolism during exercise. II. Uptake, release and oxidation of individual FFA and glycerol. Scand. J. Clin. Lab. Invest. 21:263-276, 1968.
14. HAGENFELDT, L., AND J. WAHREN. Metabolism of free fatty acids and ketone bodies in skeletal muscle. In: Muscle Metabolism during Exercise. Adv. Expl. Med. Biol. New York: Plenum Publ. Corp. 1970, this volume.
15. HAVEL, R.J., B. PERNOW, AND N.L. JONES. Uptake and release of free fatty acids and other metabolites in the legs of exercising men. J. Appl. Physiol. 23:90-96, 1967.
16. HERMANSEN, L., E.D.R. PRUETT, J.B. OSNES, AND F.A. GIERE. Blood glucose and plasma insulin in response to maximal exercise and glucose infusion. J. Appl. Physiol. 29:13-16, 1970.
17. JORFELDT, L., AND J. WAHREN. Human forearm muscle metabolism during exercise. V. Quantitative aspects of glucose uptake and lactate production during prolonged exercise. Scand. J. Clin. Lab. Invest. 26:73-81, 1970.
18. KEUL, J., E. DOLL, AND D. KEPPLER. The substrate supply of the human skeletal muscle at rest, during and after work. Experientia. 23:974-979, 1967.
19. KLASSEN, G.A., G.M. ANDREW, AND M.R. BECKLAKE. Effect of training on total and regional blood flow and metabolism in paddlers. J. Appl. Physiol. 28:397-406, 1970.
20. MANNERS, D.J. Glycogen storage disease, type I. In: Control of Glycogen Metabolism. Churchill, London: Ciba Foundation Symposium, 1964, p. 320-333.
21. MOORE, S., D.H. SPACKMAN, AND W.H. STEIN. Automatic recording apparatus for use in the chromatography of amino acids. Fed. Proc. 17:1107-1115, 1958.
22. MORGAN, C.R., AND A. LAZAROW. Immunoassay of insulin: two antibody system. Plasma insulin levels of normal, subdiabetic and diabetic rats. Diabetes. 12:115-126, 1963.
23. REICHARD, G.A., B. ISSEKUTZ, JR., P.KIMBEL, R.C. PUTNAM, N.J. HOCHELLA, AND S. WEINHOUSE. Blood glucose metabolism in exercising man. J. Appl. Physiol. 16:1001-1005,1961.
24. ROSE, I.A., AND E.L. O'CONELL. The role of glucose-6-phosphate in the regulation of glucose metabolism in human erythrocytes. J. Biol. Chem. 239:12-17, 1964.
25. ROWELL, L.B., E.J. MASORO, AND M.J. SPENCER. Splanchnic metabolism in exercising man. J. Appl. Physiol. 20:1032-1037, 1965.

26. ROWELL, L.B., K.K. KRANING II, T.O. EVANS, J.W. KENNEDY, J.R. BLACKMON, AND F. KUSUMI. Splanchnic removal of lactate and pyruvate during prolonged exercise in man. J. Appl. Physiol. 21:1773-1783, 1966.
27. SANDERS, C.A., G.E. LEVINSON, W.H. ABELMANN, AND N. FREINKEL. Effect of exercise on the peripheral utilization of glucose in man. New Engl. J. Med. 271:220-225, 1964.
28. SEGAL, S., A.E. BLAIR, AND J.B. WYNGAARDEN. An enzymatic spectrophotometric method for the determination of pyruvic acid in blood. J. Lab. Clin. Med. 48:137-143, 1956.
29. SOELDNER, J.S., AND D. SLONE. Critical variables in the radioimmunoassay of serum insulin using the double antibody technic. Diabetes. 14:771-779, 1965.
30. WAHREN, J. Quantitative aspects of blood flow and oxygen uptake in the human forearm during rhythmic exercise. Acta Physiol. Scand. 67, Suppl. 269:1-93, 1966.
31. WAHREN, J., AND L. HAGENFELDT. Human forearm muscle metabolism during exercise. I. Circulatory adaptation to prolonged forearm exercise. Scand. J. Clin. Lab. Invest. 21:257-262, 1968.
32. WIELAND, O. Glycerin. In: Methoden der enzymatischen Analyse. Weinheim: Verlag Chemie, 1962, p. 211-214.

INTERRELATIONSHIP BETWEEN AMINO ACID AND CARBOHYDRATE METABOLISM

DURING EXERCISE: THE GLUCOSE ALANINE-CYCLE

Philip Felig and John Wahren

Yale University School of Medicine, New Haven, Connecticut, USA and Karolinska Institutet at Serafimerlasarettet, Stockholm, Sweden

In the past decade considerable interest and effort have been devoted to increasing our understanding of the factors regulating energy exchange in exercising muscle. Much of this effort has been directed at studying the utilization and availability of carbohydrate and lipids. Little information is available however, on the influence of exercise on amino acid metabolism. Our own interest in amino acid metabolism in exercise derives from studies which we undertook on the role of substrates in the regulation of gluconeogenesis in intact man (2,3,4).

In examining the pattern of amino acid exchange across resting forearm muscle, alanine was observed to be released to a greater extent than all other amino acids (5). Since alanine comprises only 5-8 % of the amino acid residues in muscle protein (8), the basis for the primacy of alanine output is unclear. Peripheral synthesis of alanine by transamination of glucose-derived pyruvate has been suggested (5). By this formulation, alanine formation and release from muscle would depend not only on the rate of protein dissolution but also on the rate of glycolysis and availability of pyruvate. To test this hypothesis and to characterize further the pattern and regulation of peripheral amino acid release, we have investigated amino acid metabolism during muscular exercise, a condition characterized by increased glucose utilization.

Healthy adult male volunteers were studied after an overnight fast (Fig. 1). Catheters were placed in the brachial artery, femoral vein and hepatic vein, and simultaneous blood samples were obtained with the subjects at rest and after 10 and 40 min of mild and moderately heavy exercise. The exercise was performed in the upright position on a bicycle ergometer, and resulted in a 3-9 fold increase in oxygen consumption (Fig. 1).

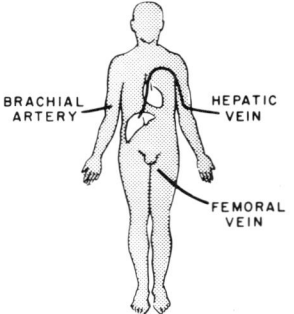

Fig. 1. Experimental protocol and oxygen consumption at rest and during exercise.

Net exchange of amino acids across leg tissue was determined by measurement of arterio-femoral venous differences, a negative A-V difference, indicating net release. As shown in fig. 2, in the resting state there was significant net release from leg tissue, of 13 out of 19 plasma amino acids. By far the largest A-V difference across the leg was that for alanine, which accounted for 40 % of total net amino acid output. A significant positive A-V difference or net uptake was observed for citrulline, serine and cystine.

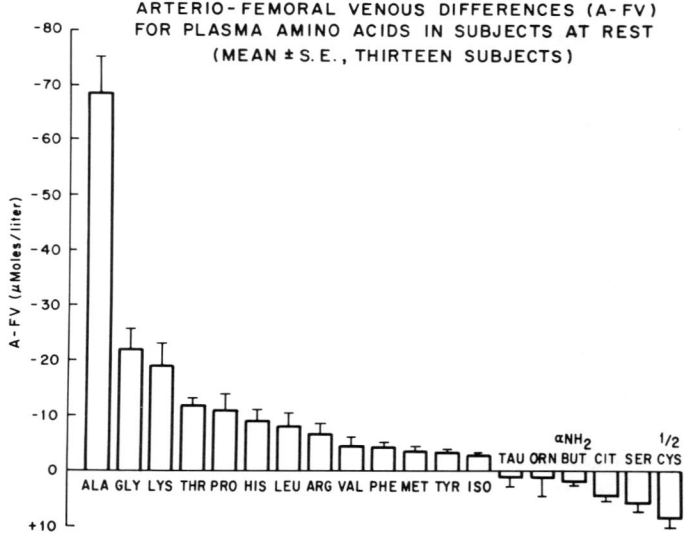

Fig. 2.

To evaluate the possible interaction of alanine and glucose metabolism, the relation between arterial alanine and arterial pyruvate was examined (Fig. 3). In resting subjects, a highly significant correlation between arterial alanine concentration and arterial pyruvate concentration was observed. No significant correlation with pyruvate was demonstrable for any of the remaining 18 amino acids measured.

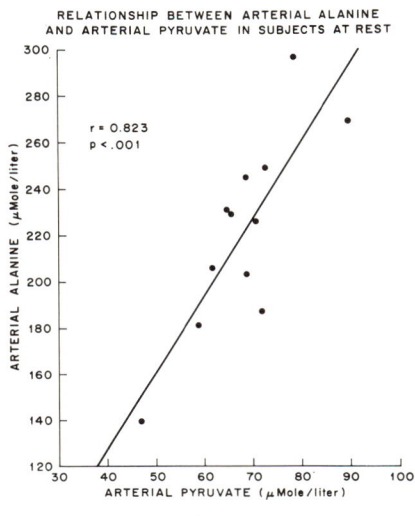

Fig. 3.

The influence of exercise on peripheral amino acid exchange is shown in fig. 4. After 10 and 40 min of exercise, significant net amino acid release from leg tissue was demonstrable only for alanine. Although the A-V difference for alanine was smaller than in the resting state, when account is taken of the increased blood flow to the leg in exercise, estimated alanine output increased approximately two fold. With the exception of serine and cystine for which small uptakes were observed, as indicated by the positive A-V differences, for all other amino acids, exchange across the leg during exercise was too small or variable to result in a significant net A-V difference. Thus the values shown for the remaining 16 amino acids were not significantly different from zero.

The pattern of amino acid exchange across the leg at heavier levels of exercise was virtually identical with that depicted for mild exercise.

The effect of exercise on the arterial concentrations of the glycolytic end-products pyruvate and lactate and on arterial alanine levels are shown in fig. 5. During both mild and heavy exercise the anticipated increases in arterial pyruvate and lactate were observed. Particularly noteworthy was the increase in arterial alanine.

With mild exercise alanine rose by 25 %, while at heavier work loads, arterial alanine increased two fold.

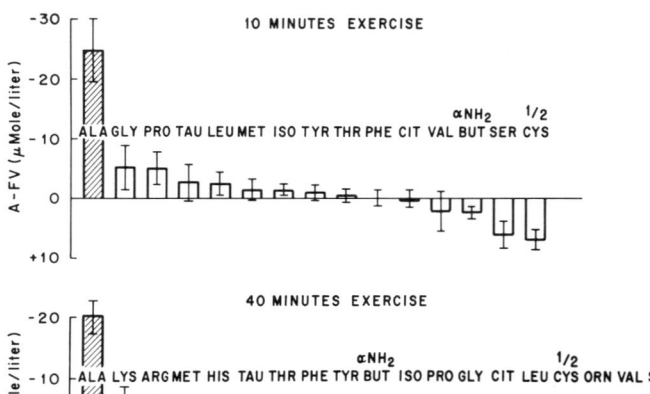

Fig. 4.

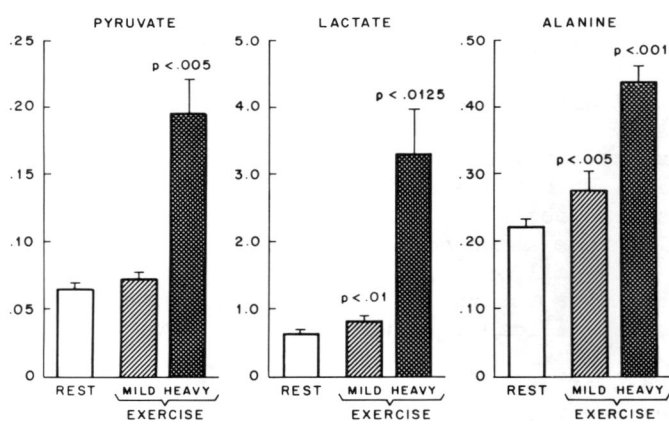

Fig. 5.

The specificity of the increase in arterial alanine concentration is evident from the response of the remainder of the plasma amino acids to exercise. Thus, during exercise the arterial concentration of all amino acids other than alanine was unchanged from resting levels or varied by less than 10 percent.

THE GLUCOSE ALANINE-CYCLE

The relation between arterial levels of alanine and pyruvate during exercise is shown in fig. 6. The values presented are those observed with both mild and heavy work loads. As in the resting state, during exercise a highly significant correlation was demonstrable between arterial pyruvate and arterial alanine. No correlation with pyruvate was observed for any of the other amino acids.

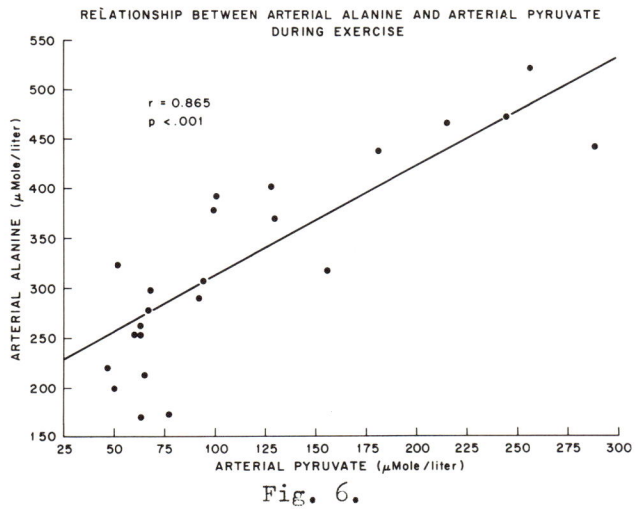

Fig. 6.

An increase in arterial alanine concentration could result not only from augmented release from exercising muscle, but also from a diminution in the rate of removal of alanine from the circulation. Since the liver is the primary site of alanine disposal, the influence of exercise on splanchnic amino acid exchange was investigated.

Splanchnic uptake of amino acids was calculated from arterio-hepatic venous differences and hepatic plasma flow, determined by the continuous infusion technique using indocyanine green. In the

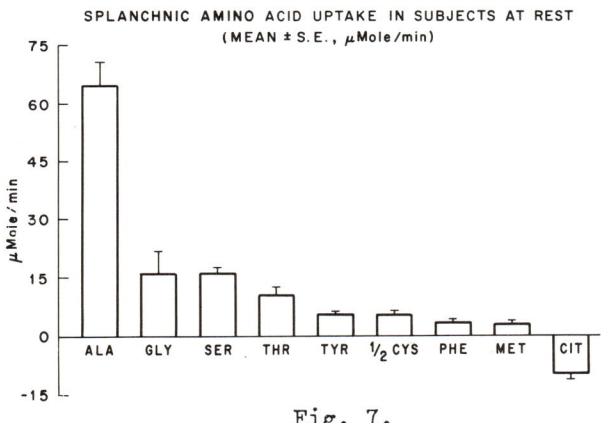

Fig. 7.

resting state, a significant net uptake was demostrable for eight amino acids (Fig. 7). However, splanchnic removal of alanine was far greater than that of all other amino acids and accounted for over 50 % of total amino acid uptake. Interestingly, splanchnic release was observed only for the urea cycle intermediate, citrulline.

During exercise, splanchnic alanine uptake again exceeded that of all other amino acids. Furthermore, during mild exercise (400 kgm/min) despite a moderate reduction in hepatic blood flow, the rate of alanine uptake rose by 60 % above resting levels (Fig. 8). This increase in alanine removal was not solely due to an elevation in arterial concentration, but was due in part to augmented fractional extraction of alanine. Whereas in the resting state, 40 % of the arterial alanine content was extracted, during exercise 70 % was removed (Fig. 8).

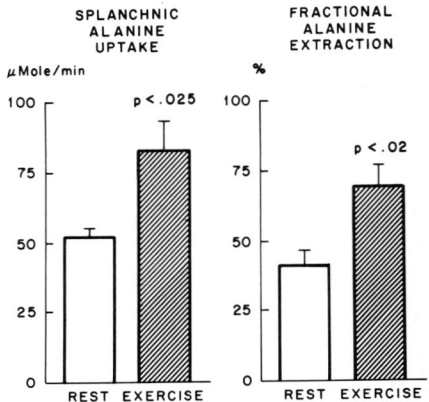

Fig. 8. Splanchnic uptake and fractional extraction of alanine during mild exercise.

The changes in alanine metabolism observed at heavier work loads (800-1200 kgm/min) are shown in fig. 9. Although estimated hepatic blood flow (EHBF) fell by 50 %, fractional alanine extraction rose from 55 % to 80 %. As a result of this augmented extraction, the absolute rate of alanine uptake increased (after forty min exercise) to slightly above resting levels, despite the reduction in blood flow. Thus exercise - induced hyperalaninemia even at heavy work loads, cannot be ascribed to diminished splanchnic uptake.

The current findings demonstrate a unique role for alanine in amino acid metabolism. Alanine release from peripheral tissue is proportionately far greater than its content in constituent muscle proteins. It is the only amino acid released in significant quantities during exercise and is unique in demonstrating as much as a two fold elevation in arterial concentration during exercise. Furthermore, this accumulation of alanine cannot be ascribed to

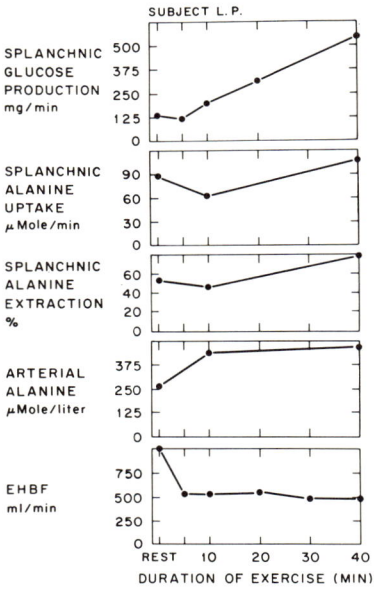

Fig. 9.

diminished hepatic removal. Finally, both at rest and during exercise alanine is the only amino acid whose concentration is directly proportional to arterial pyruvate levels. The results thus are consistent with the conclusion that alanine formation and release are not solely dependent on protein dissolution but are related to peripheral glucose utilization and pyruvate formation. Accordingly, it appears that alanine is synthesized by transamination of glucose-derived pyruvate. In as much as certain amino acids, notably leucine, isoleucine and valine are preferentially catabolized in muscle (10), a steady flow of amino groups is available for this transamination. Supporting this formulation is the recent observation that a significant proportion of muscle glycogen utilization in exercise cannot be accounted for by lactate formation or CO_2 production (6). The carbon skeleton of alanine thus may constitute an important and previously unrecognized end-product of glycolysis.

Turning to the data on splanchnic alanine uptake, it is noteworthy that *in vitro* perfusion studies have demonstrated that the rate of gluconeogenesis from alanine is far greater than that of all other amino acids (9). The primacy of alanine uptake by the splanchnic circulation suggests that alanine serves not only as a glycolytic end-product but also as a key endogenous substrate for hepatic glucose production. On the basis of these observations a glucose-alanine cycle involving muscle and liver may be constructed as shown in fig. 10.

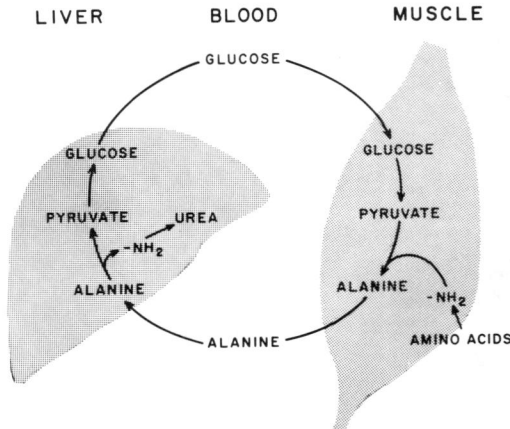

Fig. 10. The glucose-alanine cycle.

Glucose extracted from blood or released from tissue glycogen stores is converted by muscle to pyruvate. Transamination of pyruvate results in formation of alanine which is released to the circulation. Glucose-derived alanine is in turn taken up by the liver where it is reconverted to glucose and released to the circulation, thus completing the cycle. This cycle is thus analogus to that described many years ago by Cori for lactate (1).

The question arises however, as to the quantitative significance of the carbon skeleton of alanine as a glycolytic end-product. While direct measurements are currently being pursued with isotopic techniques, some crude estimates may be made from the available data (Fig. 11). Since lysine is present in muscle protein in virtually the same concentration as alanine (8), and is not significantly catabolized in muscle, the extent to which alanine release exceeds that of lysine may reflect the proportion of alanine derived from glucose. On this basis it is estimated that production of glucose-derived alanine by muscle occurs at 35-60 % of the rate at which lactate is produced (Fig. 11).

Employing the same assumptions, one may also arrive at an estimate of the proportion of glucose consumed by muscle which may be accounted for as alanine. As shown in fig. 12, on the basis of observations both in the leg and forearm, alanine release may account for 12-18 % of the glucose extracted by muscle.

As indicated, these quantitative estimates can be considered little more than crude guesses at this time. Nevertheless, the data presented in both resting and exercising man suggest a central role for alanine in both amino acid and carbohydrate metabolism.

RATIO OF ESTIMATED "GLUCOSE-DERIVED ALANINE" PRODUCTION
TO LACTATE PRODUCTION

<u>Leg Muscle</u> (Arterio-Femoral Venous Differences)
    A-FV Alanine = -70 μMole/L
    A-FV Lysine = -20 μMole/L
    ∴ A-FV "Glucose-derived Alanine" = -50 μMole/L

    A-FV Lactate = -140 μMole/L

    ∴ "Glucose-derived Alanine" : Lactate = 50:140 = 35%

<u>Forearm Muscle</u> (Arterio-Deep Venous Differences)
    A-DV Alanine = -110 μMole/L
    A-DV Lysine = -40 μMole/L
    ∴ A-DV "Glucose-derived Alanine" = -70 μMole/L

    A-DV Lactate = -120 μMole/L

    ∴ "Glucose-derived Alanine" : Lactate = 70:120 = 60%

Fig. 11.

RATIO OF ESTIMATED "GLUCOSE-DERIVED ALANINE" PRODUCTION
TO GLUCOSE CONSUMPTION (A/G)

<u>Leg Muscle</u> (Arterio-Femoral Venous Differences)
    A-FV Alanine = -70 μMole/L
    A-FV Lysine = -20 μMole/L
    ∴ A-FV "Glucose-derived Alanine" = -50 μMole/L

    A-FV Glucose = 214 μMole/L

    ∴ A/G = 25/214 = 12%

<u>Forearm Muscle</u> (Arterial-Deep Venous Difference)
    A-DV Alanine = -110 μMole/L
    A-DV Lysine = -40 μMole/L
    ∴ A-DV "Glucose-derived Alanine" = -70 μMole/L

    A-DV Glucose = 200 μMole/L

    ∴ A/G = 35/200 = 18%

Fig. 12. The data for forearm muscle in this and the previous figure are based on the studies of Felig <u>et al</u> (5), and Jorfeldt and Wahren (7). The calculations for both leg and forearm muscle are based on observations in the resting state.

ACKNOWLEDGMENTS

Supported in part by U.S. Public Health Service Grant (AM-13526) and by a grant from the Swedish Medical Research Council (B71-19X-3108-01).

REFERENCES

1. CORI, C.F., AND G.T. CORI. The mechanism of epinephrine action. II. The influence of epinephrine and insulin on the carbohydrate metabolism of rats in the postabsorptive state. J. biol. Chem. 79: 321-341, 1928.

2. FELIG, P., O.E. OWEN, J. WAHREN, AND G.F. CAHILL JR. Amino acid metabolism during prolonged starvation. J. clin. Invest. 48: 584-594, 1969.

3. FELIG, P., E. MARLISS, O.E. OWEN, AND G.F. CAHILL JR. Blood glucose and gluconeogenesis in fasting man. Arch. Int. Med. 123: 293-298, 1969.

4. FELIG, P., E. MARLISS, AND G.F. CAHILL JR. Amino acid metabolism in the regulation of gluconeogenesis in man. Amer. J. clin. Nutr. 23: 986-992, 1970.

5. FELIG, P., T. POZEFSKY, E. MARLISS, AND G.F. CAHILL JR. Alanine: Key role in gluconeogenesis. Science. 167: 1003-1004, 1970.

6. HULTMAN, E. Studies on muscle metabolism of glycogen and active phosphate in man with special reference to exercise and diet. Scand. J. clin. Lab. Invest. 19: Suppl. 94, 39-40, 1967.

7. JORFELDT, L., AND J. WAHREN. Human forearm muscle metabolism during exercise. V. Quantitative aspects of glucose uptake and lactate production during prolonged exercise. Scand. J. clin. Lab. Invest. 26: 73-81, 1970.

8. KOMINZ, D.R., A. HOUGH, P. SYMONDS, AND K. LAKI. The amino acid composition of actin, myosin, tropomysin and the meromysins. Arch. Biochem. Biophys. 50: 148-159, 1954.

9. MALLETTE, L.E., J.H. EXTON, AND C.R. PARK. Control of gluconeogenesis from amino acids in the perfused rat liver. J. biol. Chem. 244: 5713-5723, 1969.

10. MILLER, L.L. The role of liver and the non-hepatic tissues in the regulation of free amino acid levels in the blood. In: Amino Acid Pools. New York, ed. J.T. Holden, Elsevier, 1962, 708-721.

UPTAKE OF SUBSTRATES IN ISOLATED CONTRACTING SLOW AND

FAST MUSCLES IN SITU IN RELATION TO FATIGUE

    O. Hudlicka

    Department of Physiology, Medical School

    University of Birmingham

Although it has been established that carbohydrates are not the only energy source in skeletal muscle (Andres et al., 1956; Havel, et al., 1963; Bass & Hudlicka, 1964; Paul, 1970) it appears that a large part of energy supply for the performance of work is derived from the breakdown of glycogen (see e.g. Ahlborg et al., 1967; Corsi et al., 1969). In a dog gastrocnemius muscle glycogen is used mainly during the first 10-15 min of rhythmic contractions (Hirche et al., 1970) at which time the work performed by this muscle is diminishing. An early onset of fatigue has been described in fast muscles (see e.g. del Pozo, 1942; Eberstein & Sandow, 1963) which have a high content of glycogen, while slow muscles where the content of glycogen is lower can work for a long time without signs of fatigue (Kugelberg & Edström, 1968; Edström & Kugelberg, 1968). In order to elucidate these discrepancies consumption of some substrates has been followed in isolated cat soleus (slow) and gastrocnemius (mainly fast) muscles in situ measuring the blood flow and A-V differences while the muscles have been performing isotonic contractions.

The resting consumption of oxygen has been found seven times higher in soleus than in gastrocnemius in agreement with a previous report (Hudlicka, 1969) and could be explained by much higher blood flow in this muscle (Hilton & Vrbova, 1968; Hilton et al., 1970), since the A-V differences of oxygen were similar in both muscles (Fig. 1). The resting glucose consumption has been found to be ten times higher; the A-V differences and extraction ratio for glucose were significantly higher in soleus than in gastrocnemius. The differences in lactate output were somewhat less pronounced.

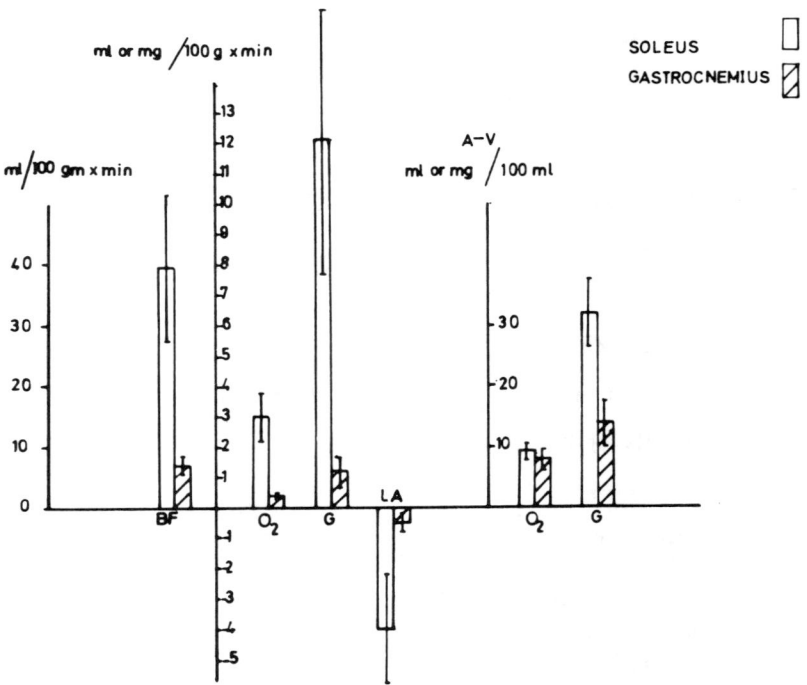

Fig. 1. Blood flow (BF), consumption of oxygen (O_2) and glucose (G) and output of lactic acid (LA) in cat soleus and gastrocnemius at rest. Mean values $\pm$ S.E.

The higher the uptake of glucose in soleus the higher the output of lactic acid. This would suggest that energy in a resting soleus was mainly derived from a breakdown of blood glucose; no similar relation has been found in gastrocnemius.

During rhythmic contractions at 2/sec gastrocnemius showed signs of fatigue after the fourth minute, and the work performed by this muscle has fallen to 45% of the peak value after 10–15 min of stimulation while the performance of soleus did not change over the same period of time. The oxygen consumption in gastrocnemius increased from 0.42 ± 0.075 ml/100 g x min at rest to 2.42 ± 41 ml/100 g x min at the end of the second min, and to 3.08 ± 0.57 ml/100 g x min at 10–15 min after the beginning of stimulation (Fig. 2). Consumption of oxygen in soleus increased from 3.0 ± 0.81 ml/100 g x min at rest to 6.93 ± 0.96 and 7.52 ± 1.33 respectively. While the increase of oxygen consumption in gastrocnemius was ensured both

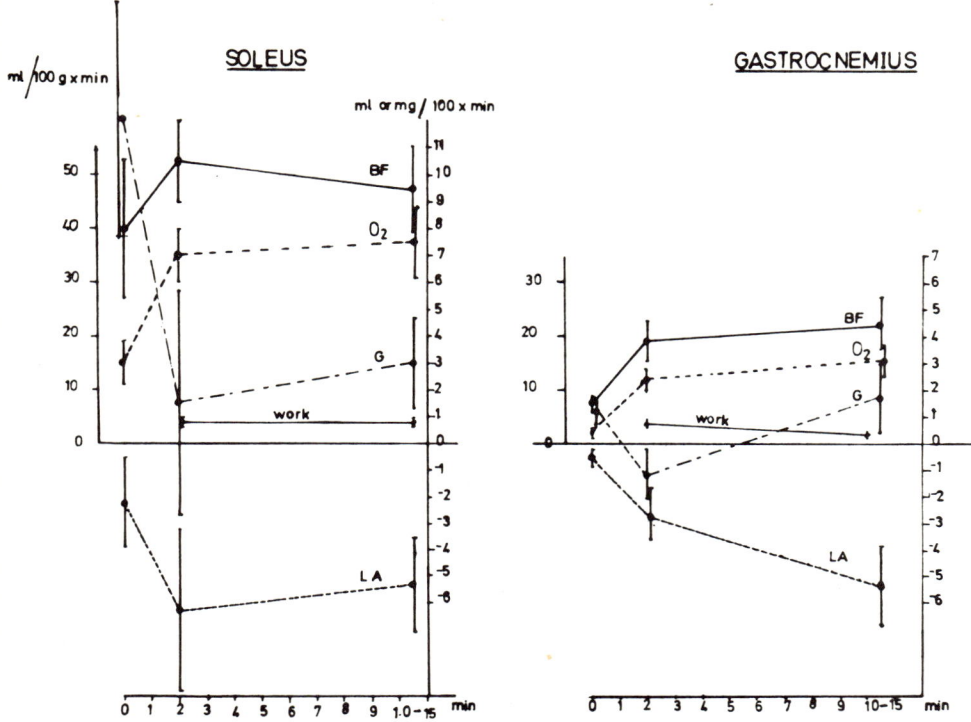

Fig. 2. Changes in blood flow (BF), oxygen (O_2) and glucose (G) consumption and lactic acid (LA) output in cat soleus and gastrocnemius during muscular contractions (stimulation of the cut sciatic nerve) at 2/sec. 0 time = values immediately before stimulation.

by the increase in blood flow and the A-V differences, the increased consumption in soleus was mainly due to the increase in A-V differences, since blood flow in this muscle does not increase very much during muscular contractions, if it increases at all (Hilton & Vrbova, 1968; Hilton et al., 1970). There was no significant change in lactate output from the soleus (Fig. 3), while lactate output from the gastrocnemius increased from 0.48 ± 0.31 mg/100 g × min at rest to 2.73 ± 0.91 at the end of the second min and to 5.29 ± 1.16 mg/100 g × min after stimulating for 10-15 min.

Glucose consumption decreased both in soleus and in gastrocnemius at the end of the second min, and remained low in soleus while in gastrocnemius a small increase, as compared to resting values, has been observed at 10-15 min after the beginning of stimulation. In fact, most values of A-V differences of glucose in gastrocnemius were negative at the end of the

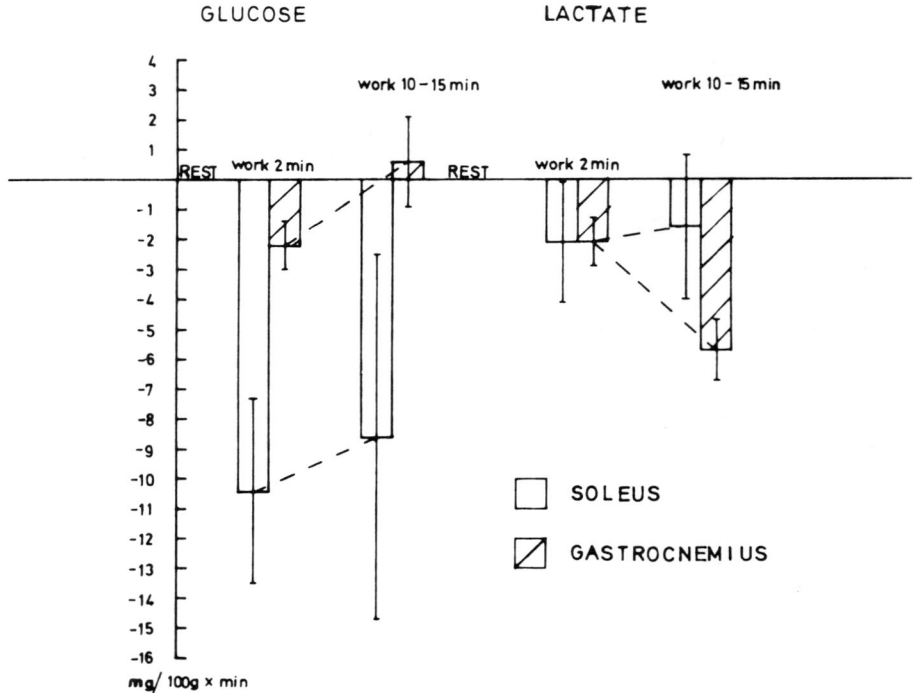

Fig. 3. Glucose consumption (or output) and lactic acid output during muscular contractions expressed as difference against values at rest (=0).

second min. Negative A-V differences of glucose during work have been described in dog gastrocnemius by Corsi et al. (1969) and in human forearm (Wahren, 1970) and leg muscles (Bergström & Hultman, 1966). The explanation is not quite clear. Since the negative A-V differences occurred in Corsi's experiment at the same time when glucose-6-P had been accumulated in the muscle it is possible to assume that glucose-6-P was split by nonspecific phosphatases (Manners, 1964) into organic phosphate and glucose.

The decreased consumption of glucose in soleus during muscular contractions might be due to accumulation of free fatty acids which has been described in pigeon breast muscles (with similar biochemical characteristics as soleus) by George & Vallyathan (1964) and which is known to decrease uptake of glucose in the heart and diaphragm (Randle et al., 1964).

Similarly to the experiments on slow and fast muscles in vitro (see e.g. Bocek et al., 1966) no appreciable glycogenolysis seems to take place in soleus, and yet this muscle can work for a long time without signs of fatigue. The energy supply to this muscle does not therefore seem to be derived from carbohydrates. On the other hand, the actual work performed by this muscle seems to be more dependent on oxygen than in gastrocnemius: (Fig. 4) the work performed by individual muscles was higher in muscles

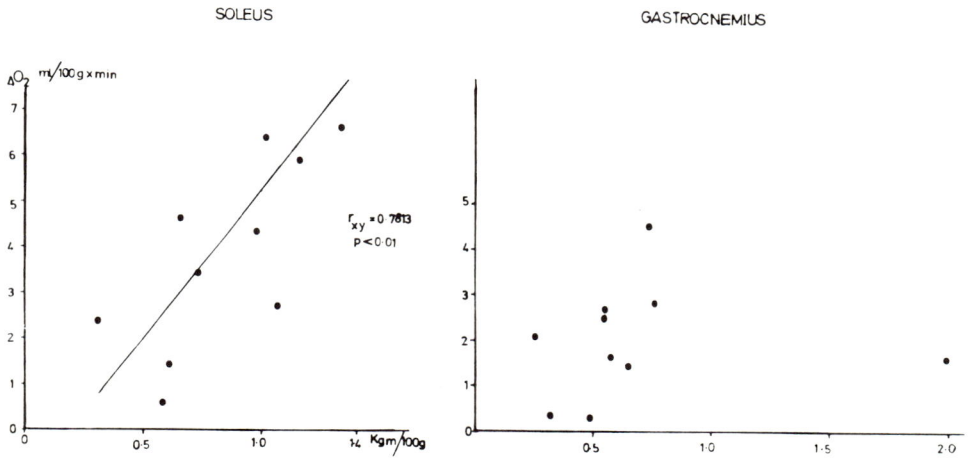

Fig. 4. Correlation between work performed and increase in oxygen consumption (O_2).

where the increase in oxygen consumption was greater.

High oxygen consumption together with low glycogenolysis seems to create favourable conditions for rephosphorylation of active phosphate in soleus, and thus to ensure continuing work without fatigue. Consequently, there might be a smaller accumulation of inorganic phosphate in this muscle. Smaller accumulation of inorganic phosphate has already been described in rabbit semitendinosus, as compared with biceps femoris by Cohn (1921). Hilton & Vrbova (1970) have found no liberation of inorganic phosphate into the venous blood from soleus.

In gastronemius, on the other hand, where the glycogenolysis takes place, the amount of phosphocreatine falls very quickly (Bergström, 1967; Wilson et al., 1967), inorganic phosphate accumulates in the extracellular space (Sacks & Cleland, 1960), and diffuses into the venous blood.

Hilton & Vrbova (1970) have described a liberation of inorganic phosphate in proportion to the rate of contractions, and the output of phosphate has still been increased at 10–15 min of stimulation (Fig. 5).

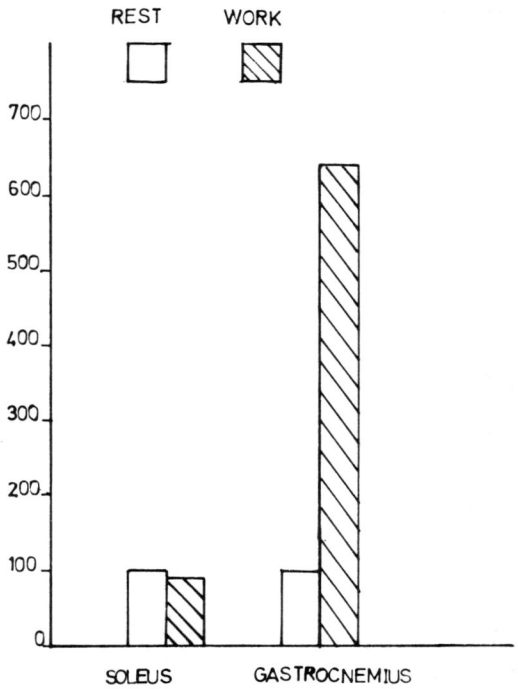

Release of PO_4 during work.
Expressed as % of rest level.

Fig. 5. Release of inorganic phosphate (calculated from V-A differences and blood flow) in the course of isotonic contractions. Values expressed as % of rest level.

It is possible that fatigue in gastrocnemius may result from the loss of inorganic phosphate from this muscle, and in fact it diminished in some experiments when inorganic phosphate has been infused by close arterial injection during muscular work.

I wish to express my gratitude to the Wellcome Foundation for their financial support which enabled me to carry out this work.

REFERENCES

Ahlborg, B., Bergström, J., Ekelund, L. G., Hultman, E.
Muscle glycogen and muscle electrolytes during prolonged exercise.
Acta physiol. scand. 70: 122-142, 1967.

Andres, R., Cader, G., Zierler, K. L. The quantitative minor role of carbohydrate in oxidative metabolism by skeletal muscle in intact man in the basal state. J. clin. Invest. 35: 671-682, 1956.

Bass, A., Hudlicka, O. Interrelations between metabolism and blood flow in normal and denervated dog gastrocnemius muscle at rest and during stimulation. Physiol. bohemoslov., 13: 48-61, 1964.

Bergström, J. Local changes of ATP and phospocreatine in human muscle tissue in connection with exercise. Circul. Res. Suppl. I vo. XX, 1 - 91-96, 1967.

Bergström, J. Hultman, E. The effect of exercise on muscle glycogen and electrolytes in normals. Scand. J. Clin. Lab. Invest 18: 16-20, 1966.

Bocek, R. M., Peterson, R. D., Beatty, C. H. Glycogen metabolism in red and white muscle. Amer. J. Physiol. 210: 1101-1107, 1966.

Cohn, F., Ueber den Einfluss der Muskelarbeit auf den Lactacidogen- gehalt in den roten und weissen Muskulatur des Kaninchens. Ztschr. f. physiol. Chemie, 113: 253-262, 1921.

Corsi, A., Midrio, M., Granata, A. L. In situ utilization of glycogen and blood glucose by skeletal muscle during tetanus. Amer. J. Physiol. 216: 1534;1541, 1969.

Edström, L., Kugelberg, E. Histochemical composition, distribution of fibres and fatiguability of single motor units. J. Neurol, Neurosurg, Psychiat. 31: 424-433, 1968.

Eberstein, A., Sandow, A. Fatigue mechanisms in muscle fibres. In: Effect of Use and Disuse on Neuromuscular Functions, ed. E. Gutmann & P. Hnik, Prague, p. 515-526, 1963.

George, J. C., Vallyathan, N. V. Effect of exercise on the free fatty acid levels in the pigeon. J. Appl. Physiol. 19: 619-622, 1964.

Havel, R. J., Naumark, A., Borchgrevink, C. F. Turnover rate and oxidation of free fatty acid of blood plasma in man during exercise: studies during continuous infusion of palmitate - 1 - C^{14}. J. clin. Invest. 42: 1054-1063, 1963.

Hilton, S. M., Vrbova, G. Absence of functional hyperaemia in the soleus muscle of the cat. J. Physiol., 194: 86P, 1968.

Hilton, S. M., Vrbova, G. Inorganic phosphate - a new candidate for mediator of functional vasodilatation in skeletal muscle. J. Physiol. 206: 29-30P, 1970.

Hilton, S. M., Jeffries, M. G. & Vrbova, G. Functional specialisations of the vascular bed of soleus. J. Physiol. 206: 543;562, 1970.

Hirche, Hj., D. Grüm, Waller, W. Utilization of carbohydrates and free fatty acids by the gastrocnemius of the dog during longlasting rhythmical exercise. Pflüger's Archiv f. gesamte Physiologie, in press.

Hudlicka, O. Resting and postcontraction blood flow in slow and fast muscles of the chick during development. Microvascular Research, 1: 390-402, 1969.

Kugelberg, E., Edstrom, L. Differential histochemical effects of muscle contractions on physphorylase and glycogen in various types of fibres: relation to fatigue. J. Neurol, Neuorsurg. Psychiat. 31: 415-423, 1968.

Manners, D. J. Glycogen storage disease, type I. In Whelan, W. J. (ed). Control of glycogen metabolism, Ciba Foundation Symposium, Churchill, London, 1964.

Paul, P. FFA metabolism of normal dogs during steady-state exercise at different work loads. J. appl. Physiol. 28: 127-132, 1970.

del Pozo, E. C. Transmission fatigue and contraction fatigue. Amer. J. Physiol. 135: 763-771, 1942.

Randle, P. J. Newsholme, E. A., Garland, P. B. Regulation of glucose uptake by muscle. Biochem. J. 93: 652-665, 1964.

Sacks, J., Cleland, M. G. Absence of phosphate interchanges in repetitive muscular contractions. Amer. J. Physiol. 198: 300-302, 1960.

Wahren, J. Human forearm metabolism during exercise IV. Glucose uptake at different work intensities. Scand. J. Clin. Lab. Invest. 25: 129-135, 1970.

Wilson, J. E., Sacktor, B., Tiekert, G. G. In situ regulation of glycolysis in tetanized cat skeletal muscle. Arch. Biochem. Biophys. 120: 542-546, 1967.

UPTAKE AND OXIDATION OF SUBSTRATES IN THE INTACT
ANIMAL DURING EXERCISE*

Pavle Paul

Division of Research, Lankenau Hospital

Philadelphia, Pennsylvania, U.S.A.

INTRODUCTION

Exercise and the biochemical mechanisms involved in muscle contraction have been of interest to the scientist for many decades. The results of studies have led to the use of exercise in therapy and rehabilitation of post-trauma and post-infarction patients and, even more recently, as a possibly preventive measure against atherosclerosis.

The recognition of FFA in plasma, together with the availability of radioactive ^{14}C-labeled compounds, was the stimulus for much investigation of the role of FFA as metabolic substrates in animals and in man (1). In spite of the small size of the plasma pool, it was shown that FFA has a rapid turnover rate and is oxidized to CO_2 in vitro as well as in vivo (2-6). The old concept that glucose is the prime fuel for muscular work has been revaluated in the last few years, and a new concept has developed in which both plasma glucose and FFA play an important metabolic role during rest and exercise. Muscle has the ability to store small amounts of glucose in the form of glycogen and FFA in the form of triglycerides (7). The two major energy storage depots, the liver and adipose tissue, supply substrates to the blood in the form of glucose and albumin-bound FFA, respectively (Fig. 1). During

---

*This work was supported in part by Grants HE-07687 and FR-5585 from the United States Public Health Service.

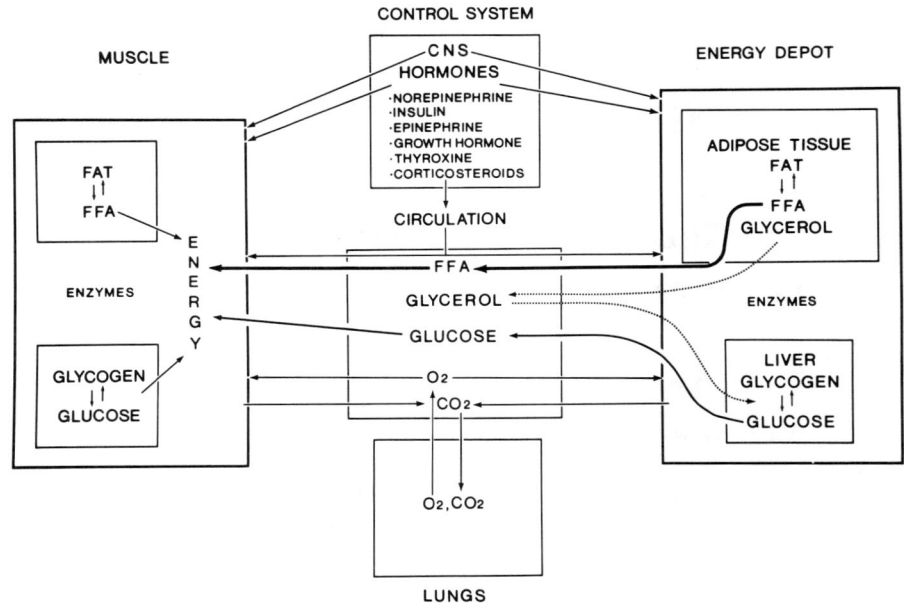

Fig. 1. Energy stores and the factors influencing their mobilization.

heavy exercise of short duration, the intramuscular energy depots are more readily available and provide the immediate energy requirement. Work of long duration, however, requires the extensive participation of intra- and, even more so, of extramuscular energy depots (6,8).

Exercise has been shown to stimulate the mobilization of substrates from these depots; however, many factors both physiological and pathophysiological play a role. To cite a few, the energy supply to the working muscle is influenced by nutrition, ambient temperature changes, hormones, disease states, hereditary factors, and physical work capacity (9-12). The circulatory system and the lungs provide the necessary mechanisms for transport of metabolic fuels, hormones, and O_2 between the energy-supply depots and the working muscle. The central nervous system (CNS) is the essential regulatory mechanism involved in the coordination of metabolic processes. Each of these five components: muscle, energy depots, circulation, lungs, and CNS could be a rate-limiting factor in work endurance.

As pointed out by Dr. Issekutz (13), there is an important difference between the electric stimulation of a relatively small muscle group and the work of almost the entire musculature of the intact organism. Whereas the former type of experiments may provide valuable information concerning the local changes of metabolism due to stimulation, experiments on exercising dogs deal with the entire interplay of nervous and hormonal mechanisms controlling work metabolism.

The purpose of the investigations discussed herein was, first, to study the contribution to metabolism of the two major energy substrates, fat and carbohydrates, in intact animals under different work loads, and second, to investigate the role of hormonal regulation in controlling the metabolic interrelationship between fat and carbohydrate metabolism.

METHODS

Preparation of Dogs

Normal Dogs. Normal dogs of both sexes weighing 10-16 kg were used. All animals were trained to lie quietly for 90-110 min on a treadmill before exercise. One group of specially selected dogs (good runners) received exercise training for a 3-4 week period, during which time the slope and duration of exercise was gradually increased. Experiments with these animals consisted of a 110-min rest period followed by 4 hr of exercise at a slope of 0-15% and at a speed of 80 or 100 m/min. The untrained dogs were used for short (30-40 min) heavy exercise at a slope of 15% and a speed of 100 m/min.

Pancreatectomized Dogs. Well-trained dogs were pancreatectomized under morphine-pentothal anesthesia 3-4 weeks prior to the experiment. These dogs were maintained on Purina dog chow with meat, methionine, pancreatine, and insulin (10-20 U/day).

Houssay Dogs. Following pancreatectomy and hypophysectomy the Houssay dogs were kept on the same food regimen as pancreatectomized dogs but without insulin. The experiments were performed after a few days of mild training 30-60 days after surgery.

Thyroidectomized Dogs. Well-trained dogs were surgically thyroidectomized (THY), and used for experiments at periods not less than 4-6 weeks later. Training was continued during this period. Before dogs were considered to be THY, the following factors were taken into consideration: gain in body weight, decreased O_2 consumption, low PBI, plasma cholesterol levels in excess of 300 mg/100 ml, high triglycerides, high phospholipids, and an increase in low and very low density lipoproteins.

Experimental Procedure

Two days before an experiment, indwelling polyethylene catheters were placed into the right carotid artery and jugular vein. The catheters were filled with saline and no anticoagulant was used. All experiments were carried out after 18-24 hr of fasting, and insulin was withheld from the diabetic dogs for 48 hr prior to the experiment. Tracer experiments were conducted with albumin-bound palmitate-1-^{14}C or uniformly labeled glucose ^{14}C (UL). After a priming dose, the tracer was infused intravenously at a constant rate throughout the entire experiment. In a number of normal dogs and also in the Houssay dog, unlabeled glucose was infused throughout the entire experiment. During the resting period a plastic hood with a rubber collar was placed over the animal's head to enable collection of expired air. A pump maintained the airflow through the hood at 20-30 liters/min during rest and at 60-100 liters/min during exercise depending on the work load. After each air collection the plastic hood was removed; it was replaced 5 min before the next sampling without interrupting the run. The samples of diluted expired air were collected for 4 min in Douglas bags by means of a second pump and analyzed for O_2 and CO_2 content by means of a calibrated Noyons diaferometer. The specific activity (SA) of $^{14}CO_2$ was measured according to the method of Fredrickson and Ono (14). Arterial blood samples were taken immediately following collection of air and analyzed for FFA and glucose SA, and blood lactate, as described in our previous studies (15,16).

Calculations

The turnover rate (release or entry = uptake) of FFA or glucose in a dynamic steady state of exercise was calculated according to the principles of isotope-dilution techniques. The infusion rate of the tracer (mμc/kg per min) divided by the SA of the substrate (mμc/μEq of FFA or mμc/μmole of glucose) represents the rate by which the unlabeled substrate enters the circulation. The glucose pool was estimated from the priming dose (t = 0) and the average glucose SA obtained during the resting period.

When a difference existed between subsequent samples, the rate of FFA or glucose entry was calculated according to the equation of Steele (17). To calculate the percentage of CO_2 derived from the immediate oxidation of plasma FFA or glucose, the following calculations were used. The CO_2 SA was divided by the FFA SA and multiplied by 1700 (assuming 17 carbons as an average of C_{16} and C_{18} fatty acids), and in the case of glucose, the CO_2 SA was divided by the glucose SA and multiplied by 600 (6 CO_2 from the uniformly labeled glucose). The product of CO_2 SA and of CO_2 output (mmole/kg per min) divided by the FFA SA or glucose SA is the amount of plasma FFA or of plasma glucose oxidized during exercise.

RESULTS AND DISCUSSION

Metabolism of FFA in the Normal Dog

Three typical experiments on the metabolism of FFA during exercise are shown in Fig. 2. During the short period of heavy exercise the O_2 uptake of the untrained dog (A) increased by a factor of 7 to 57 ml/kg per min with a concomitant 4-fold increase in blood lactate to 40 mg/100 ml. During exercise a slight increase was found in FFA level from 0.42 to 0.53 μEq/ml. Calculation showed that the FFA uptake rose from 12.2 μEq/kg per min at rest to 18.4 μEq/kg per min during work, of which about 77% was immediately converted to CO_2, representing not more than 10% of the total exhaled CO_2 and requiring 13.3% of the total O_2 consumption during exercise. In contrast, a well-trained dog running at an O_2 uptake averaging 43 ml/kg per min (seven times the resting value) for 4 hr (Fig. 2B), showed only a slight initial increase in blood lactate followed by a decrease during the exercise. The low FFA level at rest (0.32 μEq/ml) increased progressively during exercise to a final level of 1.7 μEq/ml. The rate of FFA entry to plasma increased from 14.1 to 99 μEq/kg per min at the end of exercise. At rest 20% of the CO_2 was derived from FFA. As a result of the increased oxidation of FFA from 2.84 to 59 μEq/kg per min, FFA contributed 60-70% of the exhaled CO_2. In this experiment, 63% of the total O_2 uptake was required to oxidize this amount of FFA. Fig. 2C shows the effect of superimposing a continuous infusion of nonradioactive glucose on FFA metabolism. The plasma glucose level rose during rest from 105 to 150 mg/100 ml, and the insulin level increased from 20 to 71 μU/ml. Glucose infusion decreased the FFA level from 0.746 to 0.236 μEq/ml and caused a low FFA entry of 14.2 μEq/kg per min. A high RQ of 0.93 indicates predominantly carbohydrate oxidation. During

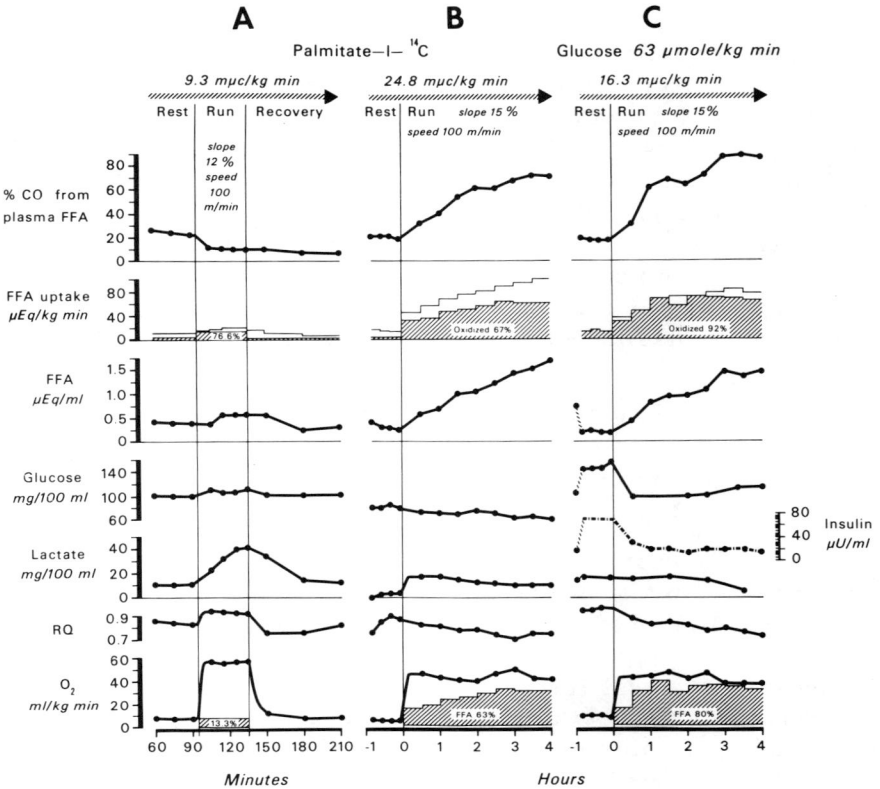

Fig. 2. Effect of exercise on the rates of uptake and oxidation of plasma FFA in normal dogs: (A) untrained, (B) trained, (C) trained with continuous (nonradioactive) glucose infusion. Priming dose of palmitate-1-^{14}C at t = -110 min. Hatched areas: upper, the rate of oxidation of FFA in μEq/kg per min; lower, its O_2 equivalent in ml/kg per min.

exercise the plasma glucose decreased sharply to 100 mg/100 ml while the insulin level dropped to an average value of 20 μU/ml. The FFA level rose constantly from the beginning of exercise and reached a level of 1.47 μEq/ml. Of the 79 μEq/kg per min FFA entry, 92% was immediately oxidized, requiring 80% of the total O_2 uptake. The RQ decreased to a level of 0.74. These results show that in spite of a constant glucose infusion, shifting of metabolism from carbohydrate to fat occurs during work.

UPTAKE AND OXIDATION OF SUBSTRATES

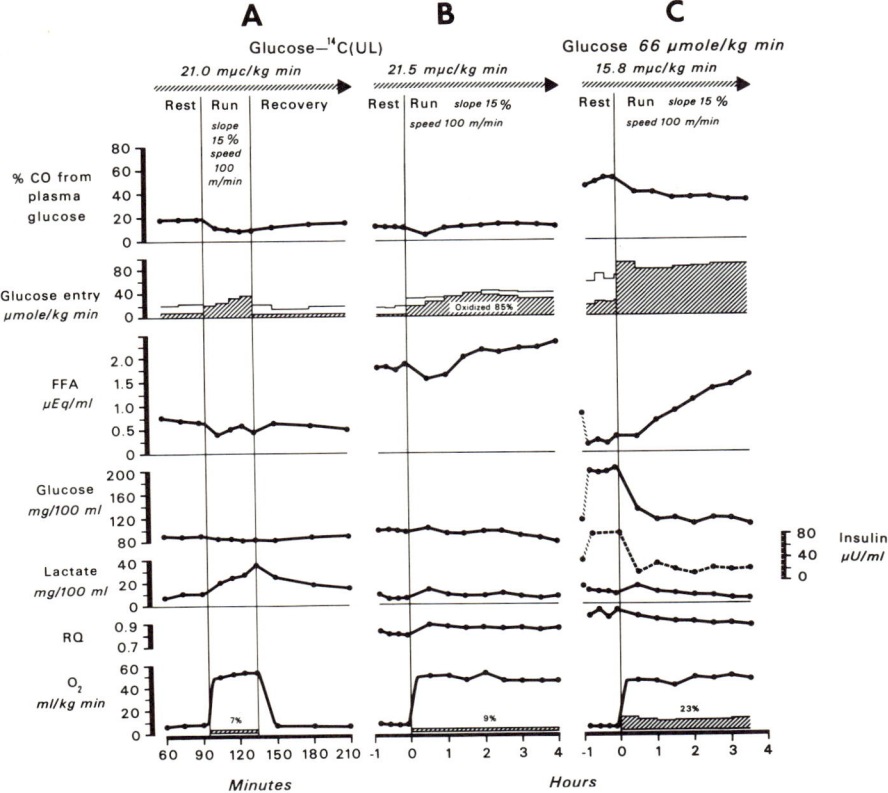

Fig. 3. Effect of exercise on the rates of uptake and oxidation of plasma glucose in normal dogs: (A) untrained, (B) trained, and (C) with continuous (nonradioactive) glucose infusion. Priming dose of glucose ^{14}C (UL) at t = -110 min. Hatched areas: upper, the rate of glucose oxidation in μmole/kg per min; lower, its O_2 equivalent in ml/kg per min.

Metabolism of Glucose in the Normal Dog

When the heavy exercise load used during the palmitate-1-^{14}C infusion experiment with an untrained animal was duplicated with glucose ^{14}C (Fig. 3A) the following observations were made. The total glucose entry, which rose from 21 to 34 μmole/kg per min, was immediately oxidized. The percent CO_2 from plasma glucose decreased from 18-19% at rest to 9% during exercise, requiring not more than 7% of the total O_2 consumption during exercise. A well-trained dog (Fig. 3B) exercising with an O_2 consumption averaging 48 ml/kg per min showed only a slight increase in plasma lactic acid and a small decrease in glucose level. At the same time, the plasma FFA level rose from 1.81 to 2.33 µEq/ml. Glucose entry and oxidation were increased from a resting level of 16.5 and 4.7 µmole/kg per min, respectively, to 36.6 and 31.1 µmole/kg per min during exercise. The glucose oxidized represented 28.5% of the uptake at rest and 85% during exercise. The percent CO_2 derived from plasma glucose was the same during rest and exercise and represented a utilization of not more than 9% of the total O_2 uptake. Administration of nonradioactive glucose (Fig. 3C) increased the resting plasma level from 116 to 200 mg/100 ml with a simultaneous 3-fold increase in glucose entry to 62 µmole/kg per min. The average RQ was 0.88 and 50% of the CO_2 was derived from plasma glucose. The concentration of plasma glucose decreased nearly 50% during exercise, indicating that its uptake exceeded its rate of entry into the pool. Glucose output was increased somewhat and was maintained at an average of 76.7 µmole/kg per min, thus contributing 32% of the exhaled CO_2. The entire glucose entry was oxidized, requiring 23% of the total O_2 uptake.

Effect of Work Load on FFA Metabolism

The effect of increasing work load on FFA metabolism is given in Fig. 4 which shows the correlation between (1) FFA concentration in the plasma and its turnover rate, (2) FFA level and percent CO_2 derived from plasma FFA, and (3) FFA turnover rate and the percent CO_2 derived from plasma FFA. Table I summarizes the regression equations of the data in Fig. 4-1. All regression coefficients are significant ($p < 0.001$), based on the null hypothesis.

On the basis of their O_2 consumption during exercise the trained dogs were classified into three work load groups: A, B, and C, while the untrained dogs were designated as group D. The regression equations describing the interrelationships in Fig. 4 and the significant differences between the regression coefficients of work loads A, B, and C have already been described (18). For comparison, the correlation between FFA concentration in the plasma and the FFA turnover rate at rest is presented here in Fig. 4-1.

When the energy demand is increased, as the trained animal

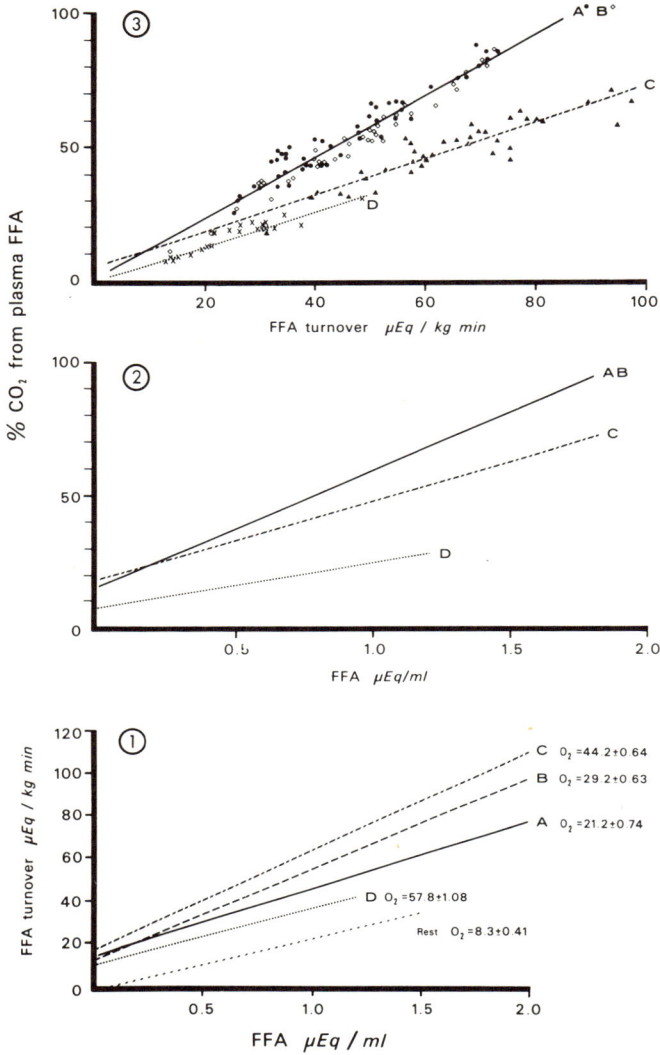

Fig. 4. 1: Interrelationship between plasma FFA level and FFA turnover rate at rest and during work loads A, B, C, and D. 2: Interrelationship between FFA concentration and percent CO_2 derived from plasma FFA during work loads A and B combined (continuous line) and work loads C and D (broken lines). 3: Interrelationship between FFA turnover rate and percent CO_2 derived from plasma FFA during work loads A (●), B (◇), C (▲) and D (X).

Table I

Regression equations of FFA turnover rate at rest and under four different work loads

Work Load	$\dot{V}O_2$ ml/kg per min *	FFA Turnover Rate µEq/kg min	SE of Regression Coefficient	% of Uptake Oxidized *	Blood Lactate * Resting mg/100 ml	Running mg/100 ml	n
Rest (11)**	8.3 ± 0.41	Y = 23.8 x − 1.5	± 2.5	22.8 ± 1.31			
A (6)**	21.2 ± 0.79	Y = 32.89 x + 13.63	± 3.0	53.4 ± 1.28	11.2 ± 0.59	10.4 ± 0.79 P = N.S.	48
B (5)**	29.2 ± 0.63	Y = 42.79 x + 12.98	± 2.7	64.7 ± 1.40	12.8 ± 0.69	10.6 ± 0.65 P = N.S.	38
C (6)**	44.2 ± 0.69	Y = 46.38 x + 16.97	± 2.7	70.4 ± 1.18	8.1 ± 0.77	17.2 ± 0.99 P = <0.001	38
D (12)**	57.8 ± 1.08	Y = 26.36 x + 10.49	± 4.0	88.2 ± 5.72	5.9 ± 0.84	45.3 ± 3.11 P = <0.001	25

*   Values are Mean ± SE   n = number of determinations

**   Number of dog experiments

x =   FFA µEq/ml

goes from rest to work loads A, B, and C, at any plasma FFA concentration the FFA turnover rate is higher at the heavier work load (Fig. 4-1). In contrast, the FFA turnover rate in untrained dogs (D) at any equivalent FFA concentration is far below that observed at work loads A, B, and C. It would appear that as long as the O_2 supply is adequate (no drastic increase in lactic acid level, Table I) the FFA level, its turnover rate, percent of uptake oxidized, and the oxidation rate will all be increased. The percentage of CO_2 derived from plasma FFA will be the same at increased work loads A and B (Fig. 4-2). The same observation applies when comparison is made between FFA turnover rate and percent CO_2 derived from plasma FFA (Fig. 4-3).

When the O_2 supply is inadequate and the lactic acid level rises significantly (Fig. 4-2,3, work load C), the percent of CO_2 derived from FFA oxidation is lower when compared to work loads A and B in spite of a higher turnover rate and larger percent of FFA uptake oxidized. Experiments on untrained dogs with high lactate (Table I) demonstrate that althought FFA uptake is almost completely oxidized (88.2%) it accounts for only a small percentage of the total CO_2 output (Fig. 4-2,3D).

It is seen from Fig. 4-3 that in well-trained dogs, A and B, the FFA turnover provided 30-90% of the total CO_2 output, work load C, 30-70%, and in untrained dogs, work load D, 5-25%.

It will be recalled from Figs. 2 and 3 that in trained dogs the FFA level, turnover rate, and the participation of FFA as an energy substrate increase during exercise. When comparing exercise to pre-exercise levels in untrained dogs, the FFA concentration and uptake in some cases remained unchanged but in others either decreased or showed a slight increase. These differences depend on the lactic acid level which reflects the anaerobic capacity of the working organism (19).

A more detailed analysis of the data in Fig. 4 is presented in Table II, in which FFA metabolism is examined at successive intervals of time during exercise at different work loads.

During the first hour of exercise at work loads A, B, and C, an increase in FFA level and in FFA uptake was found. However, even though the amount of FFA oxidized ranged from 18.8-37.5 µEq/kg per min, the percent of total O_2 uptake used for FFA oxidation was essentially the same (39-45%). During the second hour of exercise at all work loads, a further rise in the level, uptake, and oxidation of FFA was observed. Here again the participation of FFA at the three work loads remained constant.

An additional rise in all three parameters of FFA metabolism was found at all three levels of exercise during the last stage of the experiment. The amount of FFA oxidized at work load A was 25.6, B, 35.7, and C, 52.1 µEq/kg per min. Despite these appreciable differences in FFA oxidation there was no significant difference in the percent of total O_2 consumption used for this oxidation.

The untrained dogs (D) showed a decrease of the plasma FFA from the resting values of 0.80 to 0.59 µEq/ml during exercise. In spite of the fact that almost the total FFA uptake was oxidized (Table I) only 22% of the elevated O_2 consumption could be attributed to the transformation, a value far below that found in trained dogs during the first hour.

Although protein can be utilized to a limited extent as an energy source, the two major energy fuel depots are fat and glycogen. Fat is stored abundantly in the form of triglycerides in adipose tissue primarily in the peripheral and splanchnic areas, to a lesser

Table II

The effect of different work loads on FFA metabolism at successive intervals of time*

Work Load	FFA µEq/ml during rest	1 hour $\dot{V}_{O_2}$ ml/kg per min	1 hour FFA µEq/ml	1 hour FFA µEq/kg min uptake oxidized	1 hour % of total O_2 for FFA oxidation	2 hour $\dot{V}_{O_2}$ ml/kg per min	2 hour FFA µEq/ml	2 hour FFA µEq/kg min uptake oxidized	2 hour % of total O_2 for FFA oxidation	3 + 4 hours $\dot{V}_{O_2}$ ml/kg per min	3 + 4 hours FFA µEq/ml	3 + 4 hours FFA µEq/kg min uptake oxidized	3 + 4 hours total O_2 for FFA
A (6)	0.56 ±0.07	22.8 ±0.87	0.61 ±0.04	(n = 12) 34.4 ± 2.36	45.6 ± 3.50	20.4 ± 0.85	0.90 ±0.04	(n = 12) 43.7 ± 2.68	21.8 ± 1.09	58.8 ± 3.49	20.8 ± 0.71	1.13 ±0.06	(n = 24) 47.6 ± 2.42	25.6 ± 1.09	68.3 ± 3.51
B (5)	0.35 ±0.02	31.1 ±1.33	0.47 ±0.04	(n = 10) 31.8 ± 2.13	38.8 ± 2.81	28.7 ± 1.06	0.79 ±0.002	(n = 10) 46.7 ± 1.87	30.4 ± 0.88	58.0 ± 2.33	27.7 ± 0.89	1.04 ±0.06	(n = 18) 58.5 ± 2.24	35.7 ± 1.39	70.5 ± 3.05
C (6)	0.45 ±0.03	45.4 ± 1.05	0.76 ±0.06	(n = 12) 51.0 ± 2.58	44.7 ± 2.94	42.6 ± 1.15	1.01 ±0.05	(n = 12) 64.2 ± 2.54	44.4 ± 1.41	56.7 ± 1.80	45.0 ± 0.86	1.28 ±0.07	(n = 15) 77.3 ± 3.27	52.1 ± 2.02	63.0 ± 2.37
D** (12)	0.80 ±0.04	57.8 ± 1.08	0.59 ±0.06	(n = 25) 26.2 ± 1.87	23.1 ± 1.65	21.8 ± 1.56									

* Values are mean ± SE of n number of determinations.

** 30 - 40 min exercise

Number in parenthesis column 1 = number of experiments

extent around the muscle and other tissues, and finally intramuscularly. Glucose is stored to a limited extent in the form of glycogen, predominantly in the liver and muscle. An excess of carbohydrate is readily converted to lipids for storage as a more economic energy fuel. The supply and utilization of individual endogenous energy substrates respond precisely to exogenous signals which may also initiate changes in homeostatic control mechanisms. These changes affect the hormones which in turn influence certain enzyme activities. Under resting conditions, it was shown that an interrelationship exists between glucose and FFA metabolism and that the plasma FFA level influences the participation of carbohydrate in energy metabolism (20). During exercise, FFA mobilization has a sparing effect on glycogen depots (8,21).

On the basis of our experiments conducted with palmitate-1-^{14}C on normal dogs, it was shown that in untrained dogs during heavy work the FFA turnover rate can supply only a small portion of the energy requiremnts (22). From previous work we concluded that lactic acid interferes with FFA mobilization (19). Peterson and coworkers found that an inadequate O_2 supply to the muscle leads to accumulation of lactic acid and of α-glycerophosphate (23). They also showed the existence of a competition between pyruvic acid and dihydroxyacetone phosphate (DHAP) for $DPNH_2$, resulting in a striking accumulation of lactate and α-glycerophosphate. During heavy exercise when a reduction in splanchnic and skin blood flow occurs, it can be assumed that this competition exists also in adipose tissue. Increased lactic acid or inadequate O_2 supply could lead to an increased formation of α-glycerophosphate which in turn may inhibit the FFA mobilization from intramuscular and extramuscular fat depots in spite of an increased rate of lipolysis (24). Aerobic exercise in well-trained dogs with little or no rise in lactic acid shows a shifting of the metabolism towards fat oxidation regardless of the level of energy expenditure. It was shown that athletes have somewhat higher splanchnic blood flow during exercise when compared to the untrained individual (25). During prolonged exercise, redistribution of blood volume to the surface of the body to increase heat loss occurs. Greater vascularization of the splanchnic region and the surface of the body where the major fat depots are located may very well be of crucial importance in permitting mobilization and transport of FFA to the working muscle.

Although the hepatic glucose output is increased during exercise and, further, is almost completely oxidized (Table III), plasma glucose contributes not more than 15% to the caloric expenditure

Table III

Glucose oxidation during exercise*

	REST		EXERCISE					
	Plasma Glucose mg/100ml	Glucose Entry μmole/kg per min	$\dot{V}_{O_2}$ ml/kg per min	Plasma Glucose mg/100ml	Glucose Entry μmole/kg per min	Glucose oxidized μmole/kg per min	% of total O_2 for glucose oxidation	Lactic Acid mg/100ml
Untrained Dogs n = 4	122.5 ± 3.85	16.4 ± 0.74	56.5 ± 2.65	117.3 ± 2.49	30.1 ± 2.49	27.1 ± 5.48	6.03 ± 0.95	30.6 ± 3.95
Trained Dogs n = 4	104.8 ± 5.39	17.3 ± 0.68	41.6 ± 4.49	93.4 ± 5.65	35.2 ± 4.77	29.1 ± 5.83	9.94 ± 1.02	6.62 ± 0.60
Trained Dogs With Glucose infusion n = 3	185.2 ± 9.0	56.3 ± 4.03	42.3 ± 2.34	110.4 ± 2.55	76.2 ± 7.06	81.5 ± 16.5	24.4 ± 2.82	7.58 ± 0.55

* Values of Mean ± SE
n Number of dog experiments

(average 11%) (22). Regardless of whether the dog was trained or untrained or what percent of his caloric expenditure was derived from plasma FFA, the oxidation of plasma glucose required only a small fraction of the total O_2 consumption.

In untrained dogs 28% of total O_2 consumption during exercise was used for the oxidation of plasma substrates (FFA and glucose). In the case of trained dogs during the first hour of exercise, the metabolic fuels used more than 50% of the total O_2 consumption for their oxidation. In untrained dogs, during exercise the circulation does not deliver a sufficient amount of substrates to cover the energy expenditure of the work. As discussed above, FFA mobilization is inhibited in the untrained dog, implying that the animal has limited access to extramuscular fat energy depots, and the difference needed to cover the energy expenditure of the work has to be derived from intramuscular depots. The trained dog has the fat energy stores at its disposal, and as the work continues the contribution of intramuscular substrates to overall energy expenditure decreases (8). This difference in fuel supply between trained and untrained individuals is one of the major limiting factors in work endurance.

A decrease of circulating insulin level during exercise caused perhaps by a high norepinephrine level (26) enhances the effectiveness of lipolytic hormones and shifts the metabolism toward fat oxidation. At the same time, glucose mobilization and utilization is always increased during exercise. A decreased insulin level increases hepatic glucose output and also may limit glucose oxidation in the working muscle, in this way maintaining normal blood sugar in the face of an increased glucose mobilization and oxidation. In spite of the fact that FFA represents the major portion (up to 90%) of energy in exercise of long duration, depletion of the glycogen energy stores in liver results in a fall of blood sugar and, consequently, cessation of exercise will occur (27,28).

Studies with the Pancreatectomized Dog

A typical experiment with a pancreatectomized dog (Fig. 5A) shows a very high resting plasma FFA level (3.25 µEq/ml) and entry to the plasma pool (68 µEq/kg per min) of which 13% was immediately oxidized, providing 60% of the resting CO_2 output. During exercise the FFA level rose to 3.8 µEq/ml and FFA uptake to 100 µEq/kg per min of which 63% was immediately oxidized, providing 71% of the running CO_2 output and requiring 73% of the O_2 consumption. In a radioglucose experiment (Fig. 5B) a high glucose entry of 34.0 µmole/kg per min was observed with only a slight change (39.2 µmole/kg per min) during exercise. During rest only 11% of the hepatic glucose entry was oxidized, contributing 7.4% to the resting CO_2. Exercise increased the oxidation of hepatic glucose entry to 36%, but the percent of CO_2 derived from plasma glucose was only 4.5%, requiring not more than 3% of the total O_2 consumption.

This and other similar experiments (6) show that complete lack of insulin in diabetic dogs has a profound effect on metabolism at rest as well as during exercise. The full effectiveness of lipolytic hormones and enzymes becomes readily apparent in the absence of insulin. In these dogs, during rest the FFA and glucose mobilization was in the range of normal exercising dogs. The difference between the normal and diabetic resting dog was that a smaller portion of the fuel mobilized was oxidized. The metabolic picture of pancreatectomized dogs at any stage of exercise was the same as that of normal trained dogs during their last hour of work. In other words, metabolism did not shift toward fat oxidation in a stepwise fashion. This could be attributed to the fact that the FFA turnover was able to supply from the beginning to the end of exercise a major portion (up to 85%) of the energy requirement. Also, no difference or only a slight increase in hepatic glucose output was found in these dogs during exercise. Whereas normal dogs oxidized

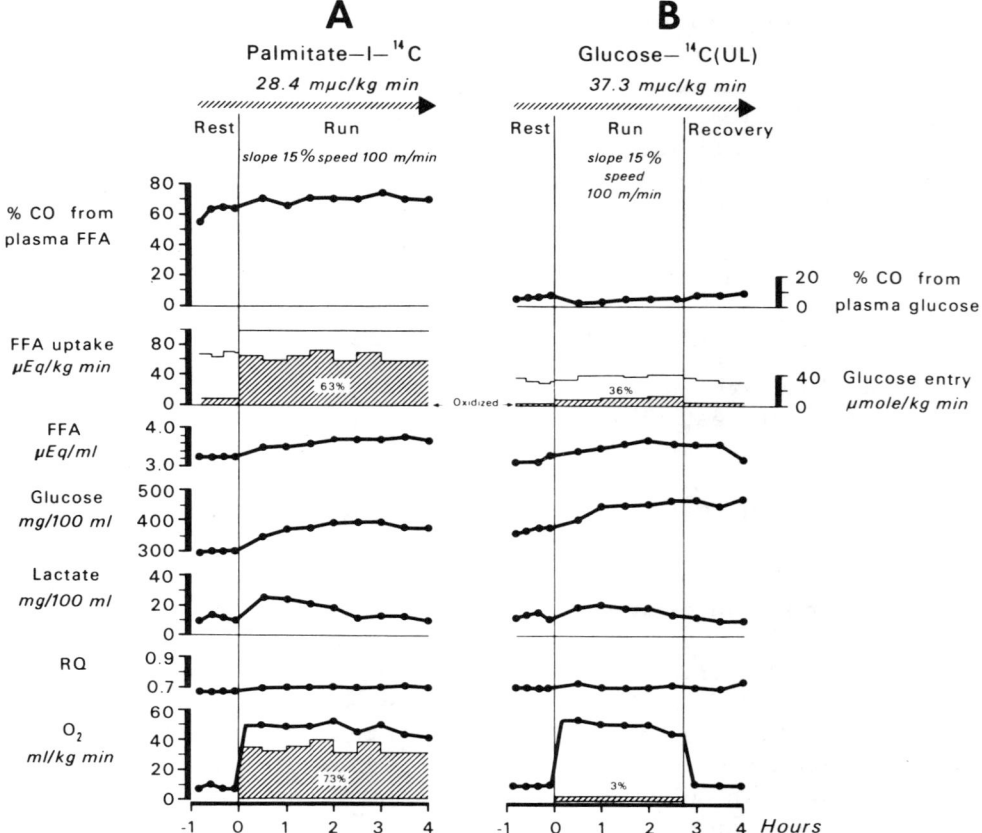

Fig. 5. Effect of exercise on the rates of uptake and oxidation of plasma FFA (A), and plasma glucose (B) in pancreatectomized dogs. Priming dose of palmitic-1-^{14}C or glucose ^{14}C (UL) at t = - 110 min, followed by constant rate infusion. Hatched areas: upper, the rate of oxidation of FFA or glucose in μmole/kg per min; lower, its O_2 equivalent in ml/kg per min.

the total glucose output providing an average of 12.7% of the energy expenditure, the diabetic dog derived only 6.3% of its energy from the oxidation of 49% of its elevated glucose output.

Studies with the Houssay Dog

Study of FFA and glucose metabolism in the pancreatectomized-hypophysectomized dog during rest and exercise is of great interest with regard to a further understanding of hormonal regulation of

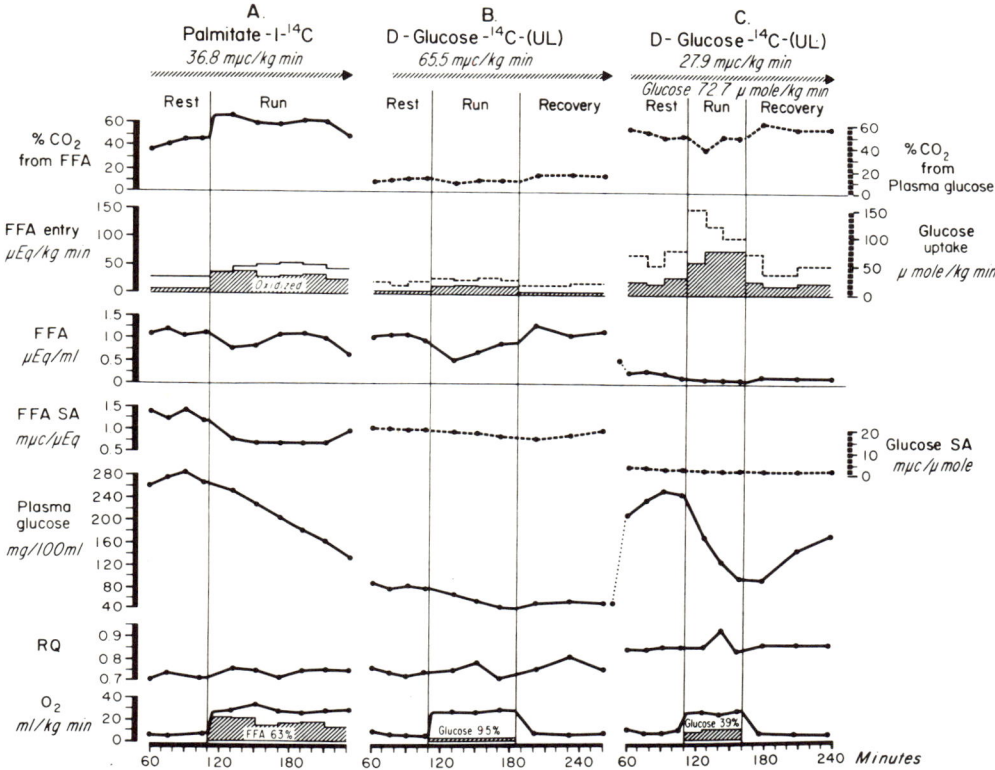

Fig. 6. Three experiments carried out on the same Houssay dog. Effect of light exercise (slope 0%, speed 100 m/min) on (A) the metabolism of FFA, (B) plasma glucose, (C) with nonradioactive glucose infusion of 72.7 μmole/kg per min. Priming dose of palmitic-1-^{14}C or glucose ^{14}C (UL) at t = -110 min. Hatched areas: upper, the rate of oxidation of FFA or plasma glucose in μmole/kg per min; lower, its O_2 equivalent in ml/kg per min.

metabolic events. These animals are essentially deprived of their hormonal regulatory system with the exception of norepinephrine production. Because of the expected decreased work capacity of these animals, a low work load was selected for the preliminary experiments shown in Fig. 6 which were carried out on the same Houssay dog. Exercise induced a 3- to 4-fold increase in O_2 uptake (26-29 ml/kg per min) over the resting values in all three experiments. The resting FFA level of 1.13 μEq/ml decreased during exercise to 0.92 μEq/ml (Fig. 6A). The turnover rate, on the other hand, increased from 28 to 48 μEq/kg per min of which

71% was immediately oxidized, contributing 60% to the exhaled CO_2 and utilizing 63% of the elevated O_2 consumption. The RQ during work averaged 0.73 and a dramatic change in plasma glucose was observed; the plasma resting level of 272 rapidly decreased to 132 mg/100 ml at the end of the exercise. It can be seen from Fig. 6B that, at rest, glucose turnover was 19 µmole/kg per min, 22% of which was oxidized, contributing 11% of the exhaled CO_2. During exercise, glucose turnover increased slightly to 26 µmole/kg per min, 57% of which was immediately oxidized, contributing 10% of the exhaled CO_2 and requiring 9.5% of the total O_2 uptake. No change in RQ was observed. The decreased glucose level resulting from the first experiment never returned to pre-exercise level. A further decrease in plasma glucose from 80 to 40 mg/100 ml occurred during the exercise phase of the second experiment (Fig. 6B). When both radioactive and unlabeled glucose were infused simultaneously throughout the third experiment at a rate approximately four times that of the hepatic sugar output (Fig. 6C), the resting plasma glucose level of 41 rose rapidly to 250 mg/100 ml. The entry rate of glucose was 68 µmole/kg per min of which 39% was oxidized, contributing 51% of the exhaled CO_2. The FFA level decreased from 0.35 to 0.19 µEq/ml. A high RQ (0.84) and the percent CO_2 from plasma glucose, in conjunction with low FFA levels, reflect a shift in the metabolism toward predominantly glucose oxidation. During exercise, glucose uptake increased to 155 µmole/kg per min in the first 15 min and then slowly decreased to 105 µmole/kg per min at the end of 45 min. Concomitantly a decrease in blood glucose from 250 to 98 mg/100 ml occurred. Of the glucose uptake, 69% was immediately oxidized, contributing 45% to the exhaled CO_2. Of the O_2 consumption, an average of 39% was used to oxidize plasma glucose. During this same period FFA levels decreased to 0.086 µEq/ml. In the postexercise period these metabolic parameters returned to pre-exercise values. Although definitive conclusions are not possible from this and the few other studies on Houssay animals, brief comments appear to be in order at this time. As long as these dogs eat constantly they can maintain a high glucose level and turnover. No differences were observed between normal and Houssay dogs with respect to FFA level, its corresponding turnover rate, and oxidation rate during rest and exercise. At rest, these dogs are similar to pancreatectomized animals, having a high blood glucose but a lower FFA level, and preferentially oxidizing FFA as an energy source. Once this equilibrium state is interrupted by starvation or stress (exercise) limited glycogen stores (due to the lack of insulin) can not maintain the glucose turnover rate and the glucose level falls.

Then these dogs become similar to normal dogs oxidizing both FFA and glucose. In contrast to pancreatectomized dogs, where the FFA and glucose levels rise during exercise, the Houssay dog did show a slight decrease in FFA and drastic decrease in glucose level. Glucose ^{14}C experiments showed that only 57% of glucose uptake was oxidized in comparison to 79-83% in normal and 40% in diabetic dogs. The most striking differences were found when unlabeled glucose was infused. In normal dogs under such conditions, the glucose entry was increased by 12.8 µmole/kg per min above the glucose infusion rate, which was 63.6 ± SE 1.13 µmole/kg per min. In experiments without glucose infusion the increase in glucose entry during exercise above the resting values was 13.7 µmole/kg per min in untrained dogs and somewhat higher in trained dogs (17.9 µmole/kg per min) (Fig. 3). In contrast, the Houssay dog increased his glucose uptake during glucose infusion from a resting value of 68 µmole/kg per min (which was practically the infusion rate of glucose) to an average of 127 µmole/kg per min, an increase of 59 µmole/kg per min above the pre-exercise value. This increase represented 3-5 times that found in normal dogs with or without glucose infusion. Glucose lowered the FFA level and shifted metabolism toward carbohydrate oxidation. This was found also in the normal but not in diabetic animals (29). During exercise, the FFA level decreased to the point where its contribution as an energy source was negligible and the plasma glucose was the only extramuscular energy source.

Oxidation of plasma glucose required 39% of the total O_2 consumption during exercise, 5 times that of normal dogs, twice as much as normal dogs with glucose infusion, and more than 10 times that of pancreatectomized dogs. The Houssay dog during exercise is not able to increase the FFA level, although it increases mobilization of FFA. These dogs are able to oxidize glucose, but they lack the ability to conserve their glycogen stores in the liver, possibly because of the low FFA level and lack of insulin.

At the point when both energy substrate levels in this dog are very low, as seen in Fig. 6C before glucose infusion, they are practically starving. It might be speculated that cells lacking hormonal control and seeking biological survival under basal conditions and even more so during work can open the door to any energy substrates given them for use to maintain metabolic functions. These experiments give insights into the metabolism of FFA and glucose during exercise as affected by lack of hormones for 30-60 days. The time factor is of crucial importance in these

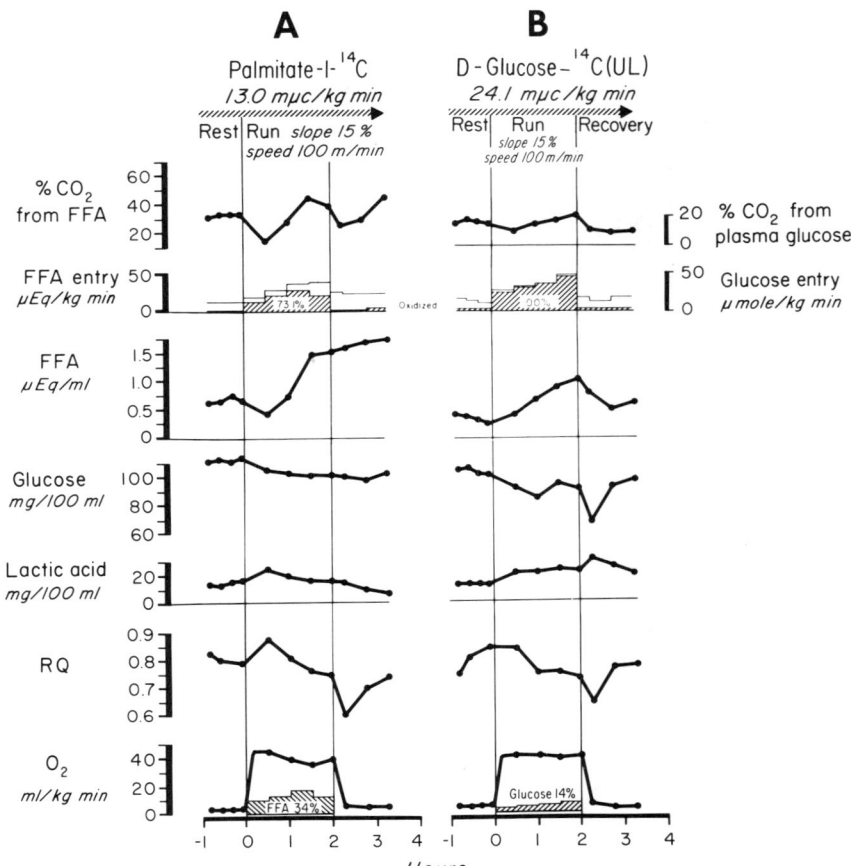

Fig. 7. Two experiments carried out on the same thyroidectomized dog. Effect of exercise on metabolism of (A) FFA and (B) plasma glucose. Priming dose at t = -110 min followed by constant rate infusion of radioactive substrates. Hatched areas: upper, the rate of oxidation of substrates in μmole/kg per min; lower, its O_2 equivalent in ml/kg per min.

dogs, as was recognized by Urgoiti and coworkers (30) and others.

Studies with the Thyroidectomized Dog

In experiments with thyroidectomized dogs the metabolic rate was increased 8 times during exercise (rest 4.82 ml O_2/kg per min, and exercise 39.8) (Fig. 7). In a larger series of experiments the thyroidectomized dogs were observed to have a significantly lower FFA turnover at rest than normal dogs at any FFA level. A smaller percentage of FFA uptake is immediately oxidized and the oxidation rate of plasma FFA is lower.

Exercise increased FFA mobilization, and the FFA level rose from an initial value of 0.68 to 1.6 µEq/ml at the end of the work period. FFA entered the plasma at the rate of 20-44 µEq/kg per min, which is approximately one-half that observed in the normal dog at the same FFA level. Of the FFA entry 73% was immediately oxidized, contributing from 17-44% of the CO_2 output. This in turn required an average of 34% of the total O_2 consumption. Unlike the normal dog, the thyroidectomized animal is unable during increased work loads to elevate FFA turnover to the same degree. The experiment with glucose ^{14}C infusion (Fig. 7B) shows that in a thyroidectomized dog exercise increases the glucose output from 16 to an average of 42.1 µmole/kg per min. At rest 18% of the CO_2 output was provided from plasma glucose oxidation. At the beginning of exercise, the percent CO_2 from plasma glucose initially decreased to 11% but reached values of 21% at the end of the work period. Of the total O_2 uptake 14% was required for glucose oxidation, indicating a greater participation of glucose in metabolism than is seen in the normal animal.

Our experiments conducted with palmitate show that FFA mobilization and its oxidation at rest is impaired in the thyroidectomized dog. Also the FFA level and FFA turnover are increased during exercise but at any equivalent FFA concentration the FFA turnover rate is far below that found in normal dogs. In spite of the fact that the percent of FFA uptake oxidized is the same as in normal dogs, the percent of total O_2 used for FFA oxidation is only a fraction of that seen in normal dogs. It would appear that in thyroidectomized dogs the rate-limiting factor in endurance during exercise probably is the rate of mobilization of FFA from adipose depots and not a deficiency of oxidative mechanisms in working muscle. These findings are in agreement with those of other investigators as described in a recent review (31) of the thyroid hormone and its regulatory function on metabolism.

In general we have concluded from these studies that under any circumstance investigated thus far the plasma glucose mobilization and oxidation represents a minor portion of the energy supply carried by the circulation to the working muscle when compared to the participation of plasma FFA.

ACKNOWLEDGEMENT

I am gratefully indebted to Drs. Anna and Otakar Sirek, Department of Physiology, University of Toronto, for kindly supplying the first series of Houssay dogs used in this work.

REFERENCES

1. Gordon, R.S., Jr. and Cherkis, A. J. Clin. Invest. 35:206 (1956).
2. Dole, V.P. J. Clin. Invest. 35:150 (1956).
3. Neptune, E.M., Jr., Sudduth, H.C., Foreman, D.R. and Fash, F.J. J. Lipid Res. 1:229 (1959).
4. Spector, A.A. and Steinberg, D. J. Biol. Chem. 240:3747 (1965).
5. Havel, R.J., Naimark, A. and Borchgrevink, C.F. J. Clin. Invest. 42:1054 (1963).
6. Issekutz, B., Jr., Paul, P. and Miller, H.I. Amer. J. Physiol. 213:857 (1967).
7. Carlson, L.A., Liljedahl, S.-O. and Wirsen, C. Acta Med. Scand. 178:81 (1965).
8. Issekutz, B., Jr. and Paul, P. Amer. J. Physiol. 215:197 (1968).
9. Christophe, J. and Mayer, J. J. Appl. Physiol. 13:269 (1958).
10. Young, D.R., Pelligra, R., Shapira, T., Adachi, R.R. and Skretti, K. J. Appl. Physiol. 23:734 (1967).
11. Carlson, A. and Pernow, B. J. Lab. Clin. Med. 58:681 (1961).
12. Armstrong, D.T., Steele, R., Altszule, N., Dunn, A., Bishop, J.S. and DeBodo, R.C. Amer. J. Physiol. 201:9 (1961)
13. Issekutz, B., Jr. Fat as a Tissue, eds. K. Rodahl and B. Issekutz, Jr. McGraw (1964) p. 228.
14. Frederickson, C.S. and Ono, R. J. Lab. Clin. Med. 51: 147 (1958).
15. Issekutz, B., Jr., Miller, H.I., Paul, P. and Rodahl, K. Amer. J. Physiol. 209:1137 (1965).
16. Miller, H.I., Issekutz, B., Jr. and Rodahl, K. Amer. J. Physiol. 205:167 (1963).
17. Steele, R., Altszuler, N., Wall, J.S., Dunn, A. and DeBodo, R.D. Amer. J. Physiol. 196:221 (1959).
18. Paul, P. J. Appl. Physiol. 28:127 (1970).

19. Issekutz, B., Jr., Miller, H.I., Paul, P. and Rodahl, K. J. Appl. Physiol. 20:293 (1965).
20. Paul, P., Issekutz, B., Jr., and Miller, H.I. Amer. J. Physiol. 211:1313 (1966).
21. Randle, P.J., Newsholme, E.A. and Garland, P.B. Biochem. J. 93:652 (1964).
22. Paul, P. and Issekutz, B., Jr. J. Appl. Physiol. 22:615 (1967)
23. Peterson, R.D., Gaudin, D., Bocer, R.M. and Beatty, C.H. Amer. J. Physiol. 206:599 (1964).
24. Issekutz, B., Jr., Miller, H.I. and Rodahl, K. Fed. Proc. 25:1415 (1966).
25. Rowell, L.B. Med. Sci. Sports 1:15 (1969).
26. Porte, D., Jr. and Williams, R.H. Science 152:1250 (1966).
27. Bergström, J. and Hultman, E. Scand. J. Clin. Lab. Invest. 19:218 (1967).
28. Hermansen, L., Hultman, E. and Saltin, B. Acta Physiol. Scand. 71:129 (1967).
29. Miller, H.I., Issekutz, B., Jr., Paul, P. and Rodahl, K. Amer. J. Physiol. 207:1226 (1969).
30. Urgoiti, E.J., Houssay, B.A. and Rietti, C.T. Diabetes 12:301 (1963).
31. Hoch, F.L. Postgrad. med. J. 44:347 (1968).

FAT MOBILIZATION AND BLOOD LACTATE CONCENTRATION

Bertil B. Fredholm

Department of Pharmacology, Karolinska Institutet

104 01 Stockholm 60, Sweden

Free fatty acids (FFA) are taken up by different organs proportionally to their concentration in plasma (6), which is largely determined by their rate of mobilization from adipose tissue (6, 9). It is known that increasing sympathetic nervous activity is a potent stimulus to increased FFA mobilization (3, 9, 13, 14). Therefore increasing sympathetic tone would tend to cause increased uptake of FFA by skeletal muscle for instance. As was pointed out by Häggendal (7, 8) the sympathetic nervous activity is markedly elevated when maximal or supramaximal exercise is performed, yet it is known, since the work of Christensen (2), that carbohydrate is the predominant source of energy under these circumstances. It would thus seem that unless there is some physiological break mechanism the skeletal muscles performing heavy exercise could be flooded with fatty acids that they have no use for.

The finding of Issekutz and Miller that there exists an inverse correlation between the FFA and lactate level in blood indicated that lactate might function as a physiological inhibitor of FFA mobilization in severe exercise (10). Later studies by Miller et.al. (12) demonstrated that lactate infusion will decrease the inflow of FFA into plasma in depancreatized dogs. The authors speculated that this effect of lactate was due to an increased re-esterification of fatty acids in adipose tissue, but they gave no evidence for this proposed mechanism of action. Björntorp (1), on the other hand, showed that lactate in rat epididymal fat pads has a different effect, it inhibited lipolytic activity and glycerol production.

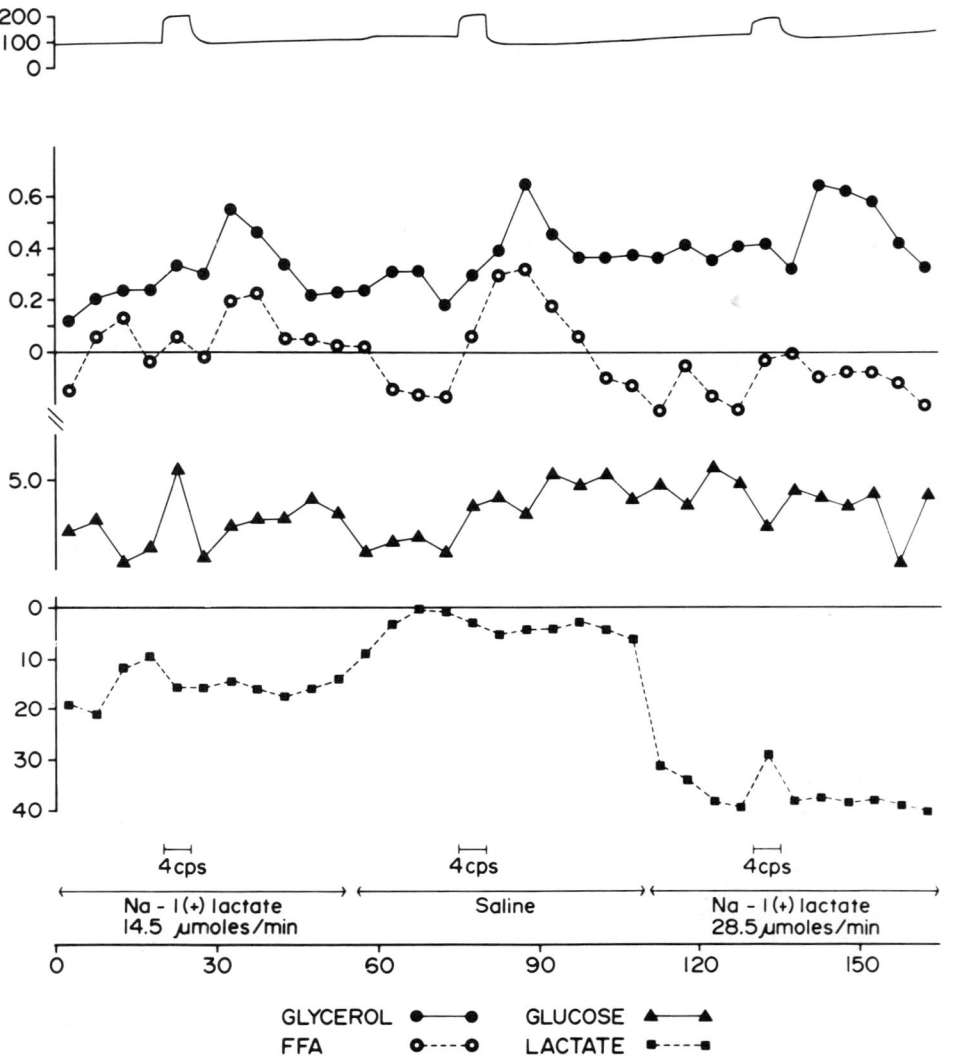

Fig. 1. Perfusion pressure (mmHg; upper panel), release of FFA and glycerol (μmoles/min/100g; second panel), uptake of glucose (μmoles/min/100g; third panel) and uptake of lactate (μmoles/min/100g; lower panel) during infusion of lactate (giving concentrations in blood of approximately 9 and 17mM) and saline. Time (min) is indicated at the bottom. From (4).

With these considerations as a background the following two questions were examined in the present report: Is it possible to antagonize the FFA mobilization induced by nerve stimulation with frequencies thought to occur during heavy exercise (see 14) by increasing the arterial lactate level to values reported for this condition? If so what is the mechanism of action of lactate?

As an experimental model the isolated, blood perfused canine subcutaneous adipose tissue described by Rosell (13) was used. The nerve to the tissue was stimulated for 5 to 10 min with 4 cps. The amount of FFA (10) and glycerol (11) released by nerve stimulation during infusion of lactate was compared with the amount released by a similar stimulation during saline infusion. In some experiments pyruvate and D(-) lactate were also infused. Further experimental details are presented elsewhere (4).

Fig. 1. illustrates the typical effect of increasing the arterial lactate concentration to about 10mM or more. During nerve stimulation the perfusion pressure increases, indicating an arteriolar vasoconstriction. Following the nerve stimulation there is an increased outflow of glycerol from the tissue. The magnitude of this response is not markedly different between stimulations. On the other hand the FFA response is definitely higher during saline than during either of the lactate infusions. There is no clearcut effect of lactate on the uptake of glucose in the experiment illustrated, and a statistical analysis of all experiments of this type failed to reveal any effect by lactate on the glucose uptake. On the other hand, it is quite clearly seen that the lactate uptake rate is dependent upon the rate of lactate administration. The cumulated results indicated an essentially linear relationship between lactate uptake and the arterial lactate concentration.

In five experiments the infusion of laevorotatory isomer had no effect whatsoever. On the other hand the infusion of pyruvate caused a decreased release of glycerol and in spite of this an increased mobilization of FFA, i.e. the opposite effect of lactate.

These results have been summarized in Fig. 2. First of all it is seen that stimulation during a second infusion of saline gave essentially identical release of glycerol and FFA as the first, indicating a good reproducibility of the nerve stimulations. The effect of lactate was clearly dose-dependent; 2 mM lactate had no effect and increasing the arterial lactate level by 5 mM had a borderline effect on the FFA release. On the other hand, lactate levels of 10 mM or more caused an

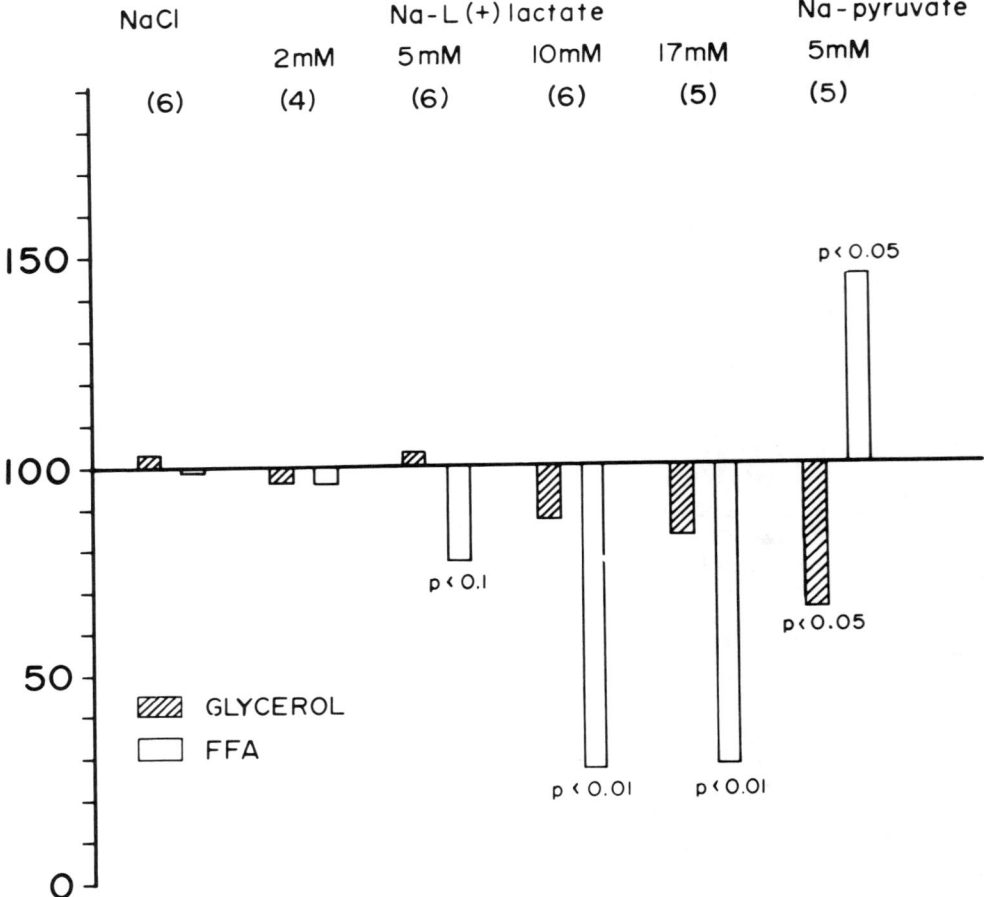

Fig. 2. Release of FFA and glycerol by nerve stimulation during infusion of NaCl, lactate or pyruvate in per cent of the amount released by a similar nerve stimulation during infusion of saline. Numbers within parenthesis denotes number of experiments in each series. The hypothesis tested by Student´s t-test for paired variates was that no change had occurred. From (4).

approximately 70 per cent inhibition of FFA mobilization during nerve stimulation.

A decreased mobilization of FFA in spite of an unchanged glycerol release is commonly equated with an increased re-esterification of fatty acids in the adipose tissue (15). It is possible that this is brought about by lactate being a substrate for glyceride-glycerol synthesis. This possibility is less likely for several reasons, most important of which is the fact that pyruvate had the opposite effect on re-esterification. If anything, pyruvate should be a better substrate for glyceride-glycerol biosynthesis (see 3).

Another possibility to explain the increased re-esterification of fatty acids brought about by lactate takes advantage of the fact that both the lactate-pyruvate and the α-glycerophosphate-dihydroxyacetone phosphate equilibria are coupled to NAD. Therefore it is possible that the lactate taken up by adipose tissue causes a reduction of NAD which shifts the second equilibrium to the left. This possibility was in fact also discussed by Miller et. al. (12).

Circumstantial evidence favouring this possibility was obtained by infusion of malate and oxaloacetate (components of the third major NAD dependent red-ox couple in the cytoplasm) which produced the expected changes in re-esterification; malate increased it and oxaloacetate caused a decrease.

It is of some interest in this connection that prolonged sympathetic nerve stimulation causes a net production and an accumulation of lactate in adipose tissue (calculated concentration in adipose tissue water more than 10 mM) and also apparently an increased re-esterification (5).

In Fig. 2. it is seen that the glycerol mobilization following nerve stimulation is inhibited by pyruvate and possibly also by high amounts of lactate. This could indicate that a direct antilipolytic effect of lactate might be present in canine adipose tissue in situ as was shown to be the case in rat fat pads in vitro (1). The fact that pyruvate was apparently more effective in this respect could be secondary to the greater uptake of pyruvate than lactate (4).

In summary, these experiments have furnished evidence that the FFA mobilization induced by increased sympathetic nerve activity can be counteracted by lactate in concentrations that are achieved in maximal exercise. Furthermore, the results demonstrate that increased re-esterification is the major mechanism of action, although some inhibition of lipolysis might also be present. The increased re-esterification might be secondary to an increased degree of reduction of the cytoplasmatic NAD.

REFERENCES

1. Björntorp, P., The effect of lactic acid on adipose tissue metabolism in vitro. Acta med. scand. 178: 253 (1965)

2. Christensen, E.H. and O. Hansen, Arbeitsfähigkeit und ernährung. Scand. Arch. Physiol. 81: 160 (1939)

3. Fredholm, B.B., Studies on the sympathetic regulation of circulation and metabolism in isolated canine subcutaneous adipose tissue. Acta physiol. scand. suppl. 79: 354 (1970)

4. Fredholm, B., The effect of lactate in canine subcutaneous adipose tissue in situ. Acta physiol. scand. In press

5. Fredholm, B. B. and J. Karlsson, Metabolic effects of prolonged sympathetic nerve stimulation in canine subcutaneous adipose tissue. Acta physiol. scand. 1970. In press

6. Fredrikson, D.S. and R. S. Gordon, Transport of fatty acids. Physiol. Rev. 38: 585 (1958)

7. Häggendal, J., Role of circulating noradrenaline and adrenaline. This volume

8. Häggendal, J., L.H. Hartley and B. Saltin, Arterial noradrenaline concentration during exercise in relation to the relative work levels. Scand. J. clin. Lab. Invest. 26: Nov (1970)

9. Havel, R.J., Autonomic nervous system and adipose tissue. In: Handbook of Physiology, sect. 5: 575 (1965)

10. Issekutz, B. Jr. and H. I. Miller, Plasma free fatty acids during exercise and the effect of lactic acid. Proc. Soc. exp. Biol. Med. 100: 237 (1962)

11. Laurell, S. and G. Tibbling, An enzymatic fluorimetric micromethod for the determination of glycerol. Clin. chim. Acta. 13: 317 (1966)

12. Miller, H.I., B. Issekutz, Jr., P. Paul and K. Rodahl, Effect of lactic acid on plasma free fatty acids in pancreatectomized dogs. Amer. J. Physiol. 207: 1226 (1964)

13. Rosell, S., Release of free fatty acids from subcutaneous adipose tissue in dog following sympathetic nerve stimulation. Acta physiol. scand. 67: 343 (1966)

14. Rosell, S., Adrenergic neuro-humoral control of lipolysis in adipose tissue. This volume.

15. Steinberg, D. and M. Vaughan, Release of free fatty acids from adipose tissue in vitro in relation to rates of triglyceride synthesis and degradation. In: Handbook of Physiology. 335 (1965)

16. Trout, D.L., H. Estes and S.J. Friedberg, Titration of free fatty acids in plasma: A study of current methods and a new modification. J. Lipid.Res. 1: 199 (1960)

Acknowledgements. These investigations were supported by grants from the Swedish Medical Research Council (40X-2553) and from Karolinska Institutet. I want to thank Miss Mona Engqvist and Miss Gunilla Wikberg for helping me in the experimental work.

CONTROL MECHANISMS FOR THE SYNTHESIS OF GLYCOGEN

IN STRIATED MUSCLE

    Sten Adolfsson and Kurt Ahrén

    Department of Physiology, University of Göteborg

    Göteborg, Sweden

    The key enzyme in the regulation of glycogen synthesis in muscle is the glycogen synthetase (UDP-glucose: α-1, 4-glucan α-4-glucosyltransferase; EC 2.4.1.11). It is well-established that many hormonal and non-hormonal factors which under physiological conditions stimulate or inhibit the rate of glycogen synthesis produce their effects by influencing, directly or indirectly, the activity of this enzyme. It is also well-known that the glycogen synthetase enzyme from muscle tissue, as well as from all other mammalian tissues heretofore investigated, exists in two forms, one (D-form) dependent upon and another (I-form) independent of glucose-6-phosphate (G-6-P) for its activity. These two forms are interconverted by a mechanism involving phosphorylation of the I-form in presence of a synthetase I kinase, and dephosphorylation of the D-form by a synthetase D phosphatase. The per cent synthetase I, observed in a tissue at any one time represents a balance between the activity of the phosphatase, catalysing D → I conversion, and the kinase, catalyzing I → D conversion. If this balance is disturbed by, for instance, a partial or total inactivation of the kinase, then the opposing action of the phosphatase becomes dominant and there is a net increase in the I-form. On the other hand, if the phosphatase is inhibited, then the action of the kinase becomes dominant with an increase in the D-form as a result. Changes in the per cent I-form produced by hormonal or non-hormonal factors can therefore be produced via influences on either the kinase or the phosphatase. It has recently been suggested that the synthetase I kinase might exist in two forms, one requiring cyclic 3´,5´-AMP for its activity and the other active in the absence of this co-factor (32, 20), and that, at least in some tissues, also the synthetase D phosphatase might exist in two forms (13). General aspects upon the synthetase enzyme system as well as references to relevant literature can be found in a recent review article (18).

In 1960-1961 Larner and co-workers (29, 30) reported that extracts prepared from rat hemidiaphragms incubated with insulin showed increased glycogen synthetase activity when measured in the absence of G-6-P ($\underline{I}$-activity). When measured in the presence of G-6-P (total activity, i. e. $\underline{I} + \underline{D}$ activity) no differences were found between extracts of control and insulin treated diaphragms. The increase in per cent $\underline{I}$-form of the synthetase enzyme after insulin treatment has since then been confirmed in many laboratories, and the observation has been extended to many types of muscles (18).

With the above-mentioned results as a background, it has become a widely accepted concept that the site of action of insulin in its stimulatory effect on glycogen synthesis is the synthetase enzyme, and that an increased $\underline{I}$-form of this enzyme is a requirement for an insulin stimulated glycogen synthesis. It has meanwhile been shown that many other hormones, which stimulate glycogen synthesis in their target organs, e. g. oestrogens in the uterus (33), also can produce an increase in the percentage $\underline{I}$-form of the glycogen synthetase, and the view is often seen in the literature that all physiological adjustments leading to an increased rate of glycogen synthesis might be explained, on the molecular level, by such an increase in the synthetase $\underline{I}$. This view seems, however, to be incorrect since there are without doubt conditions when glycogen synthesis is stimulated without a concomitant increase in synthetase $\underline{I}$. One example is the markedly increased glycogen synthesis found for several days in human skeletal muscles after vigorous exercise (12, 16). During this stimulated glycogen synthesis, the $\underline{I}$-form of the synthetase enzyme seems to be increased only during the first hours after the exercise (16). There is also one report in the literature which is not in accordance with the theory that $\underline{D} \rightarrow \underline{I}$ conversion is a requirement for an insulin stimulated glycogen synthesis, viz. that of Søvik from 1966 (27). He reported that the percentage of synthetase $\underline{I}$ was not increased in the rat diaphragm muscle after an intraperitoneal (i. p.) injection of insulin although the glycogen synthesis was highly stimulated.

We have in this laboratory during recent years studied glycogen synthesis and the synthetase enzyme in an *in vitro* preparation of the levator ani muscle of young male rats. With the technique used in this laboratory, this muscle can rapidly be prepared as an *in vitro* preparation without damage to the muscle fibres (10), and this preparation has already been shown to be suitable for several types of *in vitro* investigations of muscle metabolism (9). Effects of insulin and testosterone on glycogen synthesis in this muscle have recently been studied. Our main purpose with these investigations has been to study in detail the time-sequence of the changes in glycogen synthesis rate and synthetase enzyme activities after stimulation by these hormones. We have also, with the same muscle preparation, started to explore the effect of cyclic 3´, 5´-AMP and of electrical stimulation *in vitro*. The results of some of our experiments will be summarized in this paper, and the results will then be taken as starting point for a further discussion of the cellular mechanisms for the control of glycogen synthesis in muscle. The above-mentioned experiment of Søvik (27) on the rat diaphragm muscle has also been repeated with other time observations.

MATERIALS AND METHODS

Immature male rats weighing 50-70 g of the Sprague-Dawley strain were used. They were maintained on a semisynthetic diet but deprived of food 20-24 hr before the experiment. The intact levator ani muscle was prepared and incubated as previously described (10), i.e. the levator ani muscle was incubated in connection with the bulbocavernosus muscles in 1 ml medium in 10 ml flasks at 37°C with constant shaking. The medium was Krebs bicarbonate buffer (pH 7.4) gassed with 95 % O_2 and 5 % CO_2. The procedure for electrical stimulation of the isolated levator ani muscle has been described in detail by Arvill (8). Insulin was added to the incubation medium in a concentration of 10 mU/ml and glucose in a concentration of 2 mg/ml. A ten-fold recrystallized pig insulin was used (Novo S 23267). In some experiments, ^{14}C-labelled glucose was also added to the medium (uniformly labelled glucose-^{14}C; 2 μC/ml medium; 3 $\mu C/\mu$mole) in order to measure the rate of incorporation into glycogen.

After incubation the levator ani muscle (weighing 10-15 mg) was dissected from the rest of the bulbocavernosus complex and homogenized in an all glass homogenizer at 0°C. The synthetase enzyme activity was analysed in the supernatant after centrifugation at 5000 x g in absence (= I-form) and presence (= I + D forms) of 10 mM G-6-P, following a procedure essentially similar to that used by Larner and co-workers (31). For the determination of glycogen, the muscle was boiled in KOH and the glycogen precipitated with ethanol-Na_2SO_4. The glycogen was then split enzymatically into glucose, and the glucose was assayed by the glucose oxidase method. Glycogen from rabbit liver was used as a standard. In the experiments with ^{14}C-labelled glucose, the radioactivity of the precipitated glycogen was measured in a liquid scintillation spectrometer (Packard Tri-Carb). G-6-P was determined enzymatically in a modification of the method described by Hohorst (15).

In the experiment with the diaphragm, insulin was injected i.p. (10 mU/100 g body weight) 10 or 100 min prior to extirpation of the diaphragm. One half of the diaphragm was homogenized and analysed as described for the levator ani muscle, the other half was incubated for 60 min as an intact hemidiaphragm preparation in a medium containing ^{14}C-labelled glucose.

In the figures and the table, mean values $\pm$ standard error of the mean are given. Some of the experiments have previously been briefly reported (28, 1, 3, 4), and all methods will be given in detail in a forthcoming paper (2).

RESULTS

Insulin and the Levator Ani Muscle

The left sections of Figs 1 and 2 summarize experiments where the muscles were incubated for 30 and 210 min, respectively, in medium containing glucose. The right sections show experiments where the muscles have been incubated without glucose for 210 min.

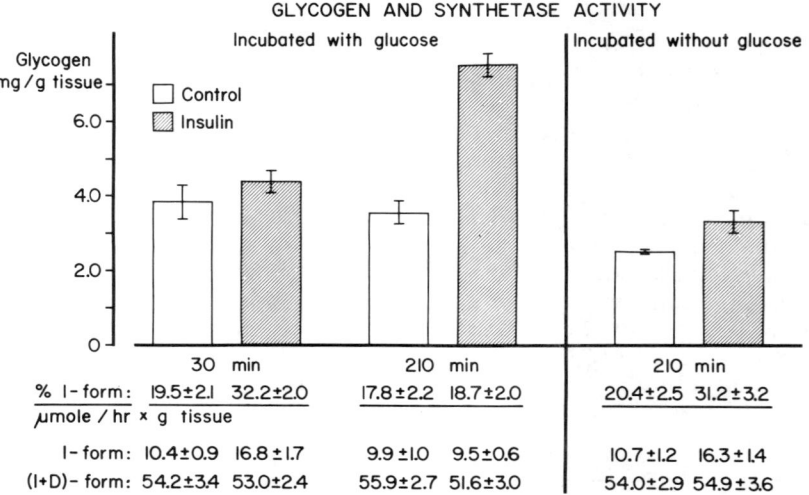

Fig. 1. Effects of insulin on glycogen content and synthetase activity in the isolated levator ani muscle. The left and the right parts of the figure show experiments with and without glucose (2 mg/ml) in the medium, respectively. The bars show the glycogen content of the muscles. The values at the lower part of the figure are the enzyme activities (μmole/hr x g tissue) in absence (I-form) and presence (I + D forms) of 10 mM G-6-P. The values for percentage of synthetase I are also given.

With glucose in medium. Fig. 1 shows glycogen contents and synthetase activities. It can be seen that the glycogen content was not yet significantly increased in the insulin group after 30 min of incubation. The per cent I-form of the synthetase was, however, markedly increased. After 210 min of incubation, insulin had doubled the glycogen content but the per cent synthetase I was now the same in the insulin and the control groups. Fig. 2 shows that the incorporation of glucose-^{14}C into glycogen was still markedly increased in the insulin group when the muscles were incubated with this radioactive isotope in the medium during a period 210-240 min after the start of incubation, indicating an increased rate of glycogen synthesis also after 210 min of incubation with insulin. In the latter experiments, glucose-^{14}C was added to the medium 210 min after the start of incubation, and the radioactivity of muscle glycogen was analysed after an additional incubation of the preparation for 30 min.

These experiments illustrate that insulin increases the per cent of synthetase I in the isolated levator ani muscle after "short-term" incubations (30 min). Quite a significant increase in the I-form has, in fact, been observed after only 5 min of incubation with insulin. The more important

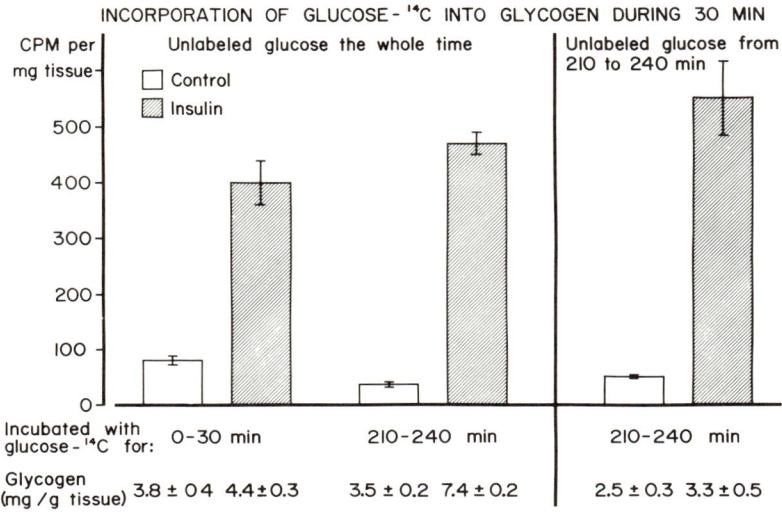

Fig. 2. Effects of insulin on incorporation of glucose-^{14}C into glycogen of the isolated levator ani muscle. The left part of the figure shows experiments where glucose (2 mg/ml) was present in the medium during the whole incubation period. The right part shows experiments where the muscles were incubated for 210 min without any glucose in the medium but where the muscles were then incubated for an additional period of 30 min (210-240 min) with both unlabelled glucose (2 mg/ml) and ^{14}C-labelled glucose present. The first pair of bars show experiments where the muscles were incubated for 30 min with glucose-^{14}C present during the whole incubation period. The next two pairs of bars show experiments where the muscles were first incubated for 210 min without ^{14}C-labelled glucose in the medium. The muscles were then transferred to new flasks with media containing glucose-^{14}C (with and without insulin), and incubated for 30 min (total incubation time thus 240 min).

observation is, however, that insulin after "long-term" incubation (210 min) still stimulates glycogen synthesis, but now without an increase in the synthetase $\underline{I}$. It can be seen from Fig. 1 that there was no increase in total synthetase activity measured in the presence of 10 mM G-6-P ($\underline{I}$ + $\underline{D}$ forms) either after 30 min of incubation or after 210 min.

In another series of experiments, G-6-P content was analysed after incubations of 30, 120 and 240 min. The G-6-P concentrations were calculated as μmole/ml intracellular water. Values for intracellular water were taken from previous studies on the levator ani muscle (10). The value for control (C) and insulin stimulated (S) muscles were: 30 min C = 0.28 ± 0.02

and S = 0.32$\pm$0.02; 120 min C = 0.23$\pm$0.02 and S = 0.33$\pm$0.02; and 240 min C = 0.16$\pm$0.01 and S = 0.35$\pm$0.04. There was thus a significant difference between control and insulin stimulated muscles after 120 and 240 min of incubation.

Without glucose in medium. The right section of Fig. 1 shows that the glycogen content of muscles incubated for 210 min with insulin but without glucose in the medium, was quite low compared to the muscles incubated for 210 min with glucose and insulin in the medium. It can also be seen that a clear increase in synthetase I was found in these muscles incubated for 210 min without glucose but with insulin in the medium.

Insulin Intraperitoneally and the Diaphragm

In one experiment, insulin was given 10 min before removal of the diaphragms in a single injection of 10 mU/100 g body weight which is the same dose as that used by Søvik (27). Glycogen content was 2.41$\pm$0.28 and 3.00$\pm$0.42 mg/g tissue in diaphragms from control and insulin injected rats, respectively. There was a marked increase in the I-form of the synthetase enzyme in the insulin group. The mean value for the control group was 16.1$\pm$3.6 per cent, and that for the insulin group 36.6$\pm$3.2 per cent. Some diaphragms were taken from both groups of rats and incubated for 60 min with glucose-^{14}C in the medium. The muscles from the insulin injected rats showed a much higher incorporation of ^{14}C into glycogen than the muscles from the control rats.

In another experiment, the same dose of insulin was injected 100 min before the removal of the diaphragms. In the control group the glycogen content was 2.20$\pm$0.35 mg/g tissue, and in the insulin group the corresponding value was 9.50$\pm$1.10. The per cent synthetase I was, however, the same in both groups (15.3$\pm$1.6 per cent in the controls, and 13.8$\pm$3.1 per cent in the insulin group). Also in this experiment a group of diaphragms were taken for incubation for 60 min with glucose-^{14}C, and the muscles from the insulin injected animals still showed a highly significant increased incorporation of ^{14}C into glycogen, indicating that glycogen synthesis was stimulated in these diaphragms although no increase in the synthetase I was detected.

Testosterone and the Levator Ani Muscle

It has been known for quite a long time that testosterone is an important regulator of carbohydrate metabolism of the levator ani muscle (6,9,11). It has previously been reported from this laboratory that an injection of testosterone proprionate 24 hr prior to removal of the levator ani muscles markedly stimulated the rate of uptake of xylose-^{14}C in the isolated preparation during a subsequent incubation period (7). No effects were seen when testosterone was added directly to the incubation medium (7).

In the present experiments, testosterone proprionate was injected intramuscularly to the rats 10 and 24 hr, respectively, before the extirpation of the levator ani muscle. Glycogen content and synthetase activities were analysed. The results are seen in the lower part of Fig. 3. The bars of the figure show the incorporation of glucose-^{14}C into glycogen in groups of

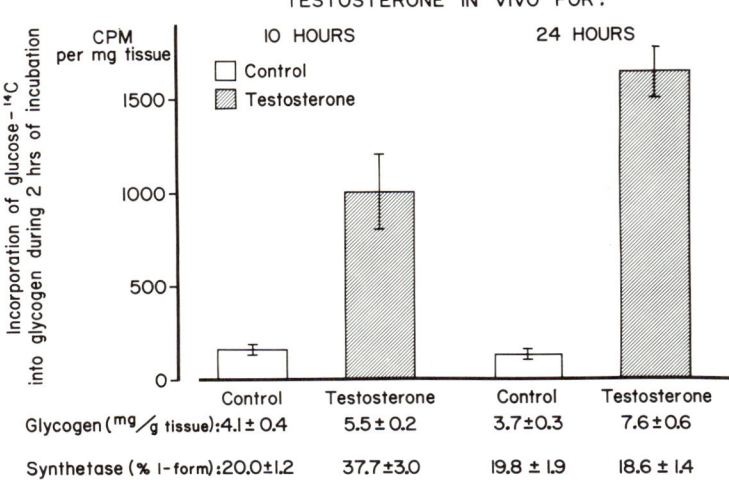

Fig. 3. Effects of testosterone on glycogen synthesis and synthetase activities in the levator ani muscle of immature male rats. Testosterone proprionate (100 mg/kg body weight) was given in one intramuscular injection 10 and 24 hr, respectively. before the extirpation of the muscles. Values for glycogen content and synthetase activity (per cent I-form) are given in the lower part of the figure. The bars in the upper part of the figure show the incorporation of glucose-^{14}C into glycogen during a 2 hr incubation period of the muscle preparations after the 10 and 24 hr in vivo stimulation by testosterone.

muscles incubated for 2 hr with glucose-^{14}C in the medium after the in vivo treatments with testosterone. It can be seen that after "short-term" stimulation by testosterone (10 hr) there was a marked increase in the synthetase I while after "long-term" stimulation (24 hr), no such increase in the I-form was seen in spite of a still highly stimulated glycogen synthesis.

G-6-P contents were also measured in this experiment. Significantly higher levels were found in the testosterone stimulated muscles both after 10 and 24 hr of testosterone treatment. Control values were 0.53 ± 0.05, and the values after 10 and 24 hr of testosterone treatment were 1.14 ± 0.09 and 1.03 ± 0.05, respectively (expressed as μmole G-6-P/ml intracellular water).

Cyclic 3´,5´-AMP and Levator Ani Muscle

The results are summarized in Table I. After 30 min of incubation with 1 mM dibutyryl cyclic 3´,5´-AMP, there was no significant decrease in glycogen content but there was a marked decrease in the I-form of the

Table I. Effects of dibutyryl cyclic 3´,5´-AMP on glycogen content and synthetase activities in the isolated levator ani muscle.

	Control	3´,5´-AMP
Exp. A: Incubation time 30 min		
Glycogen content mg/g tissue	3.2 ± 0.1	3.1 ± 0.3
Synthetase activity % I-form	23.2 ± 2.6	9.7 ± 1.3
Exp. B: Incubation time 120 min		
Glycogen content mg/g tissue	2.8 ± 0.2	1.1 ± 0.1
Synthetase activity % I-form	17.8 ± 0.9	12.8 ± 1.4

The muscles were incubated for 30 min (Exp. A) and 120 min (Exp. B) in medium containing glucose (2 mg/ml) in presence or absence of 1 mM dibutyryl cyclic 3´,5´-AMP.

synthetase enzyme. After 120 min of incubation with the same concentration of 3´,5´-AMP, glycogen content was markedly decreased but there was still a decreased per cent of the I-form of the synthetase enzyme.

Electrical Stimulation of the Levator Ani Muscle

The isolated muscles were stimulated electrically for 10 min in vitro as described by Arvill (8) with a frequency of 100 impulses/sec. Some of the muscles were taken for analyses of glycogen content and synthetase activities directly after this stimulation period (Fig. 4). Other muscles were incubated, without further stimulation, in medium containing glucose for 210 min after the stimulation period (Fig. 4).

It can be seen that synthetase I was increased in the muscles analysed directly after the period of electrical stimulation. An increased rate of glycogen synthesis immediately after the electrical stimulation was shown in experiments with glucose-^{14}C. No increase in the I-form of the synthetase enzyme was seen in the muscles analysed 210 min after the period of stimulation. The rate of glycogen synthesis was, however, increased also 210 min after the electrical stimulation. This was shown in experiments where the muscles were incubated with glucose-^{14}C for a period of 30 min, i.e. 210-240 min after the electrical stimulation.

DISCUSSION

The experiments with insulin and testosterone stimulation of the levator ani muscle illustrate two important points in the relationship between the rate of glycogen synthesis and the change in the I-form of the synthetase enzyme. First, both hormones show in an early phase of stimulation a con-

CONTROL MECHANISMS FOR THE SYNTHESIS OF GLYCOGEN

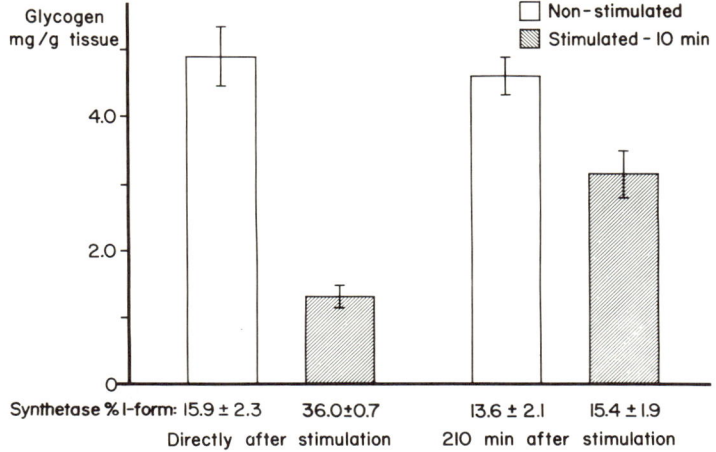

Fig. 4. Effects of electrical stimulation on glycogen content and synthetase activities in the isolated levator ani muscle. All muscles were electrically stimulated as described by Arvill (8) for 10 min with a frequency of 100 impulses/sec. Some of the muscles were analysed directly after this stimulation period (left part of the figure). Other muscles were first incubated for 210 min in medium containing glucose after the period of electrical stimulation (right part of the figure).

comitant increase in rate of glycogen synthesis and in per cent of synthetase present as I-form. This observation is thus in accordance with the theory that various hormones with an anabolic effect on glycogen metabolism act on the synthetase enzyme system by producing either an increased $D \rightarrow I$ conversion and/or a decreased $I \rightarrow D$ conversion. Secondly, both hormones show in a later phase of the stimulation a still increased rate of glycogen synthesis without any increase in synthetase I. This observation illustrates that the above-mentioned mechanism (increase in I-form) cannot be the only one for hormones stimulating glycogen synthesis. The pattern in the relationship between glycogen synthesis and the synthetase enzyme is the same after insulin and testosterone stimulation with an "early phase", characterized by increased glycogen synthesis and elevated tissue level of synthetase I, and a "late phase", characterized by increased glycogen synthesis without elevation of tissue synthetase I. The time-scale for the early and the late phases in the stimulation of glycogen synthesis is quite different for insulin and testosterone but the general pattern is the same. Our working hypothesis at present is that the cellular mechanisms behind this pattern are the same for insulin and testosterone. It is quite possible that the same pattern exists also in the stimulation of glycogen synthesis by other hormones.

It was mentioned already in the introduction that an increase in the I-form of the synthetase can be produced either by an inhibition of the synthetase I kinase or by a stimulation of the synthetase D phosphatase. Experiments of Larner and co-workers (18) indicate that the insulin action is on the kinase enzyme. Whether testosterone in its early phase of stimulation has its action on the kinase or on the phosphatase has not yet been studied.

The observation of the present study that addition of dibutyryl cyclic 3´,5´-AMP to the isolated levator ani muscle decreased the synthetase I, is in accordance with the results of various types of experiments on other tissues showing that cyclic 3´,5´-AMP can stimulate synthetase I kinase activity (26). One possibility is therefore that the effects of insulin and/or testosterone on the synthetase enzyme system are produced via a decrease in the cellular level of cyclic 3´,5´-AMP. This possibility for the action of insulin has been considered by Larner and co-workers (18) in experiments on rat diaphragms in vitro and rat skeletal muscles in vivo. Under both experimental conditions, these investigators found an increased muscle level of synthetase I without any decrease in muscle concentration of cyclic 3´,5´-AMP; in some experiments the opposite was found, i.e. a slight increase in cyclic 3´,5´-AMP. It is therefore not likely that the insulin effect on muscle synthetase is produced via alterations in the intracellular concentration of this nucleotide. That the well-known epinephrine produced decrease in synthetase activity is mediated via an increase in cyclic 3´,5´-AMP is quite likely, and experimental evidence exists in favour of such a hypothesis (18). Whether testosterone can modify the muscle metabolism of cyclic 3´,5´-AMP has not yet been studied.

The possibility that synthetase I kinase in the cells exists in two (interconvertible?) forms, one requiring cyclic 3´,5´-AMP for its activity and the other active also in the absence of this nucleotide, was mentioned in the introduction. Larner and co-workers recently reported (18) the interesting observation that insulin decreased in rat skeletal muscles the activity of the cyclic 3´,5´-AMP independent kinase, while muscles from control and insulin treated rats showed the same kinase activity when tested in the presence of cyclic 3´,5´-AMP. The authors concluded from these experiments that "insulin acts at the kinase site to bring about a greater dependence on cyclic AMP and thus inactivate the kinase, with no decrease in cyclic adenylate tissue concentrations" (18).

It may seem surprising that the "late phase" in insulin stimulation of muscle glycogen synthesis described in the present experiments on the isolated levator ani muscle and characterized by an accelerated glycogen synthesis without increase in synthetase I, has not been discovered and described earlier. The explanation for this is probably that Villar-Palasi and Larner (29, 30) made their original discovery of an increased percentage synthetase I after insulin stimulation in short-time incubation experiments (up to 30 min), and the interest has since then been focused more on the cellular mechanisms behind this change in the synthetase enzyme system than on long-time incubation experiments with insulin. Furthermore, when the insulin effect has been studied in vivo after subcutaneous or intramuscular administration, it has been considered most important to study the muscle synthetase system as short time as possible after the injection of insulin to avoid secondary adjustments due to the hypoglycemia. When given i.p., insulin produces, however, a very marked increase in diaphragm muscle glycogen without any significant effect on blood glucose, as was first shown by Rafaelsen (24). Søvik used this i.p. method of insulin administration, and studied glycogen content and the synthetase enzyme in rat diaphragm muscles 30, 60, 120 and 240 min after the insulin injection (27). He found a drastic increase in glycogen content already 30 min after injection, but there was no increase in synthetase $\underline{I}$ at any of the points studied. Søvik's interpretation of the results was that either "the increase in synthetase $\underline{I}$ might be reversed during preparation and handling of the tissue", or "the effect of insulin on synthetase $\underline{I}$, though present in vitro, might not occur in the intact rat organism" (27). We have now shown, that synthetase $\underline{I}$ is, in fact, increased in the rat diaphragm 10 min after i.p. injection of insulin. There is thus an "early phase" of accelerated glycogen synthesis in the rat diaphragm muscle with increased synthetase $\underline{I}$ activity also after i.p. injection of insulin, but this phase is, after this method of administration, shorter than that seen after in vitro addition of insulin to isolated diaphragms. This is not surprising when one considers how rapidly the glycogen content increases in the diaphragm muscle after i.p. injection compared with the conditions in vitro.

The very rapid increase in glycogen content in the rat diaphragm after i.p. injection of insulin might, in fact, have caused the rapid return of synthetase $\underline{I}$ to the control level, since a negative feed back regulation by glycogen on the synthetase enzyme system has been described. This "end product inhibition" was first shown by Danforth (14) in experiments on mouse muscles in vivo and rat diaphragm muscles in vitro. Danforth found an inverse relationship between the glycogen content and the $\underline{I}/\underline{I} + \underline{D}$ ratio of the glycogen synthetase under various experimental conditions. The same relationship has since then been reported for the perfused rat heart (18), and also for other tissues than muscles, e.g. in tissue cultures of HeLa cells (5), and in incubated ascites hepatoma cells (25). The last mentioned cells were reported (25) to lack glycogen when they were obtained from the rats, and they contained then glycogen synthetase nearly completely as $\underline{I}$-form. On incubation of these cells in medium containing glucose, glycogen synthesis was very rapid in the beginning but declined with time. In the meantime, the $\underline{I}/\underline{I} + \underline{D}$ ratio of the synthetase decreased, and the cells showed finally an almost total conversion of the synthetase to a G-6-P dependent form ($\underline{D}$-form). Villar-Palasi and Larner (18) have shown in experiments on homogenized rat muscles, that the effect of glycogen on the synthetase enzyme is brought about by an inhibition of the synthetase $\underline{D}$ phosphatase without any effect on the synthetase $\underline{I}$ kinase. It seems therefore quite likely that the cellular factor that determines the length of the "early phase" of glycogen stimulation after both insulin and testosterone is the muscle cell concentration of glycogen. This assumption is supported, for the insulin action, by the finding of the present study that levator ani muscles incubated with insulin but without glucose showed an increased synthetase $\underline{I}/\underline{I} + \underline{D}$ ratio still after 210 min of incubation, while this ratio had returned to control values at this point in muscles incubated with glucose in the medium (Fig. 1).

When Danforth (14) reported the first evidence for the above-mentioned "end product inhibition" by glycogen on muscle synthetase $\underline{I}$, then his conclusion was that this inhibition represented a very important physiological control mechanism "to favor glycogen synthesis when tissue concentrations of glycogen are low, and to slow synthesis when tissue concentrations are high". This view has since then markedly dominated the literature within this field of research, but the view has also been called in question by a few investigators, e. g. by Piras, Rothman and Cabib (22). These authors mean, that a "precursor activation" (see below) is a much more important mechanism for the control of glycogen synthesis under most physiological conditions and adjustments. We share this view of Piras, Rothman and Cabib (22), and we think that many other authors in their enthusiasm over the "end product inhibition theory" have overlooked what was shown and what was not shown in the experiments of Danforth (14) and Villar-Palasi and Larner (18), *viz.* that there exists in muscles an inverse relationship between glycogen content and the percentage of the synthetase present as $\underline{I}$-form, and that glycogen in muscle homogenates can inhibit the synthetase $\underline{D}$ phosphatase, thus decreasing the $\underline{D} \rightarrow \underline{I}$ conversion. This means that end product inhibition by glycogen has up to now been shown only for the portion of glycogen synthesis catalyzed by synthetase $\underline{I}$, but not for the portion catalyzed by synthetase $\underline{D}$.

The question then arises whether synthetase $\underline{D}$ is active under *in vivo* conditions. This question was for a long time very puzzling since it was, in fact, found that both synthetase $\underline{D}$ and synthetase $\underline{I}$ were nearly completely active when tested *in vitro* under "physiological" conditions of pH and substrate concentrations in the presence of G-6-P concentrations in the range found *in vivo* in muscles (23). It was thus difficult to understand the physiological importance of the $\underline{D} - \underline{I}$ interconversion. A partial solution to this problem was given by the finding that several metabolites (ATP, ADP, inorganic phosphate etc.) can inhibit the activity of both synthetase $\underline{I}$ and synthetase $\underline{D}$. ATP seems to be the strongest inhibitor under physiological conditions. G-6-P reverses this inhibition. At low concentration of G-6-P, within the physiological range, the $\underline{D}$-form is much more strongly inhibited by the above-mentioned metabolites than the $\underline{I}$-form. $\underline{D} \rightarrow \underline{I}$ conversion will under these conditions lead to increased total activity of the synthetase enzyme, and the differential inhibition by ATP and other metabolites provides therefore the rationale for the functionality of the D — I interconversion (21, 23).

Whether synthetase $\underline{D}$ in muscle tissue can be active under physiological *in vivo* conditions cannot yet be unequivocally answered. Some experiments, as those by Piras and Staneloni (23), indicate that the $\underline{D}$-form can be active in rat skeletal muscles under *in vivo* conditions. These investigators studied glycogen synthesis, the synthetase system and various metabolites (G-6-P, ATP, ADP etc.) in rat skeletal muscles before, during and after a short *in vivo* electrical stimulation. They also compared, in elegant but very complex experiments, the rate of glycogen synthesis *in vivo* with that obtained *in vitro* under conditions where they tried to simulate the concentrations of metabolites and the forms of the glycogen synthetase found at rest, during contraction, and during recovery in the *in vivo* experiments. It was found in these experiments, a) that the changes in the rate of glycogen synthesis registered during various phases of the in vivo experiment could not be explained solely as a consequence of changes in ratio between synthetase I and synthetase D; b) that the changes in concentration of G-6-P on the one hand and other metabolites on the other hand were of the magnitude that synthetase D might have been active during certain phases of the experiment; c) that these changes in metabolite concentrations might also have changed the activity of the synthetase I. It seems therefore likely from these experiments that synthetase D in muscle tissue can be active under physiological conditions, but the interpretation of the results is difficult since changes in the enzyme activity of synthetase I cannot be excluded. It is our opinion that the possibility that the activity of synthetase I under physiological in vivo conditions might vary with the tissue concentrations of G-6-P, must be taken into serious consideration in future studies. If this possibility can be proved to be true, then the validity of the present terminology (synthetase D and synthetase I) must be questioned, as has, in fact, already been done for the synthetase enzyme system in the liver (19).

In our own experiments with the isolated levator ani muscle, an increased rate of glycogen synthesis and an increased I/I + D ratio was found directly after a period of in vitro contractions (Fig. 4). This finding is in accordance with that found after in vivo stimulation of rat skeletal muscles (14). The finding that there was still an enhanced glycogen synthesis 210 min after in vitro electrical stimulation although the percentage of synthetase I was no longer increased (Fig. 4), is in accordance with the conditions in human skeletal muscles 1-2 days after vigorous exercise (16).

It is thus clear that there exists an "early phase" with increased I/I + D ratio and a "late phase" without such an increase, in the accelerated glycogen synthesis both after hormonal (insulin and testosteron) and non--hormonal (muscle contraction) stimulation. It is most likely that the enhanced glycogen synthesis during the "late phase" is caused by changes in the balance between intramuscular metabolites which probably can modify the activity of both synthetase D and synthetase I. Metabolite concentrations must be studied in detail in further analyses of the "late phase" of glycogen stimulation. It can be seen from the present results that G-6-P was increased in the "late phase" both after insulin and testosterone stimulation.

A mechanism causing the muscle cell to increase total glycogen synthetase activity involving neither increased D → I conversion nor increased activation of the D- or I-forms by metabolites, is an increased de novo synthesis of enzyme protein. Such a mechanism seems not to have been used during any of the phases of accelerated glycogen synthesis studied in the present experiments, since the total synthetase activity as assayed in the presence of 10 mM G-6-P was not increased during any of the experimental conditions studied. Such an increased total synthetase activity (I + D forms) has, however, been reported in skeletal muscles of trained guinea--pigs compared to muscles from untrained animals (17).

It can thus be concluded that there exist in the muscle cells, in principle, three mechanisms for increasing the glycogen synthetase activity:

1) a disturbance in the I ⎯ D interconversion favouring increased levels of the I-form; 2) changes in the metabolite concentrations (probably mainly G-6-P and ATP) in such a way that synthetase D and perpahs also synthetase I become more active; and 3) increased de novo synthesis of enzyme protein. The two first mentioned mechanisms seem to be the most important as bases for the regulation of glycogen synthesis under most physiological adjustments, and both these mechanisms seem to be active concomitantly having the effect of switching on and off glycogen synthetase activity according to the physiological situation.

ACKNOWLEDGEMENTS

This research was supported by grants from the Swedish Medical Research Council (Project 14X-27); Bergvall's Foundation, Stockholm; Nathhorst's Foundation, Stockholm; Swedish Diabetic Association; Nordisk Insulinfond, Denmark; and the Medical Faculty, University of Göteborg, Sweden. Crystalline insulin was kindly provided by Novo Research Institute, Copenhagen. Valuable technical assistance was given by Miss Marianne Dahlberg. We also wish to thank Mrs Gun Derevall and Mrs Gunilla Nordling for expert secretarial assistance in compiling this manuscript.

REFERENCES

1) ADOLFSSON, S. Effect of electrical stimulation on glycogen synthesis in an isolated muscle preparation. Acta Physiol. Scand. 79: 40A, 1970.
2) ADOLFSSON, S. To be published.
3) ADOLFSSON, S. and K. AHRÉN. Biphasic action of testosterone on muscle glycogen synthetase. Acta Physiol. Scand. 74: 30A-31A, 1968.
4) ADOLFSSON, S. and K. AHRÉN. Early and late effects of insulin on muscle glycogen synthesis. In: 7th Congr. Intern. Diab. Fed. , Excerpta Med. Intern. Congr. , Ser. No. 209: 51, 1970.
5) ALPERS, J. B. The influence of hexose and insulin on glycogen synthetase in HeLa cells. J. Biol. Chem. 241: 217-222, 1966.
6) APOSTOLAKIS, M. , D. MATZELT and K. D. VOIGT. Der Einfluss von Testosteronpropionat auf glykolytische und transaminierende Enzymaktivitäten in Leber, Musculus biceps femoris und Musculus levator ani der Ratte. Biochem. Z. 337: 414-424, 1963.
7) ARVILL, A. Effects of testosterone on the metabolism of the isolated levator ani muscle of the rat. Acta Endocr. 56: Suppl. 122: 1-14, 1967.
8) ARVILL, A. Relationship between the effects of contraction and insulin on the metabolism of the isolated levator ani muscle of the rat. Acta Endocr. 56: Suppl. 122: 27-42, 1967.
9) ARVILL, A. Hormonal effects on the in vitro metabolism of a new intact mammalian muscle preparation. Thesis. Göteborg 1967.
10) ARVILL, A. and K. AHRÉN. The levator ani muscle of the rat as an intact preparation suitable for in vitro investigations. Acta Endocr. 52: 325-336, 1966.

11) BERGAMINI, E., G. BOMBARA and C. PELLEGRINO. The effect to testosterone on glycogen metabolism in rat levator ani muscle. Biochim. Biophys. Acta 177: 220-234, 1969.

12) BERGSTRÖM, J. and E. HULTMAN. Muscle glycogen synthesis after exercise: an enhancing factor localized to the muscle cells in man. Nature 210: 309, 1966.

13) BISHOP, J.S. and K. ARMSTRONG. Rapid inactivation by glucagon of insulin-glucose activated liver glycogen synthetase D phosphatase in vivo. Fed. Proc. 29: 860, 1970.

14) DANFORTH, W.H. Glycogen synthetase activity in skeletal muscle. Interconversion of two forms and control of glycogen synthesis. J. Biol. Chem. 240: 588-593, 1965.

15) HOHORST, H.J. D-Glucose-6-phosphat und D-Fructose-6-phosphat. In: Methoden der enzymatischen Analyse. Ed. H.U. Bergmeyer. Verlag Chemie, Weinheim. 134-138, 1962.

16) HULTMAN, E. Studies on muscle metabolism of glycogen and active phosphate in man with special reference to exercise and diet. Scand. J. Clin. Lab.Invest. 19: Suppl. 94: 1-63, 1967.

17) LAMB, D.R., J.B. PETER, R.N. JEFFRESS and H.A. WALLACE. Glycogen hexokinase, and glycogen synthetase adaptations to exercise. Amer. J. Physiol. 217: 1628-1632, 1969.

18) LARNER, J., C. VILLAR-PALASI, N.D. GOLDBERG, J.S. BISHOP, F. HUIJING, J.I. WENGER, H. SASKO and N.E. BROWN. Hormonal and non-hormonal control of glycogen synthesis - control of transferase phosphatase and transferase I kinase. In: Progress in Endocrinology. Ed. C. Gual. Excerpta Med. Found. Amsterdam. 135-147, 1969.

19) MERSMAN, H.J. and H.L. SEGAL. An on-off mechanism for liver glycogen synthetase activity. Proc. Nat. Acad. Sci. (USA). 58: 1688-1695, 1967.

20) NUTTALL, F.Q. and J. LARNER. Perfusion induced increase in 3´,5´cyclic AMP dependence of glycogen transferase I kinase. Fed. Proc. 29: 616, 1970.

21) PIRAS, R., L.B. ROTHMAN and E. CABIB. Metabolite regulation of the I and D form of rat muscle glycogen synthetase. Biochem. Biophys. Res. Commun. 28: 54-57, 1967.

22) PIRAS, R., L.B. ROTHMAN and E. CABIB. Regulation of muscle glycogen synthetase by metabolites. Differential effects on the I and D forms. Biochemistry 7: 56-66, 1968.

23) PIRAS, R. and R. STANELONI. In vivo regulation of rat muscle glycogen synthetase activity. Biochemistry 8: 2153-2158, 1969.

24) RAFAELSEN, O.J. Glycogen content of rat diaphragm after intraperitoneal injection of insulin and other hormones. Acta Physiol. Scand. 61: 314-322, 1964.

25) SAHEKI, R. and S. TSUIKI. Glycogen synthesis and I to D conversion of glycogen synthetase in ascites hepatoma cells. Biochem. Biophys. Res. Commun. 31: 32-36, 1968.

26) SUTHERLAND, E.W. and G.A. ROBISON. The role of cyclic AMP in the control of carbohydrate metabolism. Diabetes 18: 797-819, 1969.

27) SØVIK, O. The action of insulin on glycogen synthesis in rat diaphragm. A comparative study of in vitro and in vivo effects. Acta Physiol. Scand. 68: 246-254, 1966.
28) SØVIK, O. and S. ADOLFSSON. Insulin-sensitive glycogen synthesis in the isolated levator ani muscle of the rat. Life Sci. 7: 549-552, 1968.
29) VILLAR-PALASI, C. and J. LARNER. Insulin treatment and UDPG glycogen transglucosylase activity. Fed. Proc. 19: 164, 1960.
30) VILLAR-PALASI, C. and J. LARNER. Insulin treatment and increased UDPG-glycogen transglucosylase activity in muscle. Arch. Biochem. Biophys. 94: 436-442, 1961.
31) VILLAR-PALASI, C., M. ROSELL-PEREZ, S. HIZUKURI, F. HUIJING and J. LARNER. Muscle and liver UDP-glucose: α-1,4-glucan α-4-glucosyltransferase (Glycogen synthetase). In: Methods in Enzymology. Eds. E. Neufeld and V. Ginsberg. 8: 374-384, 1966.
32) VILLAR-PALASI, C. and J. I. WENGER. In vivo effect of insulin on muscle glycogen synthetase. Identification of the action pathway. Fed. Proc. 26: 563, 1967.
33) WILLIAMS, H. E. and H. T. PROVINE. Effects of estradiol on glycogen synthetase in the rat uterus. Endocrinology 78: 786-790, 1966.

GLYCOGEN STORAGE IN HUMAN SKELETAL MUSCLE

E. HULTMAN, J. BERGSTRÖM, and A.E. ROCH-NORLUND

From the Department of Clinical Chemistry, S:t Eriks
Sjukhus, Stockholm, Sweden

Direct determinations of the glycogen content in skeletal muscle have been performed in numerous animal experiments and detailed studies of the biochemical events in synthesis and breakdown of glycogen in muscle tissue have been performed for many years. Both the intermediate metabolism and the enzyme systems are relatively well established to-day.

The physiological role of muscle glycogen in normal man, on the other hand, has not been studied so extensively. A prerequisite for such studies was the development of a simple method for repeated sampling of muscle tissue and which did not markedly disturb the subjects. The needle biopsy technique introduced during the last years (5, 13) meets these requirements.

The muscle glycogen content in the lateral portion of the quadriceps femoris muscle was determined in a series of 228 normal subjects (Fig. 1). The mean value was 13.9 g/kg wet muscle, measured as trichloroaceticacid-extractable glycogen, range 9.2 - 24.9. This material had an age distribution of 18-55 years. Two other series of the age groups 3-15 and 55-75 years were examined showing the same muscle glycogen content (21). Thus, no change of the muscle glycogen content with age was found.

A comparison was also made between the glycogen content in the quadriceps femoris and the deltoid muscle in the same man. The glycogen content was found to be consistently lower in the deltoid muscle, mean 9.8; range 5.3 - 14.0 g/kg wet muscle.

The variation of the muscle glycogen during the day was found to be negligible if no heavy exercise was performed. Complete

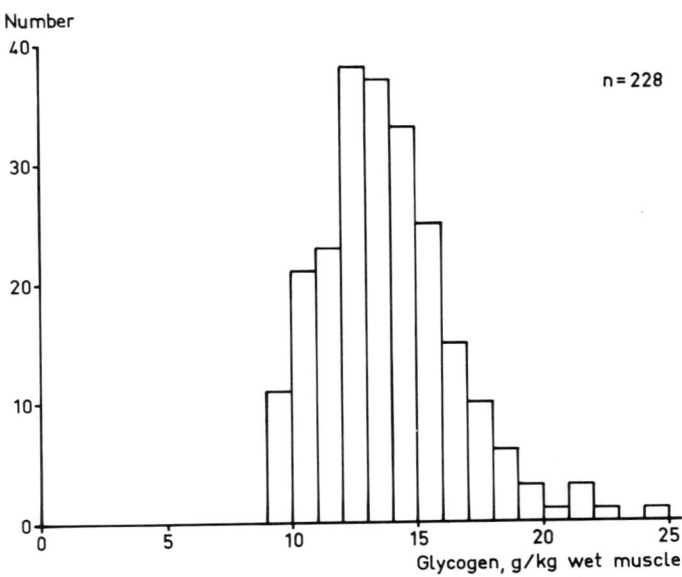

Fig. 1. The distribution of glycogen concentration values in quadriceps femoris muscle from 228 normal subjects (13).

fasting decreased the glycogen content slowly in the same way as did a carbohydrate-free diet. Over a period of 10 days, only a decrease of 30-40 % of the glycogen store in the muscles was found. Carbohydrate-rich food increased the muscle glycogen store, but only within the normal range if not preceded by glycogen depletion.

The effect of exercise on the glycogen store is dependent on the work load employed. At work loads near the subject's maximum oxygen uptake capacity, glycogen consumption is very high and the local stores are rapidly depleted, resulting in an inability to continue the exercise at the same work level. With decreasing work loads the rate of glycogen consumption decreases markedly approaching zero at a work level around 20-30 % of maximal oxygen uptake in a trained subject. Glycogenolysis during exercise will be further discussed by Bengt Saltin.

The resynthesis rate of muscle glycogen after depletion was found to be dependent on the diet given. If a carbohydrate-rich diet was given the resynthesis of the muscle glycogen store was very rapid. Resynthesis was complete after 24 hours. On a carbohydrate-free diet with the same caloric content resynthesis was complete only after 8-10 days (15). A carbohydrate-rich diet given

GLYCOGEN STORAGE IN SKELETAL MUSCLE

for more than 24 hours resulted in an increase of muscle glycogen content over the basal value, the glycogen store increasing further for up to 8 days (1) and to very high values. One experimental design used was the "one-leg" exercise, which means that the subject worked with one leg only until exhaustion of the glycogen store. The other leg rested. By this procedure it was possible to study the carbohydrate metabolism in one glycogen depleted muscle group and one with normal glycogen content in the same man. It was found that the overshoot of muscle glycogen synthesis was localized to the muscle group, where the glycogen depletion had occurred and not to other muscles in the same subject (8) (Fig. 2). Similar results have been obtained in guinea pigs by Peter and co-workers (16, 18). The effects of different diets and starvation on the muscle glycogen store in exercise and rest are summarized in Fig. 3. One-leg exercise was performed twice and the diet regime between the exercise periods was total starvation or a carbohydrate-free diet. After the second exercise, a carbohydrate-rich diet was given to all the subjects. The mean glycogen synthesis rates are shown in Fig. 4. In the rest leg, during starvation or carbohydrate-free diet, glycogen decreased at a rate of 5-10 μmol glucosyl units per kg muscle per min,

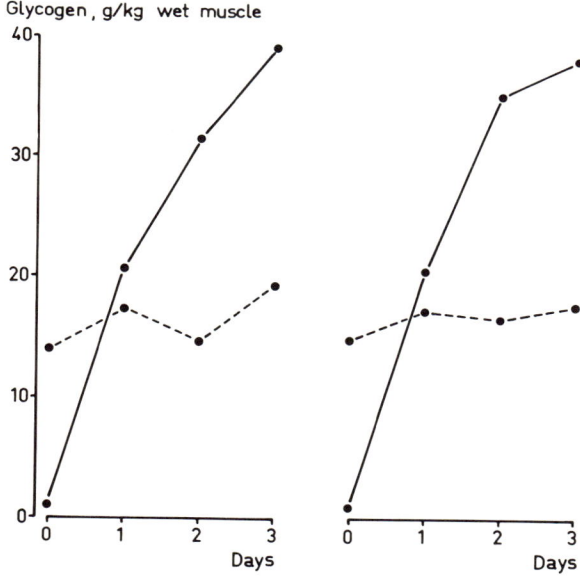

Fig. 2. Glycogen concentration of "exercise" and "rest" leg after "one-leg" exercise experiment. Biopsy specimens were taken from both legs of the subjects at approximately the same time (within 15 min) on each occasion.

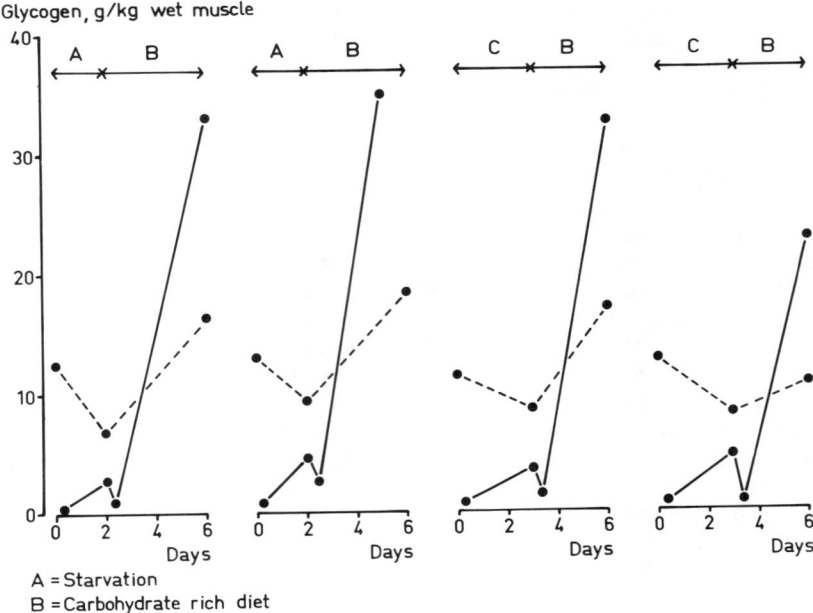

Fig. 3. Effect of starvation, carbohydrate-free and carbohydrate-rich diet on muscle glycogen concentration in the quadriceps femoris muscle of 4 normal subjects. - Exercise with one leg was performed immediately before the first biopsies; after 2 days of starvation or 3 days of carbohydrate-free diet, respectively, further exercise with the same leg was performed, and thereafter a carbohydrate-rich diet was given.
——————— work leg   - - - - - - - rest leg

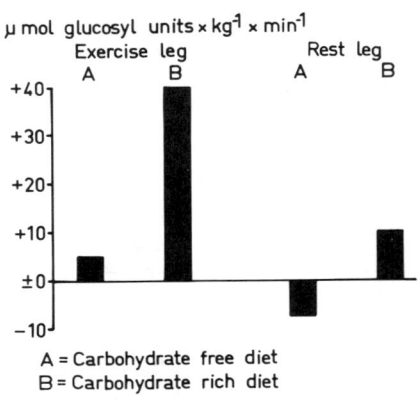

Fig. 4. Mean glycogen synthesis rate (μmol glucosyl units per kg per min) in exercise, and rest leg described in Fig. 3.

whilst a carbohydrate-rich diet gave a glycogen synthesis of 10 umol/min. In the exercised leg comparable figures for starvation or carbohydrate-free diet were + 5 µmol per kg per min and for carbohydrate-rich diet a glycogen synthesis rate of 40 µmol/min. These values for glycogen synthesis or breakdown are mean values over periods of 2-3 days. Corresponding figures for the glycogen synthesis rate during 2-4 hours of glucose infusion after one-leg exercise (9) are shown in Fig. 5. The rates were 400-500 µmol per kg muscle per min in the exercised leg and 40-50 µmol in the "rest" leg. Piras and Staneloni (20) determined the rate of glycogen synthesis in rat muscle in vivo after tetanic contraction. They found a rate of 760 µmol glucosyl units per min per 1 muscle water. The authors suggest that the rate in resting muscle is considerably lower in the range of 5-20 % of this value. Stadie et al. (23) reported a rate of 130 µmol per min in rat hemi-diaphragm.

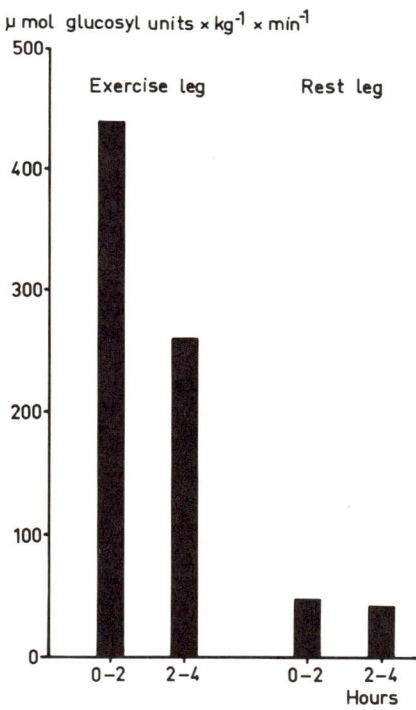

Fig. 5. Mean glycogen synthesis rate in exercise and rest leg (n = 10) during glucose infusion (infusion rate 1 g glucose per kg body weight per hour).

The maximal glycogen synthesis rates of 400-500 µmol glucosyl units per kg muscle per min are still very low in comparison with the glycogenolysis rate in the same muscle group during dynamic submaximal exercise. This rate was 3,500-4,000 µmol per min (14) and at maximal isometric contraction 40,000-43,000 µmol per min (7). Similar values were found by Piras and Staneloni (20) during tetanic contraction in rat muscle (48,000 µmol per min per 1 muscle water). The maximal rate of glycogen synthesis found in man is thus only one per cent of the maximal glycogenolysis rate.

The enzymatic mechanisms for glycogenolysis and glycogen synthesis in muscle have already been discussed by Ahrén and Adolfsson. This presentation will be limited to show some results from the determination of the glycogen synthetase I and D activity in human skeletal muscle.

The glycogen glucosyl transferase or glycogen synthetase enzyme exists in two forms, dependent on glucose 6-P (D form) and independent (I form). The transformation from I to D occurs in crude muscle extracts in the presence of ATP and magnesium (11). The reaction is stimulated by a kinase (3) which is activated by cyclic AMP (2, 3, 22).

$$\text{Synthetase I} + n\text{ATP} \xrightarrow[\text{Mg}^{2+}]{\text{kinase}} \text{Synthetase D} + n\text{ADP}$$

Two other I $\longrightarrow$ D conversion mechanisms have been reported, one involving proteolysis by trypsin (2) and another which is dependent on Ca^{2+} and a protein factor (4). The Ca^{2+} dependent reaction is not stimulated by ATP, Mg^{2+} or cyclic AMP.

The D $\longrightarrow$ I conversion is catalyzed by a specific phosphatase (11). It was shown by Danforth (10) that synthetase I activity was inversely related to the glycogen content.

The control of synthetase I activity by glycogen was shown to be dependent on the activity of the synthetase phosphatase (12); phosphatase activity being inhibited by high levels of glycogen but activated by low levels. It was suggested (19, 20) that at normal cellular levels of ATP, UDPG, P_i, and glucose 6-P only the I form is active whilst at increasing glucose 6-P or decreasing ATP levels the D form can function as an active enzyme. In normal cells, the ATP content varies only slightly and therefore the only possibility to activate the D form will be through an increase of the glucose 6-P level. This can be the case during hard exercise but in most other conditions the synthetase activity seems to depend on the I form of the enzyme. The relation between glycogen content and synthetase I activity in per cent of total was studied in a series of normal subjects (Fig. 6). The synthetase I activity increased with decreasing glycogen content from the normal mean

GLYCOGEN STORAGE IN SKELETAL MUSCLE

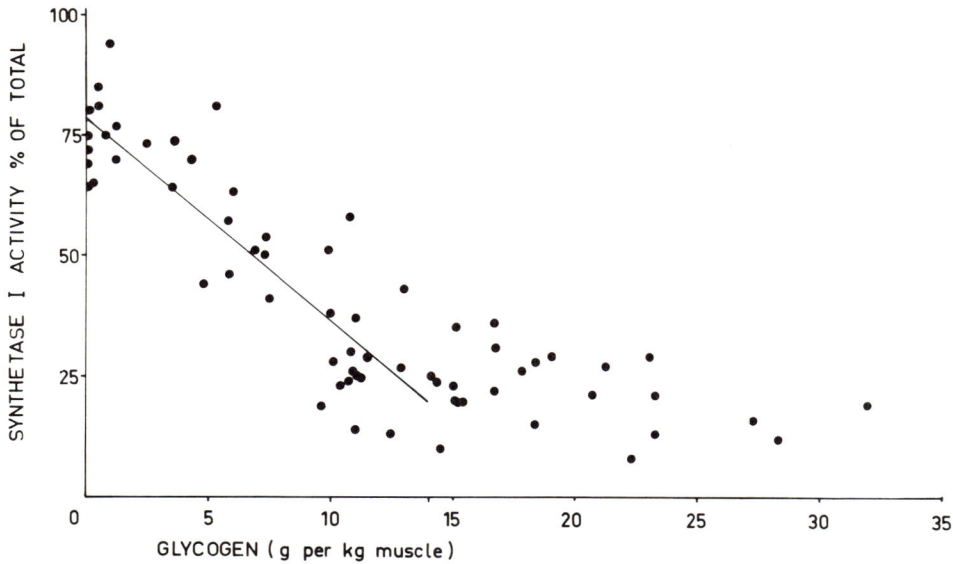

Fig. 6. Glycogen synthetase I activity in % of total, related to the muscle glycogen content. Equation for the regression line: $y = 78.8 - 42.2x$; $r = 0.88$.

value in man. Above this glycogen value, no further change in synthetase activity was found. The glycogen decrease was obtained through exercise.

In the following presentation synthetase I activity was in most cases related to glycogen content and the regression line from Fig. 6 was used. Danforth (10) found a similar correlation curve between synthetase I activity and glycogen content in rat skeletal muscle and found that the curve was displaced to the right when insulin was given and to the left when epinephrine was given. The same results were found in man (Fig. 7). There was no difference in synthetase I activity if glucose infusion (1-2 g per kg body weight per hour) was combined with insulin or not. This can, of course, be explained by the stimulating effect of glucose infusion on insulin secretion rate. In two subjects adrenaline was given intravenously during one hour. In both subjects synthetase I activity decreased in spite of a diminished muscle glycogen content. This is consistent with a postulated activation of synthetase kinase by cyclic AMP. In the insulin and/or glucose infused subjects the rate of muscle glycogen synthesis was determined and correlated to the synthetase I activity. The results are presented in Fig. 8; r - value for the regression line 0.86.

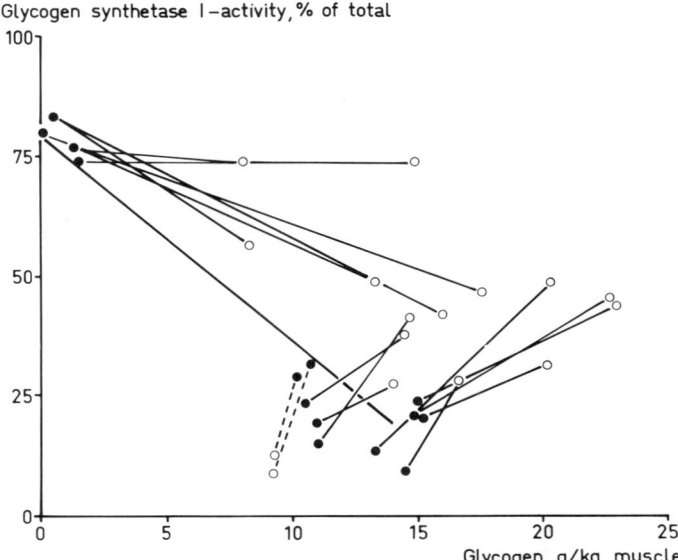

Fig. 7. Relation between synthetase I activity and glycogen content before ● and after insulin and/or glucose infusion o.

- - - - - two subjects before ● and after o adrenalin infusion
The regression line corresponds to that given in Fig. 6.

These findings that synthetase I activity is increased at low muscle glycogen levels and is further activated by insulin and/or glucose infusion would be enough to explain the rapid resynthesis of the glycogen store after exercise when carbohydrate is supplied. The carbohydrate feeding or glucose infusion will, of course, not only stimulate the synthetase activity, but also increase the amount of available substrate for glycogen synthesis in the muscle.

The glycogen synthetase activity was also determined in connection with the overshoot of the muscle glycogen store described earlier. In the morning after an overnight fast one-leg exercise was performed till exhaustion. Thereafter a carbohydrate-rich diet was given for three days. The glycogen pattern in the two legs was similar to those presented in Fig. 2. The glycogen increase during the first day in the exercise leg could be attributed to the high activity of synthetase I, but thereafter on day 2 and 3 the synthetase I activity was the same in both legs and could thus not explain the difference in glycogen synthesis. Other possible mechanisms for the difference in glycogen synthesis are a local increase in hexokinase activity after exercise or activation of the synthetase D form by ATP decrease or glucose 6-P increase.

Increased hexokinase activity was suggested as a possible mechanism for the overshoot in glycogen storage (16, 18). It was not possible, however, to confirm these results in human skeletal muscle (Fig. 9). The experiment presented is of the one-leg exercise type and determination of both hexokinase and synthetase activities concomitant with glycogen analysis was performed. The hexokinase activity varied only marginally and not in a direction which could explain the difference in glycogen synthesis in the two legs.

The second possibility discussed to-day by Ahrén inferring an activation of synthetase D was also studied. ATP, phosphocreatine, creatine, and the hexosemonophosphates were determined in muscle tissue obtained by needle biopsy technique from exercise and rest leg. Glucose was infused during one hour on the day of exercise and on the following day. Biopsies were performed before and during the glucose infusion. The subject's diet was carbohydrate-rich. A big difference in glycogen synthesis rate between the two legs was

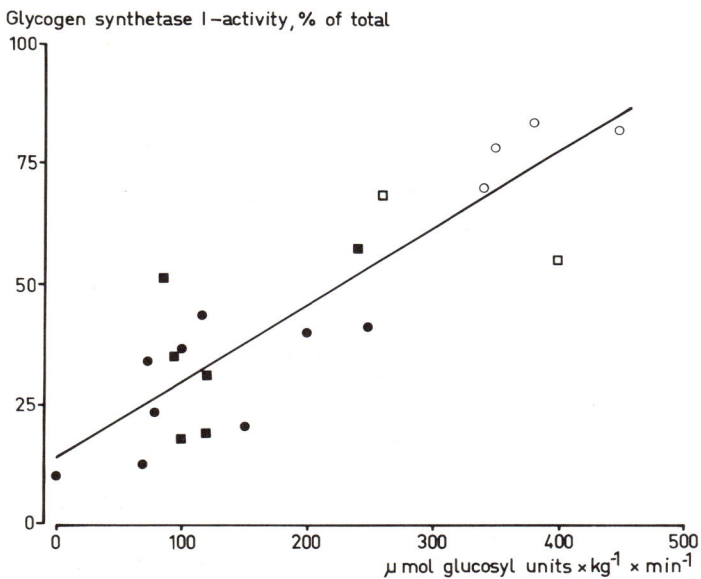

Fig. 8. Glycogen synthetase I activity in % of total related to glycogen synthesis rate in the quadriceps femoris muscle.
- ● only glucose infusion, rest leg
- ■ glucose + insulin, rest leg
- ○ only glucose infusion, exercise leg
- □ glucose + insulin, exercise leg

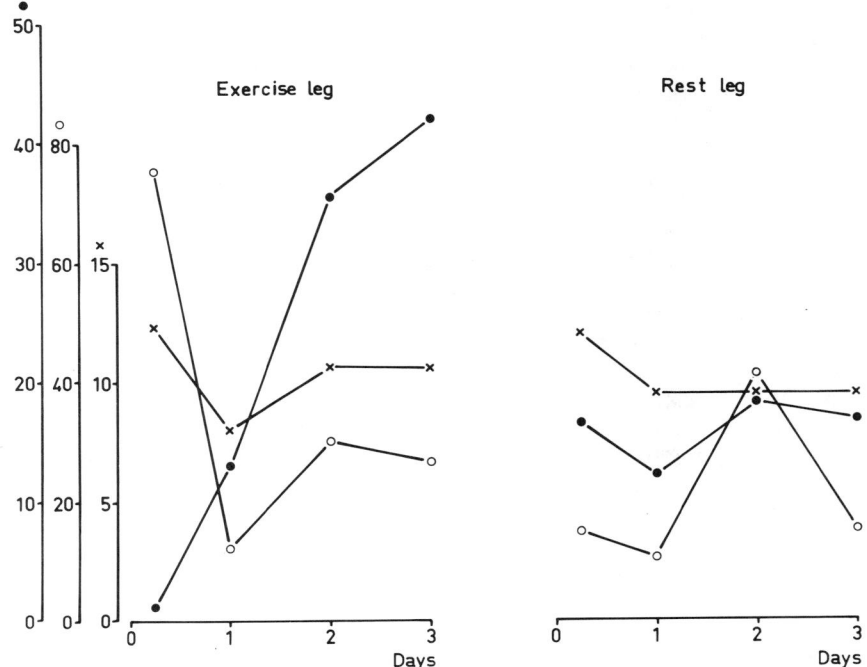

Fig. 9. One-leg exercise as described earlier. Glycogen content ●, glycogen synthetase I activity ○, and hexokinase activity ×, measured for 3 days after the exercise experiment. Carbohydrate-rich diet was given.

observed but neither ATP nor glucose 6-P showed consistent differences in concentration between the two legs. An activation of the synthetase D form in the exercise leg on day 2 is therfore not probable and cannot be explained by the observed changes in ATP or glucose 6-P concentrations (Fig. 10).

The total activity of the synthetase D form was also unchanged.

It has earlier been discussed if muscle contraction as such can give rise to transformation from D to I activity or if this is attributable to the glycogen decrease. In Fig. 11, an experiment is presented, in which 4 subjects have performed a submaximal work for 30 sec. The glycogen level was normal in two subjects and decreased by means of exercise in the two others. It can be seen that muscle contraction as such did not change the synthetase I activity measurably, but after the exercise an increased I activity occurred during a 10-min-period, whereafter the synthetase I

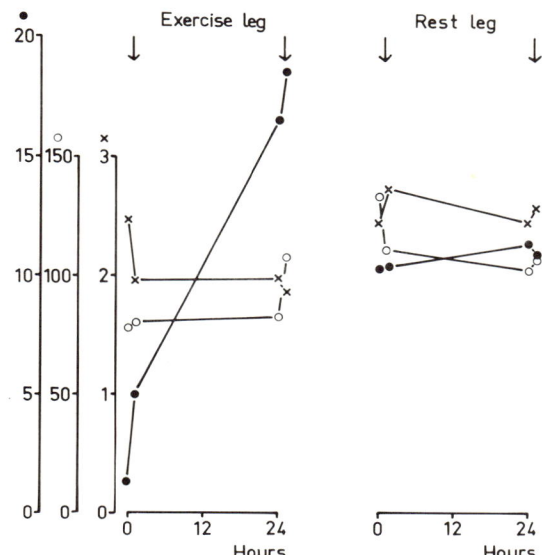

Fig. 10. One-leg experiment. Glucose infusion ↓ given during one hour after the exercise and again 24 hours later. Glycogen, ATP, and glucose 6-P content determined in biopsy specimens in both legs before and during infusion.

activity went back to the basal level, but the basal level is clearly different in the two series and related to the muscle glycogen content. It is thus possible to differentiate between the effect of muscle contraction and of glycogen depletion.

The results are in agreement with those found by Danforth (10) in rat and mouse skeletal muscle and those by Piras and Staneloni (20). The I to D transformation during repeated tetanic contraction shown by Staneloni and Piras (24) was not found in human skeletal muscle. The reason for this is not clear but could depend on the time delay in obtaining the muscle tissue after the end of contraction in our material or to the difference between tetanic contraction and dynamic work. Another reason for the difference in the results could be that the glycogen level was very low in our experiments with increased I activity before the short exercise and not in those of Piras and Staneloni (20). The synthetase D phosphatase is therefore active to transform D to I form also in connection with muscle contraction.

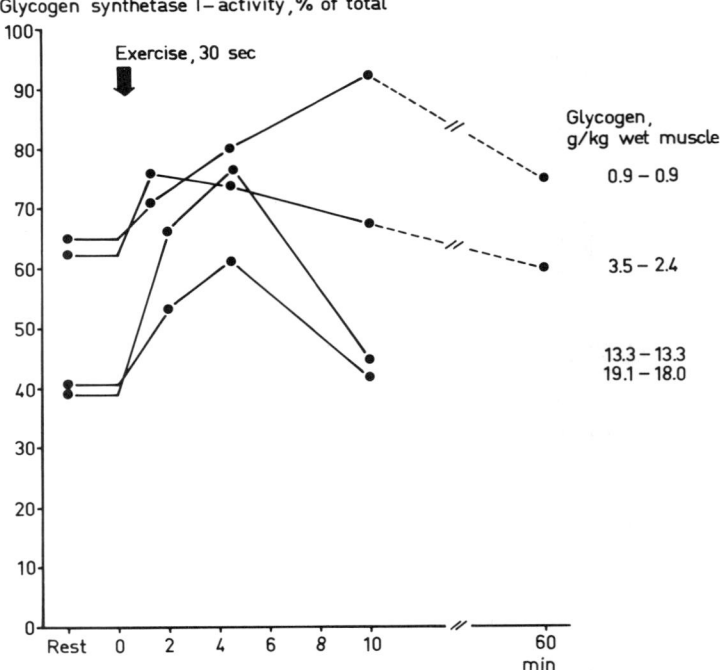

Fig. 11. Glycogen synthetase I activity in 2 subjects with normal muscle glycogen level and 2 subjects with low glycogen content. - Biopsies taken before and immediately after 30 sec exercise with submaximal work load, and thereafter at the times indicated.

To ascertain if glucose could be synthetized to glycogen during a continuous exercise, if the glycogen synthetase activity was high due to glycogen depletion a one-leg exercise experiment was again performed. After glycogen depletion in one leg the subject exercised with both legs at a moderate work load on a bicycle ergometer. During the exercise period glucose was infused continuously. The blood glucose concentration increased up to 500 mg/100 ml. In this situation glycogen was consumed in the leg with the intact glycogen store during the first hour, whilst the glycogen depleted leg synthetized glycogen during the same period despite the continuance of exercise. During the second hour both legs synthetized glycogen during continued exercise and glucose infusion (Fig. 12).

The explanation of this phenomenon must be not only an increase of synthetase activity in the glycogen depleted leg, but also an increased glucose phosphorylation in the muscle. The

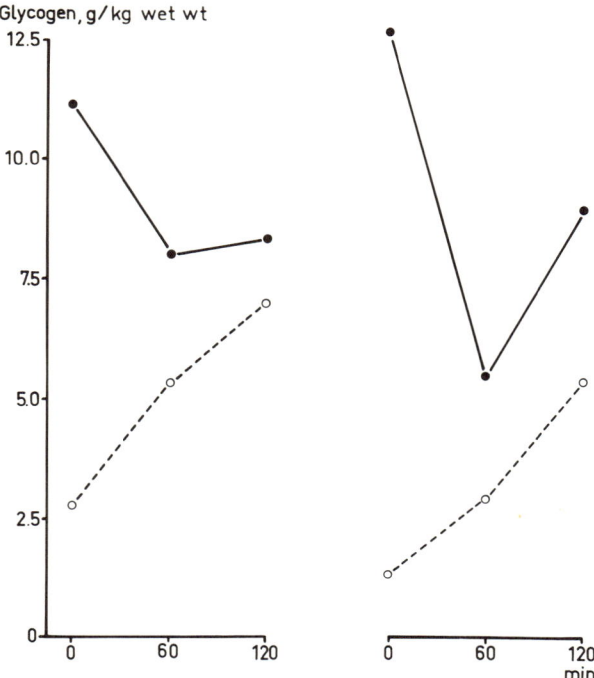

Fig. 12.   One-leg exercise is performed before the start of the study.
- - - - - -   work leg   ——————   rest leg
N.B. the different symbols for the exercise and rest leg in this figure.

Both legs are working during 2 hours at a moderate work load and glucose is continuously infused.

hexokinase responsible for this reaction is inhibited by glucose 6-P. It was earlier shown (6, 14) that during exercise the glucose 6-P level increased proportionally to the work load, but only when the glycogen store in the muscle was sufficient. At the end of the exercise when the glycogen store was depleted, the glucose 6-P level decreased, even to values lower then in the basal state. An increased glycogen synthetase activity would also tend to decrease the glucose 6-P level in the muscle. Therefore it is reasonable that the partial inhibition of hexokinase by increased glucose 6-P during exercise is abolished in the glycogen depleted leg.

Glucose penetration from plasma to muscle cells will be stimulated in both legs by the exercise and also by the increased plasma insulin level due to the glucose infusion (17). It is therefore suggested that in this situation the phosphorylation capacity and not the glucose penetration is rate limiting for the utilization of blood glucose in the muscle cells.

SUMMARY

The glycogen storage in human skeletal muscle was studied by means of a needle biopsy technique. It was found that during conditions of rest the variations in muscle glycogen due to changes in diet regime were small. Heavy exercise can empty the local glycogen stores in the working muscle rapidly. The highest glycogenolysis rate measured was 43,000 μmol glucosyl units per kg muscle per min during maximal isometric contraction.

The resynthesis of glycogen is dependent on the availability of substrate and the activity of synthetase I in the muscle tissue. Synthetase I activity is high when the muscle glycogen level is low. The highest rate of resynthesis of glycogen was found in glycogen depleted muscle during glucose infusion. The rate of synthesis was 400-500 μmol glucosyl units per kg muscle per min. The overshoot of glycogen in muscle tissue previously depleted of glycogen by exercise, cannot be explained solely by an increase in synthetase I activity, or by an activation of the D form of the enzyme by increasing glucose 6-P concentration.

This investigation was supported by grants from the Swedish Medical Research Council (Projects nos B70-19X-1002-05B, B70-19X--2647-02B, and K70-19X-2647-02BK).

REFERENCES

1. AHLBORG, B., J. BERGSTRÖM, J. BROHULT, L.-G. EKELUND, and E. HULTMAN. Human muscle glycogen content and capacity for prolonged exercise after different diets. Försvarsmedicin 3: 85 - 100, 1967.
2. APPLEMAN, M.M., E. BELOCOPITOW, and H.N. TORRES. Factors affecting the activity of muscle glycogen synthetase. Biochem. Biophys. Res. Commun. 14: 550 - 554, 1964.
3. APPLEMAN, M.M. L. BIRNBAUMER, and H.N. TORRES. Factors affecting the activity of muscle glycogen synthetase. III. The reaction with adenosine triphosphate, Mg^{2+}, and cyclic 3'5'--adenosine. Arch. Biochem. Biophys. 116: 39 - 43, 1966.
4. BELOCOPITOW, E., M.M. APPLEMAN, and H.N. TORRES. Factors affecting the activity of muscle glycogen synthesis. II. The regulation by Ca^{++}. J. Biol. Chem. 240: 3473 - 3478, 1965.
5. BERGSTRÖM, J. Muscle electrolytes in man. Determined by neutron activation analysis on needle biopsy specimens. A study on

normal subjects, kidney patients, and patients with chronic diarrhoea. Scand. J. Clin. Lab. Invest. 14: Suppl. 68, 1962.

6. BERGSTRÖM, J., G. GUARNIERI, and E. HULTMAN. Carbohydrate metabolism and electrolyte changes in human muscle tissue during heavy work. Accepted for publication in J. Appl. Physiol.

7. BERGSTRÖM, J., R.C. HARRIS, E. HULTMAN, and L.-O. NORDESJÖ. Energy rich phosphagens in dynamic and static work. Presented at the Symposium of Muscle Metabolism during Exercise, 6-9 Sept., 1970, Stockholm.

8. BERGSTRÖM, J. and E. HULTMAN. Muscle glycogen synthesis after exercise. An enhancing factor localized to the muscle cells in man. Nature 210: 309 - 310, 1966.

9. BERGSTRÖM, J. and E. HULTMAN. Synthesis of muscle glycogen in man after glucose and fructose infusion. Acta Med. Scand. 182: 93 - 109, 1967.

10. DANFORTH, W.H. Glycogen synthetase activity in skeletal muscle. Interconversion of two forms in the control of glycogen synthesis. J. Biol. Chem. 240: 588 - 593, 1965.

11. FRIEDMAN, D.L. and J. LARNER. Studies on UDPG-α glucan transglucosylase. III. Interconversion of two forms on muscle UDPG αglucan transglucosylase by a phosphorylation-dephosphorylation reaction sequence. Biochemistry 2: 669 - 675, 1963.

12. HUIJING, F., F.Q. NUTTALL, C. VILLAR-PALASI, and J. LARNER. UDPglucose: α-1,4-glucanα-4-glucosyltransferase in heart regulation of the activity of the transferase in vivo and in vitro in rat. A dissociation in the action of insulin on transport and on transferase conversion. Biochim. Biophys. Acta 177: 204 - 212, 1969.

13. HULTMAN, E. Studies on muscle metabolism of glycogen and active phosphate in man with special reference to exercise and diet. Scand. J. Clin. Lab. Invest. 19: Suppl. 94, 1967.

14. HULTMAN, E. Muscle glycogen stores and prolonged exercise. In: Frontiers of Fitness. Ed: R.J. SHEPHARD. Toronto, Canada: The Canadian Asssociation of Sports Science, 1970.

15. HULTMAN, E. and J. BERGSTRÖM. Muscle glycogen synthesis in relation to diet studied in normal subjects. Acta Med. Scand. 182: 109 - 117, 1967.

16. LAMB, D.R., J.B. PETER, R.N. JEFFRESS, and H.A. WALLACE. Glycogen, hexokinase, and glycogen synthetase adaptations to exercise. Am. J. Physiol. 217: 1628 - 1632, 1969.

17. NARAHARA, H.T. and C.F. CORI. Hormonal control of carbohydrate metabolism in muscle. In: Carbohydrate Metabolism and its Disorders. Eds: F. DICKENS, P.J. RANDLE, and W.J. WHELAN. London and New York: Academic Press. 1: 375 - 395, 1968.

18. PETER, J.B., R.N. JEFFRESS, and D.R. LAMB. Exercise - effects on hexokinase activity in red and white skeletal muscle. Science 160: 200 - 201, 1968.

19. PIRAS, R., L.B. ROTHMAN, and E. CABIB. Regulation of muscle glycogen synthetase by metabolites, differential effects on the I and D forms. Biochemistry 7: 56 - 66, 1968.

20. PIRAS, R. and R. STANELONI. In vivo regulation of rat muscle glycogen synthetase activity. Biochemistry 8: 2153 - 2160, 1969.
21. ROCH-NORLUND, A.E., J. BERGSTRÖM, H. CASTENFORS, and E. HULTMAN. Muscle glycogen in patients with diabetes mellitus. Glycogen content before treatment and the effect of insulin. Acta Med. Scand. 187: 445 - 453, 1970.
22. ROSELL-PEREZ, M. and J. LARNER. Studies on UDPG-α-glucan transglucosylase. IV. Purification and characterization of two forms from rabbit skeletal muscle. V. Two forms of the enzyme in dog skeletal muscle and their interconversion. Biochemistry 3: 75 - 88, 1964.
23. STADIE, W.C., N. HAUGAARD, and M. VAUGHAN. Quantitative relation between insulin and its biological activity. J. Biol. Chem. 200: 745 - 751, 1953.
24. STANELONI, R. and R. PIRAS. Changes in glycogen synthetase and phosphorylase during muscular contraction. Biochem. Biophys. Res. Commun. 36: 1032 - 1038, 1969.

MUSCLE GLYCOGEN UTILIZATION DURING WORK OF DIFFERENT INTENSITIES

B. Saltin and J. Karlsson

Department of Physiology, Gymnastik- och idrotts-högskolan, Stockholm, Sweden

The relative contributions of different substrates for the energy metabolism during exercise have been discussed repeatedly during the last 100 years. The greater importance for carbohydrates at heavy work intensities, which the increase in respiratory exchange ratio (RQ) implies (4), has received strong support from studies in which the glycogen breakdown in exercising muscles has been determined (7,8,13). The factors determining muscle glycogen utilization during heavy muscular work are not well understood. In this article results will be presented on glycogen utilization at different work intensities. Further, the possibility of changing the rate of glycogen utilization at a given absolute or relative work load has been investigated in conditions of lowered barometric pressure and after physical conditioning. On the basis of these results and evidence from animal experiments, it is suggested that the main determinant for the metabolic response observed during exercise in man is related to the recruitment pattern for motor units in the muscles.

METHODS AND PROCEDURE

The subjects studied were all young healthy men with different training backgrounds. They were studied at rest before, and at different times during bicycle exercise and at exhaustion. The muscle biopsies (2) were taken from the lateral portion of M. quadriceps femoris and frozen within 3-5 seconds in liquid nitrogen. The glycogen and lactate concentrations in the biopsy material were determined as described by Karlsson et al. (10). Blood lactate was analysed with an enzymatic method (14). Oxygen uptake was obtained with the Douglas bag method and the gas analyses were performed with the Haldane apparatus. Usually two bags were taken and the collection of expired air lasted for at least two minutes. The oxygen uptake measurements used for RQ calculations were always performed after at least 10 minutes of exercise. When the subjects' maximal oxygen uptake was determined the "levelling off" criterion was used.

RESULTS

Muscle Glycogen Depletion during Work

In Fig. 1 the mean glycogen depletion on work loads demanding from 30 up to 120 per cent of the individual's maximal oxygen uptake (max $\dot{V}_{O_2}$ being 3.5-4.0 l/min in the groups studied) are given. The steepness of the different curves were related to the relative work loads.

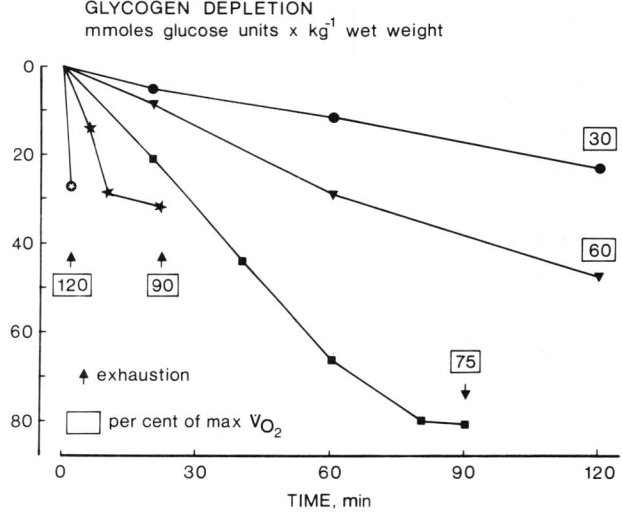

Fig. 1. Glycogen depletion in the quadriceps muscle during bicycle exercise of different intensities. The results are from ref. 7, 9, and unpublished materials.

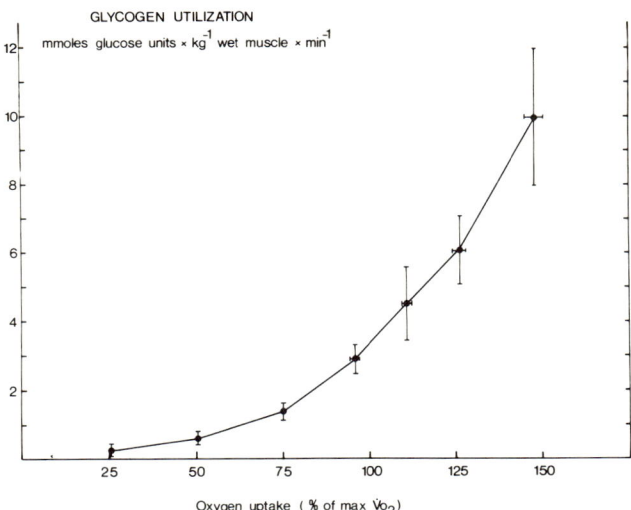

Fig. 2. Rate of glycogen utilization at different submaximal and maximal work intensities. The calculations are based on data presented in Fig. 1.

At exhaustion with work loads demanding 90 per cent of the maximal oxygen uptake or more there was ample amount of glycogen left in the quadriceps muscle. If the work demanded 70-80 per cent of the individual's maximal oxygen uptake, exhaustion coincided with muscle glycogen values approaching zero. At lower work intensities a much slower rate of glycogen utilization was observed and if the work loads were continued until the subjects gave up, the muscle glycogen was not always at a low level.

The linear part of these curves can be used to calculate the rate of glycogen depletion (Fig. 2). These results further emphasize that there was a marked enhancement in the glycogen utilization when the work intensity approached maximal and supramaximal work levels. At these very heavy work loads more than one tenth of the muscle's normal glycogen store was used per minute. This can be compared with less than half of a per cent of the muscle glycogen store utilized per minute at work loads demanding below 50 per cent of maximal oxygen uptake. The absolute values were 0.3 glucose units x kg^{-1} muscle x min^{-1} at a relative work load of 25 per cent. This increased to 0.7, 1.4, and 3.4 glucose units x kg^{-1} muscle x min^{-1}, respectively at 50, 75, and 100 per cent of the maximal oxygen uptake.

Acute Exposure to Altitude

At acute exposure to high altitude arterial oxygen tension is reduced as compared with sea level and so is the oxygen tension in femoral vein blood during exercise (Hartley, personal communication). What happens with tissue pO_2 cannot be stated. It may not be reduced during exercise but the oxygen gradient between the capillary and the mitochondria may be lowered at altitude as compared with sea level. The results for three subjects exercising at the same absolute work load at sea level and during acute exposure to an altitude of 3.000 m. are presented in Fig. 3. As the work was submaximal also at altitude there was no difference in the oxygen uptake in the two situations. It is interesting to note that RQ is approximately the same in the two situations. All three subjects behave in an identical manner in this respect. Muscle glycogen depletion was significantly more marked during the first 4 minutes of exercise at altitude as compared to sea level. During the remainder of the

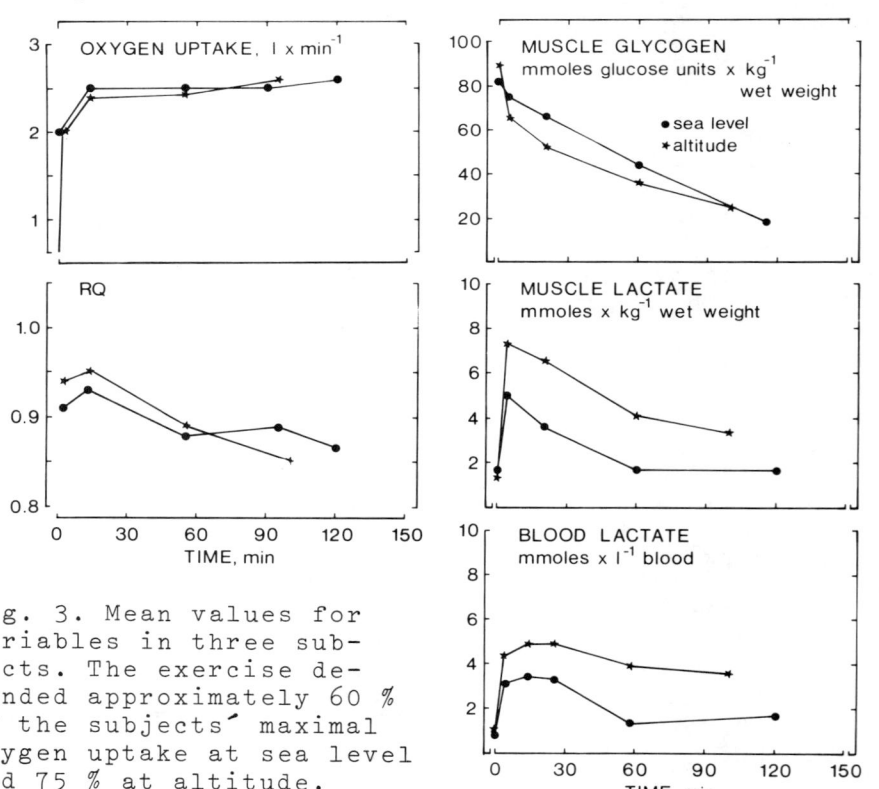

Fig. 3. Mean values for variables in three subjects. The exercise demanded approximately 60 % of the subjects' maximal oxygen uptake at sea level and 75 % at altitude.

GLYCOGEN USAGE DURING WORK OF VARIED INTENSITY

prolonged exercise period no marked differences could be observed comparing the two situations. The higher lactate production at altitude at the onset of work (indicating a lower tissue pO_2?) accounts for the accelerated glycogen breakdown during the first minutes of exercise. It is striking that the blood lactate values reflected the changes in muscle lactate.

Muscle glycogen utilization after 4 minutes of exercise agreed with the calculated value for aerobic use of carbohydrates indicating that there was no difference in the relative role of carbohydrates and fat during exercise with acute exposure to hypoxia. This is of interest not only in relation to the possibility of a reduced O_2-tension in muscle but also from the point of view that the work performed at altitude represented an approximately 15 per cent higher relative work load.

Physical Conditioning

In cross-sectional studies of untrained and well-trained subjects it has been shown that the calculated values for the aerobic use of carbohydrates were identical in both groups when they were exercising at the same relative work load (Fig. 4). This was due to the fact that the higher absolute oxygen uptake in the well-trained subjects was counterbalanced by a lower RQ during the prolonged exercise. The rate of the muscle glycogen depletion was very similar in the two groups of subjects. The slightly larger breakdown of muscle glycogen at the beginning of the exercise in the untrained subjects could explain the higher blood lactate concentration as compared with findings in the trained subjects.

Similar results were also observed in longitudinal training studies. Eight previously untrained subjects have been examined at the same absolute level of sub-maximal work before and after training. Oxygen uptake was the same throughout the exercise period but the RQ was significantly lowered at the most trained stage (Fig. 5). There was also a significantly lower rate of glycogen depletion after training and this difference could only partly be explained by the lower lactate production. The indirect calculation of the aerobic use of carbohydrates (from $\dot{V}_{O_2}$ and RQ) was in agreement with the direct determination of the glycogen used by the exercising muscles both at sea level and at altitude. On the other hand physical conditioning, in contrast to

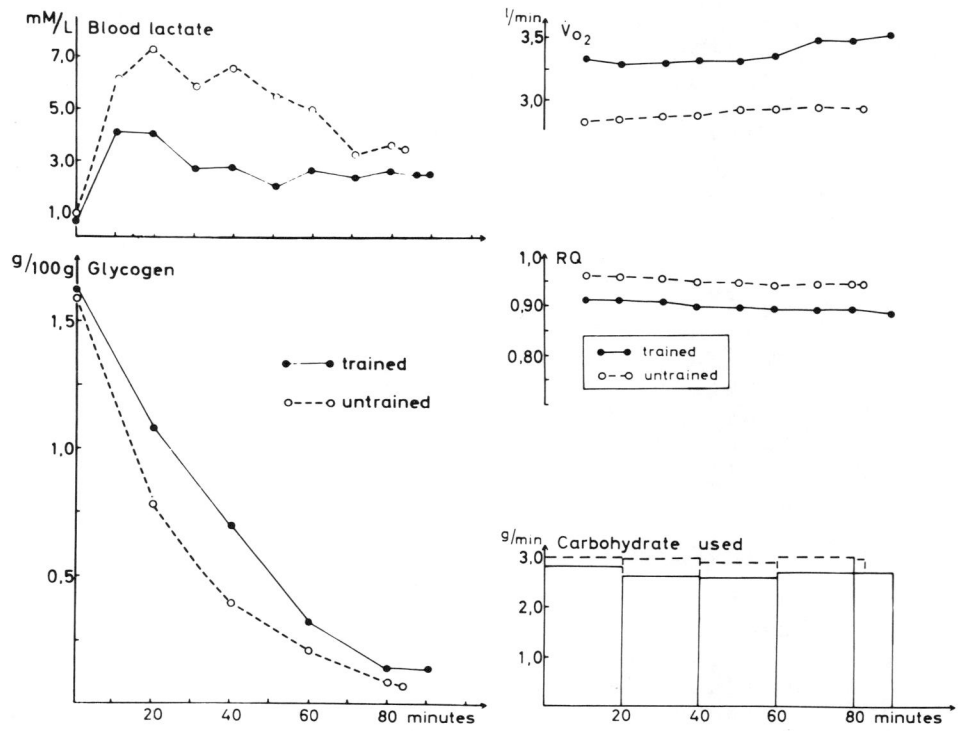

Fig. 4. Mean values for 10 untrained and 10 well-trained men at rest and during prolonged exercise to exhaustion. The work load demanded approximately 70-75 per cent of the subject's maximal oxygen uptake (7).

acute exposure to altitude, induces changes not only in lactate production but also in the relative role of the amount of fat and carbohydrates oxidized at a given submaximal oxygen uptake.

DISCUSSION

The regulatory mechanism behind the importance of the muscle glycogen for heavy muscular work is still far from solved. As demonstrated by Henneman et al. (5,6) it seems quite clear that the different sizes of the motoneurons in the spine can have a functional significance and that these motoneurons are connected to different types of fibers. These cells have different tresholds and are activated in special patterns related to the strength of

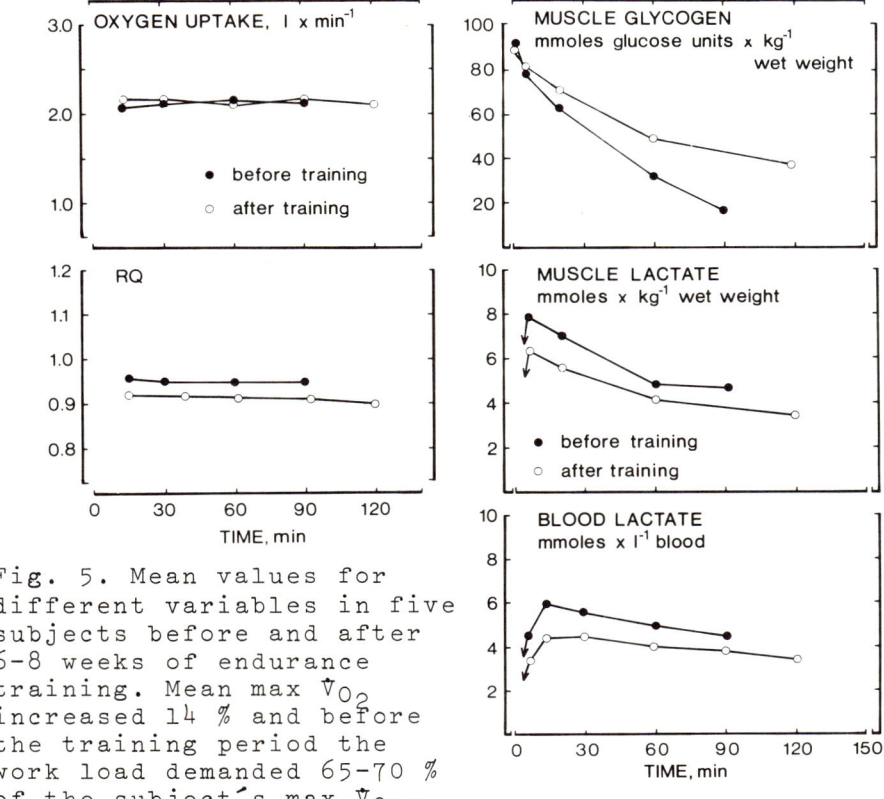

Fig. 5. Mean values for different variables in five subjects before and after 6-8 weeks of endurance training. Mean max $\dot{V}_{O_2}$ increased 14 % and before the training period the work load demanded 65-70 % of the subject's max $\dot{V}_{O_2}$.

the electrical stimuli. Further, the studies by Kugelberg and Edström (12) indicate that the fibers in the same motor unit have identical enzyme profiles. To what extent these results obtained in animals can be applied to exercising man can be questioned. If they are, the following hypothetical explanations may be advanced.

At low work intensities small motor cells innervating mainly red fibers are activated. This means a minimal amount of lactate formation and a very low rate of glycogen utilization. With increasing work rates larger motor cells are activated leading to the contraction of more glycogen containing fibers with some lactate formation. Continuation of heavy although still submaximal exercise results in a lower rate of muscle glycogen depletion after 4 minutes of exercise indicating lower or nonexistant lactate formation. Whether it also represents a change in the activation pattern of motor units cannot be settled but that possibility seems unlikely. During prolonged exercise a change in activation pattern may occur, however, as the RQ gradually become reduced on work rates lower than approxiamtely 70 per cent of maximal oxygen uptake. Of special note is the constancy of the RQ on higher work intensities (cf Fig. 4 and 5).

It has been discussed whether the oxygen tension at the mitochondrial level may play a role regulating the oxidation of the pyruvate and free fatty acids in the mitochondria. If acute exposure to 3,000 m. (a.s.l.) causes a lowering of tissue pO_2 during exercise (see above) the present results indicate that the pO_2 per se is of little or no importance for the muscle's choice of substrate. Instead the explanation of the present findings may be that bicycle exercise at identical absolute work loads brings about the same recruitment pattern with regard to the motor units that are activated, which then also results in the same pattern of metabolic changes.

After physical conditioning the relative importance of carbohydrates (muscle glycogen) at a given submaximal work load was lowered. Endurance training in animals has resulted in an increased number of red fibers (1). Morphological studies (11) and enzyme determination in human skeletal muscles (15) also indicate changes in the same direction. Endurance training should then result in an increase in the number of red fibers. Relatively more of these fibers are activated after training, during exercise at the same absolute work load. This seems also to be the case if the comparison is made at the same relative work load. In this occasion, however, the same number of glycogen containing fibers are activated.

It is well known that the diet can markedly influence not only muscle glycogen storage (3) and endurance (3,4) but also the RQ is observed during prolonged exercise (3,4). These observations do not support the concept presented above unless it is assumed that special changes occur in the muscle cell as an effect of a prolonged special diet regime or starvation. An increase in the activity of certain key-enzymes might occur, however, thereby enhancing the mitochondrias ability to oxidize a particular substrate.

To quantitate the relative role for the muscle glycogen as substrate at different work intensities the data presented in Fig. 1 and 2 can be used together with the

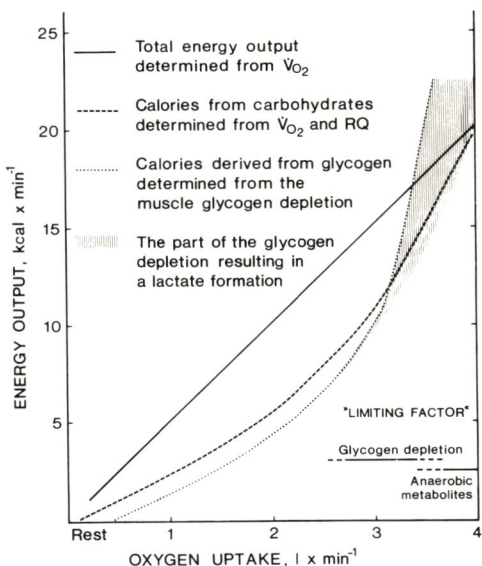

Fig. 6. The calculations are based on a subject with 4 l/min in maximal oxygen uptake. For further explanation see text.

concomitantly observed values for $\dot{V}_{O_2}$ and RQ. Two further assumptions have to be made. It is suggested that 10, 15, 20, and 25 kg of the bodies muscle mass are engaged in the bicycle exercise at work loads demanding 25, 50, 75, and 100 per cent of maximal oxygen uptake, respectively. Assuming that the glycogen depletion in the quadriceps muscle is representative for all exercising muscle groups the relative contribution of carbohydrate and fat as energy substrates can be calculated (Fig. 6). Carbohydrates accounted for less than half of the total amount of substrate used at relative work loads less than 50 per cent of maximal oxygen uptake. At these work levels a small but significant contribution of extra muscular carbohydrate, i.e. a glucose uptake, is also present. Above 50 per cent of maximal oxygen uptake carbohydrate is the dominant fuel but a tendency for a reduced absolute amount of fat to become oxidized is not found until the work loads exceed 70-80 per cent of maximal oxygen uptake. Due to the lactate production at these higher work levels the contribution of substrate from the intra-muscular carbohydrate stores in relation to external sources of carbohydrate is difficult to evaluate. A glucose uptake in the muscles during very heavy work seems to be quantitatively of minor importance, however. What has been discussed above and illustrated in Fig. 6 has not a general application as many factors may influence the choice of substrate used by skeletal muscles during exercise. As pointed out earlier the diet, the degree of training, the duration of the exercise as well as the exposure to altitude will markedly influence one or all of variables such as the lactate production, total muscle glycogen depletion and the RQ.

In summary, some comments may be made with regard to the muscle glycogen utilization and what limits prolonged exercise. At low work rates exercise can continue for many hours and no easily definable exhaustion point exists. When the subjects give up muscle glycogen may or may not be very low. Between approximately 65 and 89 per cent of maximal oxygen uptake, work time being 45-200 min, the depletion of glycogen in working muscle seems to be the limiting factor as the exhaustion point coincides with extremely low muscle glycogen values. Moreover, at these work intensities there is a need for a very high carbohydrate contribution in the energy output of the muscles, which amounts to 2-4 g $\times$ min^{-1}. At work loads demanding more than 90 per cent of the individual's maximal oxygen uptake, muscle glycogen depletion does not limit performance. The accumulation of anaerobic metabolites may, however, be the important factor at these very heavy work intensities (9).

ACKNOWLEDGEMENT

These investigations have been carried out with support from the Swedish Medical Research Council (Project No. B68-71-14X-2203-02-04).

REFERENCES

1. BARNARD, R.J., V.R. EDGERTON, AND J.B. PETER. Effect of exercise on skeletal muscle. I. Biochemical and histochemical properties. J. Appl. Physiol. 28:762-766, 1970.
2. BERGSTRÖM, J. Muscle electrolytes in man. Scand. J. clin. Lab. Invest. 14: Suppl. 68, 1962.
3. BERGSTRÖM. J., L. HERMANSEN, E. HULTMAN, AND B. SALTIN. Diet muscle glycogen and physical performance. Acta physiol. scand. 71:140-150, 1967.

4. CHRISTENSEN, E.H., AND O. HANSEN. I. Zur Metodik der Respiratorischen Quotient-Bestimmungen in Ruhe und bei Arbeit. II. Untersuchungen über die Verbrennungsvorgänge bei langdauernder, schwerer Muskelarbeit. III. Arbeitsfähigkeit und Ernährung. Skand. Arch. Physiol. 81:137-171, 1939.

5. HENNEMAN, E., G. SOMJEN, AND D.O. CARPENTIER. Functional significance of cell size in spinal motoneurons. J. Neurophysiol. 28:560-580, 1965.

6. HENNEMAN, E., AND C.B. OLSON. Relations between structures and function in the design of skeletal muscles. J. Neurophysiol. 28:581-598, 1965.

7. HERMANSEN, L., E. HULTMAN, AND B. SALTIN. Muscle glycogen during prolonged severe exercise. Acta physiol. scand. 71:129-139, 1967.

8. HULTMAN, E. Studies on muscle metabolism of glycogen and active phosphate in man with special reference to exercise and diet. Scand. J. clin. Lab. Invest. 19: Suppl. 94, 1967.

9. KARLSSON, J., AND B. SALTIN. Lactate, ATP, and CP in working muscles during exhaustive exercise in man. J. Appl. Physiol. 29:Nov. 1970.

10. KARLSSON, J., B. DIAMANT, AND B. SALTIN. Muscle metabolites during submaximal and maximal exercise in man. Scand. J. clin. Lab. Invest. 27(1), 1971.

11. KIESSLING, K.-H., K. PIEHL, AND C.-G. LUNDQUIST. Effect of physical training on ultrastructural features in human skeletal muscle. Ibidem.

12. KUGELBERG, E., AND L. EDSTRÖM. Differential histochemical effects of muscle contractions on phosphorylase and glycogen in various type of fibers in relation to fatigue. J. Neurol. Neurosurg. Psychiat. 31:415-423, 1968.

13. SALTIN, B., AND L. HERMANSEN. Glycogen stores and prolonged severe exercise. In Nutrition and Physical Activity, Almqvist & Wiksell, Uppsala, pp. 32-46, 1967.

14. SCHOLZ, R., H. SCHMITZ, T. BÜCHER, AND J.O. LAMPEN. Über die Wirkung von Nystatin auf Bäckerhefe. Biochem. Z. 331:71-86, 1959.

15. VARNAUSKAS, E., P. BJÖRNTORP, M. FAHLÉN, J. PREROVSKY, AND J. STENBERG. Effects of physical training on exercise - blood flow and succinic dehydrogenase activity in skeletal muscle. Cardiovasc. Res. 1970. In press.

THE EFFECT OF PROPRANOLOL ON GLYCOGEN METABOLISM DURING EXERCISE

R.C. HARRIS, J. BERGSTRÖM, and E. HULTMAN

From the Department of Clinical Chemistry, S:t Eriks Sjukhus, Stockholm, Sweden

It is well known that adrenalin (epinephrine) will cause activation of glycogen phosphorylase (EC 2.4.1.1.) through an increased conversion of phosphorylase __b__ to __a__ (1, 2, 3). This reaction has been shown to occur in skeletal muscle (4, 6, 7) and to be the result of an activation of phosphorylase kinase (EC 2.7.1.38.) mediated through increased levels of 3'5' AMP (5). In a series of unpublished experiments performed in this laboratory it was shown that infusion of adrenlin resulted in decreased levels of glycogen in the quadriceps femoris muscle and an increase in lactate production. In an attempt to investigate whether or not adrenalin released during moderate or heavy bicycle exercise was of importance in regulating glycogen breakdown, experiments were performed in which the beta-receptor sites in the working muscles were blocked. The blocking agent used was propranolol. The results obtained did not show any direct effects of beta-blocking upon glycogen utilization but did point to possible changes at the level of phosphofructo-kinase (EC 2.7.1.11.).

Subjects worked for one or two min at a load of 400, 1,000, or 1,200 kpm. In the simplest experiments needle biopsy samples were taken from the quadriceps femoris muscle before and after the exercise. In the experiment to be shown here further biopsies were taken 5 and 30 min after the termination of exercise. After a suitable period of rest propranolol (5 mg) was given intravenously and one hour later the experiment was repeated - the subject working at the load previously used and for the same work time. Biopsies were freeze-dried and extracted as previously described by Dr Bergström and were analysed for glycolytic intermediates and high energy phosphate co-factors. Biopsies for glycogen determinations were taken separately.

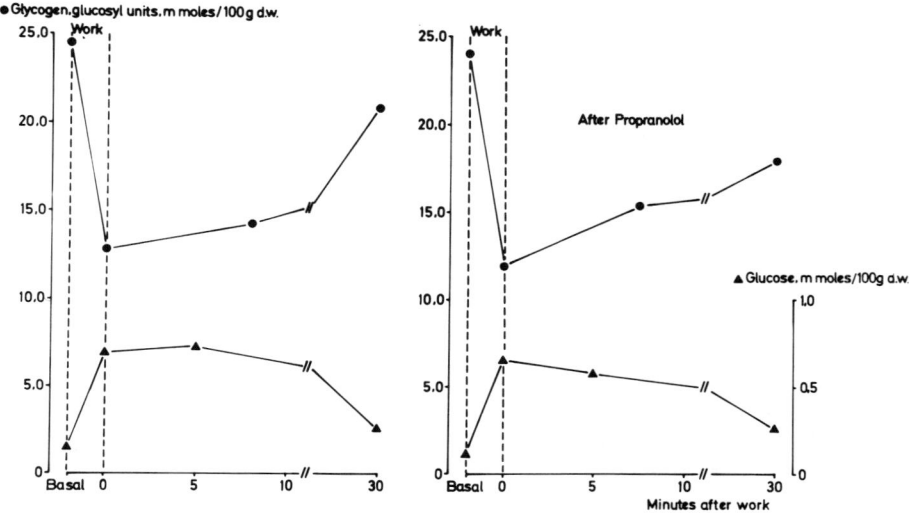

Fig. 1. The levels of glycogen and glucose in biopsy samples taken from the quadriceps femoris muscle at rest, after 2 min work at 1,200 kpm and after 5 and 30 min rest. Before and after propranolol administration.

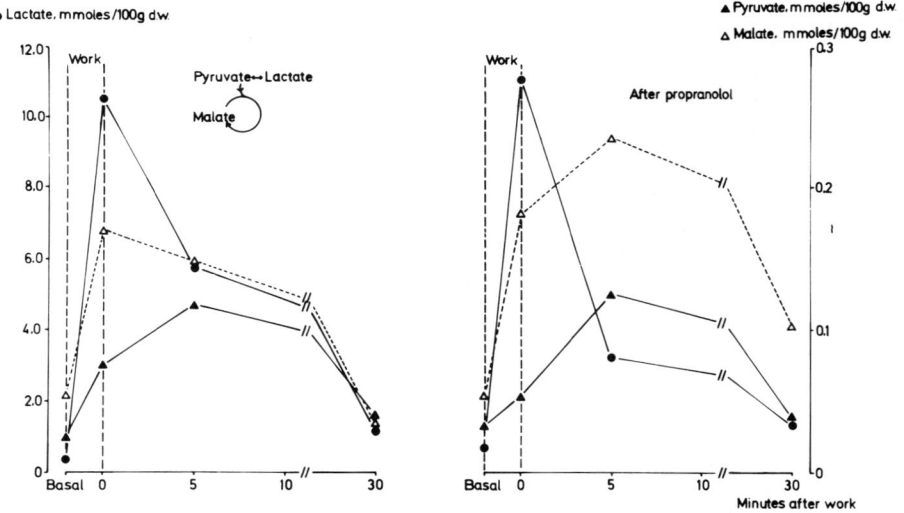

Fig. 2. The levels of pyruvate, lactate, and malate in biopsy samples taken from the quadriceps femoris muscle at rest, after 2 min work at 1,200 kpm and after 5 and 30 min rest. Before and after propranolol administration.

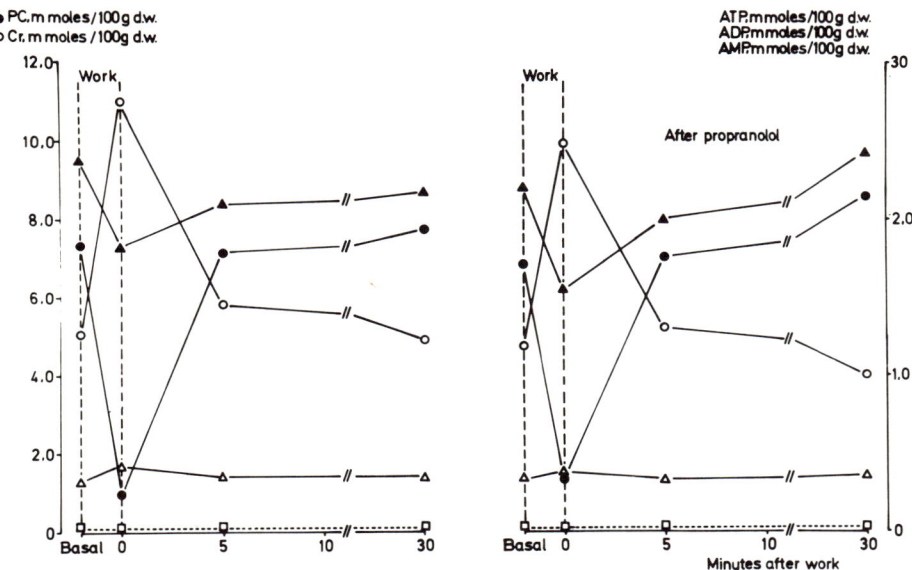

Fig. 3. The levels of ATP, ADP, AMP, creatine-P, and creatine in biopsy samples taken from the quadriceps femoris muscle at rest, after 2 min work at 1,200 kpm and after 5 and 30 min rest. Before and after propranolol administration.

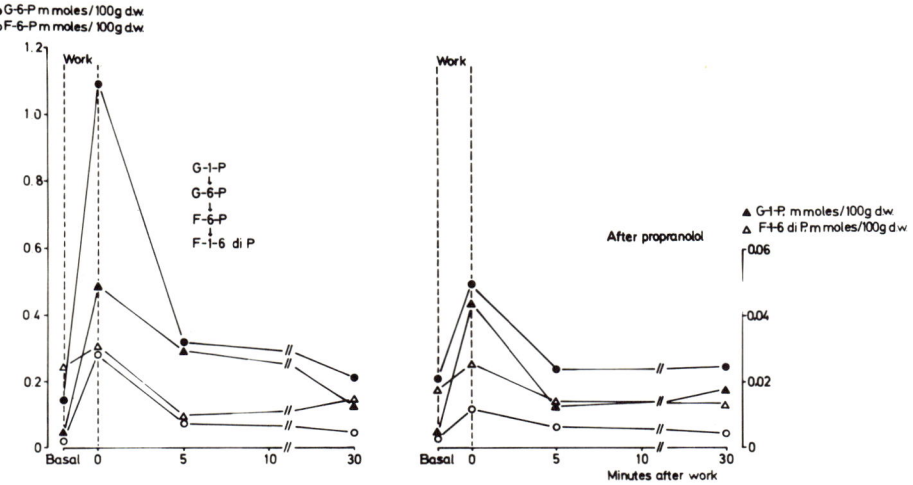

Fig. 4. The levels of glucose 1-P, fructose 6-P, and fructose 1-6-diP in biopsy samples taken from the quadriceps femoris muscle at rest, after 2 min work at 1,200 kpm, and after 5 and 30 min rest.

The results showed that after the performance of the same amount of work (before and after receiving propranolol) approximately the same amounts of glycogen were consumed. This is illustrated in Fig. 1. In this case the subject worked for 2 min at a load of 1,200 kpm after which time he had consumed 11.7 mmol glucosyl unit-equivalents of glycogen/100 g dry muscle. After receiving propranolol and after the performance of the same amount of work he was found to have consumed a further 12.1 mmol glucosyl unit-equivalents of glycogen.

Glycogen consumption following propranolol administration in 4 experiments was not significantly different from that in the control (as a mean 93 % $\pm$ SE 6.7 %). Similarly, lactate increase during work was approximately the same before and after receiving propranolol (Fig. 2). The mean increase in pyruvate + lactate following propranolol administration was, in 6 experiments, 95 % $\pm$ SE 4.8 % of the control. The amounts of α-glycerol-P accumulated by the end of exercise were unchanged by propranolol, and ATP and creatine-P breakdown were unaffected (Fig. 3). The accumulation of glucose 6-P and fructose 6-P on the other hand was decreased (Fig. 4). In 6 experiments the increase in glucose 6-P after work following propranolol administration was only 40 % $\pm$ SE 3.6 % of that in the control. The ratio of glucose 6-P : fructose 6-P was unchanged. Glucose 1-P and fructose 1-6-diP, the levels of which are low and difficult to measure, did not show any consistent pattern of change following propranolol administration.

The results obtained in this study are difficult to interpret. The apparent absence of any effect of beta-receptor blockade upon glycogen utilization does not necessarily eliminate the possibility that in the normal state adrenalin does exert a positive effect. However, the present results would indicate that either this is not of importance during the early stages of work or the muscle cells are capable of compensating for its absence. The difference in the levels of accumulated glucose 6-P and fructose 6-P after work in the two cases could be explained by an activation or de-inhibition of phosphofructokinase following propranolol administration. Other explanations are possible.

This investigation was supported by grants from the Swedish Medical Research Council (Projects nos B70-19X-1002-05B, B70-19X--2647-02B, and K70-19X-2647-02BK).

REFERENCES

1. DANFORTH, W.H. and E. HELMREICH. Regulation of glycolysis in muscle. I. The conversion of phosphorylase b to a in frog sartorium muscle. J. Biol. Chem. 239: 3133 - 3138, 1964.
2. HAMMERMEISTER, K.E., A.A. YUNIS, and E.G. KREBS. Studies on phosphorylase activation in the heart. J. Biol. Chem. 240: 986 - 991, 1965.
3. KARPATKIN, S., E. HELMREICH, and C.F. CORI. Regulation of glycolysis in muscle. II. Effect of stimulation and epinephrine in isolated frog sartorium muscle. J. Biol. Chem. 239: 3139 - 3145, 1964.
4. KREBS, E.G. and E.H. FISCHER. The phosphorylase b to a converting enzyme of rabbit skeletal muscle. Biochim. Biophys. Acta 20: 150 - 157, 1956.
5. KREBS, E.G., D.J. GRAVES, and E.H. FISCHER. Factors affecting the activity of muscle phosphorylase b kinase. J. Biol. Chem. 234: 2867 - 2873, 1959.
6. KREBS, E.G., A.B. KENT, and E.H. FISCHER. The muscle phosphorylase b kinase reaction. J. Biol. Chem. 231: 73 - 83, 1958.
7. RALL, T.W., W.D. WOSILAIT, and F.W. SUTHERLAND. The interconversion of phosphorylase a and phosphorylase b from dog heart muscle. Biochim. Biophys. Acta 20: 69 - 76, 1956.

LOCAL LIPID STORES AND EXERCISE

S.O. Fröberg, L.A. Carlson and L.-G. Ekelund

Department of Geriatrics, University of Uppsala
Uppsala, and the Department of Clinical Physiology
Karolinska sjukhuset, Stockholm, Sweden

During exercise in the fasting state it has been found that the amount of fat that is metabolized cannot be accounted for only by oxidation of plasma FFA (6,7,9). This has led to the assumption that also muscle triglycerides, as a source for fatty acids, participate in the oxidative metabolism during exercise. Direct studies on the effect of exercise on the muscle triglyceride concentration are, however, conflicting. A decrease in the concentration of muscle triglycerides has been observed in bird wing muscle after exercise (5) but not in mammalian skeletal muscle (3,4,10). In the present paper results will be presented which indicate that also in man the muscle triglycerides may serve a function as a source of fatty acids for oxidation during exercise.

RESULTS

Healthy men were exercised to exhaustion on a bicycle. The mean worktime was 99 minutes (range 40-178 minutes) and the mean work output was 7.02×10^4 kpm (range $3.27 \times 10^4 - 10.69 \times 10^4$ kpm). They started to work at about 70 per cent of the workload at a heart rate of 170 (W_{170}).

Table I shows that the mean oxygen consumption increased from 0.27 l/min at rest to 2.01 l/min during the exercise. The mean RQ increased from 0.80 at rest to 0.86 during exercise. According to the oxygen consumption and the respiratory quotient during the exercise 1575 ± 136 μmoles of fatty acids and 8524 ± 136 μmoles of glucose were oxidized per minute (Table I) which corresponds to a mean caloric expenditure of 3.94 and 6.29 KCal per minute respectively.

TABLE I

	Rest			Exercise			
	O_2 l/min	RQ		O_2 l/min	RQ	Oxidized µmole/min[1] fatty acids[2]	glucose[3]
M	0.27	0.80		2.01	0.86	1575	8524
SEM	0.01	0.02		0.13	0.01	136	986
Range	0.15-0.33	0.71-0.97		1.37-3.02	0.80-0.96	528-2226	3861-14419

Table I: Oxygen consumption and RQ during rest and exercise of estimated amount of fatty acids and glucose oxidized during the exercise (N=13)

M = mean   SEM = standard error of the mean

1) Mean of values calculated for each subject separately.

2) For oxidation of 1 mol of fatty acid (17C) 24.5 x 22.4 litres of oxygen was assumed to be needed.

3) For oxidation of 1 mol glucose 6 x 22.4 litres of oxygen was assumed to be needed.

TABLE II

	TRIGLYCERIDES µmole/g			PHOSPHOLIPIDS mg/g			GLYCOGEN mg/g		
	B	A	B-A	B	A	B-A	B	A	B-A
M	10.42	7.82	2.60	7.74	7.65	0.06	10.36	3.40	6.96
SEM	0.93	0.80	0.69	0.26	0.22	0.23	0.71	0.60	0.90
N	24	24	24	24	24	24	17	17	17
P			<0.001			>0.05			<0.001

Table II: Concentration of triglycerides, phospholipids and glycogen in the lateral vastus of the femoral muscle before and after exercise.

B = before   A = after exercise   B-A = individual difference before (B) and after (A) exercise.

M = mean   SEM = standard error of the mean
N = number of subjects   P = degree of statistical significance for the difference B-A.

In Table II the concentration of triglycerides, phospholipids and glycogen in muscle before and immediately after cessation of exercise is given. The muscle was obtained from the lateral vastus of the femoral muscle by needle biopsi technique. The concentration of triglycerides was decreased with an average of about 25 per cent after the exercise ($P < 0.01$) while no change was observed in the phospholipid concentration. Also the glycogen concentration decreased during the exercise. This decrease was about 67 per cent of the pre-exercise level.

To get some further information on the metabolic significance of the muscle triglycerides the data were analysed with regard to work performance. Some regression constant are given in Table III which also includes some regression constants on muscle glycogen.

Table III

x	y	r	p
W	TG_i	-0.53	<0.01
W	ΔTG	-0.40	<0.05

x	y	r	p
W	Glg_i	0.60	<0.02
W	ΔGlg	0.56	<0.05

x	y	r	p
TG_i	Glg_i	0.02	>0.05
ΔTG	Glg	-0.34	>0.05

W = work performed
TG_i = initial triglyceride concentration
ΔTG = change in triglyceride concentration
Glg_i = initial glycogen concentration
ΔGlg = change in glycogen concentration

Table III: Regression constant with regard to some variables.

The initial triglyceride concentration in the muscle as well as the decrease in the triglyceride concentration during the exercise were <u>negatively</u> correlated to the work performance. On the other hand, the initial glycogen concentration as well as the decrease in the glycogen concentration during exercise was <u>positively</u>

related to the work performance. No correlation was found between the initial concentration of triglycerides and glycogen in the muscle nor was the change in the two endogenous muscle substrates correlated to each other.

DISCUSSION

In the present study subjects were exercised at a work intensity of about 70 per cent of the workload at a puls rate of 170. The glycogen concentration decreased as has previously been observed under similar experimental conditions (1,2,8). The positive correlation between the pre-exercise glycogen concentration and work performance confirms previous findings (1).

A new finding was that also the muscle triglyceride concentration decreased during the exercise. In contrast to the findings with regard to muscle glycogen the muscle triglyceride concentration was negatively correlated to the amount of work performed as was also the pre-exercise triglyceride concentration.

Table IV

	Before exercise	During exercise		After exercise	
		15 min	end	15 min	60 min
M	0.65	0.56	0.86	1.39	0.87
SEM	0.06	0.07	0.14	0.16	0.13
P		>0.05	>0.05	<0.05	>0.05

For symbols see Table II.

Table IV: Concentration of plasma free fatty acids (FFA) before, during and after exercise.

In an attempt to estimate the fractional contribution of muscle triglyceride fatty acids to the energy metabolism during exercise we have used the data in Table I, II and IV which gives the mean plasma FFA concentration in five subjects during the exercise. Under the assumptions given in Table V about 75 per cent of the fatty acids oxidized were derived from muscle triglycerides and about 30 per cent from the plasma FFA. Using a similar type of calculation for the carbohydrate oxidation the muscle glycogen was estimated to account for about 70 per cent of the glucose oxidized.

The estimations in Table V indicate that not only muscle glycogen but also muscle triglycerides are of significant importance in

Table V

	Fatty acids		Carbohydrates
	Muscle[1]) triglyceride fatty acids	Plasma[2]) FFA	Muscle[1]) glycogen (as glucose)
µmole oxidized per minute	1170	500	5780
per cent of total amount of fatty acids, respectively carbohydrates, oxidized (Table I)	74	32	69

1) Assumptions: Muscle mass 40 per cent of the body weight, 50 per cent of muscle mass was active (14.8 kg) during exercise and oxidized the amount of substrates which had disappeared from muscle tissue as shown in Table II. (1 µmol of triglyceride = 3 µmoles of fatty acid).

2) Based on the data in ref. 6 where almost identical subjects were investigated. In that study, where the average FFA level in plasma was 920 µmoles/ml about 600 µmoles of FFA were oxidized per minute. Here we have used a plasma FFA value of 800 µmoles/ml (Table IV).

Table V: Estimated contribution of fatty acids from muscle triglycerides and plasma FFA and of glycogen-glucose during bicycle exercise.

the caloric homeostasis during exercise. Our data did, however, not indicate a direct relationship neither between the change in the concentration of triglycerides and of glycogen in the muscle during the exercise, nor between the pre-exercise concentration in the muscle of these muscle substrates. The physiologic implication of this is not clear. It should be emphasized, however, that the fractional contribution of fatty acids and/or glucose from endogenous muscle stores to the energy metabolism may have changed from the beginning to the end of the exercise. Thus e.g. the muscle glycogen concentration decreases rapidly in the beginning of bicycle exercise but more slowly when work is continued (8).

Support for the hypothesis that muscle triglycerides are utilized during exercise is also given from other studies on the effect of exercise on muscle triglycerides in rat and man (Table VI). It

TABLE VI

	RAT		
ELECTRICALLY INDUCED 5 twitches/sec. for 2 h basal resistance: 40 g		RED MUSCLE TISSUE[1]: WHITE MUSCLE TISSUE[1]:	24 %; $p<0.001$ 27 %; $p<0.01$
RUNNING 33 cm/sec for 3 h.		RED MUSCLE TISSUE[1]: WHITE MUSCLE TISSUE[1]: MYOCARDIUM:	44 %; $p<0.001$ 31 %; $p<0.05$ 22 %; $p<0.05$
	MAN		
SKIING mean skiing time 7 h. 45 min. mean speed 11.5 km/h.		LEG MUSCLE[2]: triglycerides glycogen	51 %; $p<0.01$ 56 %; $p<0.01$

1) Gastrocnemius muscle
2) Lateral vastus of the femoral muscle

Table VI: Per cent decrease of muscle triglycerides in rat and man and of glycogen in man during exercise (Fröberg, unpublished results).

is of interest that during skiing for seven hours the net decrease in the muscle triglyceride concentration calculated per gram of the muscle tissue obtained by needle biopsi corresponded to 69 calories and the decrease in the glycogen concentration to 35 calories. In the present study the corresponding values became 20 calories for triglycerides and 29 calories for glycogen.

SUMMARY

Fasting healthy subjects were exercised to exhaustion on the bicycle. The muscle triglyceride concentration decreased from 10.4 to 7.8 µmole per gram wet weight during the exercise. The values for the muscle glycogen concentration were 10.4 and 3.4 mg per gram wet weight before and after the exercise respectively.

The amount of fatty acids and glucose that were oxidized was estimated on the basis of the respiratory quotient and the oxygen consumption during the exercise. It was estimated that 3/4 and 1/4 of the fatty acids oxidized were derived from muscle triglycerides and plasma fatty acids respectively.

The pre-exercise concentration of triglycerides in the muscle was negatively correlated to the amount of work performed as was also the decrease in the muscle triglyceride concentration during the exercise. The pre-exercise glycogen concentration as well as the decrease during the exercise was, however, positively correlated to the work performance.

REFERENCES

1. Ahlborg, B., Bergström, J., Ekelund, L.-G. and Hultman, E.: Muscle glycogen and muscle electrolytes during prolonged physical exercise. Acta physiol. scand. 70: 129-142, 1967.

2. Bergström, J., Hermansen, L., Hultman, E. and Saltin, B.: Diet, muscle glycogen and physical performance. Acta physiol. scand. 71: 140-150, 1967.

3. Buchwald, K.W. and Cori, C.F.: Influence of repeated contractions of muscle on its lipid content. Proc. Soc. Exptl. Biol. Med. 28: 737-740, 1930-31.

4. Gemill, C.L.: The effect of stimulation on the fat and carbohydrate content of the gastrocnemius muscle in the phlorizinized rat. Bulletin Johns Hopkins Hosp. 66: 71-89, 1940.

5. George, J.C. and Berger, A.J.: Avian Myology, Academic Press, New York/London, 1966.

6. Havel, R.J., Carlson, L.A., Ekelund, L.-G. and Holmgren, A.: Turnover rate and oxidation of different free fatty acids in man during exercise. J. Appl. Physiol. 19: 613-618, 1964.

7. Havel, R.J., Naimark, A. and Borchgrevink, C.F.: Turnover rate and oxidation of free fatty acids of blood plasma in man during exercise: Studies during continuous infusion of palmitate-1-C^{14}. J. Clin. Invest. 42: 1054-1063, 1963.

8. Hermansen, L., Hultman, E. and Saltin, B.: Muscle glycogen during prolonged severe exercise. Acta physiol. scand. 71: 129-139, 1967.

9. Issekutz, Jr., B., Miller, H.I., Paul, P. and Rodahl, K.: Source of fat oxidation in exercising dogs. Am. J. Physiol. 207: 583-589, 1964.

10. Masoro, E.J., Rowell, L.B., McDonald, R.M. and Steiert, B.: Skeletal muscle lipids. II. Nonutilization of intracellular lipid esters as an energy source for contractile activity. J. Biol. Chem. 241: 2626-2634, 1966.

INFLUENCE OF INTENSITY AND DURATION OF EXERCISE ON SUPPLY AND USE OF FUELS

Richard J. Havel

Cardiovascular Research Institute and

Dept. of Medicine, Univ. of California, San Francisco, U.S.A.

In mobile animals, survival may depend upon ability to perform muscular work, often during periods of limited food supply. This work may be of high intensity, of long duration, or, occasionally, both. It is to be expected, therefore, that adaptations to meet these varying demands have developed. Because of the large energy requirements of the active as opposed to the resting state, these adaptations should include mechanisms to assure an adequate supply of fuel.

Studies performed during the last decade, including several reported at this Conference, have provided a large body of information concerning quantitative aspects of fuel supply and utilization during muscular exercise in large mammals, especially man. This paper will attempt, on the basis of these studies and certain information provided by structural and dynamic investigations on other animal species, to arrive at some generalizations concerning use of intra- and extramuscular fuels during exercise of different intensities and duration.

STRUCTURE-FUNCTION RELATIONSHIPS

Before reviewing quantitative aspects of fuel supply and utilization in man, it will be useful to point out certain relationships between structure and function of muscles in some lower forms and in muscles adapted for special uses. Particularly valuable are comparisons between classes of animals which include migratory and nonmigratory species (1-3). The former store large quantities of fat before undertaking long journeys. Examples are:

in insects, the locust; in birds, the migratory hummingbird and the goose; in fish, the salmon and eel. Their muscles are red, because of the large quantities of myoglobin and mitochondria, and are rich in fat. The latter store little fat: in insects, the honey bee and fly; in birds, the domestic chicken and turkey; in fish, the sessile "bottom" species such as the sole and cod. The main muscles of locomotion are pale, reflecting low content of myoglobin and mitochondria and are fat-poor. Muscles used primarily for rapid escape or protection from predators are generally "fast" and structurally are mainly of the white type (catch muscle of shell fish, frog gastrocnemius, dorsal muscles of tuna, pectoralis of domestic fowl). Muscles used continuously are red (deep and lateral superficial muscle of tuna, leg of domestic fowl, mammalian diaphragm). Electron microscopic studies of the diaphragm of small mammals, which contracts about twice a second, show that they are unusually rich in mitochondria and contain numerous fat drops closely associated with mitochondria in the region of the Z line (4). Since white fibers are generally larger and less well supplied by blood capillaries than red fibers (5), the structure of white fibers seems poorly suited to uptake of fuels from the blood and their aerobic oxidation. The stored fuels in white fibers, mainly phosphagen and glycogen, can yield ATP anaerobically, so that these fibers can be expected to contract rapidly without the intervention of extramuscular phenomena related to increased delivery of fuel and oxygen from the blood. The structure of red fibers seems clearly suited to uptake and aerobic oxidation of fuels supplied in the blood and to oxidation of fat stored within the muscle cells. Studies of fish muscle in vitro indicate higher oxygen consumption and capacity to oxidize fatty acids in red muscle (6, 7).

In the pigeon pectoralis, like many muscles in man composed of both white and red fibers, onset of contractions induced by electrical stimulation of its motor nerve is followed first by depletion of glycogen in the large white fibers; later, glycogen and fat are depleted in the smaller red fibers (8). In mixed mammalian muscles, there is considerable evidence that groups of muscle fibers supplied by individual motor units are structurally homogeneous (9, 10). Electrical stimulation first activates phosphorylase in white (type A) fibers (11, 12) followed in sequence by activation in two histochemically distinct types of red fibers (B and C) (12). Fatigue occurs first in A fibers, associated with loss of glycogen, while fat-rich C fibers continue to contract as their phosphorylase activity decreases. Type A fibers are innervated by high threshold motor units which discharge phasically, suggesting that they are adapted for short periods of contraction involving large work loads. Microelectrode studies of human tibial muscles indicate that rapid contraction first activates motor units with the properties of white fibers while slow contraction or prior proprioceptive stimulation activates tonic (red) fibers (13). These observations indicate that the nature of the contractile machinery and the associated use of fuels are functions of the intensity and duration of the neurogenic stimulus.

INFLUENCE OF BLOOD FLOW ON SUPPLY OF FUELS

Supply of fuels from the blood is related to their concentration and to rate of nutrient flow to the muscle. With onset of exercise, flow rapidly increases. This influences particularly the uptake of substances whose transport is largely flow-limited. For FFA, the major lipid fuel, evidence is that uptake into the legs during bicycle exercise increases from about 25% of total outflow transport at rest to 55% for light work (14) and to 75% during heavy work (15). This is accompanied by somewhat greater increases in fractional distribution of blood flow to the legs, reflecting a reduced extraction fraction with increasing flow (14). Conversely, a smaller fraction of FFA is delivered to nonmuscular tissues during exercise. The splanchnic region extracts about 40% of FFA transported in the blood at rest. Studies in rats indicate that this falls to very low levels with leg exercise (16).

MOBILIZATION OF FUELS

In the post-absorptive state in man, FFA are the chief fuel for most tissues, excepting mainly the glucose-dependent brain, erythrocyte and renal medulla (17). Their entry into the blood from adipose tissue increases rapidly with leg exercise (18, 19). Adrenergic stimulation of the adenyl cyclase system in adipose tissue, leading to activation of its hormone-sensitive lipase, appears to be mainly responsible for release of FFA from stored triglycerides (17, 20). Evidence for this includes the rapidity of onset of the mobilization, the sensitivity of fat mobilization to intravenously infused catecholamines, especially beta-mimetics, and inhibition of the process by beta-adrenergic blocking agents. Levels of growth hormone in plasma also rise with leg exercise. Increased secretion of growth hormone may contribute to fat mobilization after one to two hours of exercise, but is unlikely to be responsible for the rapid mobilization since the increase in concentration of growth hormone usually does not occur until 15 to 30 minutes after onset of exercise (21) and its effect upon fat mobilization is delayed in onset, appearing to require new protein synthesis. Secretion of growth hormone is promoted by alpha-adrenergic stimulation (22, 23), presumably through effectors in the hypothalamus which promote secretion of growth hormone-releasing factor. Thus, secretion of growth hormone as well as norepinephrine from sympathetic nerve endings during exercise may be related to central adrenergic stimulation. Such adrenergic activation could also promote fat mobilization through stimulation of alpha receptors known to inhibit secretion of insulin. Some (24, 25), but not all (26-28), studies show reduction in plasma level of insulin with moderate to heavy leg exercise. However, fat mobilization is more, rather than less, sensitive to the stimulus of exercise in insulin-dependent diabetics (29). Although the stimulus to fat mobilization appears to be basically neurogenic, the afferent pathways are not known.

Production of glucose from the liver may also increase with exercise (30, 31). This may also result from direct adrenergic stimulation, since evidence for the requisite innervation and its relation to activation of hepatic phosphorylase has been obtained (32).

SUPPLY AND USE OF FUELS IN THE POST-ABSORPTIVE STATE

At Rest

Rates of transport of the major fuels, FFA and glucose, and of the conversion of ^{14}C-labeled tracers to CO_2 have been measured by use of radioactive tracers. Additionally, regional production and uptake of tracers has been studied with the aid of appropriately placed catheters. Calculations from such data may fail to provide precise information concerning the contribution of various fuels to oxidative metabolism for several reasons. First, it must be realized that the post-absorptive state is not a steady state. Content of stored triglycerides and glycogen in tissues such as liver and muscle is changing as a result of imbalances between rates of inflow or outflow transport of glucose and FFA and their utilization. Second, measurements of conversion of ^{14}C in metabolites to CO_2 often underestimate extent of "rapid" oxidation because of isotopic exchange reactions. For example, ^{14}C in the carboxyl-carbon of palmitic acid may be transferred to glucose and to amino acids, a situation not reflecting net movement of carbon. For certain other fuels, the effective pool with which the tracer mixes rapidly may include not only the substance in question, but other metabolites. For example, ^{14}C in infused lactate may be incorporated into alanine as well as pyruvate so that apparent rates of transport may be erroneously large.

With these limitations in mind, certain generalizations are possible. Uptake of FFA-carbon into resting leg or forearm muscle is sufficient to account for production of CO_2 while uptake of glucose-carbon, corrected for release of lactate, can account for only 10 to 20 per cent (33, 34). Limited uptake of ketone-carbon (35) also occurs. Most of the FFA-carbon taken up is not rapidly released as CO_2 (14) and, from studies in experimental animals (16), it appears that an appreciable fraction is deposited as lipid-esters. Careful measurements of R.Q. across regions supplying mainly extremity muscles are consistent with almost exclusive fatty acid oxidation (14, 33, 34). These data can best be interpreted to indicate that FFA and stored lipids contribute the bulk of fuel and that there is continuous synthesis and breakdown of ester bonds so that triglyceride and phospholipid fatty acids constitute part of the pool from which the fuel is derived without appreciable net use. The fraction of glucose taken up which is oxidized is not established. Part is clearly converted to lactate (33). Limited oxidation of glucose is also to be expected from known rates of splanchnic production and of utilization in brain, erythrocytes and kidneys (17). Resting muscles also release amino acids from breakdown of protein (36, 37), but there is no evidence for appreciable oxidation locally. Transaminations, however, appear to give rise to interchange of carbon from carbohydrates with amino acids, especially alanine (38).

Exercise of Short Duration and High Intensity

Immediate fuels for this type of work include preformed high energy phosphate, mainly phosphagen, and conversion of glycogen to lactate (31). They are presumably utilized mainly in white fibers, which, as already noted, are activated first. ATP is thus resynthesized until blood flow increases. The importance of glycogenolysis is supported by the limited ability of individuals lacking myophosphorylase (McArdle's syndrome) to perform this type of work (39).

Exercise of Short Duration and Low Intensity

Use of stored glycogen, evidenced by release of lactate, occurs at the onset of light work (40). This appears to diminish after a few minutes, presumably as blood flow increases and red fibers are activated (41). This sequence fits well with the observed functional disability of individuals with McArdle's syndrome at the onset of exercise and the subsequent occurrence of a "second wind phenomenon". The disappearance of painful swelling of working muscles when these individuals continue to exercise, heralded by falling heart rate, is not accompanied by alterations in plasma concentration of major fuels (39). Increase in blood flow produced by arterial infusion of ATP eliminates their initial symptoms, presumably bypassing the phase of anaerobic metabolism (39).

Exercise of Long Duration and Low Intensity

As already noted, release of lactate rapidly decreases when light work continues for more than a few minutes. The R.Q. for leg exercise then varies from 0.70 to 0.85 and seems to be lower in trained subjects who accumulate little or no lactate in body fluids (18, 19). Measurements of lactate release from muscles of reasonably fit subjects indicate that metabolism is essentially aerobic (14). Mobilization of FFA increases rapidly during the first hour. Uptake of FFA from the blood after 1 to 4 hours accounts for from 25 to 50 per cent of CO_2 production and seems to be higher in trained subjects (18, 19). Virtually all of ^{14}C-labelled FFA entering muscle is rapidly oxidized (14), confirming the substantial contribution of FFA to oxidative metabolism. Some use of glycogen continues, its fate being complete combustion (31). Measurements of uptake of substrates, including FFA, ketones, triglycerides and glucose, from the blood and of R.Q. across leg tissues suggest that fat as well as glycogen stores in muscle are used during such exercise (14). Uptake of blood-borne substrates into leg tissues was also grossly insufficient to account for oxidative metabolism in a subject with McArdle's syndrome (39). This strongly suggests substantial use of fat stored within the muscle since muscle glycogen presumably could not be catabolized. The extent to which adipose tissue interspersed between muscle fibers rather than fat within muscle cells contributes to local use of fat is uncertain. Failure of measurements of lipid content of muscle obtained during exercise to provide clear evidence for use of fat (42) may be attributed to difficulties in sampling, related in part to the variable content of adipocytes in tissue obtained by needle biopsy. When light exercise is continued beyond the first hour, use of FFA continues to increase until its concentration reaches a value of about 2mM after 9 hours (43). Meanwhile, blood glucose concentration falls gradually to about 4mM. At this time, a metabolic steady state is reached with a ventilatory R.Q. near 0.70. Estimates of transport of FFA (20) suggest that FFA then supply virtually all the fuel of working muscle, and almost 90% of all oxidative metabolism. Glucose oxidation is twice the resting value (30). Since glycogen stores presumably are depleted, this may be attributed to increased gluconeogenesis, mainly from glycerol released during mobilization of fat from adipose tissue. That muscle glycogen is not essential for such exercise is confirmed by the ability of subjects with McArdle's syndrome to perform such work for hours (39).

After a few minutes of light exercise of forearm muscles produced by flexion and extension of the hand at the wrist, uptake of FFA-carbon can account for all of CO_2 production (34). This difference from leg exercise may reflect different work intensities or differences in blood flow, but could also result from more fundamental differences in structure or metabolism of the muscle groups. However, at somewhat higher work loads, produced by contraction against a load, the metabolic pattern in forearm muscles resembles more closely that observed in leg tissues during exercise on a bicycle ergometer (44).

Exercise of Long Duration and High Intensity

With increasing intensity of leg work, the fraction of needed fuel that can be supplied from the blood decreases and use of stored fuel increases. Glycogen content of leg muscle falls rapidly and work can generally not be sustained for long, when levels fall below 0.1 per cent of wet weight (31). Capacity to perform such work correlates well with initial glycogen content (31). Highly trained subjects can continue to exercise for at least two hours at levels requiring an eight-fold increase in oxygen consumption without accumulating appreciable lactate in the blood. Kinetic analysis of production of $^{14}CO_2$ after pulse injections of ^{14}C-palmitate and $NaH^{14}CO_3$ in such individuals demonstrates that almost all palmitate is rapidly oxidized, but an appreciable fraction transiently enters pools that appear to represent lipid esters (15). This is consistent with rapid turnover of lipid stores within the working muscles under such conditions. Lipid within muscle cells may also contribute to the metabolic mixture during such exercise, but these stores apparently cannot be utilized to an extent sufficient to obviate the need for glycogenolysis. It seems likely that endurance for such exercise in the post-absorptive state is related not only to initial stores of glycogen, but also to ability to deliver fuels, especially FFA, from the blood.

CONCLUSIONS

Skeletal muscle is omniverous. However, available data indicate that FFA are its major fuel in post-absorptive humans at rest and during at least some forms of light work. It, therefore, seems likely that they are preferred to glucose or glycogen under these conditions. Support for this proposition is provided by evidence for increased use of blood glucose and muscle glycogen when fat mobilization is inhibited by nicotinic acid (45, 46) or 5-methylpyrazole-3-carboxylic acid (35) during leg exercise. Use of stored glycogen, mainly in white fibers, occurs at onset of all levels of exercise. Its use subsequently, presumably mainly in red fibers, seems to depend upon the extent to which other fuels, particularly FFA from the blood, can supply energy needs. Thus, more glycogen is used as intensity of work increases and it appears that less is used at a given work load in trained subjects who can deliver more fuels from the blood. With prolonged work of high intensity, endurance may be limited by availability of fuels and generally cannot be sustained after stores of glycogen in muscle have been depleted. In the post-absorptive state, use of other fuels carried in the blood is limited, although small amounts of ketones and triglyceride fatty acids may be used. Little is known about use of fuels in other nutritional states. Blood glucose is almost certainly an important fuel after ingestion of carbohydrate. It is probable that triglycerides can supply appreciable fuel during alimentary lipemia and acetate after ingestion of alcohol, while ketones may be an important source during starvation. Muscular exercise has major effects upon the fate of transported fuels and the extent of muscular activity must be taken into account in evaluating the relationship between nutrition and health.

REFERENCES

1. Weis-Fogh, T. Fat combustion and metabolic rate of flying locusts. Royal Soc. of London (Phil. Trans.) Series B 237:1, 1952.
2. Drummond, G.I. and Black, E.C. Comparative physiology: fuel of muscle metabolism. Ann. Rev. Physiol. 22:169, 1960.
3. George, J. C. and Berger, A.J. Avian Myology. Academic Press, New York. 1966.
4. Gauthier, G.F. and Padykula, H.A. Cytological studies of fiber types in skeletal muscle. A comparative study of the mammalian diaphragm. J. Cell. Biol. 28:333, 1966.
5. Romanul, F.C.A. Capillary supply and metabolism of muscle fibers. Arch. Neurol. 12:497, 1965.
6. Gordon, M.S. Oxygen consumption of red and white muscles from tuna fishes. Science 159:87, 1968.
7. Bilinski, E. Utilization of lipids by fish. I. Fatty acid oxidation by tissue slices from dark and white muscle of rainbow trout(Salmo Gairdnerii). Can. J. Biochem. Physiol. 41:107, 1963.
8. George, J.C. and Jyoti, D. The lipid content and its reduction in the muscle and liver during long and sustained muscular activity. J. Animal Morphol. Physiol. 2:37, 1955.
9. Romanul, F.C.A. and Van Der Meulen, J.P. Slow and fast muscles after cross innervation. Arch. Neurol. 17:387, 1967.
10. Edström, L. and Kugelberg, E. Histochemical composition, distribution of fibres and fatiguability of single motor units. J. Neurol. Neurosurg. Psychiat. 31:424, 1968.
11. Stubbs, S. St.G. and Blanchaer, M.C. Glycogen phosphorylase and glycogen synthetase activity in red and white skeletal muscle of the guinea pig. Can. J. Biochem. 43:463, 1965.
12. Kugelberg, E. and Edström, L. Differential histochemical effects of muscle contractions on phosphorylase and glycogen in various types of fibres: relation to fatigue. J. Neurol. Neurosurg. Psychiat. 31:405, 1968.
13. Grimby, L. and Hannerz, J. Recruitment order of motor units on voluntary contraction: changes induced by proprioceptive afferent activity.

J. Neurol. Neurosurg. Psychiat. 31:565, 1968.
14. Havel, R.J., Pernow, B. and Jones, N.L. Uptake and release of free fatty acids and other metabolites in the legs of exercising men. J. Appl. Physiol. 23:90, 1967.
15. Havel, R.J., Ekelund, L-G. and Holmgren, A. Kinetic analysis of the oxidation of palmitate-1-^{14}C in man during prolonged heavy muscular exercise. J. Lipid Res. 8:366, 1967.
16. Jones, N.L. and Havel, R.J. Metabolism of free fatty acids and chylomicron triglycerides during exercise in rats. Am. J. Physiol. 213:824, 1967.
17. Havel, R.J. Lipid as an energy source. Chap. 29. In: Physiology and Biochemistry of Muscle as a Food. (E.J. Briskey, ed.). Univ. of Wisconsin Press. In press. 1970.
18. Havel, R.J., Naimark, A. and Borchgrevink, C.F. Turnover rate and oxidation of free fatty acids of blood plasma in man during exercise: studies during continuous infusion of palmitate-1-C^{14}. J. Clin. Invest. 42:1054, 1963.
19. Havel, R.J., Carlson, L.A., Ekelund, L-G. and Holmgren, A. Turnover rate and oxidation of different free fatty acids in man during exercise. J. Appl. Physiol. 19:613, 1964.
20. Havel, R.J. The fuels for muscular exercise. In: Science and Medicine of Exercise and Sport. (W.R. Johnson and E.R. Buskirk, eds.) Harper and Bros. In press.
21. Hartog, M., Havel, R.J., Copinschi, G., Earll, J.M. and Ritchie, B.C. The relationship between changes in serum levels of growth hormone and mobilization of fat during exercise in man. Quart. J. Exp. Physiol. 52:86, 1967.
22. Blackard, W.G. and Heidingsfelder, S.A. Adrenergic receptor control mechanism for growth hormone secretion. J. Clin. Invest. 47:1407, 1968.
23. Imura, H., Kato, Y., Ikeda, M. and Morimoto, M. Increased plasma levels of growth hormone during infusion of propranolol. J. Clin. Endocr. Metab. 28:1079, 1968.
24. Hunter, W.M. and Sukkar, M.Y. Changes in plasma insulin levels during muscular exercise. J. Physiol. 196:110p, 1968.
25. Pruett, E.D.R. Glucose and insulin during prolonged work stress in men living on different diets. J. Appl. Physiol. 28:199, 1970.
26. Rasio, E., Malaisse, W., Franckson, J.R.M. and Conard, V. Serum insulin during acute muscular exercise in normal man. Arch. int. Pharmacodyn. 160:485, 1966.
27. Earll, J.M., Copinschi, G., Hartog, M., and Havel, R.J. Serum levels of insulin during prolonged exercise in man. Clin. Res. 15:109, 1967.

28. Nikkilä, E.A., Taskinen, M-R., Miettinen, T.A., Pelkonen, R. and Poppius, H. Effect of muscular exercise on insulin secretion. Diabetes 17:209, 1968.
29. Carlström, S. and Karlefors, T. Studies on fatty acid metabolism in diabetics during exercise. Acta. Med. Scand. 181:747, 1967.
30. Young, D.R., Pelligra, R., Shapira, J., Adachi, R.R. and Skrettingland, K. Glucose oxidation and replacement during prolonged exercise in man. J. Appl. Physiol. 23:734, 1967.
31. Hultman, E. Studies on muscle metabolism of glycogen and active phosphate in man with special reference to exercise and diet. Scand. J. Clin. Lab. Invest. 19, suppl. 94:7, 1967.
32. Shimazu, T. and Fukuda, A. Increased activities of glycogenolytic enzymes in liver after splanchnic-nerve stimulation. Science 150: 1607, 1965.
33. Andres, R., Cader, G. and Zierler, K. The quantitatively minor role of carbohydrate in oxidative metabolism by skeletal muscle in intact man in the basal state. Measurements of oxygen and glucose uptake and carbon dioxide and lactate production in the forearm. J. Clin. Invest. 35:671, 1956.
34. Zierler, K.L., Maseri, A., Klassen, G., Rabinowitz, D. and Burgess, J. Muscle metabolism during exercise in man. In: Trans. Ass. Amer. Phys. 81:266, 1968.
35. Havel, R.J., Segel, N. and Balasse, E.O. Effect of 5-methylpyrazole-3-carboxylic acid (MPCA) on fat mobilization, ketogenesis and glucose metabolism during exercise in man. In: Drugs Affecting Lipid Metabolism, p. 105. Plenum Press. 1969.
36. London, D.R. and Foley, T.H. Evidence for the release of individual amino acids from the resting human forearm. Nature 208:588, 1965.
37. Pozefsky, T., Felig, P., Tobin, J.D., Soeldner, J.S. and Cahill, G.F. Amino acid balance across tissues of the forearm in postabsorptive man. Effects of insulin at two dose levels. J. Clin. Invest. 48:2273, 1969.
38. Felig, P. and Wahren, J. Evidence for a glucose-alanine cycle: amino acid metabolism during muscular exercise. J. Clin. Invest. 49:28a, 1970.
39. Pernow, B., Havel, R.J. and Jennings, D.B. The second wind phenomenon in McArdle's syndrome. Acta Med. Scand. Suppl. 472:294, 1967.
40. Pernow, B. and Wahren, J. Lactate and pyruvate formation and oxygen utilization in the human forearm muscles during work of high intensity and varying duration. Acta. Physiol. Scand. 56:267, 1962.
41. Jorfeldt, L. Metabolism of L (+) -lactate in human skeletal muscle during exercise. Acta Physiol. Scand. Suppl. 338:5, 1970.

42. Fröberg, S.O. Determination of muscle lipids. Biochim. Biophys. Acta 144:83, 1967.
43. Young, D.R., Pelligra, R. and Adachi, R.R. Serum glucose and free fatty acids in man during prolonged exercise. J. Appl. Physiol. 21: 1047, 1966.
44. Hagenfeldt, L. and Wahren, J. Human forearm muscle metabolism during exercise. II. Uptake, release and oxidation of individual FFA and glycerol. Scand. J. Clin. Lab. Invest. 21:263, 1968.
45. Carlson, L. A., Havel, R. J., Ekelund, L-G. and Holmgren, A. Effect of nicotinic acid on the turnover rate and oxidation of the free fatty acids of plasma in man during exercise. Metabolism 12:837, 1963.
46. Bergström, J., Hultman, E., Jorfeldt, L., Pernow, B. and Wahren, J. Effect of nicotinic acid on physical working capacity and on metabolism of muscle glycogen in man. J. Appl. Physiol. 26:170, 1969.

ENERGY-RICH PHOSPHAGENS

R. E. Davies

School of Veterinary Medicine
University of Pennsylvania
Philadelphia, Pa. 19104 U.S.A.

It has, of course, been known for a long time that when you work you get tired, and that if you work hard you breathe hard, and when you stop working you keep breathing hard for a while as you recover. It is also clear that walking up a mountain is more tiring than walking down a mountain, although the reason for this is not so obvious as it might appear. Chemical changes in the muscles during exercise have been studied for a remarkably long time. For instance, an increase of lactic acid in the muscles of deers which have been run to exhaustion was discovered by Berzelius in 1841, and as far back as 1871, Weiss showed that the glycogen content of muscles decreased with work. Even the fact that creatine was formed by working muscles was observed by Monari in 1889 and the liberation of inorganic phosphate from an organic compound during activity was recorded by Salkowski in 1890. The work of Fletcher & Hopkins (1907) supported the lactic acid theory of muscle contraction which was based on the belief that the breakdown of glucose to lactic acid was the immediate energy source.

It became known largely through the work of Meyerhof (1920) and his school that inorganic phosphate was required for the metabolism of muscle extracts, but it was not until 1930, when Lundsgaard investigated the metabolism of muscles treated with iodoacetate, that it became clear that muscles could contract and work without making lactic acid.

These treated muscles that did not make lactic acid did, however, liberate inorganic phosphate. The source of this phosphate was called phosphagen. This was quite soon identified as N-phosphorylcreatine in vertebrates and N-phosphorylarginine in most invertebrates. A variety of other phosphate-liberating compounds

have also been found in many different animal species especially
worms and leeches (see Ennor & Morrison, 1958). These compounds
such as N-phosphoryltaurocyamine, N-phosphorylglycocyamine, N-phos-
phoryllombricine, etc., are all closely related. They contain a
substituted guanidine group which can react with ATP to make a
phosphoryl derivative.

Investigations starting 40 years ago showed that these compounds have virtually no separate metabolism of their own and were
not broken down enzymically by dialysed extracts of muscle. The
material in the muscle which allowed these phosphogens to break
down was identified first as adenosine monophosphate (AMP) and then
finally as adenosine diphosphate (ADP) by Lohmann (1934). The
original incorrect identification occurred because of the presence
of the myokinase reaction in muscle which can cause an interchange
of the adenosine mono, di and triphosphates. Thus it seemed likely
that the real energy source and the phosphagen which liberated
inorganic phosphate was adenosine triphosphate (ATP). Phosphorylcreatine was thus an energy store used to replenish ATP as quickly
as it was used.

Under conditions of poisoning with iodoacetate, a breakdown of
phosphorylcreatine was found to be related to long continued muscle
work (Lundsgaard, 1934). However, it proved exceedingly difficult
to find any change in a single contraction and impossible to observe
significant changes in ATP at that time (Davies, Cain & Delluva,
1959; Cain, 1960).

Much work with myosin isolated from muscle by Engelhardt &
Lyubimowa (1939) and Szent-Györgyi (1953) with mixtures of myosin
and actin showed that the myosin, the main structural protein of
muscle, is also an enzyme and can break down adenosine triphosphate.
In the presence of actin this reaction is accompanied by mechanical
changes which could be related to changes in intact muscle. About
the same time, studies on intermediary metabolism showed that ATP
is a source of energy. It is made as a result of the breakdown of
glycogen by the pathway called anarobic glycolysis. Later it was
also found to be the end product of oxidative phosphorylation. The
problem was to show directly that it was the source of inorganic
phosphate in working muscle, was directly related to the work done
and whether it was used for contraction or relaxation.

The situation was so unsatisfactory in 1950 that Professor A.
V. Hill issued his famous "Challenge to Biochemists" to prove that
ATP really was associated with contraction in muscle. He pointed
out that "---no change in the ATP has ever been found in living
muscle except in extreme exhaustion, verging on rigor." Many
attempts were made to solve this challenge. Many other sources of
inorganic phosphate besides ATP were investigated such as the
carnosine di- and mono-phosphates, all the inosine, guanosine and

cytosine phosphates, phosphoenolpyruvate, all the phosphoglyceric acids, the phosphorus in the total acid-insoluble residue of muscle measured either as inorganic phosphate, phosphatidopeptide phosphorus, phospholipid phosphorus, ribonucleic or deoxyribonucleic acids (Cain, Delluva & Davies, 1958; Seraydarian & Williams, 1960; Cain, Kushmerick & Davies, 1963, 1964). All of them were excluded.

Many workers have tried to isolate phosphorylated proteins as intermediates in the actomyosin ATPase reaction. Exchange reactions certainly occur but no discrepancy between the rate and amount of inorganic phosphate production and the breakdown of known high-energy phosphagens has been demonstrated. The results of Cheesman & Whitehead (1969) could be explained by differential binding of radioactive calcium phosphate. Direct measurements of the difference between the amounts of acid-labile phosphate in resting and contracting muscles (Table 1) showed that there was no significant difference (Delluva, Larson, Haynes & Davies, 1968). It only became possible to observe changes in ATP in a single contraction with the use of 2,4-dinitrofluorobenzene (DNFB), an agent which is quite aggressive, but we found that at 0°C, at the right concentration and for the right time it can inhibit completely the enzyme which allows ADP to be turned back into ATP by reaction with phosphorylcreatine. This enzyme, ATP:creatine-phosphotransferase, can be blocked without interfering with the main reactions leading to contraction. The muscle can still be stimulated

Table 1. CHANGES IN INORGANIC PHOSPHATE ($\equiv$ATP) IN A 180msec ISOMETRIC TETANUS

Sartorius (Rana pipiens) at l_o supramaximally stimulated at 25/sec at 0°C. Internal work $\simeq$ 55g-cm/g

Sequential extractions of the muscles by two methods

Extraction with 48% MeOH
(7 days, -35°)

P_i Extracted from resting muscle	ΔP_i (Stimulated-Control)
1.180µmol/g	0.282±0.093µmol/g (17)

Re-extraction with 0.5M-$HClO_4$
(1.5 min, 0°C)

P_i Extracted from resting muscle	ΔP_i (Stimulated-Control)
0.250µmol/g	0.0003±0.009µmol (6)

electrically, can contract and relax, but can only use energy from
ATP. Oxidative phosphorylation is stopped completely and the
initiation of glycolysis is so delayed that short time experiments
can be done with the knowledge that only ATP is used (Cain &
Davies, 1962).

After 1.5 seconds of activity in a frog sartorius muscle,
myokinase activity becomes important. This enzyme, although
partially inhibited by DNFB during the time it takes to prepare
the muscle, is still able to act effectively. This means that
both terminal phosphates of ATP can be used by the muscle, even
though only one of them is split directly by the myosin molecules
in the muscle. Thus, the inorganic phosphate released in living,
activated muscles comes from ATP. This finding then made it
possible to investigate all the changes going on at different
stages in single contraction-relaxation cycles.

Interestingly, it was found that the rate of splitting of ATP
in a muscle inhibited by DNFB in which both ATP breakdown and
inorganic phosphate breakdown can be monitored simultaneously was
quite similar to the rate of inorganic phosphate production in a
completely unpoisoned muscle. The efficiency of the mechanism by
which the energy from ATP splitting is turned into external work
is virtually the same in both poisoned and unpoisoned muscles
(Table 2). This means that the liberation of inorganic phosphate
can be used as a monitor for the breakdown of ATP.

The first thing to be investigated was activation. Muscles at
0°C were stimulated and immediately frozen less than 50 milli-
seconds after the stimulus. They were found to have used no
detectable amount of ATP when compared with their paired controls.
There is no burst of ATP breakdown on the initiation of activity
as a result of the action potential or of the liberation of calcium
from the sarcoplasmic reticulum. It is quite certain that muscles
do not split a constant amount of ATP at the beginning of a
contraction which is then used more or less efficiently to do more
or less work.

When muscles in (3x) hypertonic solutions were investigated,
the action potentials were found to be developed normally but the
amount of ATP used was quite minuscule. It is so small that none
is available for pumping calcium. The amount was $0.006 \pm 0.008 \mu mol$
ATP/g/pulse, and is the upper limit for the amount of ATP used
under these conditions for pumping sodium ions in the first two or
three seconds following membrane depolarization (Kushmerick, Larson
& Davies, 1969).

The muscles could also be studied at a length either so short
that no further tension is developed or so long that there is no

Table 2. EFFICIENCY OF ATP UTILIZATION FOR EXTERNAL WORK

Single maximally working isovelocity contractions of frog sartorius (Rana pipiens) at 0°

Treatment and Number of Pairs	Velocity (cm/sec)	Total ATP Used (µmol/g)	External Work (g-cm/g)	Efficiency $\left(\dfrac{\text{External Work}}{\text{µmol ATP}}\right)$
Aerobic and untreated (14)	2.2	1.49±0.100	374±23	261±18
1 mM Iodoacetate, 60 min, anaerobic (12)	2.3	1.30±0.049	346±12	269±9
0.38 mM Dinitro-fluorobenzene, 40 min, anaerobic (16)	2.0	1.32±0.126	323±15	280±27

Muscles shortened a distance of 2.5 cm from 1.3 ℓ_o to 0.6 ℓ_o against an ergometer at the constant velocity indicated while being stimulated by electrical pulses at 25/sec.

270 g-cm/g external work per µmole ATP is 63% efficient at -10 kcal/mol for the free energy of ATP splitting. No allowance for ATP used for pumping calcium. (Results from Kushmerick & Davies, 1969)

overlap between the thick and thin filaments. Under these conditions the amount of ATP used following a stimulus is thus not related to the development of an external force. The ATP needed was similar in both cases and was 0.057±0.006μmol ATP/g/twitch. This depended on the rate at which the electric stimuli were passed through the muscle. If they were separated by a second or more, the amount remained constant. However, if the rate at which the stimuli were sent to the muscle was increased, then the ATP used as a result of each pulse decreased. However, the amount of ATP used per second increased and approached a maximum of 0.25μmol ATP/g/sec at the fusion frequency at 0°C. This is apparently the amount of ATP required by the calcium pumps in the muscle when they were working at a maximum rate as a result of continuous tetanic stimulation (see Infante, Klaupiks & Davies, 1964).

Muscle can shorten when stimulated and there is a heat output associated with this shortening which is known as shortening heat but this heat output is largely independent of the work done. In 1966 Professor A. V. Hill issued a "Further Challenge to Biochemists" which was to find out whether shortening heat could be accounted for by the breakdown of ATP. This challenge was taken up in several ways and the results show conclusively that shortening heat is not degraded free energy from ATP splitting (Davies, Kushmerick & Larson, 1967). In fact, shortening heat itself has rather special properties. It is seen during the shortening of the muscle but disappears in a whole series of contraction-relaxation cycles. There is still much to be learned from further studies of this particular heat transfer.

Whereas the heat of shortening was largely independent of the work done, the amount of ATP split was not and, in fact, is rather directly related to the amount of work the muscle does. The more work that is done, the more ATP is required. The efficiency by which the energy from ATP is used by the contracting muscle can be remarkably high and, after allowing for the amount of ATP used for pumping calcium and for other processes not directly related to the work output, is in the range of 70-80% of the calculated value for the maximum thermodynamic free energy available from the ATP. The efficiency is not a constant, however, since muscles are found to be less efficient at high and low speeds. More ATP is broken down under these conditions for the production of the same amount of work done by the muscle (Kushmerick & Davies, 1969).

This is a most interesting finding. At low speeds internal work associated with to and fro "dither" in the sarcomere would lower the efficiency based on external work (Larson, Kushmerick, Haynes & Davies, 1968). At high speeds the chance of forming a link would be decreased and it could form at less than the full extension of the flexible part of the crossbridge (Davies, 1963),

Table 3. ATP USAGE DURING A CONTRACTION-RELAXATION
CYCLE OF FROG SARTORIUS AT 0°C

1 μmol ATP at 10 kcal/mol at 100% efficiency
$\equiv$ 426 g-cm $\equiv$ 10 mcal $\equiv$ 4.2 × 10^{-2} Joules
(Newton-meter) $\equiv$ 4.2 × 10^5 ergs (dyne-cm)

	ATP used (μmol/g)
Stimulation (freeze before movement)	0.0
Activation (when pre-shortened, delayed freezing) 1 electric stimulus/sec	0.054±0.008
10 electric stimuli/sec	0.021±0.010
50 electric stimuli/sec	0.007±0.002
10 electric stimuli/sec (hypertonic)	0.006±0.008
External work 0 → 950 g-cm/g (at 0.3 cm/sec)	0 → 6.3
Shortening (per se)	0.0
Relaxation (depends on force developed)	0 → 0.15
Isometric (12 stim./sec) during first 1.0 sec at 0.50 l_o	0.26
at 1.00 l_o	0.64
at 1.35 l_o	0.25
When isometric at P_o (65 g), rate at 0.5 → 1.0 sec is	0.74
Stretch at 0.4 cm in 0.5 sec (1.7 P_o) rate at 0.5 → 1.0 sec is	0.25
Chemical power usage per sec Isovelocity contractions 0.6 cm/sec	0.6
2.8 cm/sec	1.4
6.0 cm/sec	1.0

Average overall efficiency for external work for 0.6 → 2.8 cm/sec isovelocity contractions = 55%. For mechanical generator efficiency = 70% at 10 kcal/mol ATP.

∴Minimum free energy for ATP at 100% efficiency is -7 kcal/mol.

since the thick and thin filaments would be moving so quickly relative to each other. Thus, all the free energy could not be utilized and the efficiency would fall. This finding excludes many theories of muscle contraction.

Relaxation of tension is associated with an extra breakdown of ATP (Nauss & Davies, 1966; Kushmerick et al., 1969). Although this has been disputed by Mommaerts & Wallner (1967), it has been confirmed by Wilkie (1970) and his colleagues.

It is interesting that muscles can perform excentric contractions, that is when they are subject to a force greater than they themselves can develop when stationary they stretch and absorb work. During this stretching of the muscle, when a large force is developed which moves through a long distance, the splitting of ATP is dramatically reduced though no net resynthesis occurs (Table 3). This is very convenient because muscles could be driven by a variety of mechanisms. For example, if the mechanism were of the type used by a rocket motor, as much energy must be used to cause motion in one direction as the other. Per pound of lunar excursion module, exactly the same amount of rocket fuel is needed to land slowly on the moon as to get back into orbit from the moon. A mechanism like an automobile uses energy to accelerate to a given speed but slows down by use of the braking system in which the kinetic energy of the moving automobile is dissipated as heat in the brake blocks. Another sort of motor is the one in streetcars or electric trains. The acceleration of the vehicle requires electrical energy but the deceleration could be used to put electric energy back into the power lines.

The problem is: what sort of mechanism occurs in muscle? It is quite clear that the mechanism is not that of the rocket motor. The absorption of work requires virtually no ATP, there is no net synthesis occurring. On the other hand, there seems to be no separate braking system within the muscle. The contractile mechanism itself must be able to be reversed without splitting ATP. This conclusion is important, because many theories of muscle contraction require that ATP be split before a link can be made between actin and myosin, i.e. between the thick and thin filaments of the sarcomeres. The development of tension and contraction then follow after. If the splitting of ATP were required to develop a force, then this splitting would still occur whatever the direction of movement. It is necessary to have a mechanism that allows for the formation of links between the filaments and for these links to be formed and mechanically broken many times whilst developing large forces but without ever splitting ATP. This can occur according to a theory of the type published by me (Davies, 1963) where the link between actin and the myosin actually contains the molecule of ATP. This is only split when the muscle contracts and

Mechanism For Developing Tension Without Using ATP In An Activated Muscle During Stretching

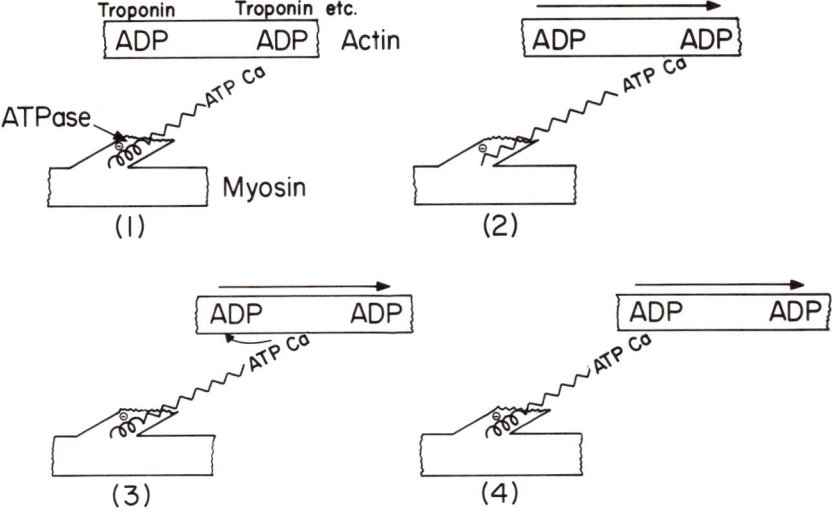

Figure 1.

This mechanism is based on the molecular theory of muscle contraction of Davies (1963). Troponin on the thin (actin) filament is taken to confer specificity for calcium as the activator in the ATPase reaction occurring on the thick (myosin) filament. The theory is based on the view that in normal contraction the liberation of calcium causes the formation of an α-helix in a small flexible polypeptide which, in each crossbridge in resting muscle, is in a randomly-coiling, though largely extended, form due to electrostatic repulsion between its two ends. This helix formation causes force development and the interdigitation of the thick and thin filaments. During stretch the linkage would go through the cycle, (1) to (2) to (3) to (4) which is identical to (1) except that the sarcomeres have been lengthened. Any α-helices already formed would be fully extended to the β- conformation and the link would then be torn apart, only to re-form further along the thin filament. Since the ATP would not get near the ATPase on the myosin it could not get split. For further details see Davies (1963).

gets shorter since this allows the ATP to come into the region where the ATPase of the myosin is so it can be split. When, however, a force of such strength is imposed on the muscle that the muscle is extended and becomes longer, then the links would be mechanically split between the ATP and the thin filament. The ATP could then link again further along the thin filament by a series of ratchet-like actions resisting stretch at all times and effectively acting as a brake but without the molecule of ATP itself ever becoming broken (Fig. 1). Of course, some ATP is separately used by the calcium pumps since the muscle is continuously activated during this time.

This accounts for the otherwise very strange observation that if you walk upstairs and then walk backwards downstairs, the muscles have moved through exactly the same distances and exactly the same forces have developed in them and yet, to walk upstairs is tiring but to walk backwards downstairs is rather easy. ATP is believed to cause a conformational change in the myosin with the splitting of hydrogen bonds. The contraction of part of each crossbridge occurs spontaneously in the presence of calcium with the formation of hydrogen bonds, the ATP being necessary for the re-extension of the contractile part of each crossbridge. When a muscle is stretched, these hydrogen bonds are broken mechanically. During contraction this conformational change would be a coil-helix type of transition. During stretch this would be reversed so that hydrogen bonds are broken. This could account both for the heat of shortening and the negative heat of lengthening in activated muscle. Stretching an activated muscle would then reverse the final mechanism which produces the external force which, I suggest, is the formation of hydrogen bonds during this conformational change in part of the myosin crossbridge. When contraction occurs, ATP is split. When the muscle is forcibly stretched, the ATP is not split. This explains why inorganic phosphate is not released during excentric contractions but is released during concentric contractions as the muscle shortens.

This work was supported by grants number 9 R01 AM 13687 and 5 T01 GM 00694 from the United States Public Health Service, and by grants from the Muscular Dystrophy Associations of America and the Heart Association of Southeastern Pennsylvania.

REFERENCES

Berzelius (1841). In Lehmann, C. G., Lehrbuch der physiologischen Chemie 1, 103, Leipzig (1850).

Cain, D. F. (1960). Ph.D. Dissertation, University of Pennsylvania. The Immediate Energy Source for Muscular Contraction.

Cain, D. F. & Davies, R. E. (1962). Breakdown of adenosine triphosphate during a single contraction of working muscle. Biochem. biophys. Res. Commun. 8, 361-366.

Cain, D. F., Delluva, A. M. & Davies, R. E. (1958). Carnosine phosphate as phosphate donor in muscular contraction. Nature, Lond., 182, 720-721.

Cain, D. F., Kushmerick, M. J. & Davies, R. E. (1963). Hypoxanthine nucleotides and muscular contraction. Biochim. biophys. Acta 74, 735-746.

Cain, D. F., Kushmerick, M. J. & Davies, R. E. (1964). Phosphoenolpyruvate, the phosphoglyceric acids and muscular contraction. Biochim. biophys. Acta 86, 81-90.

Cheesman, D. F. & Whitehead, A. (1969). Possible role in contraction of structurally bound phosphate of muscle. Nature, Lond., 221, 736-739.

Davies, R. E. (1963). A molecular theory of muscle contraction: calcium-dependent contractions with hydrogen bond formation plus ATP-dependent extensions of part of the myosin-actin cross-bridges. Nature, Lond., 199, 1068-1074.

Davies, R. E., Cain, D. & Delluva, A. M. (1959). The energy supply for muscle contraction. Ann. N. Y. Acad. Sci. 81, 468-476.

Davies, R. E., Kushmerick, M. J. & Larson, R. E. (1967). Professor A. V. Hill's "Further Challenge to Biochemists": ATP, activation, and the heat of shortening of muscle. Nature, Lond., 214, 148-151.

Delluva, A. M., Larson, R. E., Haynes, D. H. & Davies, R. E. (1968). Is protein-bound phosphate an energy source in muscle contraction? Fed. Proc. 27, 821.

Engelhardt, W. A. & Ljubimowa, M. N. (1939). Myosine and adenosinetriphosphatase. Nature, Lond., 144, 668-669.

Ennor, A. H. & Morrison, J. F. (1958). Biochemistry of the phosphagens and related guanidines. Physiol. Rev. 38, 631-674.

Fletcher, W. M. & Hopkins, F. G. (1907). Lactic acid in amphibian muscle. J. Physiol. 35, 247-309.

Hill, A. V. (1950). A challenge to biochemists. Biochim. biophys. Acta 4, 4-11.

Hill, A. V. (1966). A further challenge to biochemists. Biochem. Z. 345, 1-8.

Infante, A. A., Klaupiks, D. & Davies, R. E. (1964). Length, tension and metabolism during short isometric contractions of frog sartorius muscles. Biochim. biophys. Acta 88, 215-217.

Kushmerick, M. J. & Davies, R. E. (1969). The chemical energetics of muscle contraction. 2. The chemistry, efficiency and power of maximally working sartorius muscles. Proc. Roy. Soc. B,174, 315-353.

Kushmerick, M. J., Larson, R. E. & Davies, R. E. (1969). The chemical energetics of muscle contraction. I. Activation heat, heat of shortening and ATP utilization for activation-relaxation processes. Proc. Roy Soc. B,174, 293-313.

Larson, R. E., Kushmerick, M. J., Haynes, D. H. & Davies, R. E. (1968). Internal work during maintained tension of isometric tetanus. Biophys. Soc. Abstr. 11th Ann. Mtg., A-8.

Lohmann, K. (1934). Über die enzymatische Aufspaltung der Kreatinphosphorsäure; zugleich ein Beitrag zum Chemismus der Muskelkontraktion. Biochem. Z. 271, 264-277.

Lundsgaard, E. (1930). Weitere Untersuchungen über Muskelkontraktionen ohne Milchsäurebildung. Biochem. Z. 277, 51-83.

Lundsgaard, E. (1934). Phosphagen- und Pyrophosphatumsatz in jodessigsäurevergifteten Muskeln. Biochem. Z. 269, 308-328.

Meyerhof, O. (1920). Die Energieumwandlungen im Muskel. I. Über die Beziehungen der Milchsäure zur Wärmebildung und Arbeitsleistung des Muskels in der Anaerobiose. Pflüg. Arch. ges. Physiol. 182, 232-283.

Mommaerts, W. F. H. M. & Wallner, A. (1967). The break-down of adenosine triphosphate in the contraction cycle of the frog sartorius muscle. J. Physiol. Lond., 193, 343-357.

Monari, A. (1889). Jahresber. Tierchem., 296. Seen in von Muralt, A. (1950), The development of muscle-chemistry, a lesson in neurophysiology. Biochim. biophys. Acta 4, 126-129.

Nauss, K. M. & Davies, R. E. (1966). Changes in inorganic phosphate and arginine during the development, maintenance, and loss of tension in the anterior byssus retractor muscle of Mytilus edulis. Biochem. Z. 345, 173-187.

Salkowski, T. (1890). Z. klin. Med. 17, Suppl. 21. Seen in von Muralt, A. (1950), The development of muscle-chemistry, a lesson in neurophysiology. Biochim. biophys. Acta 4, 126-129.

Seraydarian, M. W. & Williams, E. B. (1960). Studies on the acid-insoluble fraction of frog's muscle. Biochim. biophys. Acta 41, 352-355.

Szent-Györgyi, A. (1953). Chemistry of muscular contraction. Academic Press, New York, N. Y.

Weiss, S. (1871). Sitzber. Akad. Wiss., Wien, 64, 1. Seen in von Muralt, A. (1950). The development of muscle-chemistry, a lesson in neurophysiology. Biochim. biophys. Acta 4, 126-129.

Wilkie, D. R. (1970). Personal communication.

ENERGY RICH PHOSPHAGENS IN DYNAMIC AND STATIC WORK

J. BERGSTRÖM, R.C. HARRIS, E. HULTMAN, and L.-O. NORDESJÖ

From the Departments of Clinical Chemistry, S:t Eriks
Sjukhus, and Clinical Physiology, Karolinska Sjukhuset,
Stockholm, Sweden

The understanding of muscle metabolism and its regulation in situ has been greatly advanced in recent years by the analysis of the levels of metabolic intermediates under differing conditions. Only recently has it been possible to perform similar studies in man. In the present studies muscle samples have been obtained using the needle biopsy technique (1, 10, 11) and the levels of a number of metabolites in these have been determined by enzymatic micro-methods.

A few years ago Dr Hultman and I with the assistance of Mr Norman Mc Lennan Anderson, studied the effect of bicycle work upon the levels of creatine phosphate and adenosine triphosphate (ATP) in man (11). This study confirmed earlier in vitro and in vivo studies in experimental animals and emphasized the central role of the phosphagens as the immediate energy sources for muscle contraction. More recently we have extended the range of our analyses to include a number of the glycolytic intermediates and have begun to study muscle metabolism during isometric or static work. Isometric work offers several advantages over dynamic work in that during isometric muscle contraction of moderate to high intensities, little or no blood-flow to the muscle occurs. As a result, reduced amounts of oxygen and substrates will be taken up from the circulating blood and loss of metabolites from the muscle should be at a minimum. By determining the metabolite levels in the muscle at the end of an isometric contraction it should therefore be possible to calculate the relative contributions of the different substrates for energy production. In the following we will present a brief review of our earlier experiments on dynamic work and a more detailed report on those concerned with isometric work.

BICYCLE OR DYNAMIC WORK

The details of this investigation have been given earlier (11). Muscle samples were obtained by needle biopsy from the quadriceps femoris muscle, before and after different periods of bicycle exercise performed with different work loads. The muscle samples were immediately frozen in either dry ice or alcohol or in liquid nitrogen, freeze-dried and analysed for ATP and PC by the method of Fawaz and co-workers (6). It was found that a rapid breakdown of PC occurred directly after work had begun. Following this, the level of PC remained relatively constant up to the end of the exercise period. Upon the termination of work PC was rapidly resynthetized back to its former level (Fig. 1). ATP showed a

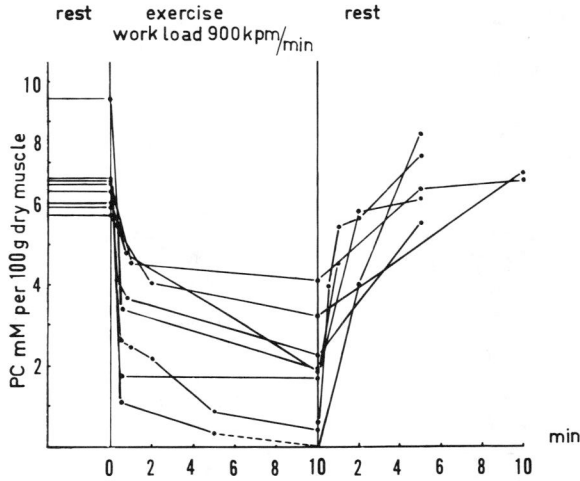

Fig. 1. Breakdown and resynthesis of PC in connection with exercise. Work load 900 kpm/min.

similar decrease following the commencement of work though to a much smaller extent than in the previous case. In an experiment in which eight subjects worked for periods of 5 min at different loads a correlation was found between the level of PC as recorded at the end of the exercise period and the work load (Fig. 2). Thus, the steady level of PC in the muscle cells bears an inverse relationship to the work load. The constancy of the PC level during exercise at a given work load was confirmed in a long term experiment of one hour's duration, during which biopsies were taken every 15 min.

When the work load was very high (1,500 - 2,400 kpm/min) the

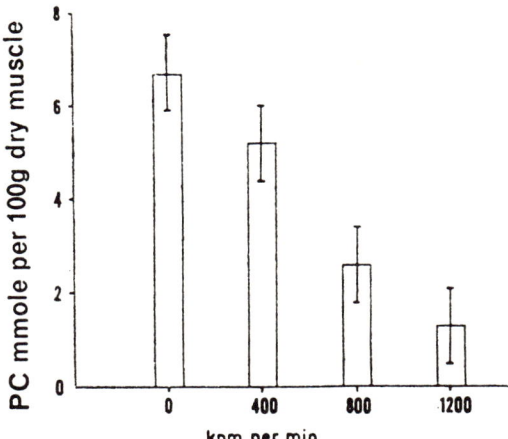

Fig. 2. PC concentration versus work load. Mean values ± standard deviation.

level of PC decreased from a rest value of 7 mmol/100 g dry weight to 0.21 mmol. The corresponding values for ATP were 2.48 mmol and 1.43 mmol, respectively. These changes in PC and ATP were highly significant. The relationship between PC and ATP in these experiments is shown in Fig. 3.

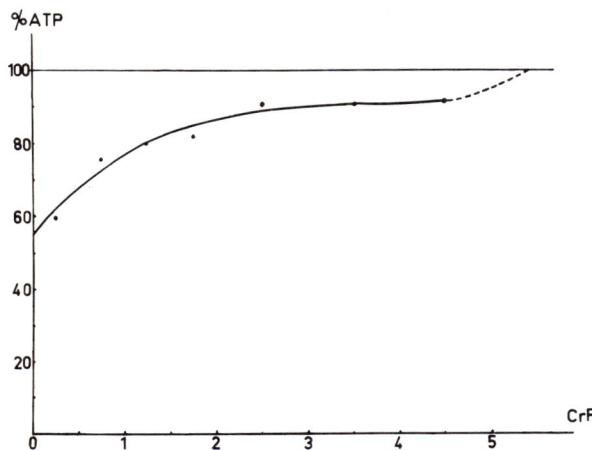

Fig. 3. The relation between ATP concentration (in percent of basal value) and PC concentration in human muscle during exercise.

At exhaustive exercise when the PC level is less than 10 % of the basal value, the level of ATP is reduced by 40 % on an average. Thus, following maximal consumption of energy there is still a large intact store of ATP in human muscle. This is in accordance with the generally accepted view that the ATP-pool in muscle is compartmentalized and that only a small fraction of it is in equilibrium with PC (9).

The oxygen debt due to the breakdown of PC and ATP upon exhaustion was estimated to be about 70 ml oxygen/kg muscle tissue. Assuming that 15 kg muscle tissue is working, this corresponds to a total oxygen debt of more than 1 litre. Other workers have found that part of the total oxygen debt after exercise cannot be accounted for by anaerobic glycolysis with lactate formation. We have concluded that this alactic oxygen debt is in part at least due to the decrease in ATP and PC in the working muscle.

Unfortunately these experiments on muscle metabolism during dynamic work did not show whether or not there was any initial period of alactic, anaerobic energy release as proposed by Di Prampero and Margaria (5) or whether lactate production began immediately. Similarly, no information was gained in these studies on how the muscle cells met the increased energy demands made upon them during heavy work, i.e. what proportion of the cells' energy needs was met by consumption of active phosphate or by glycolysis, or what factors may limit the cell's capacity to meet these energy requirements. Our more recent studies on muscle metabolism during isometric work were made in an attempt to answer these questions.

ISOMETRIC OR STATIC WORK

Normal subjects were placed in a sitting position with the legs 90° to the thighs in a special chair of the type described by Tornvall (21). A waist band was used to secure the subject to the chair. Work was done by extention of the knee, and the force generated was measured through a dynamometer secured to the leg at the level of the lateral malleolus. The force generated over the work period was recorded on a strip chart and simultaneously shown on a dial placed in front of the subject. This enabled the subject to maintain a predetermined force over the required period of time. A few days before the start of each experiment the maximal working capacity for each subject was determined. This was defined to be the greatest force which the subject could sustain for a period of 5 to 10 sec, and was determined for both legs on at least two separate occasions.

Two subjects, A and B, worked at their maximum intensity during each of three work periods and with alternate legs. In the first instance for 6.6 sec, in the second for approximately 12 sec and in the third to a point shere they would no longer maintain the required work load. Between each work period the subjects were rested for a period of at least 1 h. Three subjects, C, D, and E, worked with a load of 40 % of their maximum. Again, each subject worked three times, in the first instance for 20-30 sec, in the second for 50-72 sec, and in the third, until they could no longer maintain the required load. Needle biopsies were taken from the quadriceps femoris muscle at rest and immediately after each work period with the leg relaxed. The muscle samples were immediately frozen in liquid freon at its melting point and were later freeze-dried. The time gap between the taking of the biopsy and the freezing of the sample was in the order of 2 sec. The freeze-dried muscle samples were extracted with ice-cold 0.5 M perchloric acid containing 1 mM EDTA and the extracts were neutralized with $KHCO_3$. All metabolites were assayed by enzymatic procedures based on those described by various authors in "Methods of Enzymatic Analysis" (ed: H.U. BERGMEYER). The oxidation and reduction of pyridine nucleotides was followed at 334 mµ using an Eppendorf photometer.

Table 1. Metabolite levels in biopsy samples taken from the quadriceps femoris muscle at rest.

METABOLITE	n	µmol/100 g dry muscle weight (mean ± SE)
Creatine-P	20	7,682 ± 180
Creatine	20	4,766 ± 128
Creatine-P + Creatine	20	12,397 ± 226
ATP	20	2,459 ± 59
ADP	18	307 ± 11
AMP	18	8 ± 1
Glucose	19	197 ± 24
Glucose 1-P	18	6 ± 1
Glucose 6-P	18	153 ± 12
Fructose 6-P	18	25 ± 3
Fructose 1-6-diP	15	25 ± 3
Dihydroxyacetone-P	15	16 ± 4
Glyceraldehyde 3-P	15	8 ± 4
Glycerol 1-P	16	84 ± 14
Pyruvate	20	51 ± 5
Lactate	20	663 ± 47
Malate	16	35 ± 7

Table 1 shows the levels of the metabolites measured in biopsy samples taken from the quadriceps femoris muscle at rest together with the standard error of the mean; Table 2 gives the ratios of

Table 2. The ratios of some of the metabolites listed in Table 1. The values presented here were calculated from the same set of data as used in the previous table.

RATIO	n	mean ± SE
Creatine P : Creatine	20	1.620 ± 0.060
ATP : ADP	18	8.300 ± 0.057
Glucose 6-P : fructose 6-P	16	6.500 ± 0.386
Fructose 1-6-diP : Fructose 6-P	14	1.165 ± 0.141
Dihydroxyacetone-P : Glycerol 1-P	13	0.292 ± 0.066
Pyruvate : Lactate	20	0.085 ± 0.009

some of these metabolites to each other. The results presented in Table 2 were derived from the same set of data as used in Table 1. The ratios of PC to creatine and ATP:ADP are comparable with values obtained from experimental animals with more rapid freezing techniques (3, 9, 24) indicating that the present method preserves the <u>in vivo</u> condition of the muscle. (In earlier experiments where metabolites were extracted by the method of Schulz and co-workers (20) using 3.3 M perchloric acid, considerable losses of PC were noted to occur and a much lower ratio of ATP:ADP was found (mean 3.35 ± 0.22).)

CHANGES WITH EXERCISE

Figs 4, 5, 6, and 7 show the changes in the metabolite levels in subject A (working at max intensity) and C (working at 40 % of max intensity). In both subjects there was a rapid decrease in PC and a corresponding increase in free creatine during work. Total creatine (creatine-P + creatine) remained constant. The lowest values of PC were seen at exhaustion when the work load was 40 % of maximum.

ATP in subject A showed an immediate and steady decrease following the commencement of work, whereas in C it remained constant for the first period of work and only showed a decrease towards exhaustion. Small increases in the levels of ADP were found in both subjects, though these were not comparable to the decreases in ATP and no increases in the levels of AMP were detected. The summation, ATP + ADP + AMP, thus showed a decrease throughout the course of work.

Continuous increases occurred in both subjects in the levels of the hexose-monophosphates (glucose 1-P, glucose 6-P, and fructose 6-P), glycerol 1-P, pyruvate, and lactate. In general, the rates of increase in subjects working at max intensity were 2 to 3

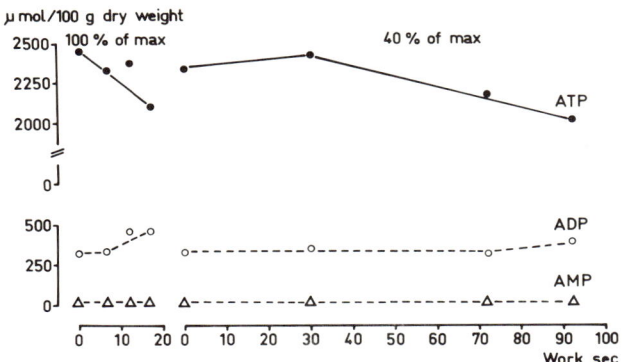

Fig. 4. The levels of ATP, ADP, and AMP in biopsy samples taken from the quadriceps femoris muscle of subjects A and C. Subject A worked at 100 % of his max intensity for 6.6 sec, 11.9 sec and to a point where he could no longer maintain the required work load - 17.1 sec. Biopsy samples were taken at the beginning of the experiment and at the end of each work period. - Subject C worked at 40 % of his max intensity for 30 sec, 72 sec and to a point where he could no longer maintain the required work load - 92 sec. Biopsy samples were taken as before. - In both subjects the first, second, and fourth biopsy sample was taken from the same leg and only the levels of the individual metabolites in these have been joined together in the Fig.

times greater than in subjects working at 40 % of max. The final levels of lactate reached in subjects C, D, and E were very high (in one instance, 17,000 µmol/100 g dry weight). It was apparent that during work of max intensity lactate production began within 7 sec. No increases in the levels of dihydroxyacetone-P and glyceraldehyde 3-P were detected during work; fructose 1-6-diP showed an initial increase, but decreased thereafter.

From the observed changes in the levels of the phosphagens and pyruvate and lactate, estimates of the rates of high energy phosphate (∼P) utilization which had occurred during work at either intensity were made. These are shown in Table 3 together with the proportions of ∼P flux, in each case, derived from glycolysis (calculated from lactate and pyruvate increase). In all cases the rate of ∼P utilization during the first work period was approximately the same as that during the longest, showing there was a near constant rate of ∼P usage throughout work. The rates of ∼P utilization in subjects A and B working at max intensity were some

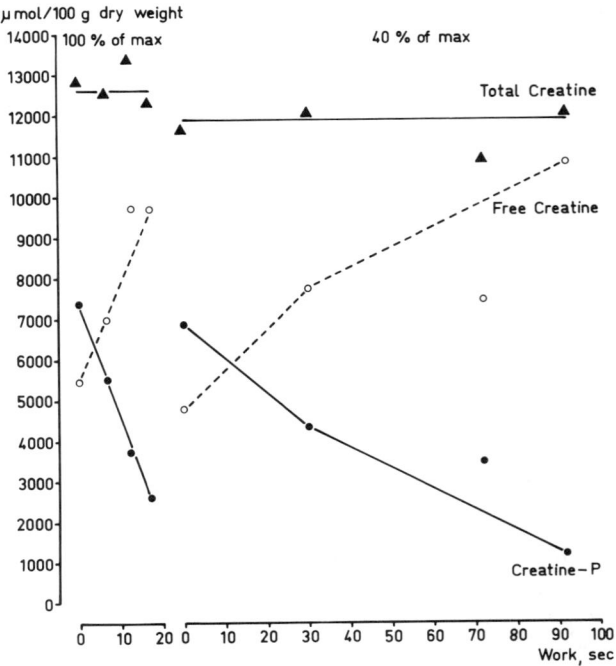

Fig. 5. The levels of creatine-P and creatine in biopsy samples taken from the quadriceps femoris muscle of subjects A and C, before and after the performance of different periods of work. The conditions of work were as described in the legend to Fig. 4.

3 to 4 times greater than those in subjects C, D, and E. In subjects A and B the proportion of $\sim$P derived from glycogen metabolism was as high during the first seconds of work as during the rest of the exercise, indicating that glycolysis must begin almost immediately at the start of exercise.

In Table 3 are also shown the rates of glycogen consumption as calculated from the increases in all of the glycolytic intermediates. The proportion of glycogen accounted for by pyruvate and lactate increase in these experiments varied between 52 and 78 %.

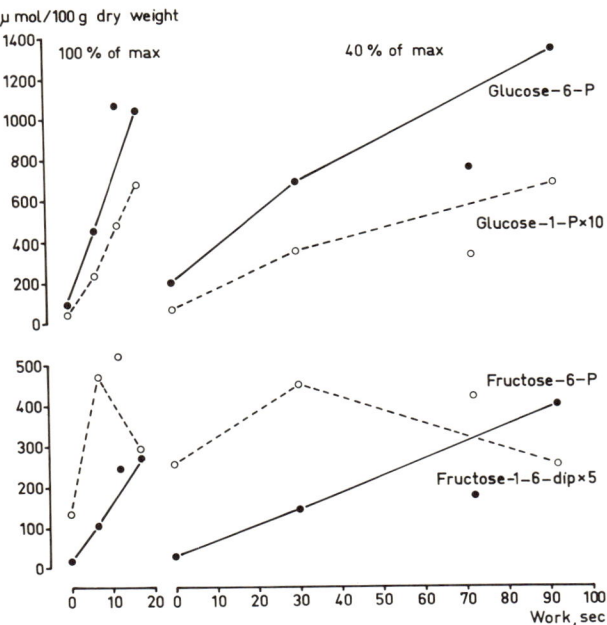

Fig. 6. The levels of glucose 6-P, glucose 1-P, fructose 6-P, and fructose 1-6-diP in biopsy samples taken from the quadriceps femoris muscle of subjects A and C before and after the performance of different periods of work. The conditions of work were as described in the legend to Fig. 4.

DISCUSSION ON THE ISOMETRIC WORK EXPERIMENTS

The 3 to 4 fold difference in the rates of ∼P utilization between subjects A and B working at max intensity and C, D, and E working at 40 % of max, points to a direct relationship between relative work load and rate of energy utilization. The apparent 3 fold difference in the glycolytic rates between the two groups as judged by the rates of pyruvate + lactate formation and glycerol 1-P formation points to a synchronization between energy utilization and regeneration.

During isometric work of this type the only substrate available to the cells for ∼P regeneration will be glycogen. The two rate limiting steps in glycogenolysis are reportedly those reactions catalyzed by phosphorylase and phosphofructokinase

350   J. BERGSTRÖM et al.

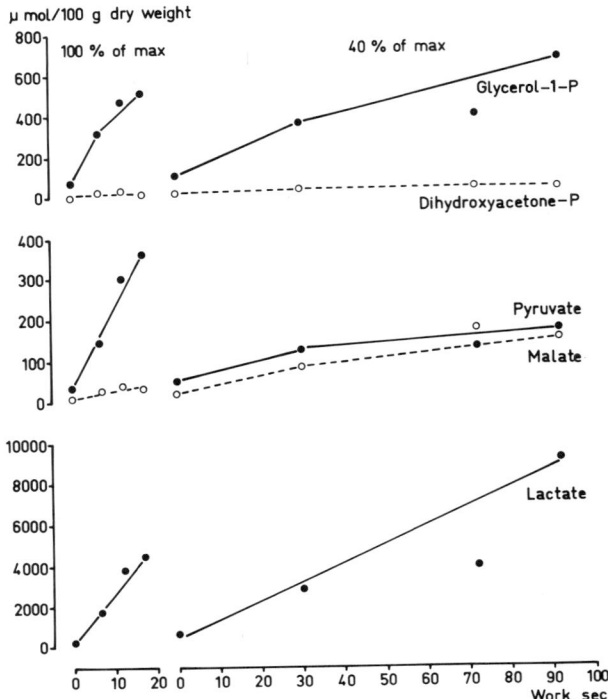

Fig. 7. The levels of dihydroxyacetone-P, glycerol 1-P, pyruvate, lactate, and malate in biopsy samples taken from the quadriceps femoris muscle of subjects A and C, before and after the performance of different periods of work. The conditions of work were as described in the legend to Fig. 4. - The levels of dihydroxyacetone-P in either subject were too low to be measured by the methods used.

(4, 12, 16, 24). In the resting muscle phosphorylase is in the relatively inactive b form but upon contraction part of this will be converted to the more highly active a form (19), thereby initiating glycogen breakdown. The rate of glycogen breakdown at this stage will be dependent upon the total phosphorylase a and b activity.

The very large increases in the levels of the hexose monophosphates during work compared to those of fructose 1-6-diP are consistent with phosphofructokinase being limiting to glycolysis in human muscle, as in other systems (3, 24). The kinetics of PFK

Table 3. The rates of pyruvate + lactate formation, glycogen consumption and $\sim$P utilization calculated for subjects working at max and at 40 % of max intensity. Glycogen consumption was calculated from the changes in the levels of all the glycolytic intermediates measured and not from direct measurements. These figures are therefore at best minimal estimates. The rates of $\sim$P utilization were calculated from the changes in the recorded levels of ATP, ADP, AMP, creatine-P, and pyruvate + lactate. It was assumed that all pyruvate and lactate formed was derived from glycogen breakdown rather than from glucose utilization and that for every mol formed 1.5 mol of $\sim$P were generated.

TIME INTERVAL sec	RATE OF PYRUVATE + LACTATE FORMATION µmol/100 g d.w./sec	RATE OF GLYCOGEN CONSUMPTION (µmol glucosyl units /100 g d.w./sec)	% OF GLYCOGEN CONSUMED RECOVERED AS PYRUVATE + LACTATE	RATE OF $\sim$P UTILIZATION µmol/100 g d.w./sec	% OF $\sim$P UTILIZED GLYCOGEN METABOLISM
Subject A					
0 - 6.6	233.5	230.5	50.7	668.0	52.4
6.6- 17.1	282.5	222.0	64.3	744.5	56.9
0 - 17.1	263.6	225.3	58.5	714.9	55.3
Subject B					
0 - 6.6	243.9	241.5	53.1	746.3	48.8
6.6- 17.1	405.6	384.4	52.8	955.7	63.7
0 - 17.1	342.8	329.2	52.1	874.9	58.8
Subject C					
0 - 30	75.0	66.7	56.2	189.1	59.5
30 - 92	102.5	71.0	72.1	217.3	70.7
0 - 92	93.5	69.6	67.1	208.1	67.4
Subject D					
0 - 30	98.5	86.0	57.2	251.1	58.7
30 -133	134.2	86.0	78.0	236.5	85.1
0 -133	126.2	86.0	73.3	239.9	78.9
Subject E					
0 - 22	71.7	62.5	57.4	231.3	46.5
22 -108	89.5	68.7	65.1	186.0	72.1
0 -108	85.8	67.4	63.7	195.2	66.0

Table 4. Factors known to affect mammalian muscle and/or brain phosphofructokinase activity in vitro.[**]

INHIBITORS OR CO-INHIBITORS	ACTIVATORS OR DEINHIBITORS
ATP	ADP
Citrate	AMP
Creatine-P[*]	3'5'-AMP
3-P glycerate[*]	Inorganic phosphate
2-3-P_2 glycerate[*]	Glucose 1-6-diP
2-P glycerate[*]	Fructose 6-P
P-enolpyruvate[*]	Fructose 1-6-diP
An increase in H^+	NH_4^+

[*] Inhibitory to PFK at neutral pH but not at 8.1 and then only in the presence of inhibitory levels of ATP and non-activating levels of fructose 6-P.

[**] Refs 2, 7, 8, 13, 14, 15, 17, 18, 22, 23.

are complex. In Table 4 are summarized the effects of several factors known to affect mammalian PFK in vitro. Present theories of glycolytic control envisage that in the resting muscle PFK is inhibited by high levels of ATP; upon the commencement of work ATP falls whilst the levels of ADP, AMP, P_i, fructose 6-P, and possibly glucose 1-6-diP increase, thereby bringing about activation. Activation will in turn result in an increase in fructose 1-6-diP and the process will for a time be autocatalytic. In the present series of experiments changes in the levels of several of the factors listed in Table 4 were observed to have occurred by the end of the first work period and these were such as to favour activation. In both series of experiments the levels of fructose 1-6-diP increased initially during work but decreased thereafter. The ratio of fructose 1-6-diP, fructose 6-P showed a progressive decrease throughout work from a mean of 1.3 in the basal state to 0.2 upon exhaustion. These observations are not in accord with activation of PFK during work of this type but point instead to a progressive inhibition. One possible mechanism by which this could have been brought about is, that the rapid build up in lactate in these experiments resulted in a pronounced decrease in the intracellular pH. A decrease in pH is known to inhibit PFK activity and to enhance the inhibitory effect of ATP (8, 22, 23). In Fig. 8 is shown a plot of the fructose 1-6-diP: :fructose 6-P ratios, vs. lactate as found in these experiments and in three others of the same type. It shows that there is at least a coincidental relationship between these two parameters which points to a possible inhibition of PFK activity in muscle cells resulting from increasing lactate concentrations.

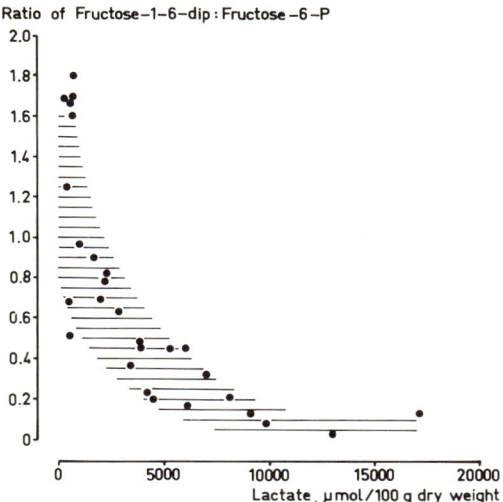

Fig. 8. The relationship between the ratio of fructose 1-6-diP:fructose 6-P in biopsy samples from the quadriceps femoris muscle to the lactate content. Subjects performed isometric work at either 40 or 100 % of their max intensity for different periods of time. Biopsies were taken at the beginning of each experiment and at the end of each work period.

In all experiments the last and longest work was performed until the subject was unable to maintain the required load. If the load had been decreased at this point the subjects could continue to work showing that cessation of work had not been due to the build up of any metabolite to toxic levels. The end point of the longest hold in all subjects was characterized by a low PC level (31.5 % of basal in subjects A and B, and 17.5 % of basal in subjects C, D, and E) and a decreased (85-90 % of basal) ATP level. It has already been mentioned that probably only a fraction of the total cell ATP is available for work. Thus, it would seem that cessation of work in these subjects was brought on by the lack of available P rather than by any other means. It can readily be calculated from the data given in Table 3 that at exhaustion following work at either intensity only a small part of the muscle glycogen store had been utilized, thereby eliminating substrate depletion as a causing factor. The lack of P therefore must have been due to a too low rate of glycolysis. The possible inhibition of PFK by decreasing cell-pH, as previously described, may well have then been the determining factor, since the resulting high levels of glucose 6-P would in turn inhibit phosphorylase b activity (16), thereby bringing about a reduction in the glycolytic rate.

This investigation was supported by grants from the Swedish Medical Research Council (Projects nos B70-19X-1002-05B, B70-19X--2647-02B, and K70-19X-2647-02BK).

REFERENCES

1. BERGSTRÖM, J. Muscle electrolytes in man. Determined by neutron activation analysis on needle biopsy specimens. A study on normal subjects, kidney patients, and patients with chronic diarrhoea. Scand. J.Clin.Lab.Invest. 14: Suppl. 68, 1962.
2. BUEDING, E. and J.M. MANSOUR. The relationship between inhibition of phosphofructokinase activity and the mode of action of trivalent organic antimonials on schistosoma mansoni. Brit.J.Pharm. 12: 159 - 165, 1957.
3. CORSI, A., M. MIDRIO, and A.L. GRANATA. In situ utilization of glycogen and blood glucose by skeletal muscle during tetanus. Am. J. Physiol. 216: 1534 - 1541, 1969.
4. DANFORTH, W.H. and J.B. LYON Jr. Glycogenolysis during tetanic contraction of isolated mouse muscles in the presence and absence of phosphorylase $\underline{a}$. J. Biol. Chem. 239: 4047 - 4050, 1964.
5. DI PRAMPERO, P.E. and R. MARGARIA. Mechanical efficiency of phosphagen (ATP + CP) splitting and its speed of resynthesis. Pfluegers Arch. Ges. Physiol. 308: 197 - 202, 1969.
6. FAWAZ, E.N., G. FAWAZ, and K. VON DAHL. Enzymatic estimation of phosphocreatine. Proc. Soc. Exp. Biol. Med. 109: 38 - 41, 1962.
7. GARLAND, P.B., P.J. RANDLE, and E.A. NEWSHOLME. Citrate as an intermediary in the inhibition of phosphofructokinase in rat heart muscle by fatty acids, ketone bodies, pyruvate, diabetes and starvation. Nature 200: 169 - 170, 1963.
8. HOFER, H.V. and D. PETTE. Wirkungen und Wechselwirkungen von Substraten und Effektoren an der Phosphofructokinase des Kaninchen-Skeletmuskels. Z. Physiol. Chem. 349: 1378 - 1392, 1968.
9. HOHORST, H.J., M. REIM, and H. BARTELS. Studies on the creatine kinase equilibrium in muscle and the significance of ATP and ADP levels. Biochem. Biophys. Res. Commun. 7: 142 - - 146, 1962.
10. HULTMAN, E. Studies on muscle metabolism and active phosphate in man with special reference to exercise and diet. Scand. J. Clin. Lab. Invest. 19: Suppl. 94, 1967.
11. HULTMAN, E., J. BERGSTRÖM, and N. MC LENNAN ANDERSON. Breakdown and resynthesis of phosphorylcreatine and adenosine triphosphate in connection with muscular work in man. Scand. J. Clin. Lab. Invest. 19: 56 - 66, 1967.
12. KREBS, E.G. and E.H. FISCHER. Phosphorylase activity of skeletal muscle extracts. J. Biol. Chem. 216: 113 - 120, 1955.

13. KRZANOWSKI, I. and F.M. MATSCHINSKY. Regulation of phosphofructokinase by phosphocreatine and phosphorylated glycolytic intermediates. Biochem. Biophys. Res. Commun. 34: 816 - 823, 1969.
14. LARDY, H.A. and E. PARKS Jr. in: GAEBLER, O.H. ed. Enzymes: Units of biological structure and function. New York: Academic Press, 1956, p. 584.
15. MANSOUR, T.E. Studies on heart phosphofructokinase: purification, inhibition and activation. J. Biol. Chem. 238: 2285 - 2292, 1963.
16. MORGAN, H.E. and A. PARMEGGIANI. Regulation of glycogenolysis in muscle. II. Control of glycogen phosphorylase reaction in isolated perfused heart. III. Control of muscle glycogen phosphorylase activity. J. Biol. Chem. 239: 2435 - 2439, 1964.
17. PARMEGGIANI, A. and R.H. BOWMAN. Regulation of phosphofructokinase activity by citrate in normal and diabetic muscle. Biochem. Biophys. Res. Commun. 12: 268 - 273, 1963.
18. PASSONNEAU, J.V. and O.H. LOWRY. P-fructokinase and the control of the citric acid cycle. Biochem. Biophys. Res. Commun. 13: 372 - 379, 1963.
19. POSNER, J.B., R. STERN, and E.G. KREBS. Effects of electrical stimulation and epinephrine on muscle phosphorylase, phosphorylase b kinase and adenosine 3'5' phosphate. J. Biol. Chem. 240: 982 - 985, 1965.
20. SCHULZ, D.W., J.V. PASSONNEAU, and O.H. LOWRY. An enzymic method for the measurement of inorganic phosphate. Anal. Biochem. 19: 300 - 314, 1967.
21. TORNVALL, G. Assessment of physical capabilities with special reference to the evaluation of maximal voluntary isometric muscle strength and maximal working capacity. Acta Physiol. Scand. 50: Suppl. 201, 1963.
22. TRIVEDI, B. and W.H. DANFORTH. Effect of pH on the kinetics of frog muscle phosphofructokinase. J. Biol. Chem. 241: 4110 - 4114, 1966.
23. UI, M. A role of phosphofructokinase in pH-dependent regulation of glycolysis. Biochim. Biophys. Acta 124: 310 - 322, 1966.
24. WILSON, J.E., B. SACKTOR, and C.G. TIEKERT. In situ regulation of glycolysis in tetanized cat skeletal muscle. Arch. Biochem. Biophys. 120: 542 - 546, 1967.

ISOMETRIC EXERCISE - FACTORS INFLUENCING ENDURANCE AND FATIGUE

R.H.T. EDWARDS[*], L.-O. NORDESJÖ, D. KOH, R.C. HARRIS, and E. HULTMAN
From the Departments of Clinical Physiology, Karolinska Sjukhuset, and Clinical Chemistry, S:t Eriks Sjukhus, Stockholm, Sweden

In two healthy male subjects we have made some preliminary observations of the factors influencing endurance and fatigue during successive sustained contractions at two thirds the intensity of a maximal voluntary contraction of the Quadriceps Muscle. In all experiments subjects sat in an adjustable chair (5), measurements being made with a strain gauge attached to the ankle when the knee joint was flexed to 90°. Fig. 1 shows the relationship between the duration of successive contractions and the interval between contractions, both under normal conditions and when the circulation to the leg was occluded using a sphygmomanometer cuff inflated to 250 mm Hg, immediately prior to starting the series. Each subject was allowed at least 24 hours recovery between each experiment. In the absence of circulatory occlusion more than seven holds were possible, the duration of successive holds being significantly related ($p = 0.016$) to the recovery interval. With circulatory occlusion no more than three holds could be sustained, the duration of the second and third holds being unrelated to the recovery interval. As observed by others (3, 4) the duration of the first hold was not significantly reduced by prior circulatory occlusion suggesting that muscle blood flow was practically arrested at this relative load. The quadriceps thus behaved as an isolated muscle preparation, convenient for studying muscle metabolism in man. The metabolites, assayed spectrophotometrically using NAD linked enzymic reactions - modified after Lowry et al. (2) - in percutaneous needle biopsy samples (1) taken at the start or immediately at the end of contractions, are shown in Fig. 2. A recovery interval of 20 sec was used in both experiments. In subject 1, ATP was maintained close to the pre-exercise control

[*] In receipt of a Swedish Welcome Travelling Research Fellowship

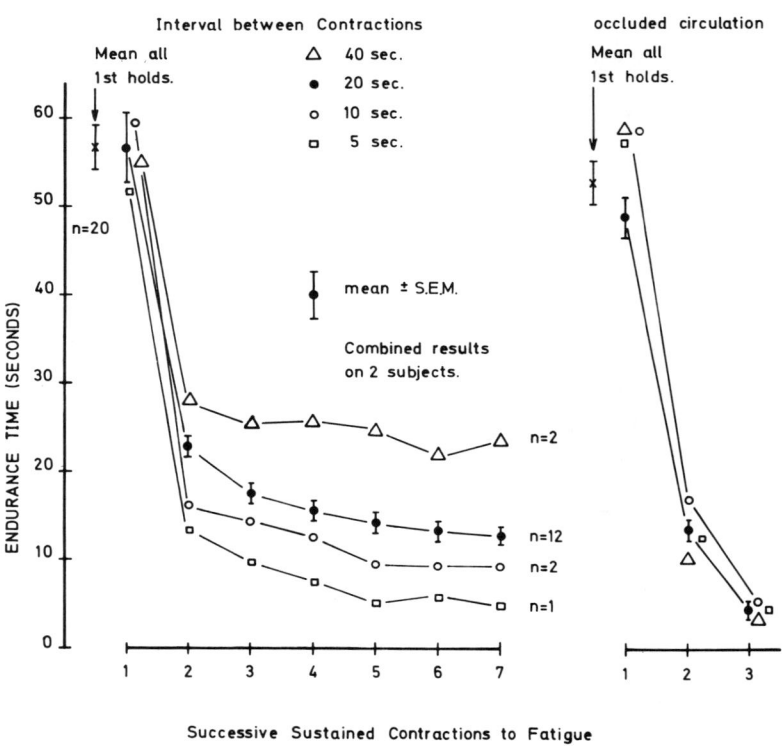

Fig. 1. Endurance times of sustained isometric contractions with and without circulatory occlusion. (MVC = Maximum Voluntary Contraction)

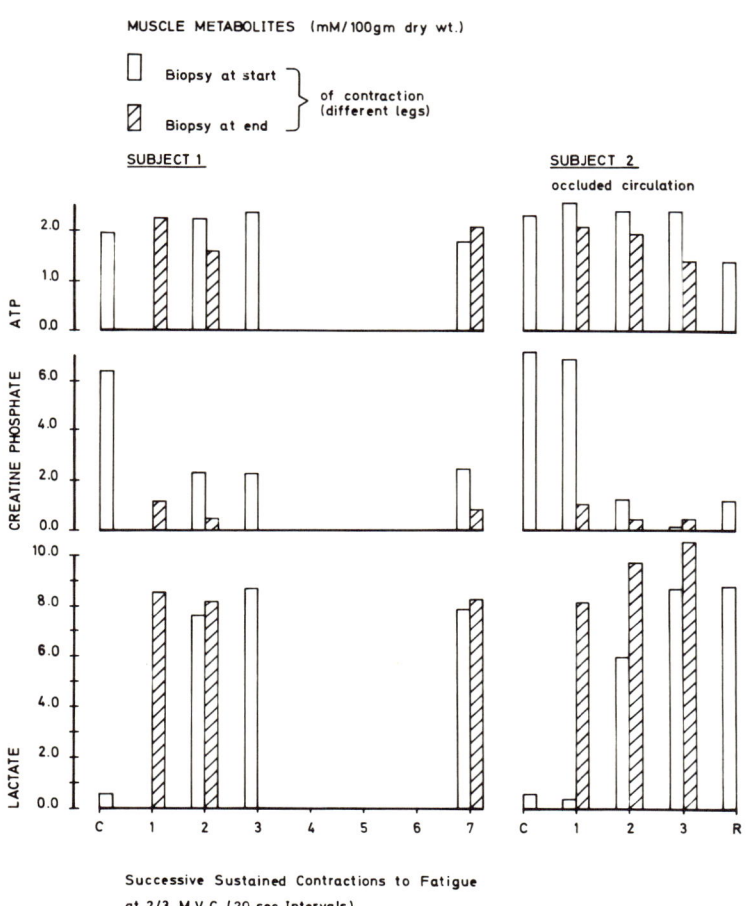

Fig. 2. Muscle metabolites in biopsy samples of quadriceps in two subjects. (C = resting control, R = sample taken 20 sec after end of last contraction, the circulation having been restored)

level (C) throughout the 7 contractions but in subject 2, in whom the circulation was occluded, the concentration fell to 60 % of the pre-exercise control level by the end of the final contraction and was unchanged after 20 sec recovery with the circulation restored. Creatine phosphate was maintained at 37 % of control at the start of successive contractions (15 % at the end of contractions) in the limb with the open circulation, but fell to about 7 % by the end of the third contraction with the circulation occluded. Lactate rose similarly during the first contraction in the two subjects and was maintained at a constant level in the one in whom there was an open circulation, in contrast to the progressive increase with successive contractions found in the subject in whom the circulation was occluded. No quantitative conclusions can be drawn as to the influence of the recovery interval on muscle lactate concentration as biopsies were taken from different legs.

The end point of any sustained isometric contraction to fatigue is determined by several factors. These experiments suggest the importance of i) the availability of ATP and creatine phosphate as immediate sources of energy and ii) a normal circulatory adaptation during recovery intervals, possibly to allow aerobic resynthesis of high energy phosphates.

REFERENCES

1. HULTMAN, E. Studies on muscle metabolism and active phosphate in man with special reference to exercise and diet. Scand. J. Clin. Lab. Invest. 19: Suppl. 94, 1967.
2. LOWRY, O.H., J.V. PASSONNEAU, F. HASSELBERGER, and D.W. SCHULZ. Effect of ischaemia on known substrates and cofactors of the glycolytic pathway in brain. J. Biol. Chem. 239: 18 - 30, 1964.
3. ROYCE, J. Isometric fatigue curves in human muscle with normal and occluded circulation. Res. Quart. 29: 204 - 212, 1958.
4. START, K.B. and R. HOLMES. Local muscle endurance with open and occluded intra-muscular circulation. J. Appl. Physiol. 18: 804 - 807, 1963.
5. TORNVALL, G. Assessment of physical capabilities. Acta Physiol. Scand. 58: Suppl. 201, 1963.

LACTATE AND OXYGEN DEBT: AN INTRODUCTION

Howard G. Knuttgen

Department of Biology, Boston University, 2 Cummington Street, Boston, 02215 Massachusetts, USA

The relationship of the formation of lactate acid to skeletal muscle energy release in exercising humans was first explored by A.V. Hill and co-workers (21, 22). The term "oxygen debt" was suggested by them to describe the excess oxygen consumption of recovery which they felt was closely related. A combination of their work and the earlier work of Krogh and Lindhard (35) resulted at that time in the belief that a certain amount of energy release during the transition from rest to exercise was provided by a non-aerobic source, glycolysis. The resulting accumulation of lactic acid (as lactate) in the body required an extra consumption during recovery for its oxidative removal.

Jervell (24) subsequently showed that, in exercise, the greatest accumulation in blood took place during the first few minutes. He felt that the blood lactate increase was due to a shortage of oxygen during the transition period. The observation was also made for the first time that the increased level of lactate due to exercise could be made to fall faster if mild exercise was employed by the subjects in place of sedentary recovery.

The work of Margaria, Edwards and Dill (40) appeared in 1933. They observed that exercise (treadmill running) could be carried on at low levels without significant changes in resting levels of blood lactate. They employed one research subject and they found lactate increase above a running speed eliciting an oxygen consumption above 2/3´ of maximal $\dot{V}_{O_2}$, with a direct and curvilinear relationship existing between lactate concentration and oxygen consumption level. Work was never carried on for longer than 10 min. As regards the oxygen debt, they concluded that (a) the initial fast component of oxygen debt was not due to lactate removal (and called it the "alactacid portion"), (b) the slow component, which could last in excess of an hour, was due to the oxidation of the accumulated lactate (as described by Hill et al.), and (c) an increase in resting metabolism resulting from the exercise. The latter has been referred to in later work (9,31) as a disturbance to resting state.

The hypothesis that the initial, rapid-repayment portion of the oxygen debt was not due to lactate removal was not contested until the work of Huckabee and the advancement of his concept of "Excess Lactate" (23). According to this concept, there was no alactacid portion: The debt could be satisfactorily explained by the oxygen necessary to oxidize the portion of the accumulated lactate which was due solely to tissue hypoxia. Subsequent studies (19,30,39,46,51,57) have failed to confirm the relationship as proposed by Huckabee. Further, the "excess lactate" concept has been objected to on theoretical grounds (3,19,42).

During the last 35 years, bits and pieces have been added to the puzzle of lactate formation and removal and the interrelationships with oxygen consumption --- but the problem remains far from solved. An increased production of lactate occurs in skeletal muscle during exercise under certain conditions. Work intensity, especially as assessed with relation to a person's capacity and/or state of training is the most important factor. Small changes are seen during work eliciting oxygen consumptions below 50-60 per cent of aerobic capacity. At higher submaximal loads, lactate production is seen to increase during the first minutes of exercise (6,15,24) and in direct proportion to the work intensity. At very high submaximal and at maximal loads, a constant production is observed. Other factors, such as lowered blood oxygen tension (e.g. at altitude) and circulatory disorder also affect the relationship by causing increased lactate production at the same absolute work loads. Whether hypoxia is the only factor has been questioned on the basis of experiments with exercising humans (28) and animal experiments (25). It has been suggested that pyruvate is formed (glycolysis is taking place) at a rate faster than the remaining steps (oxidative phosphorylation) can tolerate and, thus, the pyruvate is reduced to lactate.

The lactate formed then diffuses out of the muscle and is removed principally by circulating blood. The serious question then appears as to the extent of the diffusion of the lactate into various parts of the body and in the various fluid compartments The answer to this question is crucial to the attempts to estimate total lactate production by employing data concerning concentrations observed in blood, muscle tissue, etc. It has been reported for example that, in experiments where lactate was infused in the blood of resting dogs, the distribution space for lactate after 40 min was 75 per cent of estimated total body water (4).

Increased quantities of lactate found in the blood are then observed to decrease during sedentary recovery or in recovery involving submaximal exercise (15,18,24,41,44,53). The rate of disappearance during exercise is related to the intensity of the exercise, with maximal removal rate reported to occur at approximately 60-70 per cent of aerobic capacity (53). The locations of removal and the fate of the lactate and the various removal sites are but partially understood. The amounts found in urine (24,26,54) or excreted in sweat appear to be negligible. The kidneys might, however, remove a significant quantity by gluconeogenesis. Removal has been observed in the heart (8,29), in the liver (20,45,46,47), and in the resting limbs with resting skeletal muscle tissue presumably playing the dominant role in the latter.

Use of lactate as substrate would appear to be its fate in the heart and skeletal muscle. It is generally held that gluconeogenesis is impossible in mammalian skeletal muscle due to the lack of appropriate enzymes for conversion of pyruvate to phosphopyruvate (33,34,52). The fate of lactate in the liver would presumably be gluconeogenesis.

The relationships among lactate formation, the oxygen deficit observed at the beginning of exercise, lactate removal, and the excess oxygen consumption of recovery then become complex. One would anticipate a good quantitative relationship between oxygen deficit (oxygen either not available or not employed) and lactate formation. Experiments employing blood lactate values (38,39) and both blood and muscle values (27) for estimation of lactate production have found this to be true. The variety of disappearance routes then complicates the relationship.

If lactate is taken up by the liver and resynthesized to glucose or glycogen, an appropriate increase in body oxygen consumption during exercise and/or recovery would be expected. That such an increase does exist has been shown in animal experiments employing agents (tryptophan and quinolinic acid) for blocking gluconeogenesis (7). Lactate removal in this fashion costs a greater amount of ATP than was furnished by lactate formation (see discussion in ref. 37). Therefore, if all or most of the lactate is removed by gluconeogenesis, recovery oxygen (debt) would be greater than the original deficit. On the other hand, if all lactate formed during the deficit period is taken up by various tissues and employed as substrate (to pyruvate and then the tricarboxylic acid cycle and respiratory chain), no extra oxygen consumption would be observed. The lactate would furnish the energy for the basic needs of the cells where it was employed. In this case, there would be a measurable deficit, a quantity of lactate formed and observed in the extracellular fluid, but no oxygen debt. Under real conditions of sport and labor, it is probable that an all-or-none condition of these two removal routes never occurs but, rather, various combinations. Therefore, the understanding of the debt (the excess oxygen consumption of recovery), is made extremely difficult. Also, it points up the difficulty which arises when investigators employ the terms deficit and debt interchangeably. It would be desirable if the term "deficit" could refer to the quantity of oxygen not employed during the transition to a higher intensity of activity (e.g. from rest to exercise) and the "debt" to the excess oxygen consumption of recovery. The debt would involve the increase in oxygen utilization due to gluconeogenesis, resynthesis of depleted high-energy phosphates and restoration of electrolyte balance. It would also involve the increase due to body temperature elevation, increased sympathetic activity and increased hormone level.

Studies with animals have served to both lead the way in the understanding of lactate metabolism in exercise and, also, to confirm observations on exercising humans. Dating back to the studies of Sacks et al. (49) in 1937 and Flock et al. (16) in 1939, the formation of lactate in submaximal exercise has been observed as an initial process. It has been generally held that lactate could pass across the muscle cell membrane readily. In a large number of studies, lactate was found to enter skeletal muscle cells readily but considerable evidence exists that little or no carbohydrate can be synthesized from this lactate (1,14,32,33,34,48,52,56,58). There exists certain dissension from the free movement point of view. Tranquade (55) and Kübler et al. (36) have presented evidence suggesting the existance of an active transport mechanism. Tranquade also concluded that blood lactate, pyruvate, and L/P ratios did not accurately reflect changes in specific tissues. In addition to studies demonstrating glycogen formation from lactate in the liver (10,11), numerous studies have presented evidence of tagged lactate either being employed directly as substrate, or, at least, appearing subsequently as CO_2 (10,12,13,43,58).

In closing, I would observe that there are enough questions involved in the understanding of lactate metabolism in exercise to keep many laboratories busy for many years. Perhaps the most important questions could be summarized as follows:
1) What are the various conditions which cause lactate formation in the muscle cell?
2) What quantities of energy can be supplied by glycolysis and at what rates to produce tension in muscle?
3) How does lactate pass across cell membranes and to what extent does it become distributed throughout the body?
4) What controls the routes and the rates of lactate removal or disappearance?

5) What is the quantitative relationship of lactate removal by gluconeogenesis to the excess oxygen consumption of recovery (oxygen debt)?

REFERENCES

1. ABRAMSON, H.A., M.G. EGGLETON & P. EGGLETON. The utilization of intravenous sodium r-lactate. III. Glycogen synthesis by the liver. Blood sugar. Oxygen consumption. J. Biol. Chem. 75:763-778, 1927.

2. AGNEVIK, G., J. KARLSSON, B. DIAMANT & B. SALTIN. Oxygen debt, lactate in blood and muscle tissue during maximal exercise in man. Medicine and Sport 3:62-65, 1969.

3. ALPERT, N.R. Lactate production and removal and the regulation of metabolism. Ann. N.Y. Acad. Sci. 119:995-1011, 1965.

4. ALPERT, N.R. & W.S. ROOT. Relationship between excess respiratory metabolism and utilization of intravenously infused sodium racemic lactate and sodium L(-) lactate. Am. J. Physiol. 177:455-462, 1954.

5. ÅSTRAND, I. Lactate content in sweat. Acta physiol. scand. 58:359-367, 1963.

6. BANG, O. The lactate content of blood during and after muscular exercise in man. Skand. Arch. Physiol. 74, Suppl. 10:51-82, 1936.

7. BARNARD, R.J. & M. L. FOSS. Oxygen debt: effect of beta-adrenergic blockade on the lactacid and alactacid components. J. Appl. Physiol. 27:813-816, 1969.

8. CARLSTEN, A., B. HALLGREN, R. JAGENBURG, A. SVANBORG & L. WERKÖ. Myocardial metabolism of glucose, lactic acid, amino acids and fatty acids in healthy human individuals at rest and at different work loads. Scand. J. clin. Lab. Invest. 13:418-428, 1961.

9. CHRISTENSEN, E.H., R. HEDMAN & I. HOLMDAHL. The influence of rest pauses on mechanical efficiency. Acta physiol. scand. 48:443-447, 1960.

10. CONANT, J.B., R.D. CRAMER, A.B. HASTINGS, F.W. KLEMPERER, A.K. SOLOMON & B. VENNESLAND. Metabolism of lactic acid containing radioactive carboxyl carbon. J. Biol. Chem. 137:557-566, 1941.

11. CORI, C.F. & G.T. CORI. Glycogen formation in the liver from d- and l-lactic acid. J. Biol. Chem. 81:389-403, 1929.

12. DEPOCAS, F., Y. MINAIRE & J. CHATONNET. Rates of formation and oxidation of lactic acid in dogs at rest and during moderate exercise. Can. J. Physiol. Pharmacol. 47:603-610, 1969.

13. DRURY, D.R., A.N. WICK & T.N. MORITA. Metabolism of lactic acid in extrahepatic tissues. Amer. J. Physiol. 180:345-349, 1955.

14. ELIAS, H. & E. SCHUBERT. Über die Rolle der Säure im Kohlenhydratstoffwechsel. III. Säure und Muskelglygogen. Biochem. Z. 90:229-243, 1918.

15. ESKILDSEN, P.P. Arbejdsfysiologiske undersøgelser hos nogle hjertepatienter og patienter med asthma-emfysem. Copenhagen: Ejnar Munksgaards Forlag, 1945.

16. FLOCK, E.V., D.J. ANGLE & J.L. BOLLMAN. Formation of lactic acid, an initial process in working muscle. J. Biol. Chem. 129:99-110, 1939.

17. FREYSCHUSS, V. & T. STRANDELL. Limb circulation during arm and leg exercise in supine position. J. Appl. Physiol. 23:163-170, 1967.

18. GISOLFI, C., S. ROBINSON & E.S. TURRELL. Effects of aerobic work performed during recovery from exhausting work. J. Appl. Physiol. 21:1767-1772, 1966.

19. HARRIS, P., M. BATEMAN & J. GLOSTER. Relations between the cardio-respiratory effects of exercise and the arterial concentration of lactate and pyruvate in patients with rheumatic heart disease. Clin. Sci. 23:531-543, 1962.

20. HARRIS, P., M. BATEMAN & J. GLOSTER. The regional metabolism of lactate and pyruvate during exercise in patients with rheumatic heart disease. Clin. Sci. 23:545-560, 1962.

21. HILL, A.V., C.N.H. LONG & H. LUPTON. Muscular exercise, lactic acid, and the supply and utilization of oxygen. Proc. Roy. Soc., London, Ser. B 97:84-138, 1924.

22. HILL, A.V. & H. LUPTON. Muscular exercise, lactic acid, and the supply and utilization of oxygen. Quart. J. Med. 16:135-171, 1923.

23. HUCKABEE, W.E. Relationships of pyruvate and lactate during anaerobic metabolism. II. Exercise and formation of O_2 debt. J. Clin. Invest. 37:255-263, 1958.

24. JERVELL, O. Investigation of the concentration of lactic acid in blood and urine. Acta Med. Scand. Suppl. 24:1-135, 1928.

25. JÖBSIS, F.F. & W.N. STAINSBY. Oxidation of NADH during contractions of circulated mammalian skeletal muscle. Respiration Physiol. 4:292-300, 1968.

26. JOHNSON, R.E. & H.T. EDWARDS. Lactate and pyruvate in blood and urine after exercise. J. Biol. Chem. 118:427-432, 1937.

27. KARLSSON, J. & B. SALTIN. Lactate, ATP, and CP in the working muscles during exhaustive exercise in man. J. Appl. Physiol. In press.

28. KEUL, J., E. DOLL & D. KEPPLER. The substrate supply of the human skeletal muscle at rest, during and after work. Experientia 23:974-979, 1967.

29. KEUL, J., E. DOLL, H. STEIM, U. FLEER & H. REINDELL. Über den Stoffwechsel des menschlichen Herzens. III. Der oxidative Stoffwechsel des menschlichen Herzens unter verschiedenen Arbeitsbedingungen. Pflügers Archiv 282:45-53, 1965.

30. KNUTTGEN, H.G. Oxygen debt, lactate, pyruvate, and excess lactate after muscular work. J. Appl. Physiol. 17:639-644, 1962.

31. KNUTTGEN, H.G. Oxygen debt following submaximal physical exercise. J. Appl. Physiol. In press.

32. KOEPPE, R.E., N.F. INCIARDI, L.G. WARNOCK & W.E. WILSON. Some aspects of the metabolism of D- and L-lactic acid -2-^{14}C by rat skeletal muscle in vivo. J. Biol. Chem. 239:3609-3612, 1964.

33. KREBS, H. Gluconeogenesis. Proc. Roy. Soc., London, Ser. B 159:545-563, 1964.

34. KREBS, H.A. & M. WOODFORD. Fructose 1,6-diphosphate in striated muscle. Biochem. J. 94:436-445, 1965.

35. KROGH, A. & J. LINDHARD. The changes in respiration at the transition from work to rest. J. Physiol., London. 53:431-439, 1920.

36. KÜBLER, W., H.J. BRETSCHNEIDER, W. VOSS, H. GEHL, F. WENTHE & J.L. COLAS. Über die Milchsäure- und Brenztraubensäurepermeation aus dem hypothermen Myokard. Pflügers Arch. ges. Physiol. 287:203-223, 1965.

37. LLOYD, B. The energetics of running: on analysis of world records. Advancement of Science, pp. 515-560, January 1966.

38. MARGARIA, R., P. CERRETELLI & F. MANGILI. Balance and kinetics of anaerobic energy release during strenuous exercise in man. J. Appl. Physiol. 19:623-628, 1964.

39. MARGARIA, R., P. CERRETELLI, P.E. di PRAMPERO, C. MASSARI & G. TORELLI. Kinetics and mechanism of oxygen debt contraction in man. J. Appl. Physiol. 18:371-377, 1963.

40. MARGARIA, R., H.T. EDWARDS & D.B. DILL. The possible mechanisms of contracting and paying the oxygen debt and the role of lactic acid in muscular contraction. Am. J. Physiol. 106:689-715, 1933.

41. NEWMAN, E.V., D.B. DILL, H.T. EDWARDS & F.A. WEBSTER. The rate of lactic acid removal in exercise. Am. J. Physiol. 118:457-462, 1937.

42. OLSON, R.E. "Excess lactate" and anaerobiosis. Ann. intern. Med. 59:960-963, 1963.

43. OMACHI, A. & N. LIFSON. Metabolism of isotonic lactate by the isolated perfused dog gastrocnemius. Amer. J. Physiol. 185:35-40, 1956.

44. RÄMMEL, K. & G. STRÖM. The rate of lactate utilization in man during work and at rest. Acta physiol. scand. 17:452-456, 1949.

45. ROWELL, L.B., G.L. BRENGELMANN, J.R. BLACKMON, R.D. TWISS & F. KUSUMI. Splanchnic blood flow and metabolism in heat-stressed man. J. Appl. Physiol. 24:475-484, 1968.

46. ROWELL, L.B., K.K. KRANING II, T.O. EVANS, J.W. KENNEDY, J.R. BLACKMON & F. KUSUMI. Splanchnic removal of lactate and pyruvate during prolonged exercise in man. J. Appl. Physiol. 21:1773-1783, 1966.

47. ROWELL, L.B., E.J. MASORO & M.J. SPENCER. Splanchnic metabolism in exercising man. J. Appl. Physiol. 20:1032-1037, 1965.

48. SACKS, J. & W.C. SACKS. Carbohydrate changes during recovery from muscular contraction. Am. J. Physiol. 112:565-572, 1935.

49. SACKS, J., W.C. SACKS & J.R. SHAW. Carbohydrate and phosphorus changes in prolonged muscular contractions. Am. J. Physiol. 118:232-240, 1937.

50. SAIKI, H., R. MARGARIA & F. CUTTICA. Lactic acid production in submaximal work. Int. Z. angew. Physiol. 24:57-61, 1967.

51. SCHNEIDER, E.G., S. ROBINSON & J.L. NEWTON. Oxygen debt in aerobic work. J. Appl. Physiol. 25:58-62, 1968.

52. SCRUTTON, M.C. & M.F. UTTER. The regulation of glycolysis and gluconeogenesis in animal tissue. Ann. Rev. Biochem. 37:249-302, 1968.

53. STENSVOLD, I. & L. HERMANSEN. The rate of lactic acid removal during recovery while working and resting (Abstract). Acta physiol. scand. Suppl. 330:85, 1969.

54. STRÖM, G. The influence of anoxia on lactate utilization in man after prolonged muscular work. Acta physiol. scand. 17:440-451, 1949.

55. TRANQUADE, R.E. The relationship of intracellular and extracellular lactate and pyruvate concentrations. Clin. Res. 14:179, 1966.

56. WARNOCK, L.G., R.E. KOEPPE, N.F. INCIARDI & W.E. WILSONS. L(+) and D(-) lactate as precursors of muscle glycogen. Ann. N.Y. Acad. Sci. 119:1048-1060, 1965.

57. WASSERMAN, K., G.G. BURTON & A.L. VAN KESSEL. The excess lactate concept and the oxygen debt of exercise. J. Appl. Physiol. 20:1299-1306, 1965.

58. WICK, A.N. Chemistry and metabolism of L(+) and D(-) lactic acids. Ann. N.Y. Acad. Sci. 119:1061-1069, 1965.

THE ALACTIC OXYGEN DEBT: ITS POWER, CAPACITY AND EFFICIENCY

P. E. di Prampero

Department of Physiology, University of Milano, Italy

The immediate energy source for muscular contraction is generally assumed to be the splitting of ATP, which is immediately resynthesized by creatinephosphate (CP) splitting. These two reactions can be considered in series and, by calling phosphagen (PG) the sum of ATP + CP, the overall reaction can be simplified (5):

$$PG \rightleftharpoons Pi + G \qquad 1)$$

During <u>steady state exercise</u> the rate of PG splitting must be equal to the rate of its resynthesis, which in turn depends on energy yielding processes such as oxidations and/or glycolysis. However, since both lactic acid (LA) production and O_2 consumption are delayed processes in respect to the mechanical events of muscular contraction (8, 11), during the initial phase of the exercise the rate of PG splitting is higher than its rate of resynthesis. It follows that during work at steady state the concentration of PG in muscle is lower than at rest (3, 13). The net alactic O_2 debt ($V_{O_2}^{al}$) represents the oxidative energy necessary to resynthesize the amount of phosphagen split at the onset of work: this has been found to increase linearly with the O_2 uptake both in man and in isolated muscle (10, 13).

During <u>short term exercise</u> the rate of PG splitting may be higher than its rate of resynthesis, so that a steady state can not be attained. Moreover, in very heavy exercise, exhaustion may be reached within 5 to 10 seconds, a time too short to build an appreciable amount of LA, or to increase the O_2 uptake significantly above the resting value (8, 11). It may then be assumed, without appreciable error, that in these conditions the work is performed entirely at the expenses of PG splitting. Oxidations and/or LA production enter into the picture only later, when the exercise is over, to restore the PG level in muscle at its resting value.

For a full understanding of the PG splitting mechanism the following characteristics should be known: a) the rate of energy output (power), b) the total energy available (capacity) and c) the efficiency.

a) <u>Power.</u>

It has been shown by Margaria et al. (8) that in the first few seconds of muscular activity energy is drawn only from PG breakdown and no LA is formed, even during very strenuous exercise (Fig. 1). By plotting the intensity of exercise (in oxidative energy requirement) as a function of the duration of the work without the intervention of LA (Fig. 2, line "alact.") a straight line is obtained. This can be extrapolated at t = 0,

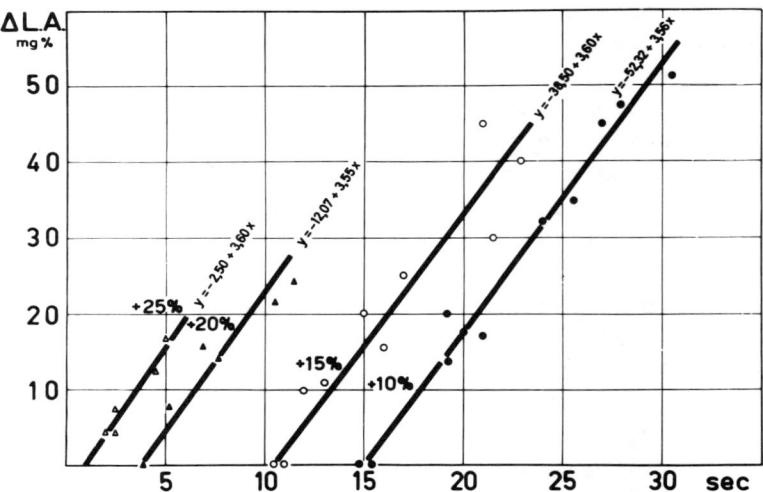

Fig. 1. Increase in lactic acid concentration in blood above resting value (ΔLA, mg %), as a function of the duration of the exercise (sec), in four non athletic subjects running on the treadmill at 18 km/h, at the incline indicated. The time during which the exercise can be carried on without LA production is given, for each work load, by the intercept on the abscissa. (After 8).

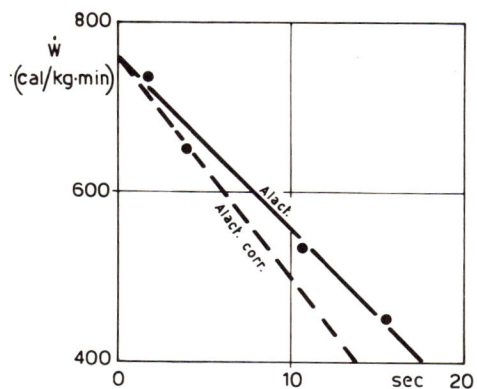

Fig. 2. Energy requirement of the exercise ($\dot{W}$, cal/kg min) as a function of the time (sec) during which the exercise can be carried on without the accumulation of LA in blood (line "alact."); this is further corrected for the small amount of O_2 consumed during the run (line "alact. corr."). Same experiments described in fig. 1. (After Margaria et al., 8).

thus giving an indication of the maximal power which can be sustained for a very short time. This appears then to be the maximal power due only to PG splitting and it amounts to about 150 ml/kg min or to 750 cal/kg min.

The time course of the maximal power due to PG splitting can be obtained, as from Margaria et al. (6), by measuring the vertical component of the speed in subjects running at top speed up a normal staircase. In these conditions the maximal speed is reached in less than two seconds and it is maintained up to the 5th second, declining thereafter. In the constant speed phase, the vertical component of the velocity (m/sec) can be identified with the mechanical power output (kgm/kg sec), all the energy being employed to lift the body (6). As the exercise is of very short duration, the power developed may be considered indicative of the phosphagen splitting mechanism alone. Moreover, since in this type of exercise the efficiency approaches 0.25 (4, 7), from the mechanical power output the corresponding energy expenditure can be calcualted. This, in normal subjects at the age of 20 - 25, attains a maximal value of 150 to 160 ml/kg min, in agreement with the data obtained by Margaria et al. (8) with the previously described technique.

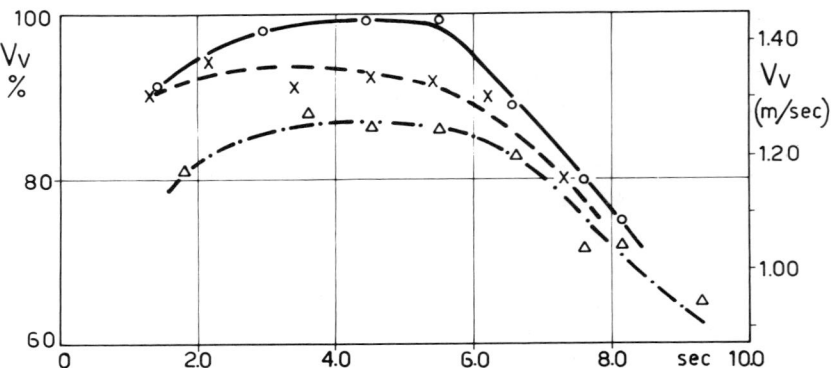

Fig. 3. Vertical component of the speed (percentage of the maximal value obtained when the test started from rest) as a function of the time (sec) from the onset of maximal run. Right ordinate gives vertical component of the velocity in m/sec. Only 3 experimental conditions are given for simplicity: maximal run starting from rest (o) and from steady state exercise (x,△) involving an O_2 uptake above basal of 18 and 33.5 ml/kg min respectively. Each curve is the average of three runs on four subjects.(After 9).

The maximal power due to PG splitting, as measured by means of this approach, is relatively constant for the first 5 to 6 seconds (fig. 3). This seems to indicate that in maximal voluntary contraction PG splitting behaves at first as a zero order reaction. As it has been recently pointed out by Margaria et al. (9), this finding can not be explained on the basis of reaction 1), according to which PG splitting should be a first order reaction. PG splitting however cannot take place spontaneously in muscle, but must be activated by the nervous impulses reaching the muscle. As a tentative explanation it may then be assumed (9) that PG must first combine with an activator (A) produced at the site of excitation to form an extremely labile compound:

$$PG + A \longrightarrow PGA \qquad 2)$$

Only than can the energy yelding reaction take place:

$$PGA \longrightarrow Pi + G + A \qquad 2')$$

If the affinity of the activator with PG is assumed to be very high, practically the total amount of activator liberated (Ao) is in the combined form (PGA):

$$Ao = PGA \qquad 3)$$

According to reaction 2') the maximal rate at which the muscle can liberate energy depends on the concentration of PGA:

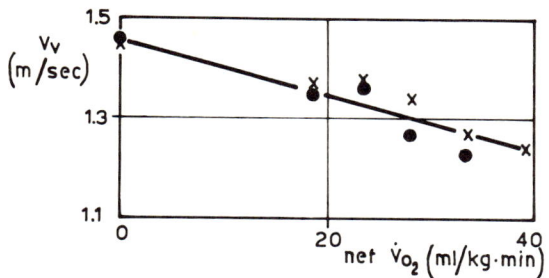

Fig. 4. Maximal vertical component of the speed (m/sec) as a function of the O_2 uptake above basal (net $\dot{V}_{O_2}$) of the exercise preceding the run, in two subjects. (After 9).

$$\frac{\overrightarrow{d \; PGA}}{dt} = k \; PGA \qquad 4)$$

From 3) and 4):

$$\frac{\overrightarrow{d \; PGA}}{dt} = k \; Ao \qquad 5)$$

The power the muscle can develop would therefore depend on the rate at which the activator is produced, which may be assumed to be constant. After 5 to 6 seconds of maximal effort the concentration of PG itself becomes presumably a limiting factor, thus explaining the observed decrease in maximal power output after the first few seconds of exercise (fig. 3). According to current theories of muscular contraction (12), the formation of the activator as an effect of nervous stimulation may be identified with the liberation of Ca^{++} at the contractile elements level.

When the test is made on subjects performing a steady state aerobic exercise (9), the maximal power, as measured from the vertical component of the maxiaml velocity (V_v, m/min), decreases linearly with increasing the O_2 uptake (ml/kg min) of the preceding exercise (fig. 4):

$$V_v = 87 - 0.318 \; \dot{V}_{O2} \qquad 6)$$

In this type of exercise the overall power output is derived from: 1) PG breakdown, 2) oxidations and 3) LA formation from glycogen. To calculate the power due to PG breakdown only, the O_2 uptake of the preceding exercise was subtracted from the total power output, expressed in equivalent energy requirement. This was calculated from the mechanical power output considering that 1 kgm = 2.34 cal = 0.47 ml O_2 and that the efficiency in this type of exercise amounts to 0.25 (4, 7); 1 kgm of mechanical work is then equivalent to 0.47/0.25 = 1.88 ml O_2 requirement. No o-

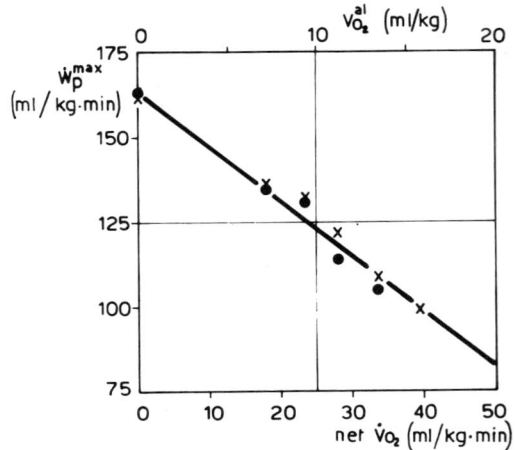

Fig. 5. Maximal power due only to extra PG breakdown ($\dot{W}_p^{max}$, ml/kg min) in equivalent O_2 requirement, as a function of the O_2 uptake of the preceding exercise (bottom abscissa) or of the corresponding alactic O_2 debt ($V_{O_2}^{al}$) (top abscissa).

ther correction was introduced, as the maximal vertical velocity was attained within 4 sec, a time too short to produce appreciable amounts of LA (8), or to increase the O_2 uptake significantly above the preceding steady state value (11). The power due to extra phosphagen breakdown ($\dot{W}_p^{max}$) decreases linearly with increasing the O_2 uptake before the test (fig. 5):

$$\dot{W}_p^{max} = 164 - 1.64 \, \dot{V}_{O_2} \qquad 7)$$

b) <u>Capacity.</u>

The maximal capacity of the alactic system has been calculated by Margaria et al. (8), by correcting the line "alact." of fig. 2 for the amount of O_2 consumed during the first phase of work, in which no LA is produced. The line "alact. corr." is thus obtained: this indicates the maximal time during which an exercise of the intensity given on the ordinate can be carried on at the expenses of PG splitting only. By extrapolating to $\dot{W} = 0$ and by calculating the integral of this function, the maximal theoretical capacity of the alactic mechanism was obtained: this amounts to 196 cal/kg (8).

The maximal theoretical capacity of the alactic O_2 debt can also be calculated from the data obtained by Margaria et al. (9) and described in formula 7). In fact the net alactic O_2 debt has been shown to be proportional to the O_2 uptake at steady state (10) (2):

$$\dot{V}_{O_2} = 2.5\ V_{O_2}^{al} \qquad 8)$$

The $\dot{V}_{O_2}$ data on the abscissa of fig. 5 can then be substituted by the corresponding $V_{O_2}^{al}$ data, the equation of the function becoming then, as from 7) and 8):

$$\dot{W}_p^{max} = 164 - 4.1\ V_{O_2}^{al} \qquad 9)$$

The alactic O_2 debt is the O_2 equivalent of the split phosphagen in muscle; eq. 9) indicates therefore that the maximal power due to PG breakdown decreases with increasing the split phosphagen in muscle. When all the PG available has been split, $\dot{W}_p^{max} = 0$. The maximal potential alactic O_2 debt amounts then to $164/4.1 = 40$ ml/kg or 200 cal/kg. This value, which is substantially equal to that previously obtained (8), is indicative of the total phosphagen content of the muscles. Since 1 mole PG requires 3.74 l of O_2 for its resynthesis (P/O = 3), 40 ml O_2 are equivalent to 10.8 m-moles PG. The fraction of muscles in the body is about 0.4: the PG concentration in muscle can then be calculated as $10.8/0.4 = 26.8$ m-moles/kg, a value not far from those obtained from direct chemical analysis of human muscle (3).

It has been shown by Margaria et al. (8) that only half the maximal capacity of the alactic system (about 100 cal/kg) is utilized in maximal exercise of very short duration (5 - 10 sec) starting from rest.

The amount of energy derived from extra PG breakdown during maximal exercise can also be calculated (9) when performing the test from a previous steady state exercise. In fact the total amount of energy spent in maximal exercise is given by the area included between the curves of fig. 3 and the coordinates, provided that the curves be corrected for the energy spent to accelerate the body during the initial phase of the run (6). As previously pointed out, in this type of exercise the total energy output is met by : 1) PG breakdown, 2) oxidations, 3) LA formation from glycogen. This last amounts to about 20 cal/Kg and since it is probably delivered when the short run (8 sec) is over (9) it has been neglected. The amount of O_2 consumed during the run can be calculated by knowing that the O_2 consumption increases as an exponential function of time ($t_{\frac{1}{2}} = 30$ sec) tending to the O_2 requirement level (11). If this is subtracted from the total energy spent, expressed in equivalent O_2 requirement, the amount of energy due to extra PG breakdown is obtained (fig.6). It amounts to about 95 cal/kg when the run starts from rest and

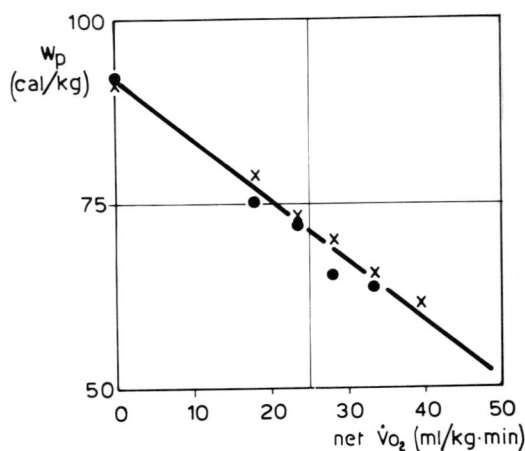

Fig. 6. Amount of energy derived from extra PG breakdown (cal/kg) during maximal run (8 sec) preceded by steady state exercise of the intensity given on the abscissa. (After 9).

to about 50 cal/kg when the run is preceded by maximal aerobic exercise.

During aerobic work at steady state the net alactic O_2 debt ($V_{O_2}^{al}$), which is the O_2 equivalent of the split phosphagen in muscle, increases according to eq. 8). Assuming that the total phosphagen in muscle (PGo) is constant and given by the sum of split (G) plus unsplit (PG) phosphagen:

$$PGo = PG + G \qquad 10)$$

by expressing the phosphagen in O_2 equivalent units (1 mole ATP = = 3.74 l O_2 = 18.7 kcal), eq. 8) can be written:

$$\dot{V}_{O_2} = 2.5\,(PGo - PG) \qquad 11)$$

As previously shown PGo amounts to about 200 cal/kg; for each $\dot{V}_{O_2}$ value the PG concentration in muscle can then be calculated: fig. 7, line "PG st. st.". By subtracting from this line the amount of PG split during the maximal run, as from fig. 6, the concentration of PG in the exhausted muscle is obtained: fig. 7, line "PG Exhausted". It appears from fig. 7 that, when starting the run from maximal aerobic exercise (about 50 ml/kg min), the total amount of PG utilized at exhaustion reaches a value of about 150 cal/kg. The concentration of PG in muscle however does not seem to reach, in these experimental conditions, very low levels, for which it would be necessary to increase the O_2 consumption much above the maximal values. In short bursts of very

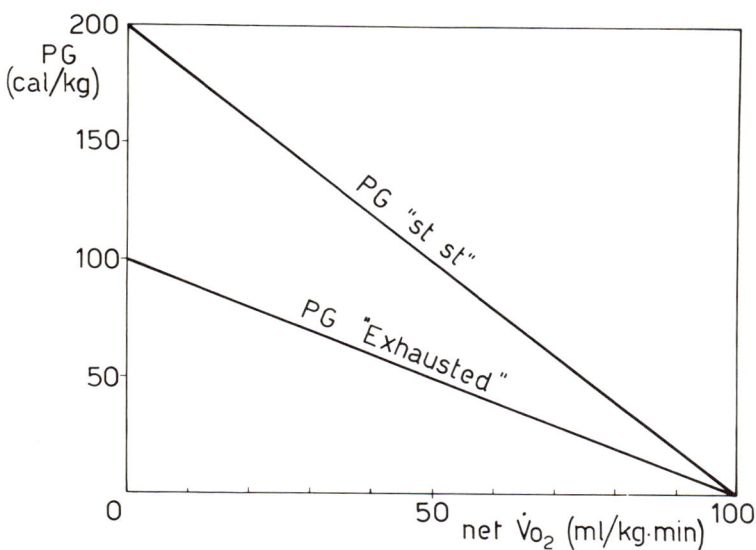

Fig. 7. PG concentration in muscle, in O_2 equivalent units (cal/kg), as a function of the O_2 consumption during steady state exercise: line PG "st. st.". The line PG "Exhausted" indicates the concentration of PG in muscle at the end of an exhaustive burst of very heavy exercise of 8 sec duration, preceded by a steady state exercise of the intensity indicated on the abscissa (schematic). Both PG and $\dot{V}_{O_2}$ are referred to 1 kg body weight.

heavy exercise the muscle is then exhausted before the PG stores have been depleted. Since in reality phosphagen is the sum of ATP and CP, this finding may be explained if it is assumed that ATP splitting is faster than its speed of resynthesis, which depends on the rate of CP splitting. If this is so the fraction of ATP in muscle which is immediately available for mechanical work performance may be reduced to levels too low to maintain an adequate power output while the CP concentration is still relatively high.

c) <u>Efficiency</u>.

The total mechanical power output, as given by the ordinate of fig. 4 can be expressed in O_2 equivalent units (1 kgm = = 0.47 ml O_2). Eq. 6) becomes then:

$$\dot{W}_p = 40.9 - 0.15 \dot{V}_{O_2} \qquad 12)$$

where both power output ($\dot{W}_p$) and O_2 consumption are expressed in ml/min. Eq. 12 indicates that the maximal power the muscle can develop decreases with increasing the O_2 consumption from which the run is started. This seems to suggest that the aerobic mechanism is less efficient than PG splitting in producing work. In fact, were the efficiency of the two systems the same, and assuming that in all cases the muscle is equally and maximally activated, the line of fig. 4 should be horizontal, since the loss in power from PG splitting, due to the fact that a fraction of the available PG is continuously involved in the oxidative turnover, should be compensated by the power developed by the oxidative turnover itself.

The overall efficiency of the exercise (E), is given by:

$$E = \frac{W}{PG} \cdot \frac{PG}{V_{O_2}} = S \cdot R \qquad 13)$$

where S indicates the efficiency of all anaerobic processes occurring during muscle contraction (PG splitting) and R indicates the efficiency of PG resynthesis at the expenses of oxidations and/or glycolysis, all quantities being given in oxygen equivalent units. The amount of energy derived from PG splitting only, is transformed in mechanical work with an efficiency of S, while the energy due to oxidations and/or glycolysis is converted into work with a lower efficiency value (E), this last depending on the efficiency of PG resynthesis (R).

The loss in maximal power due to the previous steady state exercise is then given by the power theoretically developed by oxidations, were S the overall efficiency of the exercise, $(\dot{V}_{O_2}S)$, minus the power actually developed, $(\dot{V}_{O_2} E)$. The amount of power (ΔP) that should be added to the power actually developed, at any given $\dot{V}_{O_2}$ level to obtain an horizontal line is then:

$$\Delta P = \dot{V}_{O_2} (S - E) \qquad 14)$$

It has been experimentally found that the actual loss in power (slope of the line) is $- 0.15 \dot{V}_{O_2}$ (Fig.4). By adding this value to the power effectively developed at any given $\dot{V}_{O_2}$ the line becomes horizontal; then:

$$\Delta P = 0.15 \dot{V}_{O_2} \qquad 15)$$

From 14) and 15):
$$E = S - 0.15 \qquad 16)$$
By giving to the overall efficiency of the exercise (E) the maximal value experimentally found during exercise, which is of the order of 0.25 (4, 7), S turns out to be 0.4, which seems therefore to be the maximal value of efficiency of the anaerobic processes of muscular contraction. This is of the same order of magnitude as that calculated by Whipp and Wasserman (14) for bicycle ergometer exercise and by Cerretelli et al. (1) on dog gastrocnemius. It also follows from eq. 13) and 16) and assuming E = 0.25 that the efficiency of PG resynthesis from oxidations amounts to 0.62, which also appears not to be far from the usually accepted values.

References

1) Cerretelli P., P.E. di Prampero and J. Piiper. - Energy balance of anaerobic work in the dog gastrocnemius muscle. - Am. J. Physiol. 217, 581 - 585, 1969.

2) di Prampero P.E., C.T.M. Davies, P. Cerretelli and R. Margaria. - An analysis of O_2 debt contracted in submaximal exercise. - J. Appl. Physiol. 1970 (in press).

3) Hultman E., J. Bergström and N. McLennon Anderson. - Breakdown and resynthesis of phosphorylcreatine and adenosinetriphosphate in connection with muscular work in man. - Scand. J. Lab. Invest. 19, 56 - 66, 1967.

4) Margaria R. - Sulla fisiologia e specialmente sul consumo energetico della marcia e della corsa a varia velocità ed inclinazione del terreno. - Atti Acc. Naz. Lincei 7, 299 - 368, 1938.

5) Margaria R. - Aerobic and anaerobic energy sources in muscular exercise. - In "Exercise at Altitude", Excerpta Medica Foundation, 1967 (R. Margaria editor).

6) Margaria R., P. Aghemo and Rovelli E. - Measurement of muscular power (anaerobic) in man. - J. Appl. Physiol. 21, 1662 - 1664, 1966.

7) Margaria R., P. Cerretelli, P. Aghemo and G. Sassi. − Energy cost of running. − J. Appl. Physiol. 18, 367 − 370, 1963.

8) Margaria R., P. Cerretelli and F. Mangili. − Balance and kinetics of anaerobic energy release during strenuous exercise in man. − J. Appl. Physiol. 19, 623 − 628, 1964.

9) Margaria R., P.E. di Prampero, P. Aghemo, P. Derevenco and M. Mariani. − The effect of a steady state exercise on the maximal anaerobic power. − J. Appl. Physiol. (in press).

10) Margaria R., H.T. Edwards and D. B. Dill. − The possible mechanism of contracting and paying the oxygen debt and the role of lactic acid in muscular contraction. − Am. J. Physiol. 106, 687 − 714, 1933.

11) Margaria R., F. Mangili, F. Cuttica and P. Cerretelli. − The kinetics of the oxygen consumption at the onset of muscular exercise in man. − Ergonomics 8, 49 − 54, 1965.

12) Mommaerts W.F.H.M. − Energetics of muscular contraction. − Physiol. Rev. 49, 427 − 508, 1969.

13) Piiper J., P.E. di Prampero and P. Cerretelli. − Oxygen debt and high energy phosphates in gastrocnemius muscle of the dog. − Am. J. Physiol. 215, 523 − 531, 1968.

14) Whipp B.J. and K. Wasserman. − Efficiency of muscular work. − J. Appl. Physiol. 26, 644 − 649, 1969.

MUSCLE ATP, CP, AND LACTATE IN SUBMAXIMAL AND MAXIMAL EXERCISE

J. Karlsson

Department of Physiology, Gymnastik- och idrotts-högskolan, Stockholm, Sweden

Since the work of Fletcher and Hopkins (4) and of Lundsgaard (14) it is well established from animal experiments that the splitting of the ATP and CP stores (the phosphagens) and the lactate formation are responsible for the anaerobic energy yield of the skeletal muscle. Not until lately the role of these two processes could be evaluated in man during exercise (3, 8, 10). The introduction of a muscle biopsy technique (1) and the development of methods for analysing metabolites on small pieces of skeletal muscle (11) have made it possible to perform more systematic studies also in humans. This paper will focus on the following two problems.

1. Lactate accumulation in the muscles at submaximal and maximal exercise and the concentration gradient for lactate in the muscle.
2. The relationship between oxygen deficit and muscle ATP, CP, and lactate at submaximal and maximal exercise.

SUBJECTS

The subjects were all healthy, young physical education students, well acquainted with physical exercise and the testing procedure. Nobody, however, could be defined as extremely well-trained or as an athlete. This was the case both with regard to capacity for endurance and sprint events.

METHODS

Oxygen uptake determinations were performed with the Douglas bag method and the gas samples were analysed with a Haldane apparatus. Blood lactate was determined enzymatically (15) and muscle metabolites with enzymatic methods introduced by Lowry (13) and adapted for the present needs (11). The oxygen uptake ($\dot{V}_{O_2}$) in the steady state at a submaximal work load is by definition the amount of oxygen needed for a complete aerobic energy output by the muscle. It is then possible to estimate the total amount of oxygen needed for a complete aerobic energy output (Fig. 1). The difference between this value and the determined oxygen uptake during the work will represent the oxygen lack during the work. This calculated oxygen deficit represents the amount of energy derived from anaerobic energy processes expressed in O_2-equivalents. Assuming a constant mechanical work efficiency also at supramaximal work loads a linear relation-

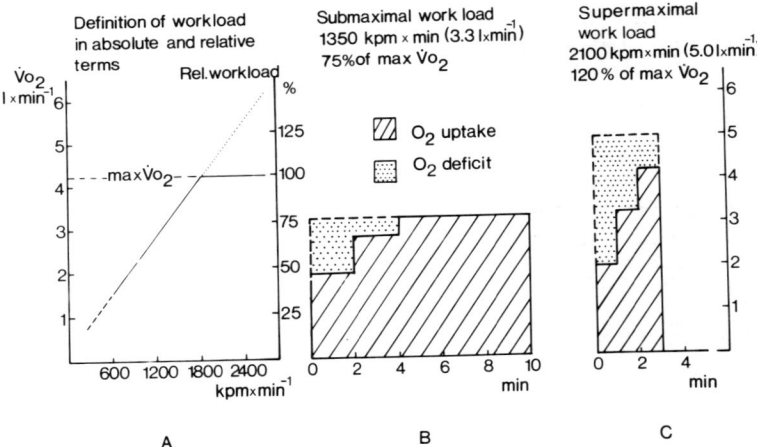

Fig. 1. Principles underlying calculation of relative work load and oxygen deficit. Left - $\dot{V}_{O_2}$ is related to work load, $\dot{V}_{O_2}$ max being defined by the levelling off criterion. Centre - oxygen deficit at a submaximal load ($\dot{V}_{O_2}$ = 3.3 l x min^{-1}, equivalent to 75 per cent of max $\dot{V}_{O_2}$). Right - oxygen deficit in supramaximal work ($\dot{V}_{O_2}$ ≈ 5.0 l x min^{-1}, for complete aerobic energy output, equivalent to 125 per cent of max $\dot{V}_{O_2}$).

ship between oxygen uptake and work load may be extrapolated as indicated in the figure. From this relationship the theoretical oxygen uptake for a complete aerobic energy output can be given and the oxygen deficit can be calculated also for these supramaximal loads.

PROCEDURE

The resting values for blood lactate (finger tip blood) and muscle metabolites were obtained after 45-60 minutes of rest in supine position. The exercise consisted of pedalling a bicycle ergometer. The exercise values for the muscle metabolites were obtained with the subject sitting on the bicycle and the samples were frozen in liquid nitrogen within 3-5 seconds after the termination of work. To secure the highest blood lactate concentration 2-3 blood samples were taken after the exercise. Exhaustion was defined as the incapability to keep up the pedal rate at a frequency of 60 revolutions x min^{-1}.

RESULTS AND DISCUSSION

Muscle Lactate Accumulation during Exercise

At rest the muscle lactate concentration was 1.4 mmoles x kg^{-1} wet muscle and there was no marked increase until the work load exceeded 50-60 per cent of the maximal oxygen uptake (Fig. 2). Above this work level a rapid increase in the accumulation of lactate was observed and reached at maximal exercise a mean value of 23 mmoles x kg^{-1} wet muscle. The blood lactate concentration reflected the values found in the muscles rather well. Expressed per liter water basis (1) a concentration gradient for lactate from the muscle to the blood existed. This gradient was enhanced the higher the work load. The exhaustive work intensities were not sustained long enough to get an equilibrium for the flux of lactate between the muscles and the blood. Thus, the blood lactate concentration increased during the initial part of the recovery period (Fig. 3).

The muscle lactate concentration per liter of water immediately after the termination of the work amounted to twice that obtained in blood. The extracellular volume in a skeletal muscle approximates one quarter of the total muscle volume (6). Further, if the blood lactate concentration is representative for the extracellular

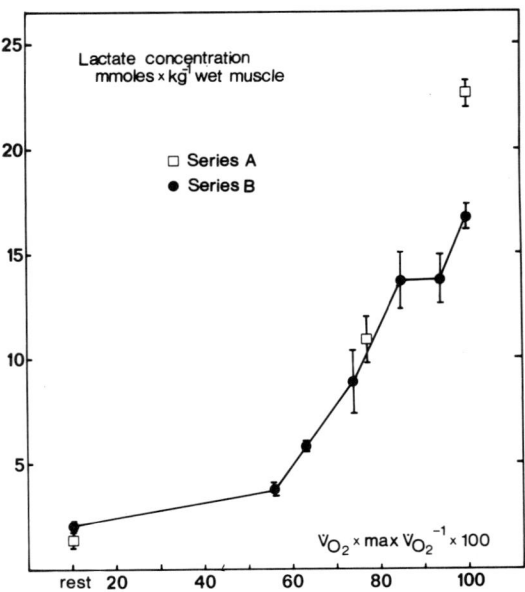

Fig. 2. Muscle lactate concentration (mean ±1 SE) in relation to relative work load. Series A and Series B denote two groups of subjects, average max $\bar{V}_{O_2}$ of 4.4 and 3.3 $l \times min^{-1}$, respectively (11).

space it can be calculated that the intracellular lactate concentration was around 29 mmoles x l^{-1} assuming that there was an equal intracellular distribution of lactate. Direct determinations of lactate in muscle fiber structures dissected from slices illustrated similar concentration gradients (Karlsson, to be published). In slices from a biopsy specimen obtained in a similar experiment the highest concentration observed was 146 µmoles x g^{-1} dry wet weight and the lowest 59 µmoles x g^{-1} corresponding to approximately 29 and 12 mmoles x kg^{-1} wet weight, respectively. After extremely heavy bicycle exercise lasting for only 50 seconds, the gradient was further enhanced and the corresponding values were 39 and 7 mmoles x kg^{-1} wet muscle respectively, indicating that ratios in the order of 3-6 might be obtained for the intracellular - extracellular lactate concentrations during short time exhaustive exercise. Within the same muscle, no marked gradients for lactate concentration has been observed. This has been studied both in stimulated canine gracilis muscles which were then frozen in liquid nitrogen (12) and in the lateral portions of M. quadriceps

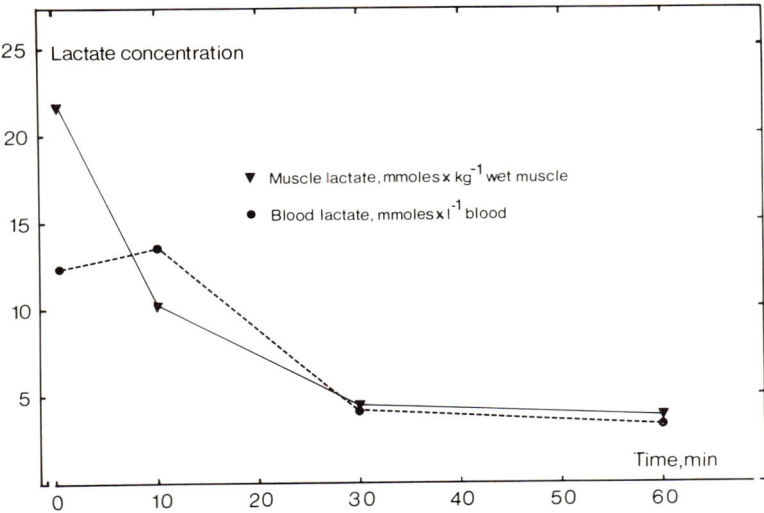

Fig. 3. Muscle and blood lactate concentration in one subject immediately, 10, 30, and 60 min after maximal bicycle exercise (3).

femoris in man with repeated biopsies at different sites during submaximal and short term maximal exercise (Karlsson, to be published). It seems reasonable therefore to conclude that although the muscle sample is very small the lactate concentration obtained in the biopsy specimen is representative for the muscle studied.

Oxygen Deficit and Muscle ATP, CP, and Lactate

Three subjects were studied when they performed exhaustive exercise on the bicycle ergometer (10). Three different work loads were applied in each subject, exhaustion time being approximately 2, 7, and 16 min. These three work loads demanded 130, 100, and 90 per cent of the subject´s maximal oxygen uptake, respectively. In separate experiments the lowest work load was interrupted after 2 and 7 minutes and the medium work load after 2 minutes of work. In each subject six experiments were performed, three exhaustive and three non-exhaustive. In each experiment the total oxygen uptake was determined and when the work was terminated a muscle biopsy was obtained.

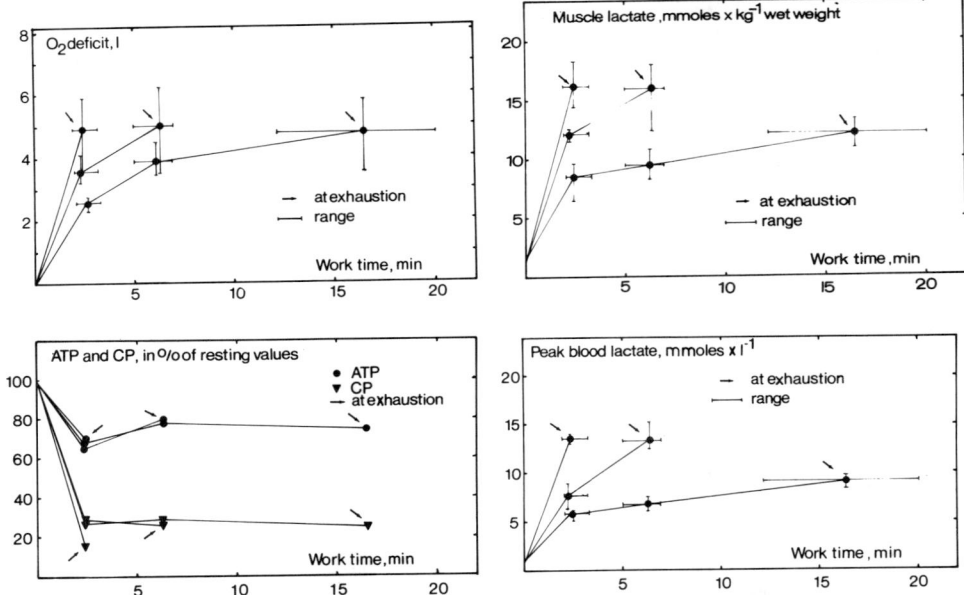

Fig. 4. Mean values and range for oxygen deficit (upper left panel), muscle lactate concentration (upper right panel), phosphagen depletion (lower left panel) and blood lactate concentration (lower right panel) versus work time in three subjects at three separate work loads representing 130, 100, and 90 per cent of max $\dot{V}_{O_2}$ (10). The six experiments were performed on separate days over a period of two months.

Almost identical values for the oxygen deficit were obtained from the onset of work to exhaustion for the three different work loads (Fig. 4). The main part of the oxygen deficit was contracted in the initial phase of the work. Two and a half minutes after start of work the mean oxygen deficit was 2.6 liters at the lowest load and 4.9 liters at the highest work load indicating an anaerobic energy output almost twice as high at the heaviest work load as during the lightest load. The muscle lactate accumulation followed the same pattern and after the same time the muscle lactate concentration was 8.5 on the lowest compared with 16.1 mmoles x kg^{-1} wet muscle on the highest work load. In contrast to these findings phosphagens attained approximately the same degree of depletion after two and a half minutes of work at all loads. This was the case regardless whether the subjects were exhausted or could continue the exercise for 14 minutes more. At this point two pertinent comments can be made. The alactacid part of the oxygen deficit can be maximally utilized although the oxygen deficit has not attained its maximal value. The results may also indicate that a depletion of the ATP and CP stores do not limit short time exhaustive exercise of the examined duration.

At the two heaviest work loads the same muscle lactate concentration ($\approx$ 16 mmoles x kg^{-1}) was attained at exhaustion, but a considerable lower value (12 mmoles x kg^{-1} wet muscle) was observed at exhaustion on the lowest work load. As the oxygen deficit as well as the splitting of the phosphagens was the same in all three situations, the total lactate production was probably the same in the longest run. The explanation for the lower muscle lactate concentration in the longest run could be that greater distribution and utilization of lactate in the body has taken place. If local lactate concentration in working muscle was the limiting factor for physical performance at the two highest work loads some additional factor must be present at the lowest load sensitizing the muscle to the lactate concentration, e.g. a decreased buffering capacity of the muscle due to a prolonged exposure to metabolic acids.

The highest blood lactate concentration in each experiment was in good agreement with the corresponding muscle lactate concentration. Thus, the highest blood lactate values were obtained after the two higher work loads and a somewhat lower concentration was found at the lowest work load.

The factors involved in the regulation of the lactate production are frequently discussed but far from solved. The pyruvate concentration, the LDH activity (H- or M-form) and the concentration of extra mitochondrial NAD and NADH are three of the more important factors. Only a very minor increase in the pyruvate concentration is observed during exercise (9) and an increased pyruvate concentration can then be excluded as an important factor for the lactate accumulation found during heavy exercise as suggested by Huckabee (7). In skeletal muscle LDH is present in its M-form and this favours the formation of lactate, but no significant changes occur in the LDH activity (5) on going from rest to short term exercise. During exhaustive exercise the oxygen demand of the muscles are higher than can be met by the circulatory system. Further, it is shown that contractions result in an accumulation of extra mitochondrial NADH (2) which leads to a lactate production. It seems reasonable to assume that there exists a causal relationship between the oxygen deficit and the accumulation of NADH at onset of heavy exercise. This may explain why such a good relationship has been found between the oxygen deficit and the muscle lactate concentration.

Fig. 5 illustrates that the phosphagen depletion both during maximal and submaximal work levels was related to the calculated oxygen deficit. A levelling off for the depletion of the ATP and CP stores was present also in this situation. The relationship between increasing oxygen deficit values, maximal phosphagen depletion and increasing muscle lactate concentrations might be used to estimate the order of magnitude of the alactacid

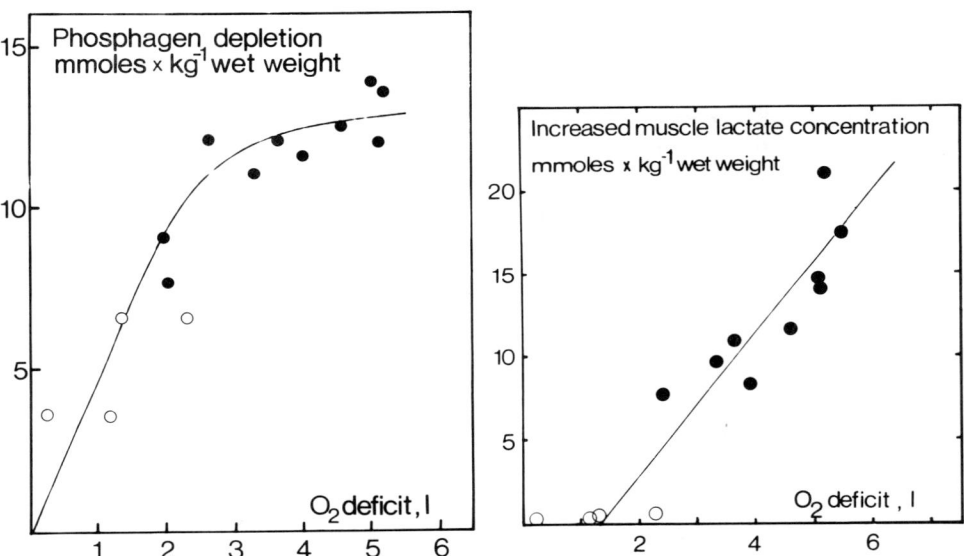

Fig. 5. Phosphagen depletion (left panel) and increase in muscle lactate concentration (right panel) in relation to oxygen deficit (10). Each point represents the mean of at least three values. Open circles represent data obtained by Knuttgen and Saltin (to be published). The equation of the regression line for the lactate data (in work loads > 50 per cent of max $\dot{V}_{O_2}$) in relation to the oxygen deficit is y = -5.2 x 4.1x; (r = 0.88).

portion of the oxygen deficit. Assuming a linear relationship between the increase in muscle lactate concentration and the oxygen deficit the equation of the regression line could be calculated (Fig. 5). The intercept on the x-axis thus obtained (1.3 l of O_2) corresponds to the maximal oxygen deficit value without any increase in muscle lactate concentration, i.e. the maximal alactacid portion of the oxygen deficit. The phosphagens, however, were not maximally depleted until the oxygen deficit amounted to 2-3 liters. This must mean that the anaerobic glycolysis is of significance as an energy delivering process already before the energy reserves of the phosphagens are fully utilized.

Above it has been discussed how the splitting of phosphagens (ATP + CP) occurred already at low work loads, but no lactate accumulation was found until the work intensity exceeded 50 per cent of the maximal oxygen uptake. At this relative work load the depletion of phosphagens amounted to approximately 5 mmoles x kg^{-1} wet muscle (Fig. 6). If the corresponding increase in the concentration of ADP and inorganic phosphate in muscle acts as a trigger mechanism for the glycolysis or in another way increases the glycolytic activity cannot be settled at present.

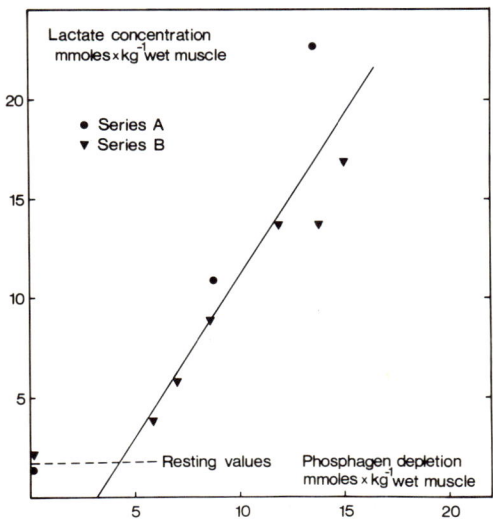

Fig. 6. Muscle lactate concentration in relation to phosphagen depletion (11). Open circles denote data obtained by Knuttgen and Saltin (to be published).

As a summary the following three points will be mentioned. The anaerobic energy output is contracted:

1. at low work intensities with a relatively small anaerobic energy output, by phosphagen splitting
2. at moderate work intensities ($>$ 50 % of max $\dot{V}_{O_2}$) partly by phosphagen splitting and partly by anaerobic glycolysis with concomitant lactate formation
3. at heavy work loads ($>$ 90 % of max $\dot{V}_{O_2}$) by maximal phosphagen depletion and maximal anaerobic glycolysis

ACKNOWLEDGEMENT

These investigations have been carried out with support from the Swedish Medical Research Council (Project No. B69-71-14X-2203-02-04).

REFERENCES

1. BERGSTRÖM, J. Muscle electrolytes in man. Scand. J. clin. Lab. Invest. 14: Suppl. 68, 1962.
2. CHANCE, B., AND F, JÖBSIS. Changes in fluorescence in a frog sartorius muscle following a twitch. Nature. 184:195-196, 1959.
3. DIAMANT, B., J. KARLSSON, AND B. SALTIN. Muscle tissue lactate after maximal exercise in man. Acta physiol. scand. 72:383-384, 1968.
4. FLETCHER, W.M., AND F.G. HOPKINS. Lactic acid in amphibian muscle. J. Physiol. 35:247-302, 1907.
5. GOLLNICK, P.D., P.J. STRUCK, AND T.P. BOGYO. Lactic dehydrogenase activities of rat heart and skeletal muscle after exercise and training. Scand. J. clin. Lab. Invest. 22(4):623-627, 1967.
6. HOLLOSZY, J.O., AND H.T. NARAHARA. Studies of tissue permeability. X. Changes in permeability to 3-methyl-glucose associated with contraction of isolated frog muscle. J. biol. Chem. 240(9):3493-3500, 1965.
7. HUCKABEE, W.E. Relationships of pyruvate and lactate during anaerobic metabolism. I. Effects of infusion of pyruvate or glucose and of hyperventilation. J. Clin. Invest. 37:244-254, 1958.

8. HULTMAN, E., J. BERGSTRÖM, AND N. McLENNAN ANDERSON. Breakdown and resynthesis of phosphorylcreatine and adenosine triphosphate in connection with muscular work in man. Scand. J. clin. Lab. Invest. 19:56-66, 1967.

9. KARLSSON, J. Pyruvate and lactate ratios in muscle tissue and blood during exercise in man. Acta physiol. scand. 1971. In press.

10. KARLSSON, J., AND B. SALTIN. Lactate, ATP, and CP in working muscles during exhaustive exercise in man. J. Appl. Physiol. 29:Nov. 1970.

11. KARLSSON, J., B. DIAMANT, AND B. SALTIN. Muscle metabolites during submaximal and maximal exercise in man. Scand. J. clin. Lab. Invest. 27(1), 1971.

12. KARLSSON, J., S. ROSELL, AND B. SALTIN. Carbohydrate and fat metabolism in contracting canine skeletal muscle. Am. J. Physiol. Submitted for publication.

13. LOWRY, O.H., J.V. PASSONNEAU, F.X. HASSELBERGER, AND D.W. SCHULZ. Effect of ischemia on known substrates and cofactors of the glycolytic pathway in brain. J. Biol. Chem. 239(1):18-30, 1964.

14. LUNDSGAARD, E. Weitere Untersuchungen über die Muskelkontraktionen ohne Milchsäurebildung. Biochem. Z. 269:308-328, 1930.

15. SCHOLZ, R., H. SCHMITZ, T. BÜCHER, AND J.O. LAMPEN. Über die Wirkung von Nystatin auf Bäckerhefe. Biochem. Z. 331:71-86, 1959.

MUSCLE ATP, CP, AND LACTATE DURING EXERCISE AFTER PHYSICAL CONDITIONING

Bengt Saltin and Jan Karlsson

Department of Physiology, Gymnastik- och idrottshögskolan

Stockholm, Sweden

After training reduced blood lactate concentration at a given submaximal work load is a common finding (2, 3, 11). A more prolonged training period results in a lower blood lactate also at the same relative work load (2, 3, 4). The problem to be focused on in this paper is therefore how the accumulation of the muscle lactate concentration and the depletion of the ATP and CP stores during submaximal exercise were influenced by training. Further, as morphological studies of these subjects' skeletal muscles were performed too (9) the relationship between biochemical and morphological changes were analyzed.

SUBJECTS

The study was performed on the same group of subjects as that of Kiessling et al. (9) which consisted of 15 conscripts (mean weight 71 kg; mean height 178 cm). They were studied at induction and after 12 and 28 weeks of endurance training. The physical training consisted of running three times a week with maximal speed over a distance of 5 km (3.5 miles). Muscle biopsies (1) were taken at rest and after 10 minutes of bicycle exercise at two different submaximal work loads and after an exhaustive work load. The biopsy material was analyzed for lactate, ATP and CP according to Karlsson et al. (7).

RESULTS

Maximal oxygen uptake increased from 3.3 at induction to 3.8 and 4.1 l/min after 12 and 28 weeks training, respectively. The

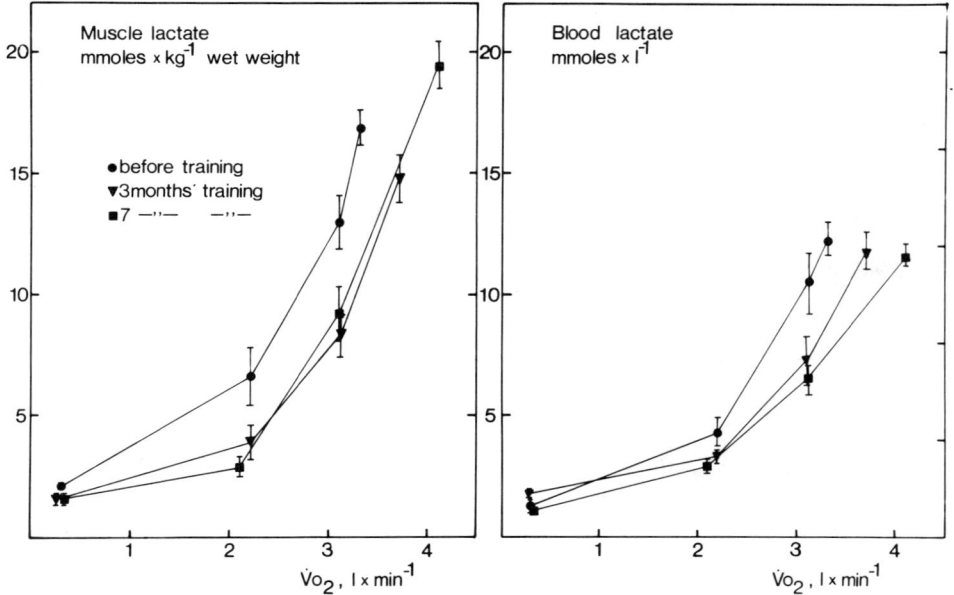

Fig. 1. Muscle and blood lactate concentrations in relation to the oxygen uptake at the different work loads before, after three and seven months of physical conditioning. Each point represents the mean value ± 1 SD.

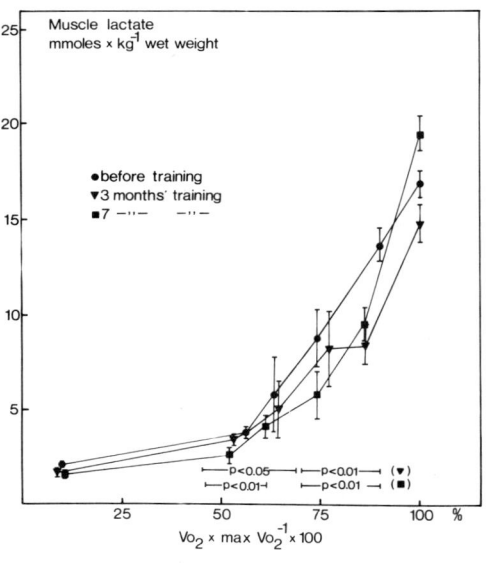

Fig. 2. The muscle lactate concentrations in relation to the relative work load. The calculations of the significance level are based on comparisons between the results obtained after physical conditioning and the pre-training data.

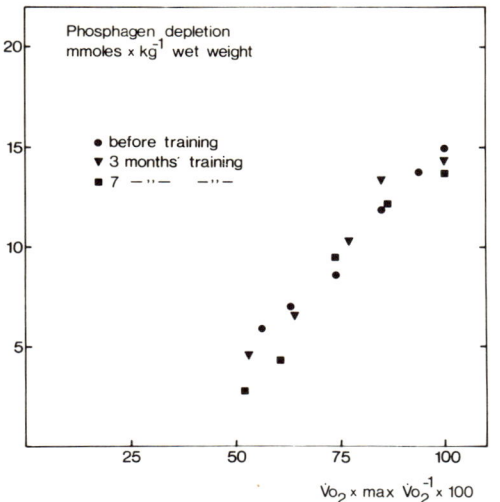

Fig. 3. The phosphagen depletion in relation to the relative work load before and after physical conditioning.

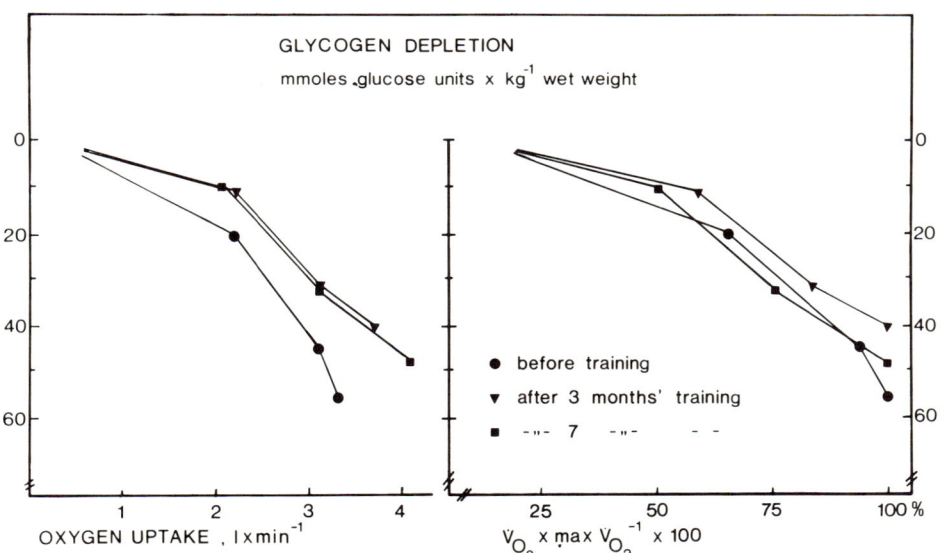

Fig. 4. The glycogen depletion during two submaximal work loads of 10 minutes duration and the maximal work load. In the left panel the work is expressed in absolute values (the oxygen uptake l/min) and in the right in relative terms.

muscle lactate was markedly reduced at the same submaximal work load after the first part of the training and a further reduction in the muscle lactate was observed also during the latter part of the training period (Fig. 1). Even if the work load was expressed in per cent of the subjects' maximal oxygen uptake a highly significant reduction in the muscle lactate concentration at the same relative work load was observed (Fig. 2).

The blood lactate concentration follows the same pattern as the muscle lactate (Fig. 1). The depletion of the ATP and CP stores were in all three test situations related to the relative work load (Fig. 3) which means that the splitting of the phosphagens was less pronounced at the same absolute work load in the most trained stage.

DISCUSSION

It may be worthwhile to speculate on how the reduction in muscle lactate concentration is brought about by training. One way to explain the findings is that the reduction in muscle lactate observed at the same absolute work load is due largely to the activation of relatively more red fibers. Similar glycogen depletion after training (Fig. 4) at the same relative work load suggests that lower lactate values are not due to a lower lactate production by glycolysis. After training, however, some of the formed lactate may be utilized immediately in other parts of the muscles. Kiessling et al's data with an increased number of mitochondria after endurance training (9) and the study by Jorfeldt (6) support this hypothesis.

In this study the oxygen deficit at the onset of work was not measured. However, in cross-sectional studies of untrained and well-trained subjects it has been documented that a good relationship exists at work loads above 50 per cent of maximal oxygen uptake between the oxygen deficit and the accumulation of muscle lactate (9). This speaks in favour of a faster acceleration of circulatory oxygen transport at the onset of work in trained subjects which also has been observed (5, 8, 10).

ACKNOWLEDGEMENT

This study was supported by grants from the Swedish Delegation for Applied Medical Defence Research (project No. 23:090/69) and the Swedish Medical Research Council (project No. B70-14X-2203-04B).

REFERENCES

1. BERGSTRÖM, J. Muscle electrolytes in man. Scand. J. Clin. Lab. Invest., Suppl. 68, 1962.

2. EDWARDS, R.H.T., A. MELCHER, C.M. HESSER, and O. WIGERTZ. Blood lactate concentrations during intermittent and continuous exercise with the same average power output. In: Muscle Metabolism during Exercise. Adv. Expl. Med. Biol., New York, Plenum Publ. Corp. 1970, this volume.

3. EKBLOM, B. Effect of physical training on oxygen transport system in man. Acta Physiol. Scand., Suppl. 328, 1969.

4. HERMANSEN, L. and B. SALTIN. Blood lactate concentration during exercise at acute exposure to altitude. In: Exercise at altitude., Amsterdam, Excerpta Medica Foundation, 1967, p. 48.

5. JONES, W.B. and T.J. REEVES. Total cardiac output response during four minutes of exercise. Amer. Heart J. 76: 209-216, 1968.

6. JORFELDT, L. Metabolism of L(+)-lactate in human skeletal muscle during exercise. Acta Physiol. Scand., Suppl. 338, 1970.

7. KARLSSON, J., B. DIAMANT, and B. SALTIN. Muscle metabolites during submaximal and maximal exercise in man. Scand. J. Clin. Lab. Invest. In press. 27: 1971.

8. KAIJSER, L. Limiting factors for aerobic muscle performance. Acta Physiol. Scand., Suppl. 346, 1970

9. KIESSLING, K.-H., K. PIEHL, and C.-G. LUNDQUIST. Effect of physical training on ultrastructural features in human skeletal muscle. In: Muscle Metabolism during Exercise. Adv. Expl. Med. Biol., New York, Plenum Publ. Corp. 1970, this volume.

10. ROBINSON, S. Experimental studies of physical fitness in relation to age. Arbeitsphysiologie, 10: 251-323, 1938.

11. SALTIN, B., L.H. HARTLEY, Å. KILBOM, and I. ÅSTRAND. Physical training in sedentary middle-aged and older men. II. Oxygen uptake, heart rate, and blood lactate concentration at submaximal and maximal exercise. Scand. J. Clin. Lab. Invest. 24: 315-322, 1969.

LACTATE PRODUCTION DURING EXERCISE

Lars Hermansen

Institute of Work Physiology

Oslo Norway

The individual's maximal oxygen uptake sets an upper limit for the amount of energy, which can be liberated during prolonged exercise (i.e. 10 min or more). However, the rate of energy can be markedly increased during shorter work periods because energy can be derived anaerobically in addition to that derived from the aerobic processes. Anaerobic processes give rise to a production of lactate, which diffuses from the muscle cell to the blood, where it can easily be measured. The lactate concentration in arterial blood is usually taken as an indication of the extent to which the anaerobic processes are involved during work.

During light to moderate work the required energy is almost exclusively derived from aerobic processes, and the blood lactate concentration remains approximately at the resting level during the entire work period. The stores of adenosinetriphosphate (ATP) and phosphocreatine (PC) provide the necessary amount of energy in the critical period during transition from rest to work before respiration and circulation are adapted to the work load. However, as the work load increases, anaerobic processes play an increasingly large role.

As will be seen in Fig.1 the blood lactate concentration shows a pronounced rise when work load is increased above a certain level both in untrained and well-trained subjects. It is well known that the blood lactate concentration is lower in well-trained than in untrained subjects when they are working on the same absolute work load (1). This observation is commonly explainable by the higher maximal oxygen uptake in well-trained subjects. However, as shown in Fig.1 the blood lactate concentration is lower in well-trained than in untrained subjects, even when the work load is expressed in relation to the individual's maximal oxygen uptake.

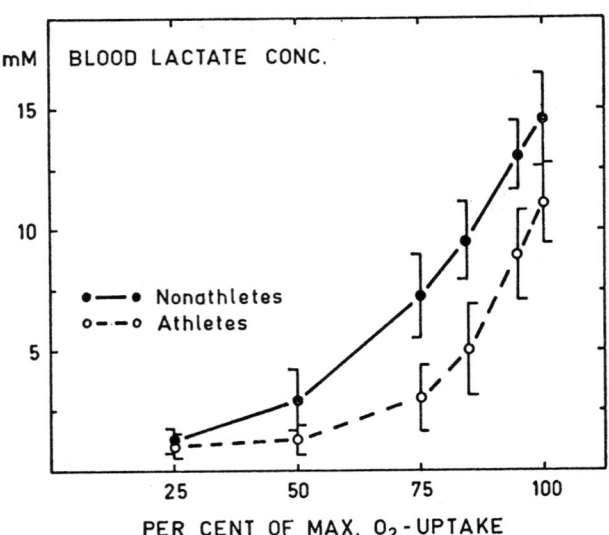

Fig. 1 Blood lactate concentration in 6 non-athletes with an average maximal oxygen uptake of 44.0 ml/kg x min, and in 8 athletes with an average maximal oxygen uptake of 66.8 ml/kg x min.

Consequently, it can be concluded that training elicits changes not only in the capacity to transport oxygen from the lungs to the working muscles, but also in the utilization of oxygen in the muscle cell itself.

The explanation for this observation is still obscure. Evidence from animal studies indicates that during exercise of light intensity, red muscle fibres are brought into action (7,10). As the work load increases, white muscle fibres are also activated. Therefore, it is possible that a larger proportion of the red fibres are activated during exercise in athletes than in untrained subjects. However, although this might explain some of the differences between well-trained and untrained subjects, it should be noted that there exists as yet no complete explanation for the increased production of lactate during submaximal exercise, either in untrained or in well-trained subjects.

The delay of an adequate oxygen supply at the onset of work is probably one of the explanations. However, as shown by Carlson and Pernow (2) and by Keul et al (6), lactate is produced regardless of a relatively high oxygen content and oxygen pressure in the venous blood draining the muscle. These observations were explained by Keul et al (6) as an imbalance between the glycolysis and the oxidative capacity of the cell. However, it should be noted that only a certain proportion of the total number of muscle cells are activated even during maximal work. Venous blood draining the muscle, therefore, consists of blood coming from activated and non-activated motor units within the same muscle group. Consequently, care should be taken when making assumptions concerning changes in intra-cellular processes from measurements in the blood.

The blood lactate concentration at rest and during muscular exercise is dependent on several factors:
1. The rate of lactate production.
2. The rate of lactate diffusion from the cells to the blood.
3. The rate of lactate removal.

Lactate is known to be a relatively small and easily diffusible molecule. It is assumed, mainly from animal studies, that lactate diffuses freely and rapidly into all water compartments of the body (8). For instance, it was shown in mammals by Evans and Eggleton (4) that within 30 seconds after a short tetanic concentration, the concentration of lactate in the blood was approximately equal to that in the muscle.

Recent studies by Diamant et al (3) have shown that the lactate concentration in human skeletal muscle was much higher than in the blood immediately after maximal exercise of approximately 3 min duration. However, shortly after the cessation of work, at the time when the peak blood lactate value was reached, the muscle and the blood lactate values were approximately the same. These results seem to indicate that there is a fairly rapid distribution also in the intact human organism.

On the other hand, measurements of the lactate concentration in the cerebrospinal fluid in human subjects before and after exercise revealed that the lactate is not evenly distributed in all water compartments of the body. Samples of blood and cerebrospinal fluid were taken at rest before a maximal exercise (3 min duration) to complete exhaustion, and 5 and 60 min after cessation of work. The blood lactate concentration was found to increase from a resting value of 1.2 mM to a peak value of 14.6 mM measured 5 min after cessation of work. The blood lactate concentration after 60 min of recovery was still above resting values (i.e. 4.3 mM). The lactate concentration in the cerebrospinal fluid, however, remained essentially unchanged, i.e. 1.7 mM at rest, 1.7 mM and 1.8 mM at 5 and 60 min of the recovery period, respectively. Although these results seem to indicate that lactate is not evenly distributed in all body compartments, it should be emphasized that the cerebrospinal fluid space represents only a small fraction of the total body water. Consequently, if the lactate has reached an equilibration in the rest of the body water, the blood lactate value will give valuable information not only concerning the changes in the blood, but also about the total amount of lactate in the body.

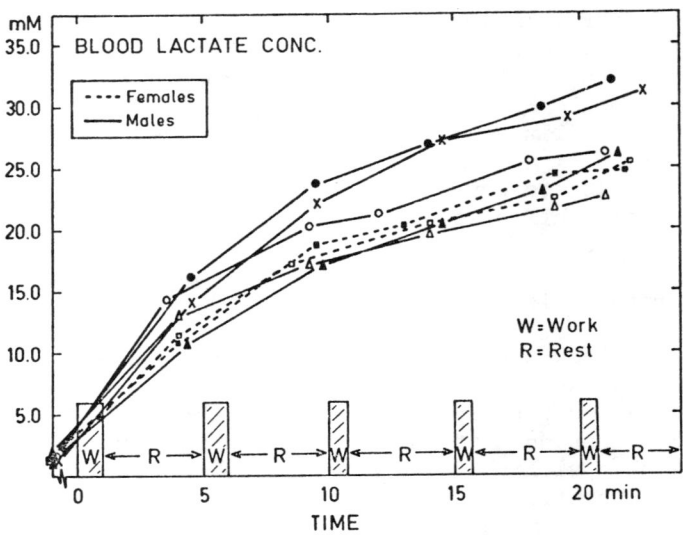

Fig.2 Blood lactate concentration in 5 male and 2 female subjects after 5 maximal work periods of approximately 1 min duration.

In order to make a further evaluation of the distribution of lactate in the body, subjects were asked to perform 5 maximal work periods, allowing only a few minutes (i.e. approximately 4 min) rest in between. As shown in Fig.2, the blood lactate concentration increased after each of the five work periods to extremely high values. The highest values obtained were 31.1 and 32.0 mM. It should be noted that the blood lactate values obtained by this procedure are even higher than the lactate values reported by Diamant et al (3) for the muscle. Thus, it seems reasonable to believe that in these experiments the blood lactate concentration will give a fairly good estimation of the lactate concentration in the total body water.

When the intensity of the work approaches or surpasses that which corresponds to the individual's maximal oxygen uptake, blood lactate concentration will increase during the entire work period. However, even under resting conditions, the glycolytic processes proceed slowly, maintaining a small but constant level of lactate in the blood and muscle (5).

During maximal work of short duration, i.e. less than 10 min, the rate of lactate production can be markedly increased. The more intensive the work, the sooner the blood lactate concentration will reach a level at which the continuation of work is impossible.

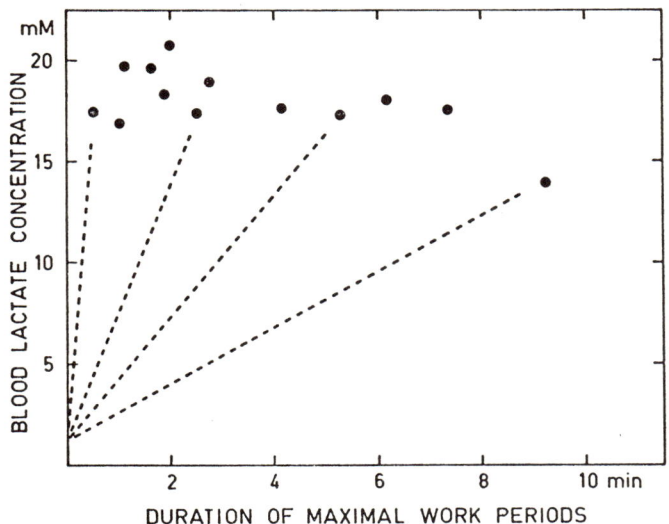

Fig.3 Peak blood lactate concentration in one male athlete after 13 maximal work periods of different duration.

Fig.3 gives the results of 13 maximal work experiments on the same individual. Each maximal run was continued until the subject was completely exhausted. Although there is a tendency toward somewhat lower blood lactate concentrations at the shortest and the longest work periods, it should be noted that the blood lactate level at the time of exhaustion was approximately the same in all experiments. In other words, when the subjective feeling of fatigue or complete exhaustion reached a stage which was incompatible with continuation of the work, the blood lactate concentration had increased to approximately 18 mM for the subject in question.

Consequently, the average rate of lactate production must be much higher during the shorter work periods than the longer ones, as indicated by the slope of the different dotted lines in Fig.3.

The values for the blood lactate concentration given in Fig.3 have been used to calculate the total amount of lactate produced during the actual work periods. These values, which were obtained in laboratory experiments, are compared with values observed in champion athletes after important competitions (Fig.4).

In making the calculations it is assumed that the lactic acid is evenly and rapidly distributed in all water compartments and also that the fraction of water was 0.6 for the body and 0.8 for the blood, as proposed by Margaria et al (9).

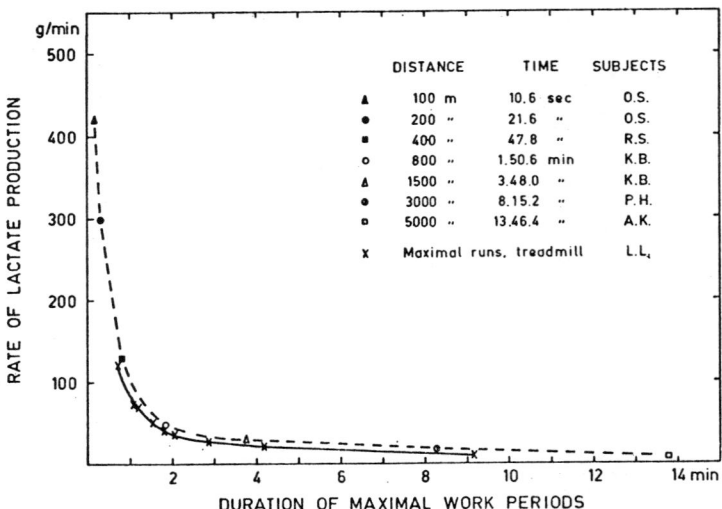

Fig. 4 Rate of lactate production in athletes during maximal running. For further explanation, see text.

The results show that the rate of lactate production is closely related to the intensity of the work, and that the production of lactate is pronounced even at work times as short as 10 sec. The peak value obtained after the 100 m race was actually 16.8 mM.

The motivation of the subjects plays an important role in establishing maximal blood lactate concentrations. This may possibly explain the fact that all blood lactate values in the laboratory experiments were lower than those obtained after participation in important competitions.

The crucial question is, of course: how is it possible to increase the rate of lactate production by 200 or 300 times or more?

The enzyme lactate dehydrogenase catalyzes the reaction as presented in the following equation:

$$\text{Pyruvate} + \text{NADH} + \text{H}^+ \xrightleftharpoons{\text{LDH}} \text{Lactate} + \text{NAD}^+$$

In the presence of pyruvate and NADH + H$^+$, the equilibrium of this reaction strongly favours lactate production. From this equation it also appears that the production of lactate is dependent on the concentration of pyruvate, NADH and H$^+$, and also on the activity of the enzyme.

Measurements at rest and immediately after maximal work using the needle biopsy technique have revealed a two-fold increase in the pyruvate concentration in the muscle (3). However, a two-fold increase in the pyruvate concentration alone cannot explain more than a small fraction of the total increase in lactate production.

Therefore it is reasonable to believe that the concentration of NADH and H^+ or the activity of the enzyme (or both) must be changed during the exercise.

REFERENCES

1. Asmussen, E.: In Handbook of Physiology Respiration II. Muscular exercise. Chapt. 36, Amer.Physiol.Soc. 1965.

2. Carlson, L.A. and B. Pernow: Studies on the peripheral circulation and metabolism in man. I. Oxygen utilization and lactate-pyruvate formation in the legs at rest and during exercise in healthy subjects. Acta physiol.scand. 52: 328-342. 1961.

3. Diamant, B., J. Karlsson and B. Saltin: Muscle tissue lactate after maximal exercise in man. Acta physiol.scand. 72: 383-384, 1968.

4. Evans, C.L.: Recent advances in physiology. J. Churchill and A. Churchill, London 1930.

5. Hohorst, H.J., F.H.Dreutz and T.Buecher: Über Metabolitgehalte und Metabolitkonzentrationen in der Leber der Ratte. Biochem. Z. 332: 18 - , 1959.

6. Keul, J., E. Doll and D.Keppler: The substrate supply of the human skeletal muscle at rest, during and after work. Experientia 23: 1-6, 1967.

7. Kugelberg, E. and L. Edström: Differential histochemical effects of muscle contractions on phosphorylase and glycogen of various types of fibres: relation ot fatigue. J. Neurol. Neurosurg. Psychiat. 31: 415-423, 1968.

8. Margaria, R., H.T.Edwards and D.B.Dill: The possible mechanisms of contracting and paying the oxygen debt and the role of lactic acid in the muscular contraction. Am.J.Physiol.106:689-715, 1933.

9. Margaria, R., P. Cerretelli, P.E.diPrompero, C.Massari and G.Torelli: Kinetics and mechanism of oxygen debt contraction in man. J. Appl. Physiol. 18: 371-377, 1963.

10. Romanul, F.C.A.: Enzymes in muscle: I. Histochemical studies of enzymes in individual muscle fibres. Arch.Neurol.11: 355-368. 1964.

TURNOVER OF ^{14}C-L(+)-LACTATE IN HUMAN SKELETAL MUSCLE DURING EXERCISE

L. Jorfeldt

From the Department of Clinical Physiology, Karolinska Institutet at Serafimerlasarettet, Stockholm, Sweden

It is well established that the exercising skeletal muscle is able to produce lactate and it has also been shown that resting (2, 11) and possibly exercising skeletal muscle (4, 13), too, like heart muscle, is able to take up lactate. It therefore seems reasonable to assume that the two processes - uptake and release - occur simultaneously. Quantitative studies of this phenomenon appear to be possible only with isotopically labelled lactate and even they have been complicated by the lack of a method for analysing labelled L(+)-lactate, which is the naturally occurring stereo-isomere.

This investigation was performed in order to elucidate the mutual relations between uptake and release of lactate in skeletal muscle during exercise, and the metabolism of the lactate taken up. L(+)-lactate-1-^{14}C was used and a method was worked out for the determination of ^{14}C-L(+)-lactate in blood. The human forearm was used as an experimental model.

SUBJECTS, PROCEDURES AND METHODS

The results are based on data from young male healthy volunteers who were in the postabsorptive state. The subjects performed a rhythmic forearm exercise for about 60 min on a handergometer at an intensity of 10 kpm/min, with 60 contractions per min. Two infusions of a solution, which besides the labelled lactate also contained dye (Indocyanine green) for blood flow determination (14), were made into the brachial artery through a teflon catheter (Fig. 1). The infusions were started after about 7 and 40 min of exercise and lasted 6 - 7 min. Before, during and after the infusions, blood samples for the determination of concentrations of dye, labelled

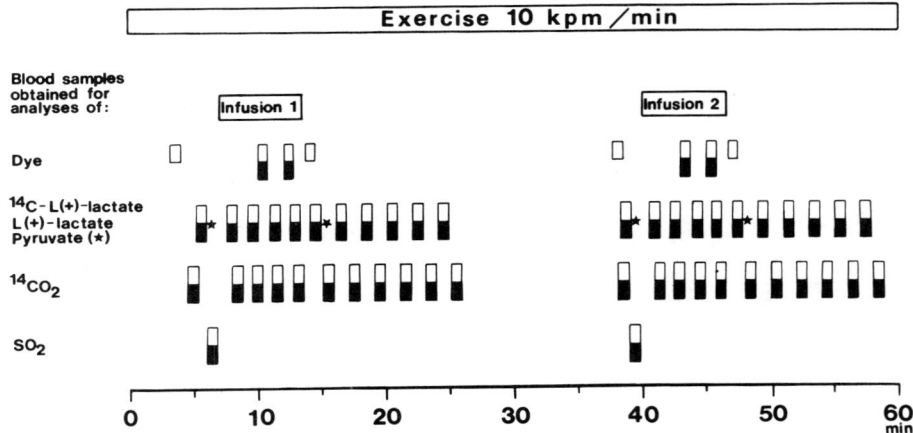

Fig. 1. The experimental design. The approximate times of the infusions and the blood sampling from the radial artery (□) and deep vein (■) are indicated. An asterisk denotes blood samples which were also used to analyse the pyruvate concentration.

and total lactate, pyruvate and $^{14}CO_2$ were collected from the radial artery and a deep forearm vein through catheters placed percutaneously in these vessels. Oxygen saturation and content were determined immediately before each infusion. The investigation was performed with a normal arterial lactate concentration in 15 subjects and in 6 subjects at an elevated concentration obtained by the intravenous infusion of sodium lactate.

The radiochemical purity of the L(+)-lactate-1-^{14}C used was more than 98 %.

<u>Oxygen saturation</u> was determined spectrophotometrically (8) and the hemoglobin concentration by the cyanmethemoglobin technique (3).

$^{14}CO_2$ <u>in blood</u> was determined as described by Hagenfeldt (6).

The analyses of the L(+)-lactate, ^{14}C-L(+)-lactate and pyruvate concentrations were made with the same perchloric acid (PCA) extract of blood. The <u>L(+)-lactate</u> (7) and <u>pyruvate</u> (1, 12) concentrations were analysed enzymatically.

The ^{14}C-L(+)-lactate concentrations were determined using an enzymatic technique, which is outlined only briefly here. For further details see Jorfeldt (9).

The lactate in the PCA extract was oxidized to pyruvate by LDH, with NAD as proton acceptor and hydrazine sulphate as pyruvate trap. The pyruvate formed was then precipitated as phenylhydrazone after the addition of phenylhydrazine hydrochloride, carrier and pH-adjustment. The precipitate was filtered off, washed and dried and the radioactivity determined by liquid scintillation. The procedure was then repeated on another sample but with the LDH and NAD solutions replaced by ammonium sulphate solution and water respectively. The difference in ^{14}C-radioactivity between the precipitates was taken to reflect ^{14}C-L(+)-lactate radioactivity.

The recovery was 85.2 ± 4.0 % (mean $\pm$ SD) and the coefficient of variation was 2.8 % in samples containing more than 1000 dpm/ml and 7.2 % in samples containing less than 1000 dpm/ml.

Unless otherwise stated the results given in the text are means $\pm$ one standard deviation (M $\pm$ SD).

Definitions

a = <u>arterial concentration</u>, v = <u>deep venous concentration</u>.

<u>Fractional uptake of ^{14}C-L(+)-lactate</u> (F_{upt}) was calculated as the (a - v) difference of labelled lactate at steady state during infusion as a fraction of the arterial concentration of labelled lactate.

<u>Uptake of lactate</u> was computed as the product of F_{upt} and arterial lactate concentration.

<u>Rate of lactate uptake</u> was calculated as the product of lactate uptake and muscle blood flow.

<u>Net uptake and net release of lactate</u> are defined as the magnitude of, respectively, a positive and a negative (a - v) difference of lactate.

<u>Fractional oxidation of ^{14}C-L(+)-lactate</u> was calculated as the (v - a) difference of $^{14}CO_2$ at steady state during the infusions as a fraction of the (a - v) difference of ^{14}C-L(+)-lactate.

RESULTS AND DISCUSSION

Normal arterial lactate concentration (Table I)

The net release of lactate was lower after 40 than after 10 min of exercise which is in agreement with earlier observations (15).

Table I. Data on metabolism (M ± SD) of L(+)-lactate in forearm muscle obtained after 10 and 40 min of exercise at normal (A) and elevated (B) arterial concentrations of lactate; p-values denote the probability that random factors are responsible for the difference between 10 and 40 min exercise (p_{10-40}) and between 10 min values and 40 min values at normal as opposed to elevated arterial concentrations of lactate (p > 0.05 values have been omitted).

Variable	A			B			Difference A - B	
	10 min n = 15	40 min n = 9	p_{10-40}	10 min n = 6	40 min n = 6	p_{10-40}	10 min p	40 min p
Arterial conc. of lactate, mmoles/l	1.09 ± 0.33	0.98 ± 0.36		3.35 ± 0.54	3.36 ± 0.54		<0.001	<0.001
Net release of lactate, mmoles/l	1.58 ± 0.96	0.42 ± 0.28	<0.01	0.65 ± 0.47	0.14 ± 0.29	<0.05	<0.05	
Rate of net release of lactate, mmoles/min	0.48 ± 0.34	0.16 ± 0.11	<0.05	0.18 ± 0.12	0.05 ± 0.11		<0.05	
Fractional uptake of ^{14}C-L(+)-lactate	0.32 ± 0.08	0.30 ± 0.07		0.23 ± 0.05	0.23 ± 0.04		<0.05	
Uptake of lactate, mmoles/l	0.33 ± 0.09	0.29 ± 0.11		0.76 ± 0.22	0.76 ± 0.14		<0.001	<0.001
Rate of uptake of lactate, mmoles/min	0.10 ± 0.03	0.11 ± 0.05		0.23 ± 0.10	0.25 ± 0.09		<0.001	<0.01
Fractional oxidation of ^{14}C-L(+)-lactate	0.38 ± 0.16	0.52 ± 0.19		0.52 ± 0.19	0.48 ± 0.10			

During the infusion periods the arterial concentration of ^{14}C-L(+)-lactate remained approximately constant. The (a - v) difference of ^{14}C-L(+)-lactate was high initially but fell successively and reached a constant level after about 4 min of infusion on both occasions (Fig. 2). This shows that lactate from the blood was taken up continuously in muscle. In spite of the higher net release

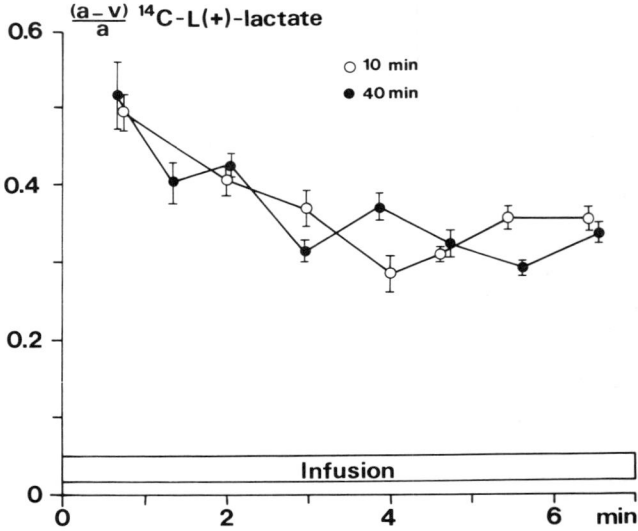

Fig. 2. Uptake of ^{14}C-L(+)-lactate during infusion, expressed as a fraction of the arterial concentration of ^{14}C-L(+)-lactate. Mean values ± SE are given from infusion 1 (open circles) and infusion 2 (filled circles).

of lactate at 10 min, there was no difference between 10 min and 40 min of exercise as regards the fractional uptake of ^{14}C-L(+)-lactate, the uptake of lactate or rate of uptake of lactate. As illustrated in Fig. 3, the release of $^{14}CO_2$ was already considerable about 1 min after the start of the infusions. This release increased further but reached a constant level during the infusions. The fractional oxidation of ^{14}C-L(+)-lactate was 0.38 ± 0.16 and 0.52 ± 0.19 after 10 min and 40 min of exercise respectively. It is thus obvious that part of the lactate taken up by the muscles was oxidized to CO_2. However, since the fractional oxidation was less than unity, direct oxidation to CO_2 cannot be the only metabolic route for the lactate taken up. About 20 per cent of the labelled carbon from the ^{14}C-L(+)-lactate taken up was in fact recovered as a release of other PCA-extractable metabolites. No release of fat-extractable ^{14}C-labelled metabolites was recovered. An incorporation of ^{14}C in a metabolic pool with slow turnover would serve to explain the discrepancy between uptake of labelled lactate and release of labelled metabolites. This hypothesis is supported by results from a supplementary study in which the release of $^{14}CO_2$ was followed up to about 50 min after the infusion (Fig. 4).

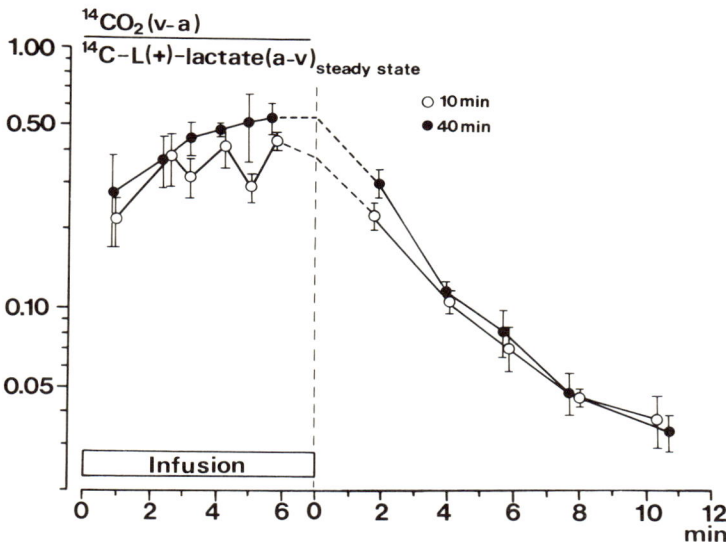

Fig. 3. Semilogarithmic representation of $^{14}CO_2$ during and after infusion, expressed as a fraction of the uptake of ^{14}C-L(+)-lactate at steady state during infusion. The calculated fractional oxidation is plotted on the y-axis and joined to the neighbouring observations by broken lines. Mean values $\pm$ SE are given from infusion 1 (open circles) and infusion 2 (filled circles).

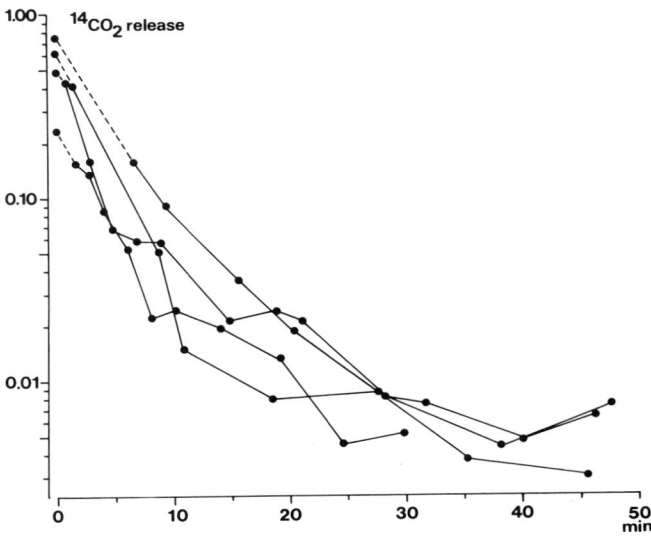

Fig. 4. Semilogarithmic representation of the post-infusion release of $^{14}CO_2$ from the ^{14}C-L(+)-lactate taken up by the forearm muscles during the infusions.

In order to be able to compensate for the formation of $^{14}CO_2$ from recirculating labelled metabolites other than lactate, the subjects performed an identical exercise with the contralateral arm as well and the release of $^{14}CO_2$ from this arm was determined too. An incorporation of ^{14}C in the glycerol fraction of a muscular triglyceride pool or in glycogen could serve to explain these results.

Elevated arterial lactate concentration (Table 1)

The arterial lactate concentration, about 3.4 mmoles/l, remained approximately constant during the exercise period. In this situation the net release of lactate was lower than at normal arterial lactate concentration but it was still higher at 10 min than after 40 min of exercise. After 55 min of exercise a shift had occurred towards a net uptake. Although there was a net release of lactate, a net uptake of pyruvate was recorded in two cases after 10 min and in four cases after 40 min of exercise. In this situation, too, the uptake of ^{14}C-L(+)-lactate and the release of $^{14}CO_2$ reached a steady state during the infusion periods. The fractional uptake of ^{14}C-L(+)-lactate was less than at normal arterial concentration but the uptake of lactate and the rate of uptake of lactate were higher. The fractional oxidation of the ^{14}C-L(+)-lactate taken up was of the same magnitude as at normal arterial lactate concentration. In the total material, the rate of lactate uptake correlated most closely to the arterial inflow of lactate to the forearm, i.e. the product of the arterial lactate concentration and forearm muscle blood flow (Fig. 5). This relation seems to be curvilinear, since the curve cannot intersect the positive y-axis, indicating saturation kinetics for the lactate uptake.

The net release of lactate from forearm muscle was positively correlated to oxygen uptake but otherwise no correlation was found between the lactate variables studied and oxygen uptake.

The results thus demonstrate that there is an exchange of lactate between blood and exercising skeletal muscle, with simultaneous release and uptake.

An uptake of lactate of this type may occur in all cells with an intracellular lactate turnover. Thus if the muscle mass studied was homogenous, the results could simply reflect a diffusion across the cell membranes. In this situation, determinations of (a - v) differences of total lactate are a sufficient basis for determining the net lactate metabolism in muscle. The use of labelled lactate makes it possible to study the metabolic processes. However, the fact that the uptake of lactate was the same after 10 and 40 min of exercise in spite of a significantly higher net release on the former

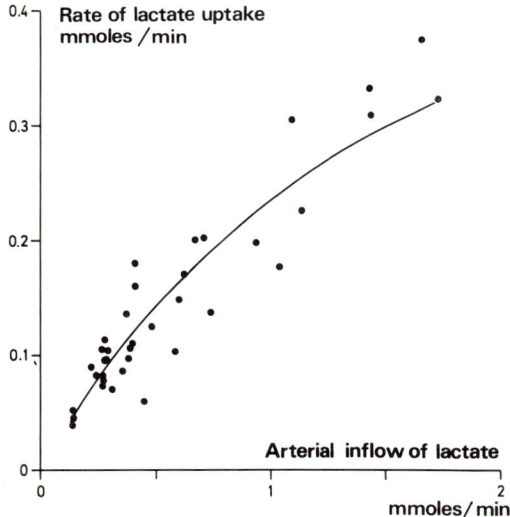

Fig. 5. Relationship between the rate of lactate uptake and the arterial inflow of lactate. The curve corresponding to the linear regression for the reciprocals of the variables is shown.

occasion makes it improbable that the results simply reflect a diffusion equilibrium. An additional explanation may be that the lactate metabolism is different in different parts of the muscle mass. It is possible, for instance, that parts of the muscle mass did not participate in the exercise and that the uptake of lactate was proportionally higher in these parts while the bulk of the release took place in the working parts. In this situation (a - v)-determinations will underestimate the total lactate production in muscle induced by exercise.

The occurrence of red and white muscle fibres within an individual muscle is of special interest since these two types of fibres have different metabolic patterns and different recruitment patterns (5, 10) during prolonged exercise. It seems possible that the release of lactate is proportionally higher in the white fibres than in the red and vice versa as regards the uptake. An activation of those fibres which predominantly release lactate at the beginning of the exercise period could explain the high net release during this phase.

SUMMARY

The results demonstrate that there is an exchange of lactate between blood and exercising skeletal muscle, with a simultaneous release and uptake, and that the lactate taken up is oxidized in part to CO_2 but that other metabolic pathways are also involved.

REFERENCES

1. BÜCHER, T., R. CZOK, W. LAMPRECHT, AND E. LATZKO. Pyruvat. In: *Methoden der Enzymatischen Analyse.* Weinheim: Verlag Chemie, 1962, p. 253-259.

2. CARLSON, L.A., AND B. PERNOW. Oxygen utilization and lactic acid formation in the legs at rest and during exercise in normal subjects and in patients with arteriosclerosis obliterans. *Acta Med. Scand.* 164:39-52, 1959.

3. DRABKIN, D.L., AND J.H. AUSTIN. Spectrophotometric studies. II. Preparations from washed blood cells; nitric oxide hemoglobin and sulfhemoglobin. *J. Biol. Chem.* 112:51-65, 1935

4. FREYSCHUSS, U., AND T. STRANDELL. Limb circulation during arm and leg exercise in supine position. *J. Appl. Physiol.* 23: 163-170, 1967.

5. GRIMBY, L., AND J. HANNERZ. Recruitment order of motor units on voluntary contraction: changes induced by proprioceptive afferent activity. *J. Neurol. Neurosurg. Psychiat.* 31:565-573, 1968.

6. HAGENFELDT, L. A simplified procedure for the measurement of $^{14}CO_2$ in blood. Clin. Chim. Acta 23:320-321, 1967.
7. HOHORST, H.J. L(+)-Lactat Bestimmung mit Lactat-Dehydrogenase und DPN. In: Methoden der Enzymatischen Analyse. Weinheim: Verlag Chemie, 1962, p. 266-270.
8. HOLMGREN, A., AND B. PERNOW. Spectrophotometric measurement of oxygen saturation of blood in the determination of cardiac output. A comparison with the van Slyke method. Scand. J. Clin. Lab. Invest. 11:143-149, 1959.
9. JORFELDT, L. Metabolism of L(+)-lactate in human skeletal muscle during exercise. Acta Physiol. Scand. Suppl., 338: 1-67, 1970.
10. KUGELBERG, E., AND L. EDSTRÖM. Differential histochemical effects of muscle contractions on phosphorylase and glycogen in various types of fibres: relation to fatigue. J. Neurol. Neurosurg. Psychiat. 31:415-423, 1968.
11. MARGARIA, R., H.T. EDWARDS, AND D.B. DILL. The possible mechanisms of contracting and paying the oxygen debt and the role of lactic acid in muscular contraction. Amer. J. Physiol. 106:689-715, 1933.
12. SEGAL, S., A.E. BLAIR, AND J.B. WYNGAARDEN. An enzymatic spectrophotometric method for the determination of pyruvic acid in blood. J. Lab. Clin. Med. 48:137-143, 1956.
13. STAINSBY, W.N., AND H.G. WELCH. Lactate metabolism of contracting dog skeletal muscle in situ. Amer. J. Physiol. 211: 177-183, 1966.
14. WAHREN, J. Quantitative aspects of blood flow and oxygen uptake in the human forearm during rhythmic exercise. Acta Physiol. Scand. 67, Suppl. 269:1-93, 1966.
15. WAHREN, J., AND L. HAGENFELDT. Human forearm muscle metabolism during exercise. I. Circulatory adaptation to prolonged forearm exercise. Scand. J. Clin. Lab. Invest. 21: 257-262, 1968.

MUSCLE GLYCOGEN, LACTATE, ATP, AND CP IN INTERMITTENT EXERCISE

B. Saltin and B. Essén

Department of Physiology, Gymnastik- och idrotts-högskolan, Stockholm, Sweden

Previous experiments with intermittent exercise showed that even during very heavy exercise little or no increase in lactate was found in the blood if the work periods did not exceed 5-15 seconds (1,2,3). From these results it was suggested that the myoglobin in the muscle acted as an oxygen store, the store being employed at the onset of exercise and refilled immediately after work. To analyse further the physiology and biochemistry of intermittent exercise some of the earlier experiments have been repeated with the addition that muscle biopsies have been included in the protocol.

Three specific questions were asked:
1. Is the reloading of the phosphagens (ATP and CP) during the rest periods of sufficient magnitude that it has a quantitative significance?
2. Is the muscle lactate concentration increased?
3. Is there a need for an oxygen store?

METHODS AND PROCEDURE

Three physical education students (average maximal oxygen uptake 4.4 l/min) were studied on four occasions during a period of four months. They performed intermittent exercise each time for 30 minutes at a supramaximal work load of 2,400 kpm/min ($\approx$ 400 Watt). The work and rest periods during the intermittent exercise were varied

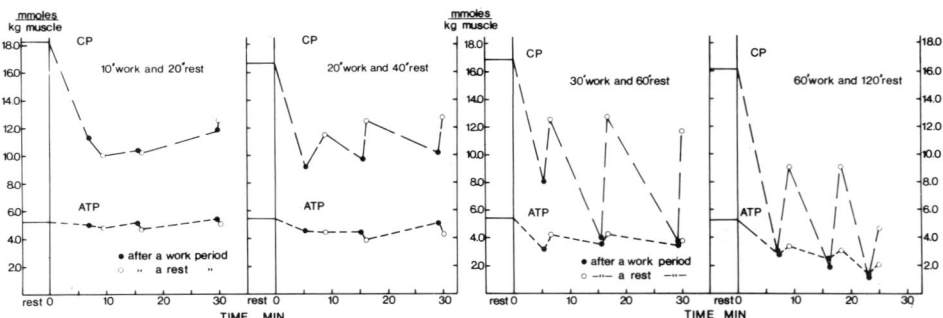

Fig. 1. Mean values for muscle creatine phosphate (CP) and adenosine triphosphate (ATP) concentration in three subjects at rest and performing intermittent exercise for 30 minutes. The results from four different experiments are given where the duration of the work period has been varied from 10 up to 60 seconds. The ratio between work and rest has in all experiments been 1:2, and the work load 2,400 kpm/min.

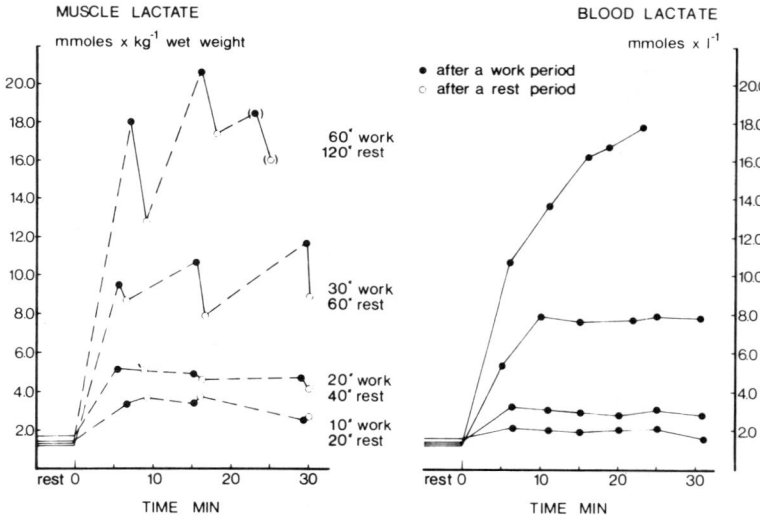

Fig. 2. Mean values for muscle lactate (left panel) and blood lactate (right panel) concentration in the same experiment as in Fig. 1.

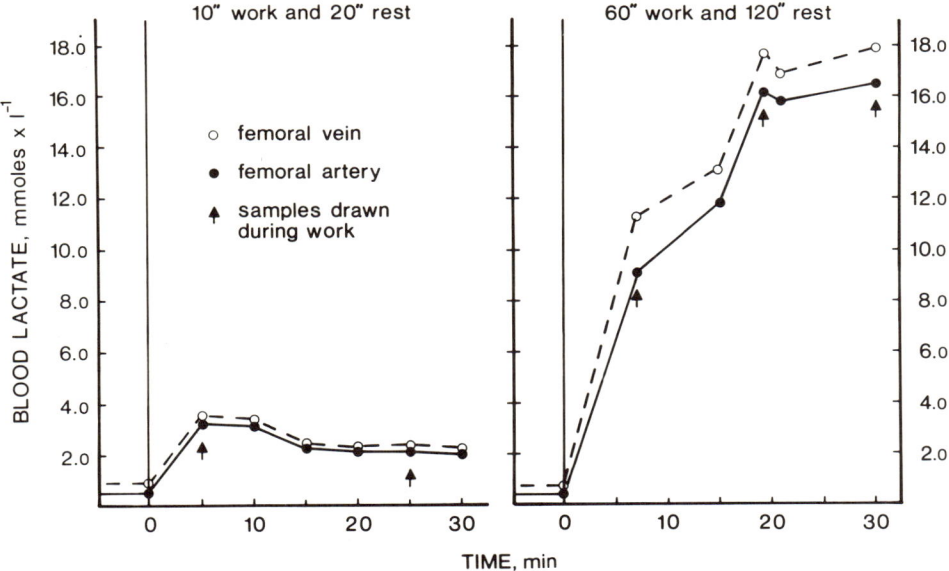

Fig. 3. The lactate concentration in the femoral artery and femoral vein in one subject at rest and during the intermittent exercise. In the left panel is given the results from experiment with 10 seconds of work and in the right panel the work periods lasted 60 seconds.

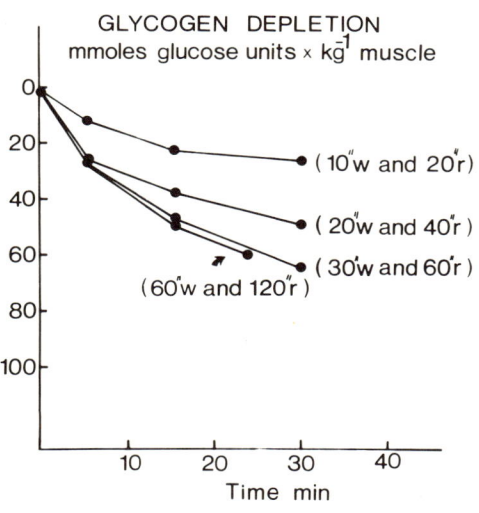

Fig. 4. Muscle glycogen depletion in intermittent exercise of 30 minutes' duration. Data are from the same experiment as described in Fig. 1.

in the following manner (a) 10 sec work, 20 sec rest, (b) 20 sec work, 40 sec rest, (c) 30 sec work, 60 sec rest, (d) 60 sec work, 120 sec rest. The ratio between rest and exercise was 1:2 and the subjects attained in each experiment a total work output of 24,000 kpm. Muscle biopsies were taken from the quadriceps muscle before work and approximately 5, 15, and 30 minutes after the work started. They were taken both immediately after a work period and at the end of a rest period. Analyses of the different metabolites in the biopsy samples were performed according to the Lowry method (4) as described by Karlsson et al. (5). Oxygen uptake, heart rate, and blood lactate were also determined at appropriate time intervals and with standard techniques used in the laboratory (3). In separate experiments cardiac output was determined with the dye-dilution technique using a Beckman densitometer with cardiogreen as indicator. In these experiments one of the venous catheters was placed in a retrograde direction in the femoral vein.

RESULTS AND DISCUSSION

The ATP and CP depletion was most marked with the longer work periods (Fig. 1), and when the rest periods lasted more than 20 seconds a significant increase in the CP and ATP was observed in the muscle specimen. The resynthesis of ATP and CP is an extremely fast process and the reason for this slow return to normal resting levels must be sought in other changes that occur in the muscle cell limiting the full resynthesis of the phosphagens. Blood lactate was only elevated above resting levels in the experiments with the longer work periods (Fig. 2). On the other hand muscle lactate was increased in all experiments which is in contrast to what has been suggested (2,6). When the work periods lasted for 30 or 60 seconds, a gradual increase was observed in the muscle lactate level during the intermittent exercise.

Whether there is a lactate production during each work period throughout the intermittent exercise also with the shortest work periods cannot be definitely answered. At the end of each rest period, lower muscle lactate values (compared with the end of a work period) were observed only in the experiments with rest periods of 60 and 120 seconds. On the other hand higher femoral venous than arterial lactate concentrations were observed throughout the intermittent exercise period also in the experiments with the shortest work periods (Fig. 3). An estimation of the amount of lactate released from the muscles in the experiment with repeated periods of 10 seconds' work has been done from the data given in Table 1. The calculation is based on the assumption that the oxygen uptake and the cardiac output during the work periods above resting (basal) oxygen uptake is equally divided between the legs. The results thus obtained indicate that even if there is a lactate formation with each work period of 10 seconds the magnitude of the energy derived from this anaerobic glycolysis is of a very minor importance.

The glycogen depletion found in intermittent exercise (Fig. 4) is less than that found with continuous heavy exercise (cf 7). This is especially the case in the intermittent exercise with the shortest work periods, which further supports the belief that very little lactate is produced. The low rate of utilization of the muscle glycogen observed during intermittent exercise is in good agreement with the findings of very low RQ values during the work periods in the intermittent exercise (Table 1).

Subject S.L.		10 sec work, 20 sec rest			60 sec work, 120 sec rest		
		REST before	WORK 10"	RECOVERY last 10"	REST before	WORK 10"	RECOVERY last 10"
Lactate mmoles/kg muscle	muscle	1.1	$3.5^{x)}$	$3.2^{x)}$	1.1	$21.1^{x)}$	$19.8^{x)}$
	femoral vein	0.9	$2.4^{x)}$	$2.2^{x)}$	0.5	$16.6^{x)}$	$16.9^{x)}$
	artery	0.5	$2.3^{x)}$	$2.1^{x)}$	0.2	$16.1^{x)}$	$15.8^{x)}$
Oxygen uptake l/min		0.37	2.52	1.86	0.36	3.40	0.99
RQ		0.85	0.92	0.87	0.81	0.84	1.37
Cardiac output l/min		6.5	19.8	16.6	6.1	21.5	16.2
Heart rate beats/min		75	$146^{x)}$	$136^{x)}$	75	$185^{x)}$	$144^{x)}$

$^{x)}$ These values were obtained at the end of the period.

Table 1. Some circulatory and metabolic data in the same experiment as in Fig. 3.

It is concluded then that reloading of ATP and CP during the rest periods, lasting not longer than 20 seconds plays – from a quantitative standpoint – an insignificant role in energy supply during intermittent exercise. The same is the case for the lactate production in intermittent exercise with very short ($<$ 15 sec) work periods. The amount of oxygen transported during each work period is not large enough either, to meet the actual demand, thus supporting earlier studies (2,3) claiming a need for an oxygen store (myoglobin?) in the muscle which is very quickly refilled during each intervening rest period.

ACKNOWLEDGEMENT

This investigation has been carried out with support from the Swedish Medical Research Council (Project No. B71-14X-2203-04B).

REFERENCES

1. ÅSTRAND, I., P.-O. ÅSTRAND, E.H. CHRISTENSEN, AND R. HEDMAN. Intermittent muscular work. *Acta physiol. scand.* 48:448-453, 1960.

2. ÅSTRAND, I., P.-O. ÅSTRAND, E.H. CHRISTENSEN, AND R. HEDMAN. Myohemoglobin as an oxygen store in man. *Acta physiol. scand.* 48:454-460, 1960.

3. CHRISTENSEN, E.H., R. HEDMAN, AND B. SALTIN. Intermittent and continuous running. *Acta physiol. scand.* 50:269-286, 1960.

4. LOWRY, O.H., J.V. PASSONNEAU, F.X. HASSELBERGER, AND D.W. SCHULZ. Effect of ischemia on known substrates and cofactors of the glycolytic pathway in brain. *J. Biol. Chem.* 239(1):18-30, 1964.

5. KARLSSON, J., B. DIAMANT, AND B. SALTIN. Muscle metabolites during submaximal and maximal exercise in man. *Scand. J. clin. Lab. Invest.* 27(1), 1971.

6. MARGARIA, R., R.D. OLIVIA, P.E. Di PRAMPERO, AND P. CERRETELLI. Energy utilization in intermittent exercise of supramaximal intensity. *J. Appl. Physiol.* 26:752-756, 1969.

7. SALTIN, B., AND J. KARLSSON. Muscle glycogen utilization during work of different intensities. Ibidem.

BLOOD LACTATE CONCENTRATIONS DURING INTERMITTENT AND CONTINUOUS

EXERCISE WITH THE SAME AVERAGE POWER OUTPUT

R.H.T. EDWARDS[*], A. MELCHER, C.M. HESSER, and O. WIGERTZ

From the Departments of Aviation and Naval Medicine
Karolinska Institutet, and Clinical Physiology
Karolinska Sjukhuset, Stockholm, Sweden

We have recently been interested in the physiological adaptations accompanying the performance of a given total amount of work by a variety of patterns of intermittent and continuous exercise.

Three healthy male volunteers exercised in the upright position on an electrically braked cycle ergometer. Fig. 1 shows

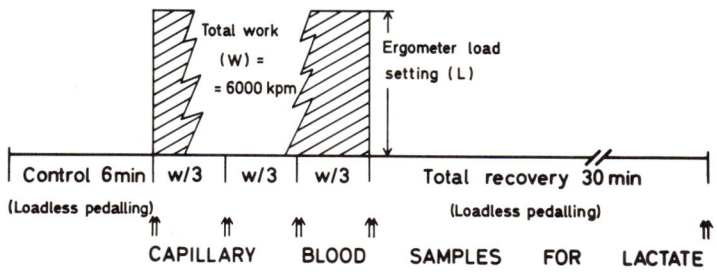

Fig. 1. Times of blood sampling for lactate relative to total work performed.

---

[*] In receipt of a Swedish Welcome Travelling Research Fellowship

the protocol used in all experiments. Duplicate capillary blood samples were taken for lactate analysis by a micro enzymic method - based on Hohorst (3) - from the warmed finger tips at the end of the 6 min control period, on completing one third, two thirds and the whole of the standard work target (6000 kpm), and after a total recovery period of 30 min. The pedalling frequency was constant at 60 rpm throughout the experiment; control and recovery were periods of loadless pedalling. The four continuous and four intermittent exercise patterns are illustrated in Fig. 2. The total work goal

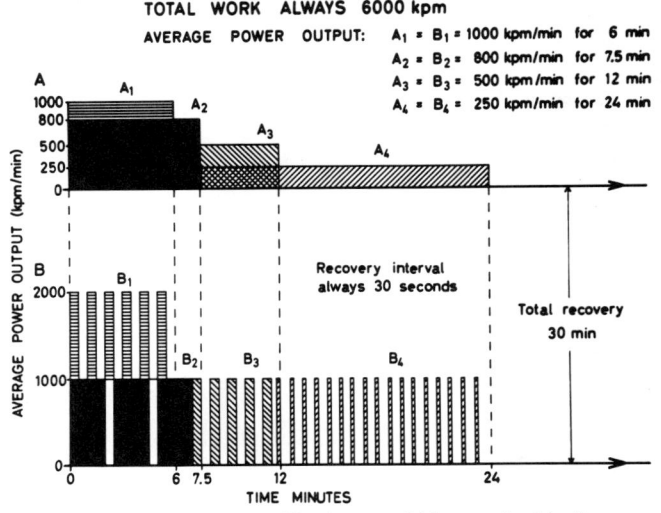

Fig. 2. Work profiles studied.

was the same in all instances, accurately achieved in the case of intermittent exercise, with square wave work profiles of 10 sec (B_4), 30 sec (B_1 and B_3) or 120 sec (B_2) alternating with 30 sec of recovery, with the use of a digital programmer controlling the ergometer load. Care was taken to ensure that the average power outputs (work per min) were the same for paired patterns of intermittent and continuous exercise; $A_1 = B_1 = 1000$ kpm/min for 6 min, $A_2 = B_2 = 800$ kpm/min for 7.5 min, $A_3 = B_3 = 500$ kpm/min for 12 min and $A_4 = B_4 = 250$ kpm/min for 24 min. Comparing first the effect of intermittent with continuous exercise at the same ergometer setting namely 1000 kpm/min (Fig. 3), it is seen that blood lactate concentration (the mean results from three subjects) rises progressively during the 6 min of continuous exercise but with intermittent exercise the increase is noticeably less until when the work periods are only 10 sec long there is no increase at all, in agreement with the observations of others (1, 2, 4). It should be noted, however, that the average power output has been markedly reduced from 1000 kpm/min to 250 kpm/min with the shortest work periods. Blood was sampled for lactate estimation at exactly the same time in relation to the start of exercise and in relation

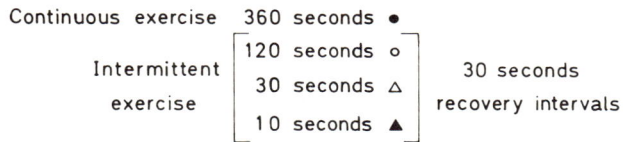

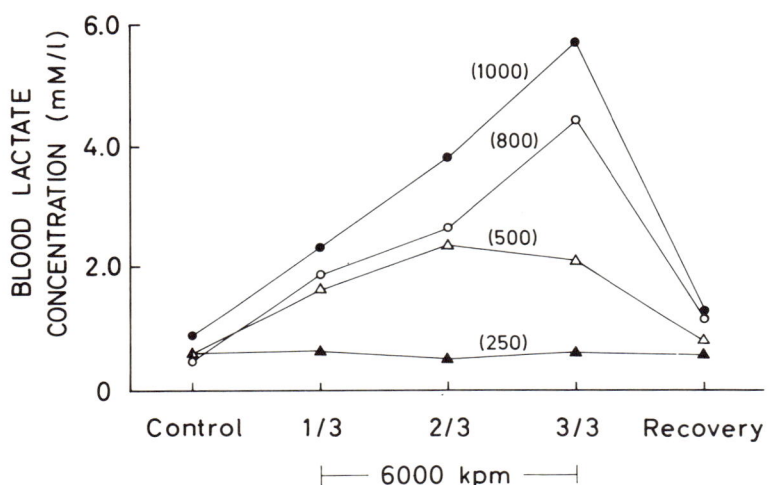

Fig. 3. Changes in blood lactate concentration resulting from splitting up continuous exercise into alternating periods of work and recovery. (Mean results in 3 subjects)

to the amount of the total work target completed and the results are compared in a concordance plot (Fig. 4). Blood lactate concentrations were slightly higher with intermittent exercise compared with continuous exercise of the same average power output. Previous studies have emphasized the reduction in blood lactate concentration which occurs when work is done intermittently. However, it appears from this study that if a given quantity of work has to be done in a set time a somewhat lower rise in blood lactate concentration is to be found if the work is performed at a continuous low rate than if performed intermittently with a higher work load.

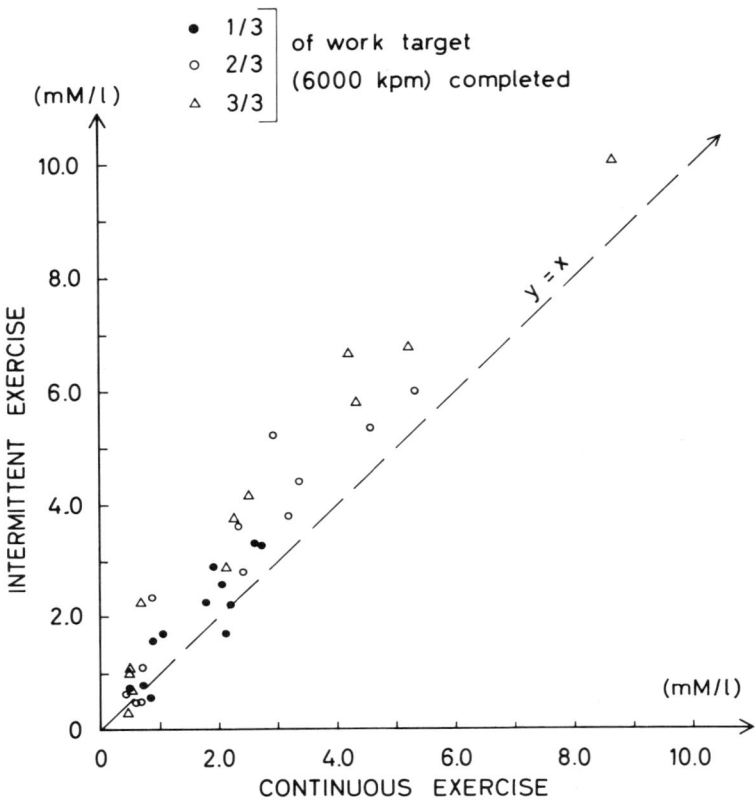

Fig. 4. Concordance plot for blood lactate concentrations at the same times during intermittent and continuous exercise (not including control and recovery values). Results from 3 subjects.

REFERENCES

1. ÅSTRAND, I., P.-O. ÅSTRAND, E.H. CHRISTENSEN, and R. HEDMAN. Intermittent muscular work. Acta Physiol.Scand. 48: 448 - 453, 1960.
2. CHRISTENSEN, E.H., R. HEDMAN, and B. SALTIN. Intermittent and continuous running. Acta Physiol.Scand. 50: 269 - 286, 1960.
3. HOHORST, H.J. Enzymatische Bestimmung von L(+) - Milchsäure. Biochem. Z. 328: 509 - 521, 1957.
4. MARGARIA, R., R.D. OLIVIA, P.E. DiPRAMPERO, and P. CERRETELLI. Energy utilization in intermittent exercise of supramaximal intensity. J. Appl. Physiol. 26: 752 - 756, 1969.

UTILIZATION OF ENDOGENOUS LIPIDS BY THE ISOLATED PERFUSED RAT HEART

Robert E. Olson and Catherine Bauer

Department of Biochemistry, St. Louis University School of Medicine, St. Louis, Mo. U.S.A.

The isolated perfused rat heart, which beats at rates commensurate with those observed in situ, is a reasonable model for the study of metabolic changes occurring during muscular exercise. We have used this preparation to study the utilization of endogenous lipids in the absence of oxidizable substrates in the perfusing medium (9). Fisher and Williamson (6) observed that hearts perfused without exogenous substrate would continue to beat for 1-2 hours, despite the disappearance of glycogen in 15-30 minutes. We have observed a limited availability of cardiac lipids to serve as fuel for contraction, the biggest contribution being made by neutral glycerides, and that under these conditions the heart fails to maintain its initial force of contractility. Today I should like to discuss 1) the general effects of perfusion of the isolated rat heart with a substrate-free medium upon contractility and the labile phosphate stores; 2) the changes which occur in intermediary metabolites of glycolysis and in the lipids over a two hour period of perfusion; and 3) the behavior of cardiac mitochondria isolated from perfused rat hearts at various times during depletion of the available endogenous substrates.

MATERIALS AND METHODS

Male albino rats of the Wistar strain from the St. Louis University colony weighing between 200 and 275 grams were used in this study. They were fed a commercial stock diet prior to use and were not fasted. They were killed by decapitation, the heart quickly removed and chilled in Krebs-Henseleit bicarbonate buffer at 4°C before being mounted on an aortic cannula and perfused through the coronary arteries by the Langendorff technique as employed by Bleehen and Fisher (1). In most experiments the effluent perfusion was not recirculated. The gas mixture was 95% O_2:5% CO_2 and the temperature 37°C. The perfusion medium was Krebs-Henseleit medium (8) containing 1.25 mM calcium. Glucose was added at the concentration of 8 mM in some experiments. In other experiments fatty acid-free bovine serum albumin was added at a concentration of 0.2%; in some others bovine serum albumin complexed to palmitate-1-^{14}C (0.85 mM in the perfusate) was added at a concentration of 0.5%.

The oxygen consumption of the perfused heart was measured in some experiments by measuring the pO_2 of the effluent perfusate with an oxygen electrode and the rate of flow with a bubble flowmeter. The heart was sampled by freezing at -70C with a Wollenberger clamp. Glycogen, glycolytic intermediates, glycerol, intermediates of the tricarboxylic acid cycle, creatine phosphate and the adenine nucleotides were measured enzymatically using DPN or TPN linked reactions measured in an Eppendorf fluorimeter (5). In some experiments reflected fluorescence from the surface of the perfused heart attributable to reduced pyridine nucleotides was measured (3). Lipids were extracted with the Folch reagent, separated on Silica gel-G thin layer plates, and their acyl content determined by the method of Rapport and Alonzo (11). Free fatty acids were measured in the neutral lipid fraction by an automated microtitration employing a Radiometer titrimeter. Radioactivity of isolated compounds was determined in a Packard scintillation spectrometer.

Mitochondria were isolated from normal and perfused hearts by homogenizing in a solution containing 0.225 M sucrose, 0.075 M mannitol, 0.0005 M EDTA, 0.004 M Tris buffer at pH 7.6. Nuclei and cell debris were removed by centrifuging at 300 x g for 10 minutes and the mitochondria by centrifuging at 8,500 x g for 20 minutes. They were washed once and suspended in the original medium to give a concentration of 10 mg/ml. P/O ratios were determined by the method of Chance and Williams (2) using an oxygen electrode. The reactions were carried out at 28°C in a medium containing 0.225 mannitol, 0.075 M KCl, and 0.0055 M phosphate at pH 7.3. The substrates were 250 μM pyruvate and 100 μM malate. Rapid or State 3 respiration was induced from the resting condition by the addition of 400 µM ADP. Respiratory control ratios (R.C.R.) were computed by dividing the rate of respiration during State 3 by the rate observed during the subsequent resting state or State 4.

RESULTS

Contractility and Labile Phosphate Supply. A comparison of the force of contraction developed by hearts perfused with 8 mM glucose as compared with those perfused with non-nutrient medium is presented in Figure 1. Contractility was maintained at 100 ± 5% for 2 hours in hearts perfused with glucose, whereas, those perfused without substrate maintained a normal force of contraction for only 30 minutes, at which time contractility began to decline and continued to decline to negligible values at 2 hours. On the other hand, the heart rates were comparable in hearts perfused with and without substrate. Likewise, coronary flow rates, which averaged 6 ml/minute were the same in hearts perfused with or without substrate.

Figure 2 shows the effect of perfusion of the rat heart with non-nutrient medium for two hours upon creatine phosphate and free and total creatine. Creatine phosphate begins to decline immediately and reaches levels 33% below initial values in 30 minutes, at which time no change in contractility is noted. A further fall to values of about 15% of initial concentrations is noted at 2 hours. The free creatine varied reciprocally with the creatine phosphate so that total creatine in the perfused heart remained essentially constant for the 2 hour period.

The changes in the adenine nucleotides of the isolated rat heart perfused with non-nutrient medium for 2 hours are presented in Figure 3. The ATP levels declined steadily from about 20 to 10 µmoles/gram dry weight over the 2 hour period. Part of this loss occurred into the perfusate where we have detected inosinic acid and other purine derivatives. The ADP levels of the heart remained relatively constant during the 2 hour perfusion, no doubt in part through the action of myokinase. The AMP content rose abruptly at 30 minutes and remained elevated during the remainder of the perfusion. In Figure 4, the changes in creatine phosphate content and the ATP/AMP ratio are shown in relationship to contractility. It is clear from these data that transphosphorylation of CP to ADP during

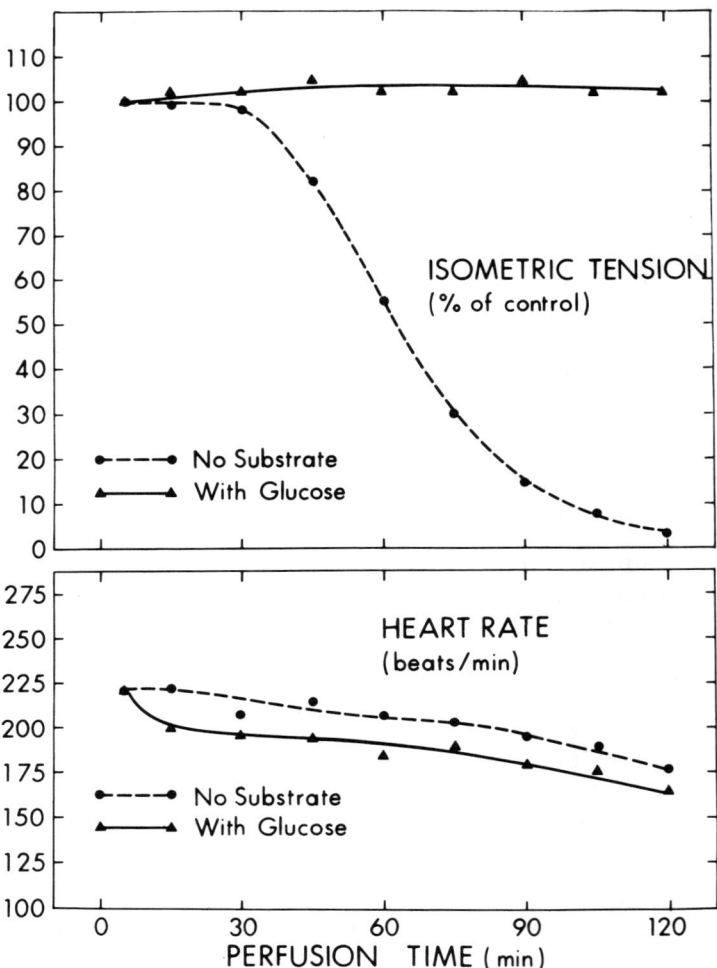

Fig. 1: Rate of change of isomeric tension development and heart rate by the isolated rat heart perfused with non-nutrient medium and glucose.

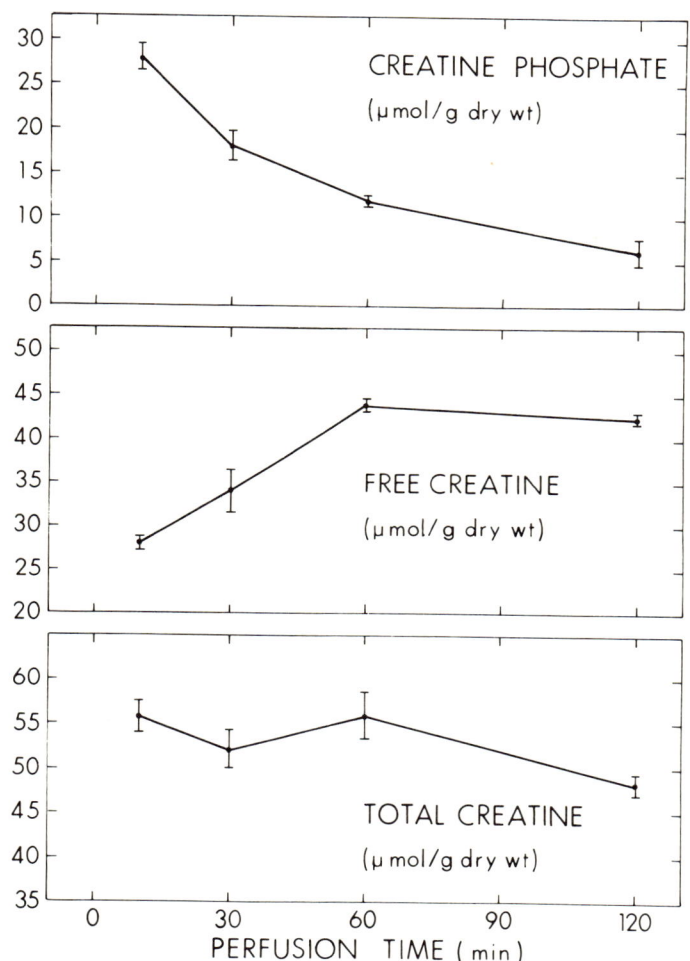

Fig. 2: Changes in creatine phosphate, free and total creatine of the isolated rat heart perfused with non-nutrient medium for 2 hours.

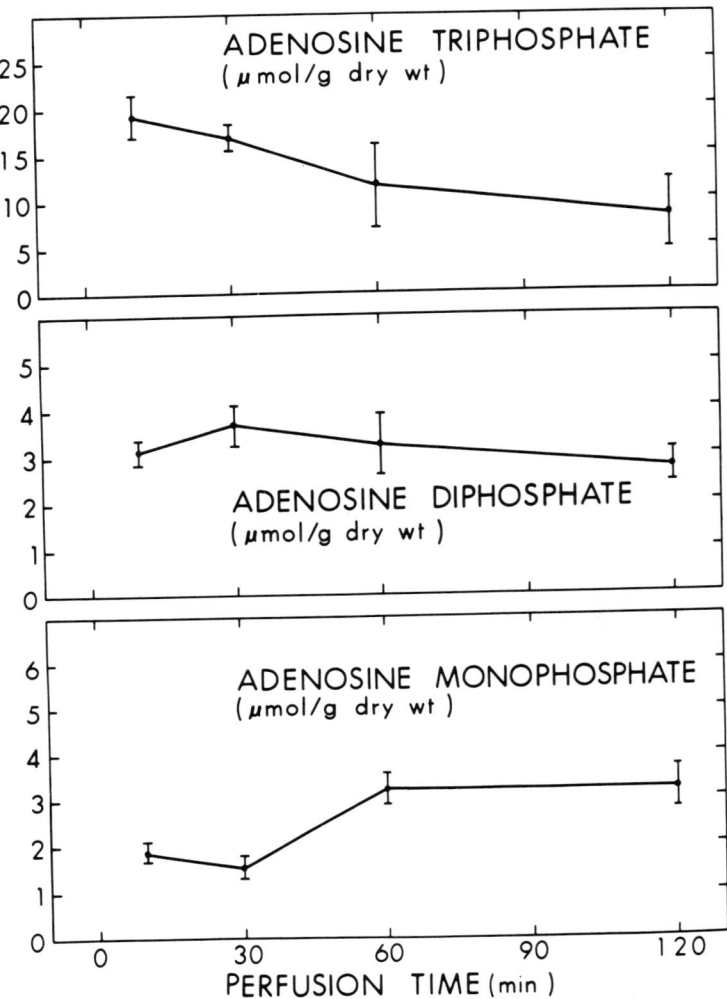

Fig. 3: Changes in the adenine nucleotide content of the isolated rat heart perfused with non-nutrient medium for 2 hours.

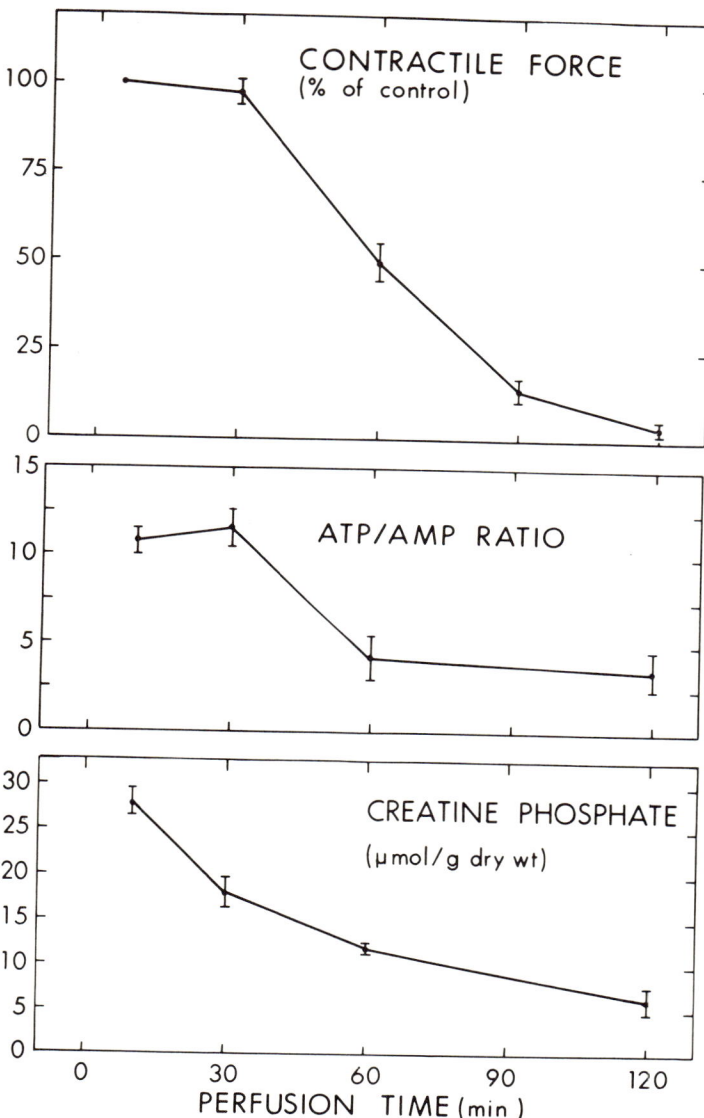

Fig. 4: Changes in the creatine phosphate content, the ratio of ATP/AMP, and the contractile force developed by the isolated rat heart perfused with non-nutrient medium for 2 hours.

the first 30 minutes maintains the ATP content in the myofibril at physiological concentrations. Since the ATP content of muscle is compartmentalized, it is not possible to state precisely what that concentration is but it is likely that it is much below the average concentration of 4 mM in fresh heart muscle. Another conclusion which may be drawn from Figure 4 is that the ATP/AMP ratio is the best index of the rate of delivery of ~P to the contractile system that we have measured.

Carbohydrate and Lipid Content of the Perfused Rat Heart. The carbohydrate content of the isolated rat heart was quickly depleted by perfusion with non-nutrient medium. Figure 5 presents the changes in glycogen and reduced pyridine nucleotide content of the rat heart during perfusion with substrate-free medium for two hours. In agreement with others (6,12), we observed that the glycogen content of the heart perfused with non-nutrient medium declined to basement values in 30 minutes. The pyridine nucleotides detected by reflective fluorescence (thought to be mainly in subepicardial mitochondria) show a gradual oxidation during the 2 hour perfusion period. Figure 6 presents the changes in three hexose-phosphates in the isolated heart perfused with non-nutrient medium for 2 hours. Glucose-6-phosphate and fructose-6-phosphate fell to negligible concentrations in 30 minutes and remained at these low values for the remainder of the perfusion. Glucose-1-phosphate reached a base line of about 25% of initial value at 30 minutes and remained at that plateau. Measurement of other intermediate metabolites of glycolysis has shown that all behave like those shown in Figure 6 (10) reaching plateau values of 5-50% of their initial values after 30 minutes of perfusion.

The lipid composition of twenty fresh rat hearts, washed free of coronary blood and immediately extracted, is shown in Table I. The total lipid of the rat heart made up 14% of the dry weight or 3.1% of the fresh weight. This myocardial lipid contained about 300 μmoles of total fatty acids/g dry weight of which only 1-3 μmoles were free. The neutral esters contained only 15% of the total fatty acids. The changes in lipid content of the rat heart during perfusion is shown in Figure 7. Phospholipid fatty acids remained constant during the entire 2 hour perfusion. Triglycerides, on the other hand, began to decline from a control value of 43 μmoles/g dry weight at 30 minutes and reached the value of 14 μmoles/g dry weight at the end of the two

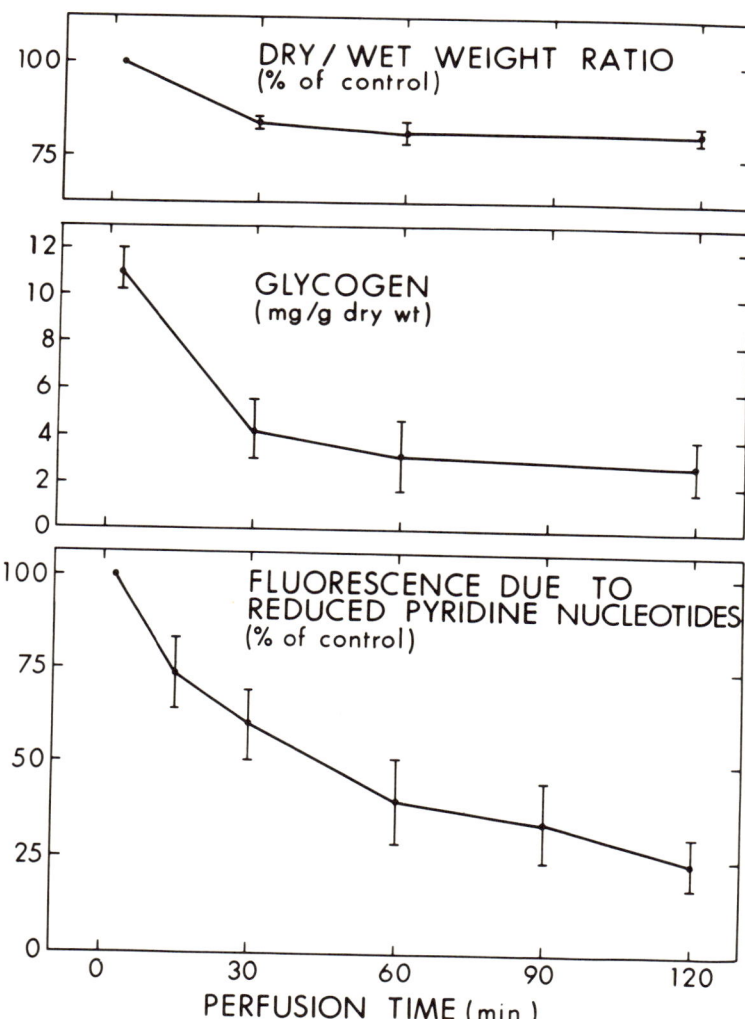

Fig. 5: Changes in dry/wet weight ratio, glycogen content, and external fluorescence due to reduced pyridine nucleotides in the isolated rat heart perfused with non-nutrient medium for 2 hours.

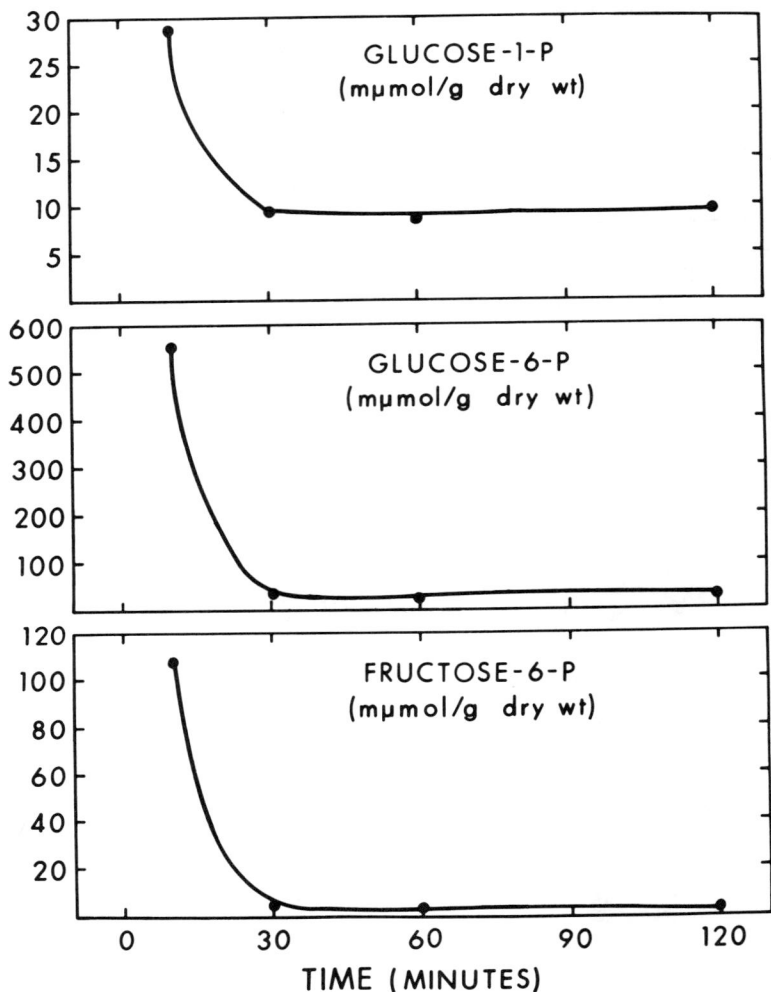

Fig. 6: Changes in glucose-1-phosphate, glucose-6-phosphate, and fructose-6-phosphate content of the isolated rat heart perfused with non-nutrient medium for 2 hours.

Table I. Lipid Composition of the Normal Rat Heart

Lipid	Concn. (µmoles of fatty acid or µg atoms of P/g dry wt)
Cholesteryl esters	1.5 ± 1.0
Free fatty acids	2.5 ± 1.0
Partial glycerides	4.0 ± 0.6
Triglycerides	40.1 ± 2.0
Phospholipids	250.0 ± 5
Lipid phosphorus	140.0 ± 6
Total lipid	298.0 ± 10

Twenty rat hearts were analyzed in six batches. The variance is presented as S.E.M.

hour perfusion. Partial glycerides and cholesterol esters did not change in concentration within analytical limits during the perfusion period. The lipids of the isolated rat heart were labeled with palmitate-1-^{14}C by

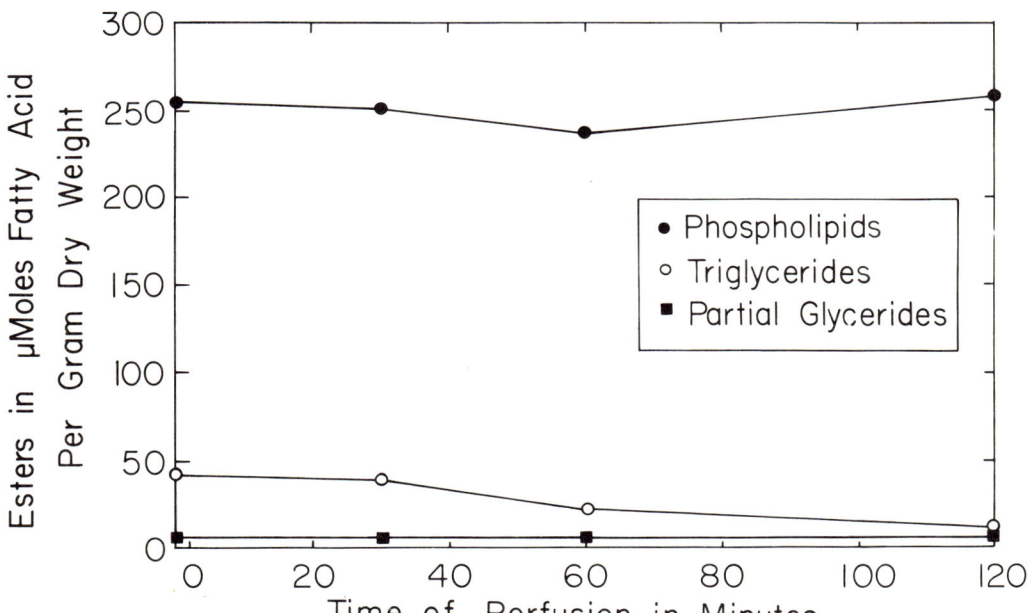

Fig. 7: Changes in phospholipid, triglyceride, and partial glyceride content of the isolated rat heart perfused with non-nutrient medium for 2 hours.

perfusing them for 30 minutes with 0.85 mM palmitate-1-^{14}C complexed to 0.5% bovine serum albumin. Four hearts (Group I) were analyzed for lipids and radioactivity at the end of this 30 minute perfusion. Four other hearts (Group II) that had been so labeled with isotopic carbon were transferred to another perfusion vessel and perfused with non-nutrient medium until exhaustion. The results of this experiment are shown in Table II. There was

Table II. Metabolism of Palmitate-1-^{14}C Labeled Endogenous Heart Lipids During Perfusion

Group	Rat No.	Total Time (Min)	PL	TG	DG	FA
1	4	30	1.80 ± .27	10.28 ± 1.48	1.33 ± .29	1.17 ± .58
2	4	93	1.84 ± .20	2.78 ± .92	0.38 ± .11	0.33 ± .04

Values are µmoles of palmitate-1-^{14}C found incorporated into each lipid class.

a highly significant loss of radioactivity from triglycerides, diglycerides, and free fatty acids during the second period of perfusion with non-nutrient medium. Only the phospholipids showed no change in radioactivity, consistent with their non-utilization during substrate deprivation. This experiment is confirmatory of the one presented in Figure 7, and supports the view that only neutral glycerides and free fatty acids are available as fuels for respiration when carbohydrate stores are depleted and exogenous substrate is lacking.

Mitochondria from Substrate-Depleted Hearts. The rates of respiration of mitochondria isolated from rat hearts perfused with non-nutrient medium and with the same medium plus 8 mM glucose are presented in Table III. The respiratory rate of mitochondria from hearts perfused with substrate-free medium declined from 121 nanomoles of O_2 per min per milligram of protein to 56 over the two hour period, a decrease of about 50%. Mitochondria isolated from hearts perfused with glucose, on the other hand, showed a decrease of 20%. The fall in total oxygen

Table III. Rates of ADP-Activated Mitochondrial Respiration and Respiratory Control Ratios From Perfused Rat Hearts

Time Min.	No Substrate		8 mM Glucose	
	Rate O_2 nM/min/mg	R.C.R.*	Rate O_2 nM/min/mg	R.C.R.
10	121 ± 10	4.33 ± 0.60	118 ± 16	4.50 ± 0.39
30	105 ± 10	3.78 ± 0.27	101 ± 8	4.05 ± 0.48
60	77 ± 17	2.61 ± 0.58	102 ± 4	4.33 ± 0.94
120	56 ± 4	2.17 ± 0.37	97 ± 4	3.45 ± 0.23

*R.C.R. = Ratio State 3 respiration/State 4 respiration. All measurements made with an oxygen electrode. Variance is reported as standard deviations.

consumption of the isolated rat heart perfused with substrate-free medium was 37% (9) whereas the fall in hearts perfused with glucose was about 10%, in fairly good agreement with the observations on isolated mitochondria. The respiratory control ratios of mitochondria obtained from hearts perfused without substrate declined steadily during the 2 hour perfusion period from 4.3 at 10 minutes to 2.2 at 2 hours. Control hearts given glucose showed only a small change during the last hour from an average of 4.3 to 3.5. The change in the respiratory control ratios signify loosening of the coupling of electron transport to phosphorylation of ADP, most marked in the hearts deprived of substrate. Measurement of the P/O ratio under these conditions of respiration showed that mitochondria from substrate deprived hearts showed a steady decline in P/O ratio from 2.5 at 10 minutes to 1.8 at two hours (Table IV). The decrease was significant at 30 minutes ($P<0.05$) in these deprived hearts, whereas no significant changes occurred in the P/O ratios of mitochondria from hearts nourished with glucose during the 2 hour perfusion. Since free fatty acids are known to uncouple oxidative phosphorylation in isolated mitochondria, it occurred to us to investigate the free fatty acid content of hearts perfused with and without glucose for 2 hours. The data are shown in Table V. Perfusion without added substrate caused the cardiac free fatty acids to rise 100% above that observed when glucose was present as substrate. When fatty acid-free bovine serum albumin was added at a concentration of 0.2% to the non-nutrient medium, the changes in oxygen

Table IV. P/O Ratios of Mitochondria from Perfused Rat Hearts

Time Min.	No Substrate	8 mM Glucose
10	2.49 ± 0.22	2.74 ± 0.14
30	2.36 ± 0.14	2.55 ± 0.14
60	2.32 ± 0.14	2.67 ± 0.14
120	1.84 ± 0.17	2.57 ± 0.17

Pyruvate and malate were used as substrates. The oxygen consumption before and after addition of ADP was measured with an oxygen electrode. Five to 10 observations were made for each experimental condition. Variance is reported as standard deviations.

Table V. Free Fatty Acid Content of Rat Hearts After 30 Minutes of Perfusion

Substrate	N	FFA Content μmoles/g Dry Weight
Glucose	5	1.15 ± 0.26
None	5	2.19 ± 0.47

consumption and P/O ratio of mitochondria isolated at the end of 2 hours of perfusion were largely prevented (Table VI).

DISCUSSION

The isolated perfused rat heart cannot generate sufficient energy from its endogenous fuel supply to sustain its initial force of contraction more than 30 minutes. It is at this point that glycogen and glycolytic intermediates reach base-line values, the intracellular ATP:AMP ratio begins to fall, neutral lipids begin to disappear, and changes indicative of a loosening of coupling between electron transport and phosphorylation in the mitochondria of the heart first appear. Even during the first 30 minutes of perfusion with non-nutrient medium, creatine phosphate and glycolysis are contributing ~P to keep ATP levels adequate in the

Table VI. Effect of Fatty Acid-Free Bovine Serum Albumin Upon P/O Ratios and Respiration of Mitochondria from Perfused Rat Hearts (2°)

Exp.	Substrate	O_2 Consumption nmoles/min/mg	P/O Ratio
1	Glucose	97 ± 4	2.57 ± 0.17
2	None	56 ± 4	1.84 ± 0.17
3	None + BSA	73 ± 14	2.52 ± 0.09

Variance is reported as standard deviations.

pool interacting with the contractile proteins. Clearly, an exogenous fuel supply is essential from the beginning to sustain the work capacity of the isolated rat heart.

It is of considerable interest that not all classes of endogenous lipids are available to the heart as a source of fuel when the fuel supply is short. Triglycerides, and to a lesser extent, diglycerides supply fatty acids for oxidation by heart mitochondria, but only after exhaustion of available carbohydrate. Even though seventy percent of the neutral fatty acid esters disppeared over the two hour perfusion period, their contribution to energy requirement of the heart was insufficient to sustain contractility at initial levels. Myocardial phospholipids were not mobilized to any detectable extent, presumably because they are integral parts of vital membrane systems such as mitochondria, reticulum, and cell membrane.

Cardiac free fatty acids were found to be higher in hearts perfused with non-nutrient medium than with glucose, presumably because of the factors favoring mobilization of endogenous fatty acids in the former condition. This rise in FFA concentration appeared to be related to the diminished oxygen consumption and reduced P/O ratio observed in mitochondria isolated from hearts perfused with non-nutrient medium since the inclusion of fatty-acid free albumin in the medium prevented these changes. Free fatty acids have long been known to uncouple mitochondria when added _in vitro_ (7) and appear to be instrumental in loosening the coupling of liver mitochondria during gluconeogenesis (4).

These findings lead to the conclusion that even though endogenous fatty acid mobilization is inadequate in the isolated rat heart perfused with substrate-free media to sustain normal contractility, the free fatty acids which appear under these conditions have a further deleterious effect upon mitochondrial membranes, electron transport, and oxidative phosphorylation.

ACKNOWLEDGMENTS

The authors wish to thank their colleagues, Drs. Robert Hoeschen and Naranjan Dhalla for their valuable collaboration. Thanks are due to the U.S. Public Health Service for grants HE-09772 and PH 43-67-680. Catherine Bauer is a predoctoral research fellow supported by 2T1 GM446.

BIBLIOGRAPHY

(1) Bleehen, N.M. and Fisher, R.B., J. Physiol. 123, 260, 1954.

(2) Chance, B. and Williams, G.R., J. Biol. Chem. 217, 383, 1955.

(3) Chance, B., Williamson, J.R., Jamieson, D., and Schoener, B., Biochem. Zeit. 341, 357, 1965.

(4) Davis, E.J. and Gibson, D.M., J. Biol. Chem. 244, 161, 1969.

(5) Estabrook, R.W. and Maitra, P.K., Analyt. Biochem. 3, 369, 1962.

(6) Fisher, R.B. and Williamson, J.R., J. Physiol. 158, 86, 1961.

(7) Helinski, D. and Cooper, C., J. Biol. Chem. 235, 3573, 1960.

(8) Krebs, H.A. and Henseleit, K., Hoppe Seyl. Zeit. 210, 33, 1932.

(9) Olson, R.E. and Hoeschen, R.J., Biochem. J. 103, 796, 1967.

(10) Olson, R.E. and Dhalla, N.S., in Heart Failure, H. Reindell, J. Keul and E. Doll (Eds.), Georg Thieme Verlag, Stuttgart, 1968, p. 325.

(11) Rapport, M.M. and Alonzo, N., J. Biol. Chem. 217, 193, 1955.

(12) Shipp, J.C., Matos, O.E., Knizley, H., and Crevasse, L.E., Amer. J. Physiol. 207, 1231, 1964.

MYOCARDIAL METABOLISM IN ATHLETES

Joseph Keul

Medizinische Universitätsklinik, Freiburg in Breisgau

W. Germany

Numerous differences in heart and circulatory regulations have been described in trained as compared with untrained men (19). Physical training elicits anatomical as well as functional changes, in consequence of which the subject develops a higher working capacity. Biochemical changes in the blood and the tissues are also known to take place as a result of training (11 e.a.). The peculiarities of the cardiac metabolism in trained men are best discussed in comparison with the "untrained" heart.

Using intubation of the coronary sinus, the arterio-venous differences of the main substrates involved in the formation of available energy, as well as of the blood gases, were studied in six well-trained professional cyclists and in 27 healthy untrained subjects of the same age, at rest, during light and hard ergometric exercise, and also during recovery. At the highest work load - i.e. 200 Watt for the untrained and 300 Watt for the athletes - both groups attained a heart rate of the same magnitude, c. 180 beats/minute.

The oxygen pressure of the coronary venous blood is only slightly lower in the untrained at rest and at 100 Watt (Fig. 1), whilst at 200 Watt the difference amounts to approximately 5 mm Hg. The lowest pO_2 in athletes is found at 300 Watt; it lies higher, however, than the value found in normal subjects. The overall drop of the oxygen pressure, from the resting level to that at maximal load, is approximately the same in both groups, i.e. 3 mm Hg (3, 4, 5). Although low, the coronary venous resting oxygen pressure shows a slight but significant decrease. The critical pO_2 value of coronary venous blood, 7 mm Hg (1, 17), is attained neither in

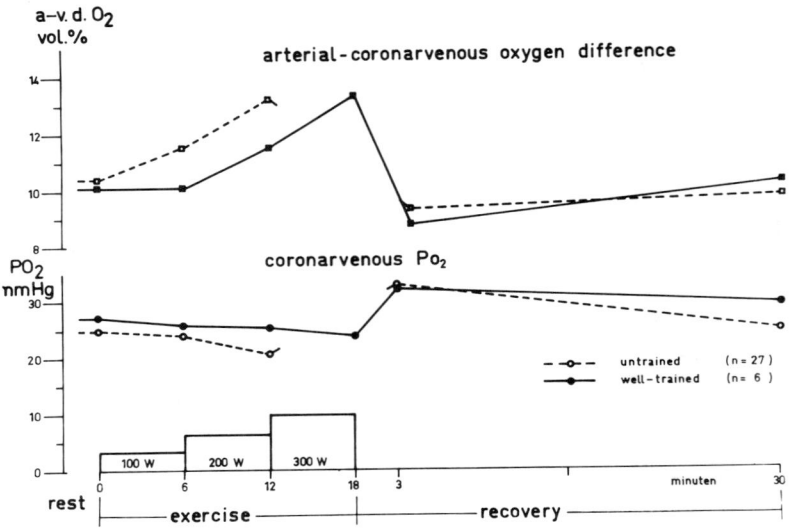

Fig. 1. The coronary arteriovenous oxygen difference and the coronary venous pO_2 in trained and untrained men at rest and during exercise (3, 4, 5).

untrained nor in athletes, even during maximal exercise. Likewise, in experiments involving a comparative oxygen deficiency effectuated at a simulated altitude up to 4250 m, the lower, critical value for oxygen pressure in the coronary venous blood was not attained (3). Thus, the supply of oxygen to the heart in healthy individuals, and particularly in trained subjects, is covered at any time, even in conditions of hard physical work. The fact that the oxygen supply to the heart is sufficient could also be deduced from the lactate/pyruvate ratio (9, 10, 11, 15, 18). After the end of the exercise, there is a significant rise of the coronary venous pO_2, as well as a corresponding decrease of the coronary arteriovenous oxygen difference; this could be interpreted as a "luxury" blood flow through the heart muscle after physical work. Such a luxury blood flow is particularly effective in intermittent exercise (3, 4, 5, 11).

The coronary arteriovenous difference of the oxygen content is essentially the same in trained as in untrained persons. At a 100 Watt work load no distinction could be found in the trained group, while the untrained showed a rise of about 10 per cent, which at 200 Watt became 30 per cent. In the trained subjects, such a rise of the arteriovenous oxygen difference occurs only at a higher work intensity. The increase of the arteriovenous oxygen difference during exercise is due mainly to hemoconcentration, a minor role being played by a higher extraction of oxygen from the coronary venous blood.

The main energy-supplying substrates for the energy metabolism of the heart muscle are glucose, lactate and free fatty acids; under physiological conditions, pyruvate, β-hydroxybutyrate and acetoacetate are of lesser importance for this process. Aminoacids are apparently not an important fuel for the energetic reactions of cardiac muscle (2, 8-15, 18). It was shown that for various substrates - particularly lactate, free fatty acids and ketone bodies - the arterial concentration represents the factor which decides their uptake by the cardiac muscle. The extraction and utilization of glucose is influenced by the insulin level of the blood, rather than by the arterial glucose level (18).

The concentration of glucose in the arterial blood falls slightly during exercise, no difference between trained and untrained subjects being observed in this respect. The lactate level rises during exercise in athletes only at 200 Watt, while this could be seen in untrained already at 100 Watt. The rise at maximal work load, with a heart rate of 180 beats/min, attains 9 µmoles/ml in untrained, while a level of only 7 µmoles/ml is found in the trained subjects (Fig. 2).

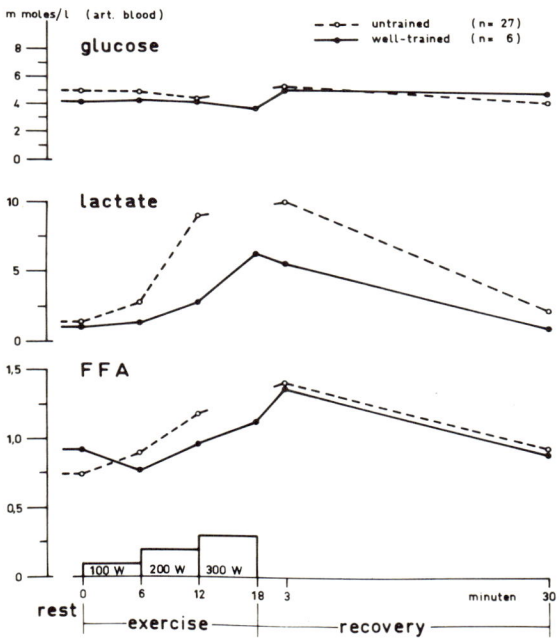

Fig. 2. The arterial concentration of glucose, lactate and free fatty acids at rest and during exercise.

The free fatty acids, too, show a higher level at submaximal and maximal load in untrained as compared to athletes. It could therefore be concluded that during this exertion, increased amounts of lactate and free fatty acids are available for the heart muscle. The same holds for less important metabolites, such as pyruvate and ketone bodies.

At rest, both the arteriovenous difference of glucose and the contribution of glucose to the energy production of the heart are similar in untrained and in athletes (Fig. 3). During exercise the glucose uptake of the heart, as well as the contribution of glucose to the oxidative metabolism are diminished in both groups, but more so in the untrained subjects. The pre-exercise values are attained after 30 minutes of recovery. The origin of the lowered utilization of glucose in the oxidative metabolism is to be sought in the strong rise of lactate in the arterial blood during exercise (Fig. 4). As a consequence, high amounts of lactate are extracted and oxidized by the heart muscle, as energy-supplying substrate. At submaximal work intensities, which are 100 Watt for the untrained and 200 Watt for the trained subjects, both the lactate level and its arteriovenous difference are the same in the two groups. During maximal load, athletes show a higher extraction of lactate, although the arterial level of this substrate is significantly lower than in untrained.

The arterial concentration may not be the sole factor determining lactate extraction by the heart muscle, since more lactate

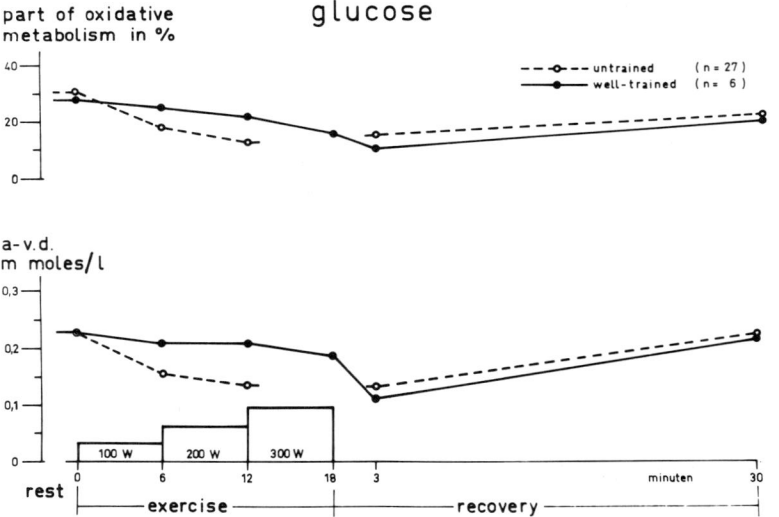

Fig. 3. The coronary arteriovenous difference of glucose and its part in oxidative metabolism of the heart during rest and exercise.

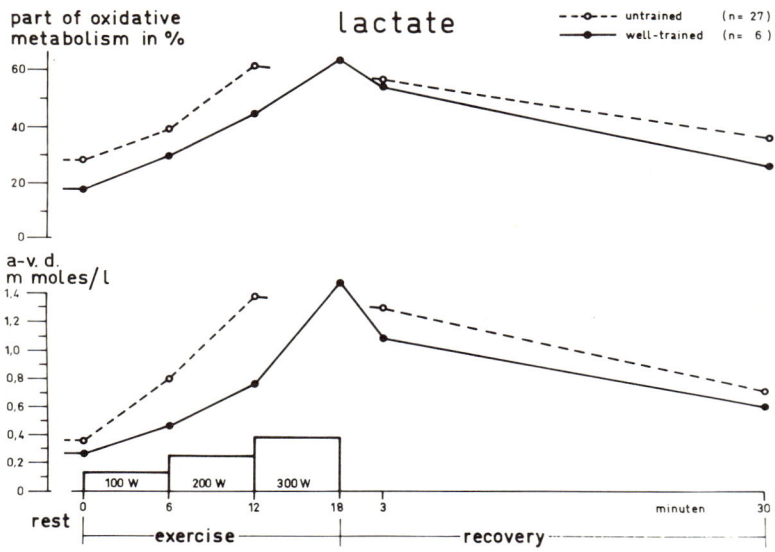

Fig. 4. The coronary arteriovenous difference of lactate and its part in oxidative metabolism of the heart during rest and exercise.

is extracted by the athlete's heart in spite of its lower arterial level at maximal work load. With an increase in work intensity, lactate becomes preferential as an energy-supplying substrate of the heart muscle. Furthermore, it was shown that patients with atrial septum defect, the cardiac work of whom was minimal during extracorporal circulation, have only low values for lactate extraction (16). In keeping with these findings, a steeper slope in the relationship: arterial level of lactate vs. arteriovenous difference, was found in athletes as compared to the untrained men.

Although the blood free fatty acids rise during exercise, the arteriovenous difference of this type of substrate sinks moderately (Fig. 5). The contribution of fats to the oxidative metabolism of the heart considerably decreased. In fact, the hematocrit value rises during exertion, thus increasing the oxygen-binding capacity of the blood; this results in a higher oxygen uptake by the heart muscle. On the other hand, the free fatty acids are only transported in the plasma, the relative volume of which decreases per ml of blood, if hematocrit rises. Consequently the free fatty acids, in spite of their increased concentration in the plasma, are not correspondingly extracted by the tissues. The preferential substrate for the energy supply in heart muscle during heavy physical work is lactate. This fact was also shown in animal experiments involving a simultaneous increase of lactate and free fatty acids (6); raising the lactate level results in lowering the free fatty acids uptake.

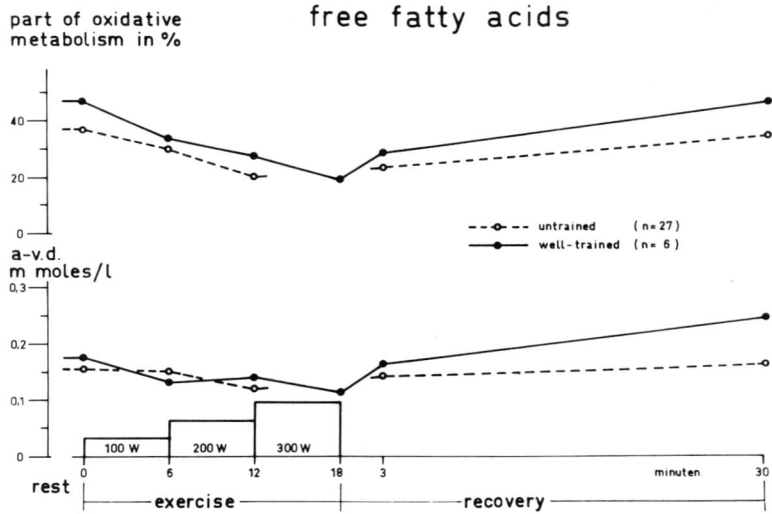

Fig. 5. The coronary arteriovenous difference of free fatty acids and their part in oxidative metabolism of the heart during rest and exercise.

During long-lasting exercise, for instance of two hours, changes of a different type are found both of the arterial concentrations of substrates and of their role in the oxidative metabolism of the heart. A marked rise of blood lactate occurs only during the first 30 minutes, while the heart rate continues to increase after this (Fig. 6). The glucose level decreases during exercise, while the free fatty acids show a strong rise. As a result of the high splitting rate of the triglycerides, the plasma glycerin markedly increases.

At the beginning of long-lasting exertion the contribution of the various substrates to the oxidative metabolism strongly resembles the previously discussed situation, i.e. a high contribution of lactate and a lower one of glucose and of the free fatty acids. However, during the later stages of exertion, the proportion of lactate as an energy-supplying substrate gradually sinks, reaching about 15 per cent, while the use of free fatty acids in the production of energy continues to increase, covering at the end of two hours of exercise up to 68 per cent of the total energy production. A higher contribution of lactate to the energy metabolism is not to be expected, since its arterial level is comparatively low. The uptake of glucose is likewise not increased, though its level does not change markedly. The free fatty acids, i.e. the substrate most

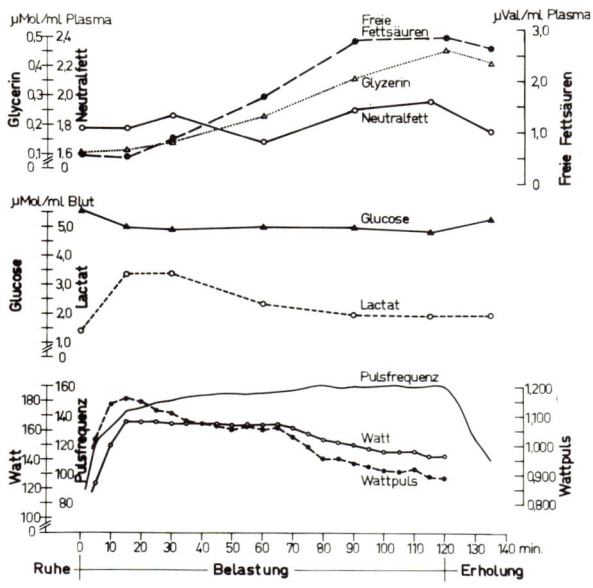

Fig. 6. Arterial concentration of energy-delivering substrates, heart rate and work load during long lasting exercise (two hours) (7).

easily available, become the main energy-supplying compounds for the human heart muscle. This is of importance, since glucose, a substrate which is indispensable for other organs, e.g. the brain, is in this way spared, while the free fatty acids, which are at hand in a much higher amount in the adipose tissue, are mobilised and used as substrates. Since in this type of exercise the maximal oxygen uptake is no longer the limiting factor, it seems unimportant that - related to one mole of oxygen - the energy formation during the degradation of the free fatty acids is about 15 per cent lower. Skeletal muscle, too, extracts in such conditions more free fatty acids and utilizes them as such. The mobilization and utilization of free fatty acids by the heart muscle is - according to experiments on which we are still working - more pronounced in athletes than in untrained subjects (7, 11).

REFERENCES

1. BRETSCHNEIDER, H.J. Sauerstoffbedarf und -versorgung des Herzmuskels. Verh. dtsch. Ges. Kreislaufforsch. 27: 32, 1961.
2. CARLSTEN, A., B. HALLGREN, R. JAGENBURG, A. SVANBORG, and L. WERKÖ. Myocardial metabolism of glucose, lactic acid, amino acids and fatty acids in healthy human individuals at rest and at different work loads. Scand. J. Clin. Lab. Invest. 13: 418, 1961.

3. DOLL, E., and J. KEUL. pO_2, pH, and pCO_2 in the coronarvenous and femoralvenous blood during exericse and hypoxia. In: Biochemistry of exercise. Basel-New York. S. Karger. 1969. p. 35.

4. DOLL, E., J. KEUL, H. STEIM, Ch. MAIWALD, and H. REINDELL. Über den Stoffwechsel des menschlichen Herzens. II. Mitt. Sauerstoff- und Kohlensäuredruck, pH, Standardbikarbonat und base excess im coronarvenösen Blut in Ruhe, während und nach körperlicher Arbeit. Pflügers Arch. Ges. Physiol. 282: 28, 1965.

5. DOLL, E., J. KEUL, H. STEIM, Ch. MAIWALD, and H. REINDELL. Über den Stoffwechsel des Herzens beim Hochleistungssportler. II. Mitt. Sauerstoff- und Kohlensäuredruck, pH, Standardbikarbonat und base excess im coronarvenösen Blut in Ruhe, während und nach körperlicher Arbeit. Z. Kreislaufforsch. 55: 248, 1966.

6. HIRCHE, H.J., and G. RÖHNER. Änderungen der Substrataufnahme des Herzmuskels bei induzierten Änderungen der arteriellen Substratkonzentration. Pflügers Arch. Ges. Physiol. 278: 408, 1963.

7. KEUL, J. (In preparation).

8. KEUL, J., and E. DOLL. The influence of exercise and hypoxia on the substrate uptake of human skeletal muscles. In: Biochemistry of exercise. Basel-New York. S. Karger. 1969. p. 41.

9. KEUL, J., E. DOLL, D. KEPPLER, H. HOMBURGER, H. KERN, and H. REINDELL. Über den Stoffwechsel des Herzens bei Hochleistungssportlern. I. Die Substratversorgung des trainierten Herzens in Ruhe, während und nach körperlicher Arbeit. Z. Kreislaufforsch. 55: 190, 1966.

10. KEUL, J., E. DOLL, D. KEPPLER, H. HOMBURGER, H. KERN, and H. REINDELL. Über den Stoffwechsel des menschlichen Herzens in Ruhe, während und nach körperlicher Arbeit. Pflügers Arch. Ges. Physiol. 282: 1, 1965.

11. KEUL, J., E. DOLL, and D. KEPPLER. Energy metabolism of muscle. München. Johann Ambrosius Barth Verlag. 1969.

12. KEUL, J., E. DOLL, D. KEPPLER, U. FLEER, and H. REINDELL. Über den Stoffwechsel des menschlichen Herzens. III. Der oxydative Stoffwechsel des Herzens unter verschiedenen Arbeitsbedingungen. Pflügers Arch. Ges. Physiol. 282: 43, 1965.

13. KEUL, J., E. DOLL, D. KEPPLER, U. FLEER, and H. REINDELL. Über den Stoffwechsel des Herzens bei Hochleistungssportlern. III. Der oxydative Stoffwechsel des trainierten menschlichen Herzens unter verschiedenen Arbeitsbedingungen. Z. Kreislaufforsch. 55: 477, 1966.

14. KEUL, J., E. DOLL, D. KEPPLER, U. FLEER, and H. REINDELL. Über den Stoffwechsel des Herzens bei Hochleistungssportlern. Das Verhalten der arteriocoronarvenösen Differenzen von Aminosäuren und Ammoniak in Ruhe, während und nach körperlicher Arbeit. Klin. Wschr. 44: 881, 1966.

15. KEUL, J., D. KEPPLER, and E. DOLL. Lactate-pyruvate ratio and its relation to oxygen pressure in arterial, coronarvenous and femoralvenous blood. Arch. Int. Physiol. Biochem. 75: 573, 1967.

16. KRAUS, H., W. OVERBECK, E. DOLL, and U. FLEER. Klin. Wschr. 42: 890, 1964.

17. LOCHNER, W., and M. NASSERI. Über den venösen Sauerstoffdruck, die Einstellung der Coronardurchblutung und den Kohlenhydratstoffwechsel des Herzens bei Muskelarbeit. Pflügers Arch. Ges. Physiol. 269: 407, 1959.

18. OPIE, L.H. Metabolism of the heart in health and disease. I-III. Amer. Heart J. 76: 685, 1968; 77: 100, 1969; 77: 383, 1969.

19. REINDELL, H., H. KLEPZIG, H. STEIM, K. MUSSHOFF, H. ROSKAMM, and E. SCHILDGE. Herz, Kreislaufkrankheiten und Sport. München. 1960.

MYOCARDIAL METABOLISM IN MAN AT REST AND DURING PROLONGED EXERCISE

B.W. Lassers, L. Kaijser, M.L. Wahlqvist and L.A. Carlson

Department of Geriatrics, University of Uppsala, Uppsala, Department of Clinical Physiology, Karolinska Sjukhuset and King Gustaf V Research Institute, Stockholm, Sweden

The technique of coronary sinus catheterisation with measurement of arterio-venous differences across the heart has provided considerable information about the uptake and utilization of plasma substrates by the myocardium in resting man. There is general agreement that in the resting fasting state plasma free fatty acids (FFA) and glucose are the principle substrates for myocardial oxidation with smaller contributions being made by plasma lactate, pyruvate and ketone bodies (1,2,3,4,5,6). However, despite the finding of triglyceride uptake by the isolated animal heart (7,8,9,10,11,12), and the demonstration of the presence of lipoprotein lipase activity in the human heart (13,14), coronary sinus catheterisation studies in man have failed to demonstrate satisfactorily the uptake of plasma triglycerides by individual hearts.

It is likely that this inability to demonstrate the disappearance of triglycerides across the heart by coronary sinus sampling in man has been due to the methodological difficulties of measuring the very small amounts of triglyceride which could provide a substantial proportion of the heart's energy supply. Thus, in an individual with a normal arterial triglyceride concentration of 1000 μmol/l plasma, there would need to be an arterio-venous difference of 150 μmol/l plasma in order for it to be detected using one of the current chemical methods of triglyceride estimation with a standard error of around five per cent for a determination. This difference is equal to about 90 μmol/l blood if the haematocrit is normal. Since the oxygen equivalent of triglyceride fatty acid is about 73.5, complete oxidation of 90 μmol of triglyceride fatty acid would require 6660 μmol of oxygen. However, the average arterio-venous difference of oxygen across the heart is about 5000 μmol/l blood. Thus, oxidation of 90 μmol of triglyceride fatty acid would need 133 per cent of the heart's oxygen extraction per litre of blood: that is, triglyceride would have an oxygen extraction ratio (OER) of 133 per cent. Since it is known that plasma FFA and carbohydrate substrates already account for a substantial fraction of the myocardial oxygen uptake, this minimum arterio-venous difference of triglyceride detectable using a method of estimation with the usual error of around five per cent, is far greater than could be expected. Even using a very much more precise method with a standard error of only one per cent, an arterio-venous difference of 19 μmol/l blood would be required and oxidation of this amount of triglyceride fatty acid would produce an OER of 28 per cent. Thus, because of the very high oxygen equivalent of triglyceride fatty acid, a very precise method of triglyceride estimation is required in order to be able to detect the quantities of triglyceride which it might be reasonable to expect the heart to extract.

The first purpose of the present study was, therefore, to examine the problem of the uptake of plasma triglyceride by the myocardium in resting fasting man by using both a radio-isotope technique and a precise chemical method of triglyceride estimation. The second purpose was then to use these techniques to investigate the utilization of plasma triglycerides and other substrates in man during prolonged exercise in the fasting state. For, although myocardial substrate utilization has been studied in man performing exercise of four to 20 minutes duration (4,15,16), there have been no reports of the effects of prolonged exercise.

METHODS

Twenty-eight unsedated, healthy male subjects between the ages of 23 and 53 years were studied in the resting state after an overnight fast with catheters in the coronary sinus and a brachial artery. In nine of these subjects measurements were made both at rest and during the last five minutes of a two hour period of supine leg exercise on a bicycle ergometer at a load of approximately 50 per cent of the rate of work which the subject was able to perform at a heart rate of 170/min. Arterio-venous differences in concentration across the heart of FFA (17), triglyceride (18), glucose (19), lactate (20), pyruvate (21) and glycerol (22) were measured chemically. Plasma triglycerides were estimated in triplicate on 10 extracts from each blood sample. The average figure for the standard error of the mean value for the 10 extracts of each sample was 0.8 per cent.

In addition to the chemical measurements, arterio-venous differences in FFA and triglyceride were determined by using a continuous infusion of albumin-bound ^{3}H palmitate as a tracer for plasma FFA and to produce endogenous labelling of plasma triglycerides (23). FFA and triglyceride were separated by thin layer chromatography and then counted in a Packard 3375 liquid scintillation spectrometer (24).

The oxygen content of the blood samples was calculated from the haemoglobin concentration, the oxygen saturation measured spectrophotometrically (25), and the oxygen tension measured with a polarographic electrode.

ARTERIAL-CORONARY SINUS CONCENTRATION DIFFERENCES IN THE RESTING FASTING STATE

Table I presents the findings in the resting fasting state. The arterial concentrations of all substrates were within normal limits. The myocardial uptake of the substrates is presented in the table both as the arterial-coronary sinus difference in concentration and as the OER. Carbohydrate substrates accounted for

SUBSTRATE	ARTERIAL CONCENTRATION	ARTERIAL-CORONARY SINUS DIFFERENCE IN CONCENTRATION	OXYGEN EXTRACTION RATIO
GLUCOSE	3640±107	151±24	18±3
LACTATE	645±44	127±22	4±1
PYRUVATE	50±4	6±4	0±0.2
FFA	706±38		
Chemical		171±12	49±3
Isotope		243±10	69±3
TRIGLYCERIDE	1080±88		
Chemical		16±4	14±4
Isotope		24±7	23±6

Table I: Myocardial metabolism in resting fasting man. Arterial concentrations, arterial-coronary sinus differences and oxygen extraction ratios are presented as the mean ± SEM for 28 subjects. Concentrations and arterial-coronary sinus differences are in μmol/l plasma except for glucose which is in μmol/l blood. Arterial-coronary sinus differences in concentration (A-CS) for FFA and triglyceride were measured by both chemical and radio-isotope techniques.

a total OER of 22 per cent. There was a considerable difference in the uptake of FFA as estimated by chemical and isotope techniques. This has also been observed by other workers (6) and is due to the fact that the coronary sinus FFA specific activity is lower than the arterial specific activity.

In addition to the uptake of carbohydrate substrates and FFA, a significant arterio-venous difference across the heart of triglyceride was found by chemical estimation in 13 subjects and by isotope measurements in 17 subjects. The average OER for all 28 subjects was 14 per cent chemically and 23 per cent isotopically. A possible explanation for the higher value found by the isotope technique is that in calculating the arterio-venous difference in concentration from the arterio-venous difference in triglyceride radio-activity, it has been assumed that the labelled triglyceride is evenly distributed throughout all of the plasma lipoprotein fractions. It is probable, however, that under these conditions the major part of the labelled triglyceride was located in the very low density lipoprotein fraction which contains about 50 per cent of the total plasma triglycerides in normal man (J. Boberg, personal communication). If this is the case, then the arterio-venous difference in triglyceride concentration determined from the arterio-venous difference in radio-activity would be half of that which has been calculated. The myocardial uptake of triglyceride would then be about 12 ± 4 (SEM) µmol/l plasma and the OER about 12 ± 3 (SEM) per cent. The chemical and isotopic techniques would then agree much more closely.

ARTERIAL-CORONARY SINUS CONCENTRATION DIFFERENCES DURING PROLONGED EXERCISE

After two hours of exercise, changes were found in the coronary sinus oxygen saturation, the arterial concentrations of substrates and in the pattern of myocardial substrate uptake. In addition, a number of subjects showed a significant myocardial production of free glycerol.

Figure 1 shows the changes in oxygen saturation of the coronary sinus blood after two hours of exercise. The saturation fell in all nine subjects from an average resting value of 34.6 ± 1.5 (SEM) per cent to 24.7 ± 1.2 (SEM) per cent. This fall, equal to 28 per cent of the resting value, was highly significant ($p < 0.001$). Since the arterial oxygen saturation did not change with exercise, the reduced coronary sinus saturation indicates an increased myocardial extraction of oxygen from each litre of blood. The average arterial-coronary sinus difference in oxygen content was 4875 ± 189 (SEM) µmol/l blood at rest and 5917 ± 170 (SEM) µmol/l blood during exercise.

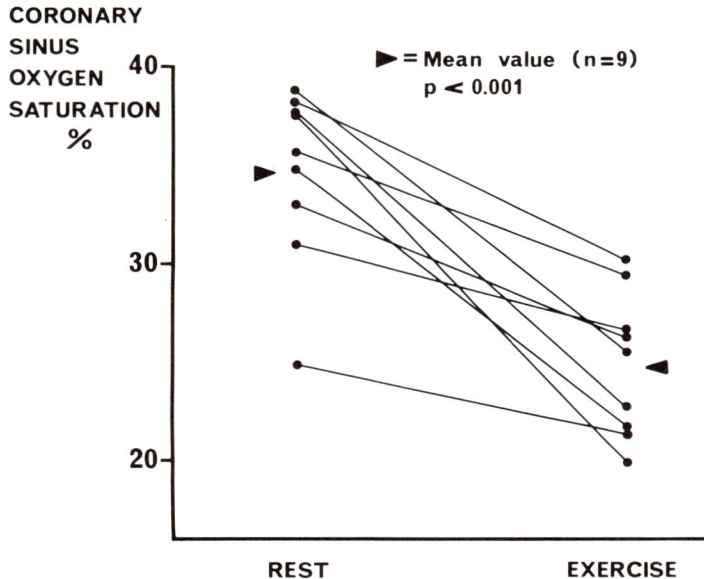

Figure 1: Changes in coronary sinus oxygen saturation after approximately two hours of supine leg exercise on a bicycle ergometer.

The effect of exercise on the arterial concentration of the various substrates is presented in Table II. Glucose concentration

SUBSTRATE	ARTERIAL CONCENTRATION	
	Rest	Exercise
GLUCOSE	3916±203	3258±167
LACTATE	603±91	1272±194
PYRUVATE	47±5	86±6
FREE FATTY ACIDS	614±49	1366±106
TRIGLYCERIDE	1127±157	1123±157

Table II: Effect of two hours of supine leg exercise on arterial concentrations of substrates. Concentrations are in μmol/l plasma except for glucose which is in μmol/l blood. Values are means±SEM for nine subjects.

fell to 83 per cent of the resting level while the concentrations of lactate and pyruvate doubled. The FFA turnover rate, which was estimated from the isotope data, increased an average of 2.8 times with exercise and the arterial concentration of FFA rose to 222 per cent of the resting level. There was no significant change in the arterial concentration of triglycerides with exercise.

Table III shows the effect of prolonged exercise on both the arterial-coronary sinus difference in concentration and the oxygen extraction ratios. The fall in arterial glucose concentration was

SUBSTRATE	ARTERIAL-CORONARY SINUS DIFFERENCES IN CONCENTRATION		OXYGEN EXTRACTION RATIOS PER CENT	
	Rest	Exercise	Rest	Exercise
GLUCOSE	191±59	97±32	24±8	9±3
LACTATE	84±32	293±113	3±1	9±3
PYRUVATE	-2±5	20±9	0±0.2	1±0.2
FFA				
Chemical	171±19	209±9	49±4	50±3
Isotope	242±18	229±13	70±4	55±4
TRIGLYCERIDE				
Chemical	15±10	17±10	14±9	12±7
Isotope	34±19	10±14	30±17	8±10

Table III: Effect of two hours of supine leg exercise on arterial-coronary sinus differences in substrate concentrations and on oxygen extraction ratios. Differences in concentration are in μmol/l plasma except for glucose which is in μmol/l blood. Values are means ±SEM for nine subjects except for the isotope estimate of triglyceride at rest where measurements were made in only five subjects.

accompanied by a fall in the OER from 24 per cent to nine per cent and the rise in arterial lactate concentration was accompanied by a rise in the OER of this substrate from three per cent to nine per cent. During exercise pyruvate had an OER of only about one per cent. Thus, during prolonged exercise in the fasting state the total OER for plasma carbohydrate substrates fell from 27 per cent to 19 per cent.

Although the arterial concentration of FFA more than doubled, there was only a small increase in FFA extraction as estimated chemically and the OER did not change. The arterio-venous difference estimated isotopically fell and the OER decreased from 70 per cent to 55 per cent.

In the case of the chemical estimates of triglyceride uptake, there was a significant increase in uptake in one of the nine subjects and a significant fall in uptake in another. Two of the five subjects in whom isotope measurements of triglyceride uptake were made both at rest and during exercise showed a significant fall in uptake. However, the average triglyceride uptakes for the group as a whole, estimated either chemically or isotopically, did not change significantly.

Figure 2 shows the effect of exercise on the average arterial concentration of free glycerol and on the arterial-coronary sinus difference in free glycerol concentration. After two hours of exercise the glycerol concentration increased from 48 ± 5 (SEM) µmol/l plasma to 278 ± 36 (SEM) µmol/l plasma. With exercise five of the nine subjects showed a significant increase in the production of glycerol across the heart, suggesting increased hydrolysis of lipid. Possible sources of this glycerol are the plasma triglycerides and cardiac lipid stores. Since no increase in the arterial-coronary sinus difference in triglyceride was detected during exercise, it seems more likely that the glycerol arose from hydrolysis of cardiac lipid stores rather than from circulating triglyceride.

SUMMARY

Myocardial substrate utilization was studied in 19 normal, fasting men at rest and in nine others both at rest and during the last few minutes of a two hour period of submaximal exercise. Arterial-coronary sinus differences in concentration were measured chemically for glucose, lactate, pyruvate, glycerol, free fatty acid (FFA) and triglycerides, and radio-isotopically for FFA and triglycerides.

At rest the total oxygen extraction ratio (OER) for plasma carbohydrate substrates averaged 22 per cent, and that for FFA, 49 per cent chemically and 69 per cent isotopically. A statistically significant arterial-coronary sinus difference in triglycerides was detected in 13 of the 28 subjects chemically and in 17 subjects isotopically.

After two hours of exercise the nine subjects showed a significant fall in coronary sinus oxygen saturation and increase in myocardial oxygen extraction per litre of blood. The arterial con-

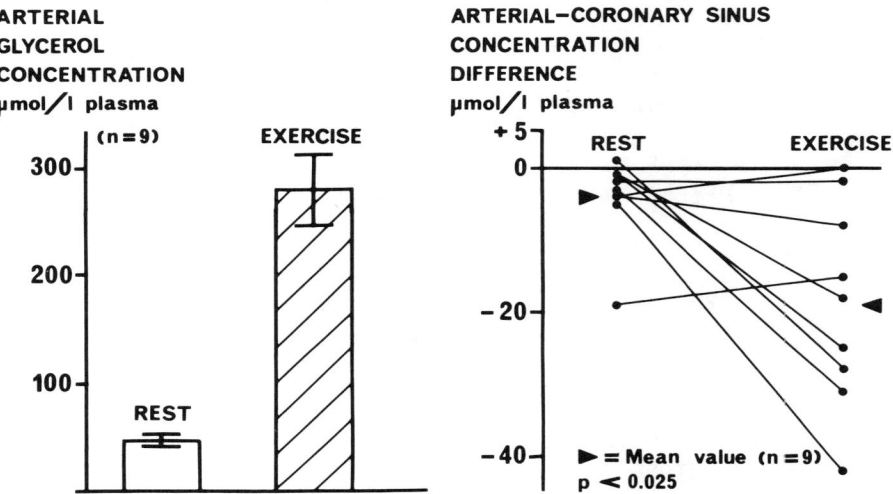

Figure 2: Effect of two hours of supine leg exercise on the average arterial concentration and the arterial-coronary sinus differences in concentration of free glycerol in nine subjects.

centration of glucose fell slightly and the concentrations of lactate, pyruvate and FFA approximately doubled, while the concentration of triglycerides did not change significantly. The total OER for plasma carbohydrate in these nine subjects fell from 27 per cent to 10 per cent during exercise. The OER for FFA estimated chemically did not change, but that estimated isotopically fell from 70 per cent to 55 per cent. No significant change in the average uptake of plasma triglycerides was detected either chemically or isotopically. During exercise five subjects showed a significant increase in the myocardial production of glycerol suggesting hydrolysis of cardiac lipid stores.

ACKNOWLEDGEMENTS

This work was supported by a grant from the Swedish Medical Research Council (No. 19x - 204 - 07). B.W.L. was in receipt of a Wellcome-Swedish Travelling Fellowship and M.L.W. was in receipt of a Life Insurance Medical Research Fund of Australia and New Zealand Overseas Research Fellowship.

REFERENCES

1. Bing, R.J., Siegel, A., Vitale, A., Balboni, F., Sparks, E., Taeschler, M., Klapper, M., Edwards, S.: Metabolic studies on the human heart in vivo. I. Studies on carbohydrate metabolism of the human heart. Amer. J. Med. 15: 284, 1953.

2. Bing, R.J., Siegel, A., Ungar, I., Gilbert, M.: Metabolism of the human heart. II. Studies on fat, ketone and amino acid metabolism. Amer. J. Med. 16: 504, 1954.

3. Goodale, W.T., Olson, R.E., Hackel, D.B.: The effects of fasting and diabetes mellitus on myocardial metabolism in man. Amer. J. Med. 27: 212, 1959.

4. Harris, P., Jones, J.H., Bateman, M., Chlouverakis, C., Gloster, J.: Metabolism of the myocardium at rest and during exercise in patients with rheumatic heart disease. Clin. Sci. 26: 145, 1964.

5. Rudolph, W., Maas, D., Richter, J., Hasinger, F., Hofmann, H., Dohrn, P.: über die Bedeutung von Acetacetat und β-Hydroxybutyrat im Stoffwechsel des menschlichen Herzens. Klin. Wchnschr. 43: 445, 1965.

6. Most, A.S., Brachfeld, N., Gorlin, R., Wahren, J.: Free fatty acid metabolism of the human heart at rest. J. Clin. Invest. 48: 1177, 1969.

7. Gousios, A., Felts, J.M., Havel, R.J.: The metabolism of serum triglycerides and free fatty acids by the myocardium. Metabolism 12: 75, 1963.

8. Delcher, H.K., Fried, M., Shipp, J.C.: Metabolism of lipoprotein lipid in the isolated perfused rat heart. Biochim. Biophys. Acta 106: 10, 1965.

9. Crass, M.F., Meng, H.C.: The removal and metabolism of chylomicron triglycerides by the isolated perfused rat heart: the role of a heparin released lipase. Biochim. Biophys. Acta 125: 106, 1966.

10. Kreisberg, R.A.: Effect of diabetes and starvation on myocardial triglyceride and free fatty acid utilization. Amer. J. Physiol. 210: 379, 1966.

11. Enser, M.B., Kunz, F., Borensztajan, J., Opie, L.H., Robinson, D.S.: Metabolism of triglyceride fatty acids by perfused rat heart. Biochem. J. 104: 306, 1967.

12. Scheur, J., Olson, R.E.: Metabolism of exogenous triglyceride by the isolated perfused rat heart. Amer. J. Physiol. 212: 301, 1967.

13. Schnatz, J.D., Ormsby, J.W., Williams, R.H.: Lipoprotein lipase activity in human heart. Amer. J. Physiol. 205: 401, 1963.

14. Muir, J.R.: The regional production of lipoprotein lipase in man. Clin. Sci. 34: 261, 1968.

15. Carlsten, A., Hallgren, B., Jagenburg, R., Svanborg, A., Werkö, L.: Myocardial metabolism of glucose, lactic acid, aminoacids and fatty acids in healthy human individuals at rest and at different work loads. Scand. J. Clin. Lab. Invest. 13: 418, 1961.

16. Keul, J., Doll, E.: The influence of exercise and hypoxia on the substrate uptake of human heart and human skeletal muscles. in Biochemistry of Exercise. Medicine and Sport. Vol. 3, Basel/New York, 1969, p. 41.

17. Trout, D.L., Estes, E.H., Friedberg, S.J.: Titration of free fatty acids of plasma: a study of current methods and a new modification. J. Lipid Res. 1: 199, 1960.

18. Kessler, G., Lederer, H.: Fluorometric measurements of triglycerides. in Automation in Analytical Chemistry. New York, 1966, p. 341.

19. Hjelm, M.: Enzymatic determination of hexoses in blood and urine. Scand. J. Clin. Lab. Invest. 18: suppl. 192, 85, 1966.

20. Lundholm, L., Mohme-Lundholm, E., Vamos, N.: Lactic acid assay with L(+) lactic acid dehydrogenase from rabbit muscle. Acta Physiol. Scand. 58: 243, 1963.

21. Bücher, T., Czok, R., Lamprecht, W., Latzko, E.: Pyruvat. in Metoden der enzymatischen Analyse. Verlag Chemie, Weinheim, 1962, p. 253.

22. Chernick, S.S.: Determimination of glycerol in acyl glycerols. in Methods in Enzymology. Vol. 19. New York, London, 1969, p. 627.

23. Boberg, J., Carlson, L.A., Freyschuss, U.: Studies on the total and splanchnic turnover of plasma free fatty acids and plasma triglycerides in man by means of isotope and chemical methods. in Progr. Biochem. Pharmacol. Vol. 5. Basel/New York, 1969, p. 149.

24. Boberg, J.: Separation of labeled plasma and tissue lipids by thin-layer chromatography. A quantitative methodological study. Clin. Chim. Acta 14: 325, 1966.

25. Holmgren, A., Pernow, B.: Spectrophotometric measurement of oxygen saturation of blood in the determination of cardiac output. A comparison with the Van Slyke method. Scand. J. Clin. Lab. Invest. 11: 143, 1959.

HABITUAL PHYSICAL ACTIVITY: AEROBIC POWER AND BLOOD LIPIDS

Gunnar Grimby, Lars Wilhelmsen, Per Björntorp,
Bengt Saltin, and Gösta Tibblin
From the Department of Clinical Physiology and
Medical Service I, University of Göteborg, Sahlgrenska
sjukhuset, S-413 45, Göteborg, Sweden

Among the different risk factors for coronary heart disease (CHD), high blood lipids and low physical activity have aroused a considerable interest and in several studies, low levels of blood lipids and high physical activity have been demonstrated to be associated with lower risk for CHD (for review see 7).

There are many reports on serum cholesterol and to a less extensive extent on serum triglycerides in populations with different degree of physical activity or with a comparison of the values before and after physical conditioning in healthy men. One problem in all these studies is that dietary habits may influence the blood lipids and be different in groups with different physical activity (15). There may also be constitutional factors which influence both the blood lipids and the degree of physical activity. The whole problem is very complex and it is therefore not astonishing that the literature about serum lipids and physical exercise is confusing. The scope of the present paper is to review some of the literature in the field and to present own recent observations, where the physical exercise factor in the populations is elucidated both with questionnaires and exercise tests.

In relating blood lipids to physical activity, two different approaches have been used, to study populations with different degree of habitual physical activity (Table I), and to follow the effect of training usually during fairly short periods (Table II).

Reduction of fasting <u>triglycerides</u> after 6 months of

Table I
Serum cholesterol and triglycerides in populations with different physical activity

	Population	Method	Effect of activity	
			Cholesterol	Triglycerides
Karvonen (14)	Lumber worker / General pop.	Caloric intake	0	
Hernberg (12)	Business men	Exercise test	– (30-49 years)	
Shane (23)	Flying personal	Exercise test	–	–
Saltin and Grimby (21)	Active athletes / Former	Questionnaire + exercise test	0	–
Carlsson and Lindstedt (4)	Health control	Questionnaire Occupation Leisure time	0 –	0 – (men)
Mc Donough et al. (18)	YMCA	Questionnaire	0	
54 years old men (present report)	General population	Questionnaire Exercise test	0 0	– –
	Athletes (52-56 years) / General pop. (54 years)	Questionnaire + exercise test	–	–

– denotes lower values with increased activity

Table II
Effect on serum cholesterol and triglycerides of a period of physical training

	Training period weeks	Effect on cholesterol	Effect on triglycerides	
Montoye et al. (19)	13	−		by decreasing body weight
Holloszy et al. (13)	26	0	−	
Fitzgerald et al. (8)	2	0		
Campbell (1)	10	−		dynamic training
		0		static training
Goode et al. (9)	2	0	−	
Naughton and Mc Coy (20)	34	0		
Campbell and Lumsden (2)	10	(−)	(−)	only obese
Mann et al. (17)	26	(−)	0	early periods
Siegel et al. (24)	15	−	(−)	not sign.
Kilbom et al. (16)	9	−	−	

− denotes reduced values after training

training was first convincingly shown by Holloszy et al. (13).
The diet and body weight were kept constant. It seems from
that study that a person have to be fairly regular in his
added activity, with training periods every two to three days,
to maintain a lower level. Shane (22) and Carlson and Lindstedt (4) showed in their population studies that groups with
increased habitual physical activity have lower triglycerides.
Comparing the triglyceride values before and after a training
period (Mann et al. (17), however, found a rise after training. They ascribed this change to an increase in food intake
during the 6 months´ training period. Their subjects expended
400-900 extra calories per day during the training and remained in weight balance. An increase was also seen in the
control subjects. Siegel et al. (24) on the other hand reported a decline in the triglyceride values after 15 weeks of
training in sedentary men, aged 32-59 years, but the difference
was not significant. In the studies by Kilbom et al. (16) and
by Siegel et al. (24), the body weight did not change significantly with training. Goode et al. (9) described decrease in
serum triglycerides with 2 weeks training in subjects on diet
free of animal fat and in caloric balance.

The literature on serum <u>cholesterol</u> is more confusing. There
are some population studies, which show somewhat lower values in
physical active (or physical fit) persons than in inactive
ones, whereas most investigations have not been able to
demonstrate such a tendency. Hernberg (12), however, studied
a large group of businessmen in Finland and assumed fairly
homogenous socio-economical conditions. He could then demonstrate a negative correlation between measured capacity
for physical work and serum cholesterol. Also when the effect
of shorter periods of training have been studied several
reports do not demonstrate a change in serum cholesterol
values with increased activity and increased physical working
performance. The body weight is in none of these studies reported to change with training. Taylor et al. (1961) in a
training study with controlled diet concluded that serum
cholesterol during moderate activity is controlled by the
per cent of fat in the diet up to approximately 40 % of total
calories. After this point a moderate physical activity
results in a lower cholesterol than expected. Holloszy and
coworkers (13) stated in their reports that physical training
at least could prevent the rise in serum cholesterol, which
accompanies increased coloric intake. The body composition of
the subjects in the conditioning program should be taken into
account. A significant reduction in cholesterol with 10 weeks
of training was found by Campbell and Lumsden (2) in obese
young men, but not in muscular or slim groups. When high
intensity of physical training with dynamic activities such
as running is used as in the study by Campbell (1), Kilbom

et al. (16) and Mann et al. (17) and in active endurance athletes, a slight reduction in cholesterol values may occur. Seasonal variation in cholesterol level is a factor, which also must be taken into account (6). In the study by Kilbom et al. (16), however, the training started in September and the subjects were restudied 2 months later.

A POPULATION STUDY ON 54 YEARS OLD MEN

A representative sample of men in Göteborg, born in 1913 on days, which numbers are equally divisible by three, was selected and studied for the first time in 1963. 88 % of the selected men (n = 855) took part in the study (26). The same population was restudied in 1967, when 803 men came. At that time an exercise test on a bicycle ergometer and a questionnaire about the physical activity during working hours and at leisure time were included in the program.

Methods

The physical activity was evaluated in ten-years periods from 20 to 50 years and thereafter between 50 and 53 years and for the previous year (Wilhelmsen et al., to be published). Four scores were used for occupational and leisure time activity respectively, according to Saltin and Grimby (21).
The activity groups are:

Occupational physical activity.

Group I	Group II	Group III	Group IV
Predominantly sedentary, sitting: desk worker, watch maker, sitting assemblyline worker light goods)	Sitting or standing, some walking: cashier, general office worker, light tool and machinery worker, foreman	Walking, some handling of material: mailman, waiter, construction worker, heavy tool and machinery worker	Heavy manual work: lumberjack, dock worker, stone mason, farm worker, ditch digger

Spare time physical activity.

Group I	Group II	Group III	Group IV
Almost completely inactive: reading, TV watching, movies, etc.	Some physical activity during at least 4 hours per week: riding a bicycle or walking to work, walking or skiing with the family, gardening	Regular activity: such as heavy gardening, running, calisthenics, tennis, etc.	Regular hard physical training for competition in running events, soccer, racing, European handball, etc. Several times per week.

In this presentation only the physical activity the year proceeding the interview (at 53-54 years of age) is considered.

Exercise test was performed with step-wise increasing work loads in the sitting position on a bicycle ergometer. In a subsample of 53 men the maximal oxygen uptake was determined (11), in the other subjects maximal heart rate was determined, if there was no contraindication to exhaustive exercise. From heart rates at rest and at submaximal and maximal work loads, the maximal oxygen uptake was then predicted according to a formula, based on the results from the subsample of the 53 men (11). In altogether 641 men a determined or predicted value of maximal oxygen uptake is available.

Serum cholesterol was determined according to Cramér and Isaksson (5) and serum triglycerides according to Carlsson (3).

Results

The distribution of the predicted maximal oxygen uptake in 641 men in different physical activity groups is shown in table III. The maximal oxygen was somewhat higher with increased activity. The difference is significant when groups 3+4 is compared with groups 1+2 for leisure time and occupational time activity, respectively. There are no differences between groups 1 and 2 or between 3 and 4. It is worth to notice that the relative number of participants in the maximal exercise test is different in different activity groups with the least percentage participation rate in the groups with low physical activity. This means that the real differences in aerobic power probably is somewhat larger. There was no difference in body weight between the different activity groups and no major difference in dietary habits. Smokers were less common in the groups with high leisure time activity.

The concentration of serum cholesterol in relation to physical activity is presented in table IV. There is no effect of the different degree of physical activity neither on leisure time nor during work. There is no significant correlation between serum cholesterol and maximal oxygen uptake.

The concentration of serum triglycerides in relation to physical activity is presented in table V. With increasing activity during occupation or leisure time, the values are lower although not significantly. However, group 4 leisure time activity comprises only 3 persons altogether. There were significantly fewer men with triglycerides in the 5:th quintile in those with leisure time activity 3-4, but no difference with respect to occupational activity. It should also be noted that the maximal oxygen uptake in group 3 leisure time activity is only a few per cent higher than in group 1+2. This value can be compared with the effect of intense physical training, which is definitely larger as discussed below.

The population was devided into quintiles with respect to maximal oxygen uptake. There is a tendency to lower serum triglyceride values with increasing maximal oxygen uptake (Table VI). However, this observation is weakened by the fact that the triglycerides were determined at the study 4 years earlier. On the other hand, we have not found evidence for large differences in training habits within these years.

Table III

Maximum oxygen uptake (ml/kg body weight) in relation to physical activity. Random sample of 54 year old men in Göteborg, Sweden. N = 803

	Score	Leisure time activity 1 - 2	Leisure time activity 3 - 4	$\sum$
Occupat. Activity	1-2 $\bar{X}$ s_x n	29.8 3.5 303	31.1 3.4 118	30.2 3.5 421
	3-4 $\bar{X}$ s_x n	30.9 3.6 162	31.3 3.8 58	31.0 3.7 220
	$\sum$ $\bar{X}$ s_x n	30.2 3.6 465	31.2 3.5 176	30.5 3.6 641

(1-2 vs 3-4 Occupat.: t=2.90, p<0.005)

t = 3.25
p < 0.005

Table IV

Serum cholesterol (mg/100 ml) in relation to physical activity. Random sample of 54 year old men in Göteborg, Sweden.
N = 855

	Score		Leisure time activity 1-2	Leisure time activity 3-4	$\sum$	
Occupat. activity	1-2	$\bar{x}$ s_x n	268 48 313	274 46 106	270 48 419	n.s.
	3-4	$\bar{x}$ s_x n	270 42 155	270 46 45	270 42 200	
	$\sum$	$\bar{x}$ s_x n	269 46 468	273 46 151	270 46 619	

t = 0.93
n.s.

Table V

Serum triglycerides (mMol/l) in relation to physical activity. Random sample of 50 year old men in Göteborg, Sweden.
N = 855

	Score		Leisure time activity 1-2	Leisure time activity 3-4	$\sum$	
Occupat. activity	1-2	$\bar{x}$ s_x n	1.30 0.74 382	1.18 0.72 192	1.26 0.73 574	t=1.13 n.s.
	3-4	$\bar{x}$ s_x n	1.21 0.64 128	1.16 0.56 61	1.20 0.61 189	
	$\sum$	$\bar{x}$ s_x n	1.27 0.71 510	1.18 0.68 253	1.26 0.82 763	

t=1.70
n.s.

PHYSICAL WELL-TRAINED MIDDLE-AGED MEN

In 1966 Grimby and Saltin (10) reported the results from a study on the cardiopulmonary function in still active middle-aged athletes with at least 20 years of hard endurance training. They were still active and successful in long-distance cross-country running (sw. orientering) and skiing. The training program consisted of at least 1-1.5 hours running 2-6 times a week during at least 8 months of the year.

Fourteen men in the age group 40-49 years had a mean maximal oxygen uptake of 4.0 (3.5-4.5) l/min = 57 ml/kg x min, serum cholesterol 222 mg/100 ml and serum triglycerides 0.85 mM/l. Fifteen men in the age group 50-52 years had a mean maximal oxygen uptake of 3.4 (2.4-4.2) l/min = 53 ml/kg x min, serum cholesterol 251 mg/100 ml and serum triglycerides 0.95 mM/l. The cholesterol value was not significantly different from the male urban population aged 50, the triglycerides, however, were significantly lower in the active athletes.

In former athletes of the same cathegory, but with at least 10 years without physical training, Saltin and Grimby (21) showed that the maximal oxygen uptake was about 25 % lower than in the still active athletes, but about 20 % higher than in sedentary men of the same ages. The serum triglycerides was higher in the formerly active than in the still active athletes and of about the same order as in group 1 and 2 in table V. The serum cholesterol, although with slightly higher mean values in the former than in the still active athletes, did not differ significantly between these two materials. The still active athletes had lower body weight than the former athletes.

To enlarge the group of middle-aged men who have trained intensively (group 4 in leisure time activity), we have a study on 52-56 years still active endurance athletes in progress. The preliminary results from 43 men are shown in table VI, together with serum cholesterol and triglycerides in the population study of 50 and 54 years old men divided in quintiles with respect to oxygen uptake. In the active athletes with definitely higher maximal oxygen uptake, the serum cholesterol and triglycerides are significantly lower than in the population sample. The athletes had in general no extreme dietary habits, but lower body weights.

Table VI

Serum triglycerides at age 50 and cholesterol at age 54 in relation to maximal oxygen uptake in randomly selected men and in still active athletes aged 52-56

54 years old men

Max $\dot{V}_{O_2}$ Quintile ml/kg × min		1 17.9-27.5	2 27.6-29.3	3 29.4-31.0	4 31.1-33.4	5 33.5-42.9
Cholesterol mg/100 ml 54 years old men	$\bar{x}$ s_x n	274 46 104	267 47 105	276 48 113	273 41 108	256 39 101

50 years old men

		1	2	3	4	5
Triglycerides mM/l 50 years old men	$\bar{x}$ s_x n	1.35 0.72 132	1.41 0.93 133	1.14 0.44 131	1.08 0.40 128	1.12 0.64 132

52-56 years old athletes

Max $\dot{V}_{O_2}$ ml/kg × min		29.4-55.3 (mean 45.1)
Cholesterol mg/100 ml	$\bar{x}$ s_x n	211 33 43
Triglycerides mM/l	$\bar{x}$ s_x n	0.76 0.28 43

EFFECT OF SHORT PERIOD OF TRAINING IN SEDENTARY MEN

Serum cholesterol values before and after 2 months of hard physical training of 42 men with sedentary jobs (age 34-50 years), were included in the study by Kilbom et al. (16). They belonged for their leisure time to activity group 1 or 2 before training and to group 3 during training. The values after training were usually lower with an average value of 258 mg/100 ml before and 235 mg/100 ml after training. The difference is small but significant. The maximal oxygen uptake had increased on an average 18 (8-44) % from 2.9 to 3.4 l/min (22). The body weight did not change with training.

CONCLUSIONS

Serum triglycerides have been demonstrated to decrease with increased physical activity specially during leisure time, and aerobic power. Serum cholesterol on the other hand shows no difference in the population study between the activity groups. However, with short periods of training, so that men belonging to group 1 or 2 is changed to group 3, cholesterol can show a slight decrease. Somewhat lower values are also seen in still active athletes, who belong to leisure time activity group 4 since many years.

REFERENCES

1. Campbell, D.E. Influence of several physical activities on serum cholesterol concentration in young men. J. Lipid Res. 6: 478, 1965.

2. Campbell, D.E. & Lumsden, T.B. Serum cholesterol concentrations during physical training and during subsequent detraining. Am. J. Med. Sci. 253: 155, 1967.

3. Carlson, L.A. Determination of serum glycerides. Acta Soc. Med. Upsalien. 64: 208, 1959.

4. Carlson, L.A. & Lindstedt, S. The Stockholm prospective study 1. The initial values for plasma lipids. Acta Med. Scand., Suppl. 493, 1968.

5. Cramér, K. & Isaksson, B. An evaluation on the Theorell method for the determination of total serum cholesterol. Scand. J. clin. Lab. Invest. 11: 213, 1959.

6. Doyle, J.T., Kinch, S.H. & Brown, D.F. Seasonal variation in serum cholesterol. J. Chron. Dis. 18: 657, 1965.

7. Fox, S.M. & Haskell, W.L. Physical activity and the prevention of coronary heart disease. Bull. N.Y. Acad. Med. 44: 950, 1968.

8. Fitzgerald, O., Heffernan, A. & McFarlane, R. Serum lipids and physical activity in normal subjects. Clin. Sci. 28: 83, 1965.

9. Goode, R.C., Firstbrook, J.B. & Shephard, R.J. Effects of exercise and a cholesterol-free diet on human serum lipids. Can. J. Physiol. and Pharmacol. 44: 575, 1966.

10. Grimby, G. & Saltin, B. Physiological analysis of physically well-trained middle-aged and old athletes. Acta Med. Scand. 179: 513, 1966.

11. Grimby, G., Wilhelmsen, L., Aurell, M., Bjure, J., Ekström-Jodal, B. & Tibblin, G. Aerobic power and related factors in a population study on men aged 54. Scand. J. clin. Lab. Invest. In press 1970.

12. Hernberg, S. Serum-cholesterol and capacity for physical work. Lancet 2: 441, 1964.

13. Holloszy, J.O., Skinner, J.S., Toro, G. & Cureton, T.K. Effects of a six month program of endurance exercise on the serum lipids of middle-aged men. Am. J. Cardiol. 14: 753, 1964.

14. Karvonen, M.J. Körperliche Tätigkeit, Cholesterinstoffwechsel und Arteriosklerose. Z. Präventivmed. 6: 269, 1961.

15. Keys, A., Andersson, J., Aresu, M., Björck, G., Broch, J., Bronte-Stewart, B., Fidanza, F., Keys, M., Malmros, H., Poppi, A., Porteli, T., Swahn, B. & Del Vecchio, A. Physical activity and the diet in populations differing in serum cholesterol. J. Clin. Invest. 35: 1173, 1956.

16. Kilbom, Å., Hartley, L.H., Saltin, B., Bjure, J., Grimby, G. & Åstrand, I. Physical training in sedentary middle-aged and older men. I Medical evaluation. Scand. J. clin. Lab. Invest. 24: 315, 1969.

17. Mann, G.V., Garelt, H.L., Farbi, A., Murray, H. & Billings, T. Exercise to prevent coronary heart disease. Am. J. Med. 46: 12, 1969.

18. Mc Donough, J.R., Kusumi, F. & Bruce, R. Variations in maximal oxygen uptake with physical activity in middle-aged men. Circulation 61: 743, 1970.

19. Montoye, H.J., Van Huss, W.D., Brewer, W.D., Jones, E.M., Ohlson, M.A., Mahoney, E. & Olsson, H. The effects of exercise on blood cholsterol in middle-aged men. Amer. J. Clin. Nutr. 7: 139, 1959.

20. Naughton, J. & McCoy, J.F. Observations on the relationship of physical activity to the serum cholesterol concentration of healthy men and cardiac patients. J. chron. Dis. 19: 727, 1966.

21. Saltin, B. & Grimby, G. Physiological analysis of middle-aged and old former athletes. Circulation 38: 1104, 1968.

22. Saltin, B., Hartley, L.H., Kilbom, Å. & Åstrand, I. Physical training in sedentary middle-aged and older men. II Oxygen uptake, heart rate and blood lactate concentration at submaximal and maximal exercise. Scand. J. clin. Lab. Invest. 24: 323, 1969.

23. Shane, S.R. Relation between serum lipids and physical conditioning. Amer. J. Cardiol. 18: 540, 1966.

24. Siegel, W., Blomqvist, G. & Mitchell, J. Effects of a quantitated physical training program on middle-aged sedentary men. Circulation 49: 19, 1970.

25. Taylor, H.L., Andersson, J.T. & Keys, A. Studies on diet, physical activity and serum cholesterol concentration. Circulation 24: 1055, 1961.

26. Tibblin, G. High blood pressure in men aged 50. Acta Med. Scand. Suppl. 470, 1967.

MYOCARDIAL METABOLISM IN PATIENTS WITH ISCHEMIC HEART DISEASE

Albert S. Most

Rhode Island Hospital, and Brown University

Providence, Rhode Island

Investigation of myocardial metabolism has been of particular interest in patients with ischemic heart disease (IHD). As a test of adequate myocardial oxygenation, metabolic studies are employed to refine the diagnosis of IHD in patients with (1) and without (2,3) coronary atherosclerosis, and to evaluate the results of surgical revascularization of ischemic myocardium (4,5). More recently the therapeutic implication of metabolic factors in acute myocardial ischemia has received considerable attention (6,7).

CORONARY SINUS CATHETERIZATION

Coronary sinus catheterization is central to metabolic studies but has inherent limitations. The patchy distribution of coronary atherosclerosis results in regions of myocardium with varying degrees of adequate perfusion. Consequently, coronary venous effluent is not homogeneous. The increased sensitivity of local venous sampling (8) not only is impossible in man but could give a distorted picture of myocardial metabolism in patients with coronary heart disease. By sampling several sites in the coronary sinus (1) one approaches the local venous technique but still samples sufficiently wide regions of myocardium to get a broadly representative sample of myocardial effluent. In practice, single-site sampling is most widely employed. This is preferred so that control and stress state coronary sinus blood samples can be drawn with certainty from the same site.

If coronary angiography has demonstrated a vascular lesion or if there is localizing ECG evidence of ischemia, that coronary sinus position may be chosen which is most likely to detect the ischemic zone during a metabolic stress test. Lack of homogeneity in the effluent does, however, make quantitative metabolic analysis very difficult.

As a coronary sinus catheter is withdrawn from the most distal site toward the orifice, the effluent becomes cumulative in nature and thereby less selective. Inferior wall ischemia is most difficult to detect selectively for this reason, its effluent entering the coronary sinus far downstream, just inside the orifice. In addition, sites far upstream cannot always be sampled owing to difficulty in catheter placement or wedging of the sampling tip. Nevertheless, clinical application has continued to confirm the usefullness of this methodology, although sensitivity is admittedly compromised owing to the aforementioned.

MYOCARDIAL LACTATE EXTRACTION

A Diagnostic Test

As a diagnostic test, reliance has largely been placed on measuring arterial and coronary sinus blood lactate levels and arbitrarily defining ischemia in terms of an expression of myocardial lactate extraction. Most commonly, a negative extraction (net production) is defined as ischemia. Other indices such as the lactate: pyruvate ratio or "excess lactate" are used less widely but in general carry the same meaning. Lactate production by the heart is indicative of ongoing anaerobic glycolysis and is therefore taken as evidence of inadequate myocardial oxygenation. Recent work (9) suggests that lactate and pyruvate observe different membrane kinetics, making the lactate: pyruvate ratio and "excess lactate" measurement more difficult to interpret.

By sampling selective sites in the coronary sinus for lactate, Gorlin and coworkers (1) were able to correlate regional metabolic ischemia with probable areas of coronary flow restriction based on cineangiographic studies. Prior to the advent of transvenous intracardiac pacing, such studies utilized pharmacologic stress to induce myocardial lactate production. Isoproterenol was used primarily but its attendent increase in the arterial lactate concentration, similar to that observed with exercise, rendered this test less sensitive than was desirable. With atrial pacing no significant elevation of circulating substrate is observed, making this the preferred method of stressing the myocardium for metabolic investigation.

Patients are now being seen who have convincing clinical evidence of myocardial ischemia but no angiographic evidence of coronary atherosclerosis (2,3,10). In these subjects, metabolic evidence of ischemia is exceedingly important if the diagnosis of IHD is to be maintained. Two laboratories (2,3) have now documented metabolic evidence of ischemia in patients without atherosclerotic involvement of the coronary arteries, thereby stimulating the search for other causes of IHD. In one series (2), five of twelve patients with exertional chest pain and normal coronary arteriograms demonstrated lactate production with isoproterenol stress. As a necessary corollary, the presence of coronary atherosclerosis need not implicate this pathology as the sole cause of myocardial ischemia in all suspect cases. That atherosclerosis is not the only explanation of the anginal syndrome is suggested by post mortem and coronary angiographic studies on patients with non-anginal complaints. In such patients, angiographic lesions indistinguishable from those in patients with angina pectoris have recently been demonstrated (11). Lactate studies, unfortunately, have not been reported on this group. The possibility exists that more subtle causes of IHD may be present even in patients with coronary atherosclerosis.

ECG Correlation

Correlating metabolic data with ECG evidence of ischemia, Kemp and coworkers (12) have shown a close correlation using isoproterenol stress. More recently, Parker and colleagues (13) have shown a similar relationship using atrial pacing. Unlike the ECG stress test, lactate studies are valid in the presence of digitalis and an abnormal resting ECG. In addition, there is less subjectivity in their interpretation.

As a diagnostic tool then, metabolic evidence of ischemia is important supportive evidence in selected patients with coronary atherosclerosis (ie., borderline anatomic lesions, atypical pain syndromes) and critically important data in patients suspected of coronary heart disease who fail to demonstrate anatomic disease on angiographic study.

Evaluating Myocardial Revascularization

In assessing the value of myocardial revascularization for coronary artery disease, evidence of metabolic improvement is the best objective proof available to substantiate claims for this therapy (4,5). Such improvement has been reported and runs counter to evidence showing that the flow contribution of an internal mammary artery pedicle implant (IMAPI) to total myocardial perfusion is too small to be of significant value (14). This disagreement suggests that the IMAPI may be of benefit through other than its direct flow contribution. In patients demonstrating IMAPI-coronary artery linkage on angiography, Kemp and coworkers (5) found lactate reversal from production pre-operatively to extraction post-operatively in eight of eleven subjects. In those failing to have coronary artery visualization after IMAPI angiography, only one of four had a similar reversal. Thus, agreement was seen between reversion to normal lactate metabolism and an IMAPI-coronary artery communication, suggesting improved myocardial oxygenation.

The diagnostic value of myocardial lactate studies is amply supported by the preceding correlations. As an objective means of judging therapeutic maneuvers in IHD, evidence favors continued and even wider use of this metabolic evaluation of adequate myocardial oxygenation.

METABOLIC INTERVENTIONS

The metabolic therapy of ischemic myocardium was first offered as a means of reducing ventricular irritability by increasing intracellular potassium through an intravenous mixture of potassium, glucose and insulin (15). More recently it has been questioned whether the glucose (16), through its contribution to anaerobic energy production, wasn't the primary factor in improving electrical stability. Since glucose has been shown to protect hypoxic muscle function experimentally (17,18), the question is raised whether hyperglycemia would be of benefit in the presence of myocardial ischemia in man.

Added glucose may protect hypoxic myocardium indirectly by reducing circulating free-fatty acids (FFA). The latter substrate has been shown to increase oxygen consumption of perfused rat heart (19) and depress contractility in rat papillary muscle (18) during hypoxia. Coupled with clinical reports implying an association between high circulating FFA levels and cardiac arrhythmias (6,7), this additional therapeutic benefit may be postulated for a glucose infusion in the presence of myocardial hypoxia. Any therapy based on this thesis must be considered experimental, however, in view of recent contrasting reports denying a relationship between FFA elevation and ventricular irritability (20,21).

MYOCARDIAL GLUCOSE EXTRACTION

In order to investigate glucose extraction by ischemic human myocardium, simultaneous myocardial glucose and lactate extraction was studied before and during atrial pacing stress in subjects with angina pectoris. Subjects were paced with increments of 10 beats/minute every 2-3 minutes until either anginal pain developed or heart rate exceeded control by 50-70%.

Three groups of subjects could be distinguished: I. Lactate extraction before and during pacing; II. Lactate extraction at rest but production during pacing; III. Lactate production both at rest and during pacing. Glucose extraction increased significantly in groups II and III. Subjects already producing lactate at rest (group III) augmented glucose extraction from already high levels. A significant correlation was noted between lactate production and glucose extraction in combined groups II and III during pacing.

No stoichiometric 2:1 ratio was observed for lactate production and glucose extraction. This is consistent with work in the ischemic dog heart (22) and may be explained by several alternative possibilities: 1. myocardial ischemia is heterogeneous in the zones sampled, 2. usable myocardial glycogen reserves were not depleted by the degree and duration of ischemia and/or 3. membrane kinetics of glucose and lactate may differ much in the same way those of lactate and pyruvate do (9). Nevertheless, myocardial ischemia in man is associated with augmented glucose extraction.

No relationship was seen between glucose extraction and the arterial glucose concentration, either before or during ischemia. This raises some doubt that elevating arterial glucose within the physiologic range studied will augment glucose uptake and possibly lactate production. Whether arterial glucose ever critically limits the capacity for anaerobic myocardial glycolysis in man is not known. Although net lactate production in human studies has been calculated to produce only 2% of the total ATP needed to sustain normal resting cardiac function (16), two factors deserve consideration. With coronary sinus sampling, net lactate production in a patchy disease such as coronary atherosclerosis may represent greater lactate production from a small focus of ischemic tissue partially masked by extraction in a much larger area. The insensitivity of coronary sinus sampling versus local venous sampling has been amply demonstrated (8). Secondly, the degree of muscle function maintained in an ischemic zone may be subnormal and consequently require less energy. Anaerobic glycolysis may therefore be sufficient to maintain marginal function in such areas until improved oxygenation can be provided. The quantitative contribution of anaerobic glycolysis to myocardial metabolism in IHD should not be discounted.

The correlation of myocardial glucose uptake with the arterial glucose concentration under resting conditions is at best a very weak one. Studies of this question have yielded conflicting results (23-26). To further explore the issue, myocardial glucose extraction was studied in 16 resting subjects before and during a 30 minute, intravenous glucose infusion. No significant relationship was discerned between glucose extraction and either the arterial glucose or insulin concentration. In spite of a 20% rise in circulating glucose and a 50% rise in insulin no significant increase in myocardial glucose extraction was noted.

Cautious attempts at substrate manipulation in the management of acute myocardial ischemia in man are justified. At this juncture in our knowledge, however, such efforts must be considered entirely experimental.

REFERENCES

1. Herman, M.V., Elliot, W.C. and Gorlin, R.: Electrocardiographic, anatomic and metabolic study of zonal myocardial ischemia in coronary heart disease. Circulation 35:834, 1967.

2. Kemp, H.G., Elliot, W.C. and Gorlin, R.: The anginal syndrome with normal coronary arteriography. Trans. Assoc. of Amer. Phys. 80:59, 1967.

3. Neill, W.A., Kassebaum, D.G. and Judkins, M.P.: Myocardial hypoxia as the basis for angina pectoris in a patient with normal coronary arteriograms. N.E.J.M. 279:789, 1968.

4. Gorlin, R. and Taylor, W.J.: Myocardial revascularization with internal mammary artery implantation: Current status. J.A.M.A. 207:907, 1969.

5. Kemp, H.G., Manchester, J.H., Amsterdam, E.A., Taylor, W.J. and Gorlin, R.: Internal Mammary artery implantation: Effect on myocardial lactate utilization. Circ. Suppl. II 41:55, 1970.

6. Gupta, D.K., Young, R., Jewitt, D.E., Hartog, M. and Opie, L.H.: Increased plasma-free-fatty-acid concentrations and their significance in patients with acute myocardial infarction. Lancet 2:1209, 1969.

7. Kurien, V.A. and Oliver, M.F.: A metabolic cause for arrhythmias during acute myocardial infarction. Lancet 1:813, 1970.

8. Owen, P., Thomas, M. and Opie, L.: Relative changes in free-fatty-acid and glucose utilization by ischemic myocardium after coronary artery occhision. Lancet 1:1187, 1969.

9. Henderson, A.H., Craig, R.J., Gorlin, R. and Sonnenblick, E.H.: Lactate and pyruvate kinetics in isolated perfused rat hearts. Am. J. Physiol. 217: 1752, 1969.

10. Eliot, R.S. and Bratt, G.: The paradox of myocardial ischemia and necrosis in young women with normal coronary arteriograms. Amer. J. Cardiol. 23:633, 1969.

11. Manchester, J.H., Herman, H.V., Amsterdam, E.A., Kemp, H.G. and Gorlin, R.: Prevalence and distribution of coronary atherosclerosis in a non-selected population. Clin. Res. 17:252, 1969.

12. Kemp, H.G., Most, A.S. and Gorlin, R.: Correlation between electrocardiographic and metabolic evidence of ischemia in man. Circ. Suppl. VI 38: 113, 1968.

13. Parker, J.O., Chiong, M.A., West, R.O. and Case, R.B.: Sequential alterations in myocardial lactate metabolism, S-T segments, and left ventricular function during angina induced by atrial pacing. Circ. 40: 113, 1969.

14. Dart, C.H., Jr., Scott, S., Fish, R. and Takaro, T.: Direct blood flow studies of clinical internal thoracic (mammary) arterial implants. Circ. Suppl. II. 41:64, 1970.

15. Calva, E., Miyica, A., Bisteni, A. and Sodi-Pallares, D.: Oxidative phosphorylation in cardiac infarct. Effect of glucose-KCl-insulin solution. Am. J. Physiol. 209:371, 1965.

16. Olson, R.E.: Metabolic interventions in the treatment of infarcting myocardium. Circ. Suppl. IV. 40:195, 1969.

17. Weissler, A.M., Kruger, F.A., Baba, N., Scarpelli, D.G., Leighton, R.F. and Gallimore, J.K.: Role of anaerobic metabolism in the preservation of functional capacity and structure of anoxic myocardium. J. Clin. Invest. 47:403, 1968.

18. Henderson, A.H., Most, A.S., Parmley, W.W., Gorlin, R. and Sonnenblick, E.H.: Depression of myocardial contractility by free fatty acids during hypoxia. Circ. Res. 26:439, 1970.

19. Challoner, D.R. and Steinberg, D.: Effect of free fatty acid on the oxygen consumption of perfused rat heart. Am. J. Physiol. 210:280, 1966.

20. Ruttenberg, H.L., Pamintuan, J.C. and Soloff, L.A.: Serum-free-fatty-acids and their relation to complications after acute myocardial infarction. Lancet 2:559, 1969.

21. Russo, J.V., Margolis, S. and Ross, R.S.: Effect of heparin on ventricular arrhythmias following myocardial infarction. Clin. Res. 18:327, 1970.

22. Brachfeld, N. and Scheuer, J.: Metabolism of glucose by the ischemic dog heart. Amer. J. Physiol. 212:603, 1967.

23. Goodale, W.T., Olson, R.E. and Hackel, D.B.: The effects of fasting and diabetes mellitus on myocardial metabolism in man. Amer. J. Med. 27:212, 1959.

24. Bing, R.J., Siegel, A., Vitale, A., Balboni, F., Sparks, E., Taeschler, M., Klapper, M. and Edwards, S.: Metabolic studies on the human heart in vivo. I. Studies on carbohydrate metabolism of the human heart. Amer. J. Med. 15:284, 1953.

25. Goodale, W.T. and Hackel, D.B.: Myocardial carbohydrate metabolism in normal dogs, with effects of hyperglycemia and starvation. Circ. Res. 1:509, 1953.

26. Barta, E., Breuser, E., Pappova, E. and Zlatos, L.: Relationship between the arterial level of basic energetic substances and their uptake in heart muscle. Cor Vasa 9:237, 1967.

METABOLISM IN PATIENTS WITH ISCHEMIC HEART DISEASE AND OBESITY AFTER TRAINING

Per Björntorp[x]

First Medical Service, Sahlgren's Hospital
University of Gothenburg, Sweden

In relation to the numerous studies concerned with the input stide of the energy balance equation in hyperlipidemic conditions comparably few studies are concerned with the effects on plasma lipids by changes in energy output. Acute effects of such changes are known (8) and physical training in relation to plasma lipids has also been studied in healthy persons (12,13).

The present work was performed in order to elucidate this problem somewhat more. In addition to plasma lipids, variables of carbohydrate metabolism were also followed as well as body composition changes.

Two studies will be reported here. The first (Part I) is concerned with physical training of patients who have suffered a myocardial infarction, and the other (Part II) with patients with extreme obesity.

PART I

Material and Methods

This study was designed to elucidate which metabolic defects were characteristic for myocardial infarction by

---

[x] These studies were performed in collaboration with Peter Berchtold, Gunnar Grimby, Anders Gustafson, Björn Lindholm, Harald Sanne, Lars Sullivan, Gösta Tibblin and Lars Wilhelmsen.

performing comparisons with a control material of the same age randomly selected from the same region as the myocardial infarction patients. In order also to see which metabolic aberrations were characteristic just for aging a control material of young men was also included. This allowed later evaluation of on which variables physical training was effective, on those due to aging or on those specific for myocardial infarction.

In these three materials body composition was determined by isotope dilution methods (15), plasma cholesterol (9), triglyceride (7), insulin (11) and blood glucose (14) after a 100 g per oral glucose tolerance test. In a random part of the materials adipose tissue fat cell diameter was also determined (4).

The myocardial infarction material consisted of all men below the age of 55 years who had suffered a myocardial infarction between 1st of January 1968 and 31st of May 1969 in the city of Gothenburg, Sweden, and survived. This was possible by the organization of a special infarction clinic. Of originally 107 men admitted during this period 3 were excluded from the present investigation because of overt clinical diabetes mellitus. The mean age of the myocardial infarction patients was 50.0 years with the range of 36-55 years.

The patients with myocardial infarction were divided into two groups by taking every second infarction to a group for physical training. The physical training program was designed individually for each patient. It consisted of 3 periods of each a half hour per week, with supervised ECG-monitored periods mainly on the ergometer bicycle. The load on the bicycle was individual and determined by the condition of the patient. The end points for the work load were either angina pectoris, ST-T depression on ECG, tendency to cardiac arrythmia or locomotor limitations. No accidents occurred. The training was started 3 months after the myocardial infarction and had a duration of 9 months. Ergometer tests with measurements of maximal oxygen uptake were performed before and after 9 months of physical training. Patients who could follow the training program sufficiently (14 men) had on an average an increase of maximal oxygen uptake of 0.3 lit/min while the remaining 14 men who were followed up after training but who could not follow the program, showed no improvement in oxygen uptake measurements.

The middle-aged control material consisted of 81 55-year old randomly selected men in Gothenburg, Sweden.

Five of these had suffered a myocardial infarction or had angina pectoris and were therefore excluded, leaving 76 men for the study. Finally 12 23-year old medical students were studied similarly.

Results

Insulin values. Figure 1 shows that middle-aged men had considerably higher insulin values than young men (55 y - 23 y). This was associated with an increase of the size of the fat cells of the middle-aged men beacuse when middle-aged men with the same fat cell diameter as

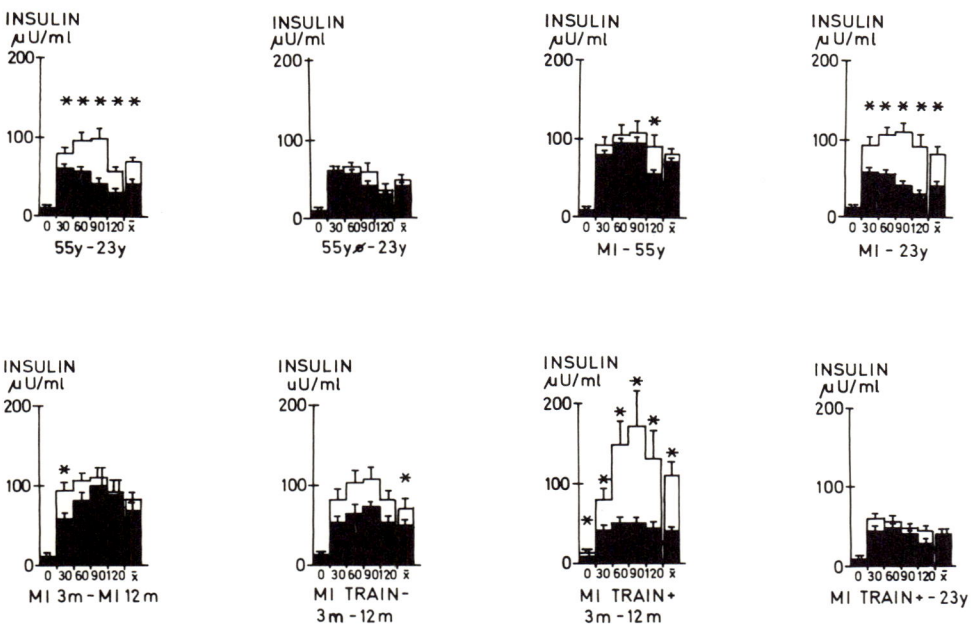

Fig. 1. Comparison of plasma insulin during glucose tolerance tests in middle-aged randomly selected men (55 y), young men (23 y) and myocardial infarction patients who were not subjected to physical training (MI). Myocardial infarction patients were trained physically whereby some were considered sufficiently trained (MI TRAIN+) and others not (MI TRAIN-) (See text!) These as well as non-trained myocardial infarction patients (MI) were examined 3 and 12 months after the myocardial infarction (3m, 12m). ø: Middle-aged men with small fat cells. * p < 0.05.

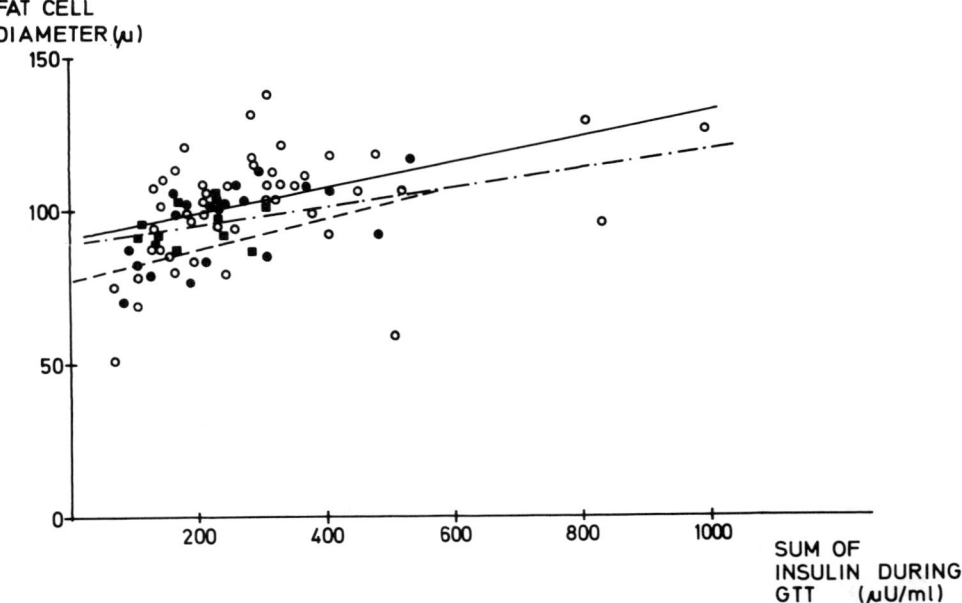

Fig. 2. Correlations between fat cell diameter and sum of insulin during glucose tolerance test in 55 year old men (open circles, whole line, y = 0.04x + 91, r:0.37, p 0.001), patients with myocardial infarction (filled circles, hatched line, y = 0.05x + 77, r:0.66, p 0.001) and 23 year old men (squares, correlation n.s.) (total material, hatched-pointed line, y = 0.03x + 89, r:0.39, p 0.001).

the young men were selected, the differences in insulin values disappeared (55 y ∅ - 23 y). This is also illustrated in Figure 2 where the rather strong correlation between sum of insulin values during glucose tolerance test and fat cell size is shown for 55-year old men and for infarction patients. This is also true within a population of obese patients (5). Figure 1 also shows that there was not much difference between myocardial infarction patients and middle-aged controls (MI - 55y), while young men had considerably lower values (MI - 23y).

At the reinvestigation after 9 months there was only a slight decrease in insulin values in the control group (MI 3m - MI 12m). This was also the case in insufficiently trained myocardial infarction patients (MI TRAIN-, 3m - 12m). The sufficiently trained patients, however, had a pronounced decrease of insulin values in all points in comparison with their own values before physical training (MI TRAIN+, 3m - 12m). As a matter of fact, they were now indistinguishable from young men (MI TRAIN+ - 23y).

<u>Body fat and plasma lipids.</u> Middle-aged men had higher blood lipids and body fat than young men (Fig. 3, 55y - 23y). The myocardial infarction patients had higher blood lipids but were not fatter than middle-aged controls (MI - 55y). In comparison with young men, however, myocardial infarctions had not only higher blood lipids but also more body fat (MI - 23y).

What happened with these variables at reinvestigation 9 months later? Controls (MI 3m - MI 12m) or insufficiently trained myocardial infarction patients (MI TRAIN-, 3m - 12m) showed no changes. The sufficiently trained myocardial infarction patients, however, had lower triglyceride and body fat after training while cholesterol was unchanged (MI TRAIN+, 3m - 12m). As was the case with insulin

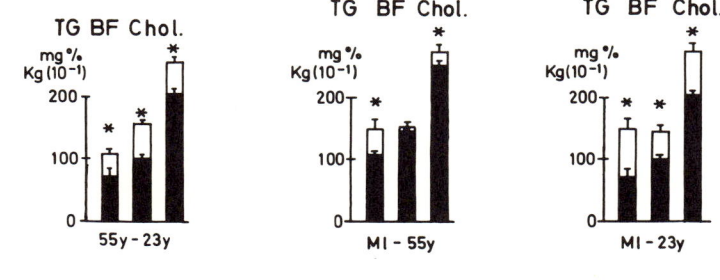

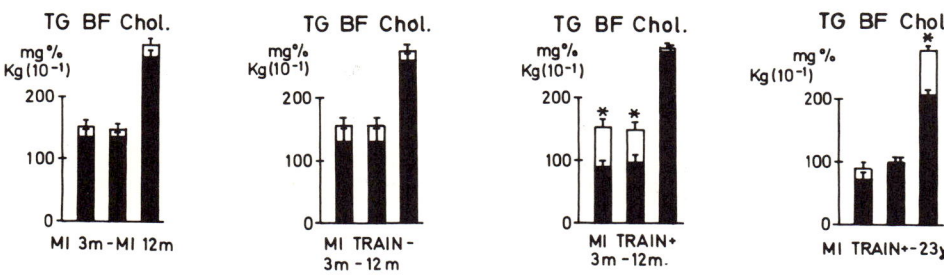

Fig. 3. Comparisons of plasma triglyceride (TG), cholesterol (chol.) and body fat (BF) in middle-aged randomly selected men, young men and myocardial infarction patients. Symbols as in Fig. 1.

described above, triglyceride and body fat were now also indistinguishable from that of young men while cholesterol remained high (MI TRAIN+ - 23y).

<u>Glucose tolerance</u>. The glucose values during the glucose tolerance test showed that 55 year old men had a decreased glucose tolerance in comparison with young men (Fig. 4, 55y - 23y). Myocardial infarction patients had a decreased glucose tolerance in comparison with controls of the same age (MI - 55y) and in comparison with young men (MI - 23y). This decreased glucose tolerance was not changed with time, nor with insufficient training and, as seen in Fig. 4 (MI TRAIN+, 3m - 12m) not by sufficient physical training either.

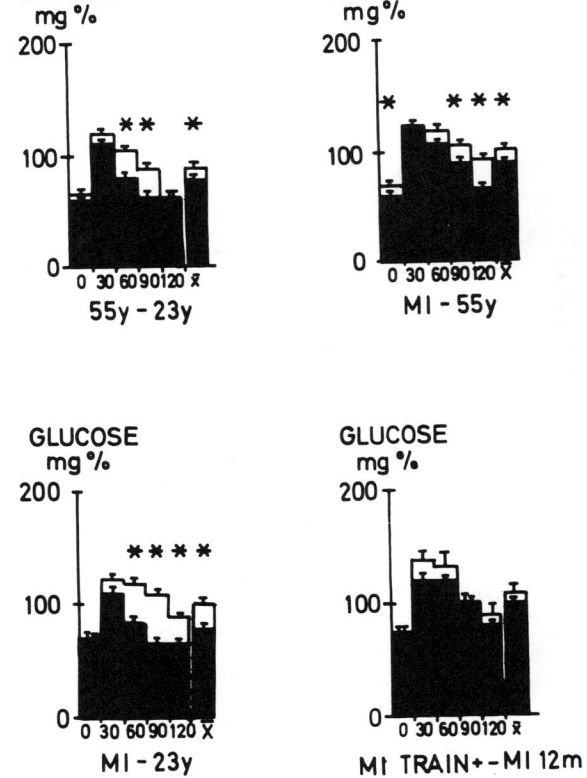

Fig. 4. Comparisons of blood glucose values during glucose tolerance test in middle-aged randomly selected men, young men and myocardial infarction patients. Symbols as in Fig. 1.

Discussion

In comparison with young men myocardial infarction patients had higher body fat, plasma lipids and insulin, increased fat cell diameter and decreased glucose tolerance. Of these findings only the decreased glucose tolerance and the increased plasma lipids remained when comparisons were performed with men of the same age. These findings thus may be considered as characteristic for patients with myocardial infarction. Increased insulin, body fat and fat cell diameter on the other hand seems to be due to aging because they were higher in older controls than in younger controls. Glucose tolerance and plasma triglyceride and cholesterol also showed an age-dependence.

The physical training period caused a decrease of plasma insulin, triglyceride, body fat and presumably therefore also of fat cell diameter (cf. 17), so that these factors were no longer different from those of young men. Physical training thus predominantly affected factors which were considered to be caused by aging. Two factors which were considered characteristic for myocardial infarction, viz. increased cholesterol and decreased glucose tolerance, were however, unchanged by training.

Increased body fat in association with increased fat cell size is due to a positive caloric balance (17). It seems possible that also insulin increase with age might be caused by such factors because when young men are compared with older men with the same fat cell size, insulin values are no longer different (4). Triglyceride increase apparently is dependent on an insulin increase in combination with a decreased glucose tolerance (2). All these age-dependent variables might thus be explained by a positive caloric balance, caused by increased caloric intake in relation to caloric output. It seems logical that these changes should be normalized by an increased physical activity and physical training, which was then also indeed demonstrated.

It is not possible to state whether some of the observed changes after the training period are due to an effect of the last working period or due to more constant adaptive changes to the physical training or both.

PART II

The previously described study thus indicated a key role for insulin in the effects of physical training on metabolism. A decrease in insulin was associated with a decrease in body fat. This association between plasma insulin and body fat is well-known (1, 16, 3, 4). In another study we wanted, however, to subject this question to a closer analysis (6). We therefore selected a clinical group with increased plasma insulin for study, viz. obese patients. With these patients we tried to avoid a change in the adipose tissue fat depot with physical training in order to study whether there is an obligatory association between these two variables.

Material and Methods

Ten extremely obese patients were selected for study. Eight were women (19 - 60 years) and two men (19 and 50 years). They were prescribed in advance not to diet during the procedures, but rather to keep their weight stable and eat ad libitum.

Since this material typically consisted of mainly middle-aged women, a randomly selected material of 25 52-year old women was also examined for comparison. One of these was excluded because of overt obesity (body fat 44.2 kg).

Physical training was then performed on the ergometer bicycle with the obese patients and also training of muscle strength. Both maximal oxygen uptake and muscle strength increased in all patients except one after 8 weeks of intensive training (3-5 days per week).

Body composition, glucose tolerance with plasma insulin and blood lipids were determined before and after training as described above.

Results

Figure 5 (Ob.-C) defines these materials as far as body fat and plasma lipids are concenrned. Body fat was of course much higher in the obese group and so was also fat cell size.

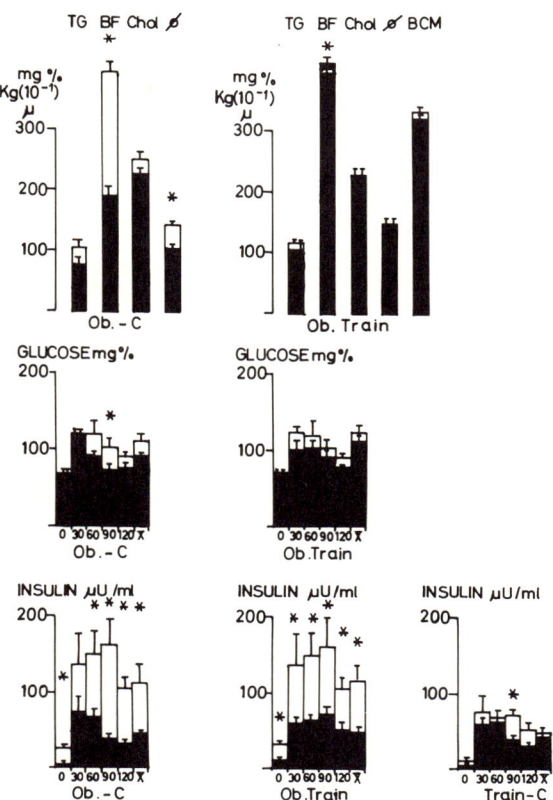

Fig. 5. Comparisons of plasma triglyceride (TG), cholesterol (chol.), body fat (BF), body cell mass (BCM) and adipose tissue fat cell diameter (⌀) and plasma insulin and blood glucose during glucose tolerance test in extremely obese patients before (Ob.) and after (Ob. Train) physical training and in middle-aged randomly selected women (C).

Glucose tolerance was almost similar in both the groups (Ob.-C) while insulin was considerably higher in the obese group (Ob.-C).

After physical training the glucose tolerance remained unchanged while insulin values now were considerably lower about a half that before training (Ob. Train) and no longer certainly abnormal (Train-C). Body fat increased somewhat, but body cell mass did not increase. Plasma lipids did not change (Ob. Train).

Discussion

This study agrees with that reported above in that insulin was considerably decreased by physical training. Now it occurred, however, without any decrease in body fat. Glucose tolerance did not change, also in agreement with the previous study. This probably means that insulin sensitivity in the periphery had increased by physical training.

Plasma lipids, which were not abnormally high, did not change by physical training. Nor did body cell mass increase other than in a few cases. This increase was not proportional to insulin decrease. The effect of physical training to produce a decreased insulin secretion thus does not necessarily go via a decrease of the adipose tissue depot. The reason for this can only be speculated on. Changes in endocrinological factors other than insulin after training are only partially known and could be one reason. Another reason might be that physical training produces changes within muscle which may facilitate the insulin effect on glucose uptake in muscle. Such effects are production of an only partially defined hypoglycermic agent by acute work (10), decrease of muscle cell substrate pools of glycogen or lipid and finally induction of enzymic changes. Aerobic enzymes increase in activity after a period of training like the one described here (18) and perhaps also other enzymes of more direct influence for muscle insulin sensitivity.

REFERENCES

1. BAGDADE, J.D., E.L. BIERMAN, AND D. PORTE JR. The significance of basal insulin in the evaluation of the insulin response to glucose in diabetic and non-diabetic subjects. J. Clin. Invest. 46:1549, 1967.

2. BERCHTOLD, P., P. BJÖRNTORP, B. LINDHOLM, G. TIBBLIN, AND L. WILHELMSEN. In preparation, 1970.

3. BJÖRNTORP, P., H. BERGMAN, E. VARNAUSKAS, AND B. LINDHOLM. Lipid mobilization in relation to body composition in man. Metabolism 18:840, 1969.

4. BJÖRNTORP, P., P. BERCHTOLD, AND G. TIBBLIN. Plasma insulin in relation to adipose tissue in men. Diabetes, in print, 1970 a.

5. BJÖRNTORP, P., L. SJÖSTRÖM, AND J. VRANA. To be published, 1970 b.

6. BJÖRNTORP, P., K. DE JOUNGE, L. SJÖSTRÖM, AND L. SULLIVAN. The effect of physical training on insulin production in obesity. Metabolism 19:631, 1970 c.

7. CARLSON, L.A. Determination of serum glycerides. Acta Soc. Med. Upsalien. 64:208, 1959.

8. CARLSON, L.A., AND F. MOSSFELDT. Acute effects of prolonged, heavy exercise on the concentration of plasma lipids and lipoproteins in man. Acta Med. Scand. 62:51, 1964.

9. CRAMÉR, K., AND B. ISAKSSON. An evaluation of the Theorell method for the determination of total serum cholesterol. Scand. J. clin. Lab. Invest. 11:213, 1959.

10. GOLDSTEIN, M.S., V. MULLICK, B. HUDDLESTUN, AND R. LEVINE. Action of muscular work on transfer of sugars across cell barriers: comparison with action of insulin. Amer. J. Physiol. 173:212, 1953.

11. HALES, C.N., AND P.J. RANDLE. Immunoassay of insulin with insulin antibody precipitate. Lancet I:200, 1963.

12. HOLLOSZY, J.O., J.S. SKINNER, G. TORO, AND T.K. CURETON. Effects of a six months' program of endurance exercise on the serum lipids of middle-aged men. Amer. J. Cardiol. 14:753, 1964.

13. KILBOM, Å., L.M. HARTLEY, B. SALTIN, J. BJURE, G. GRIMBY, AND I. ÅSTRAND. Physical training in sedentary middle-aged and older men. I. Medical Evaluation. Scand. J. clin. Lab. Invest. 24:315, 1969.

14. LEVIN, K., AND S. LINDE. Determination of glucose in blood, cerebrospinal fluid and urine with a new glucose-oxidase reagent. J. Swed. Med. Assoc. 95:3016, 1962.

15. LINDHOLM, B. Body cell mass during long-term treatment with cortisone and anabolic steroids in asthmatic subjects. Acta Endocr. (Kbh) 55:22, 1967.

16. SALANS, L.B., J.L. KNITTLE, AND J. HIRSCH. The role of adipose cell size and adipose tissue insulin sensitivity in the carbohydrate intolerance of human obesity. J. Clin. Invest. 47:153, 1968.

17. SIMS, E., AND E.S. HORTON. Endocrine and metabolic adaptation to obesity and starvation. Amer. J. Clin. Nutr. 21:1455, 1968.

18. VARNAUSKAS, E., P. BJÖRNTORP, M. FAHLÉN, J. PREROVSKY, AND J. STENBERG. Effects of physical training on exercise-blood flow and succinic dehydrogenase activity in skeletal muscle. Cardiovasc. Res., in print, 1970.

METABOLISM OF FREE FATTY ACIDS DURING EXERCISE IN PATIENTS WITH

OCCLUSIVE ARTERIAL DISEASE OF THE LEG

Lars Hagenfeldt, Bengt Pernow and John Wahren

Departments of Clinical Chemistry and Clinical Physiology, Karolinska Institutet at Serafimerlasarettet Stockholm, Sweden

Several studies on turnover rate and oxidation of free fatty acids (FFA) have clearly shown that utilization of FFA accounts for a substantial fraction of the oxidative metabolism in exercising muscle (for a recent review see Carlson, Boberg and Högberg, (2)). The fraction of FFA entering the muscle that is directly oxidized may be determined from the extraction of ^{14}C-FFA together with the output of $^{14}CO_2$ and has been reported to be 75-100 per cent during exercise with the legs (10,11). For exercise with the forearm the corresponding range is 40-100 per cent; the fractional oxidation correlated negatively to the lactate/pyruvate ratio and positively to the oxygen consumption (5,6). The present study was undertaken in order to characterize the relationship between oxygen supply and oxidation of the FFA taken up by the exercising muscles. For this purpose patients with impaired blood flow capacity of the leg due to occlusive arterial disease were studied during bicycle exercise and compared with a group of healthy subjects.

METHODS

Subjects and Procedure

The patient group consisted of 12 male subjects (age 43-64 years) with angiographically verified occlusive vascular lesions at various levels of the leg arteries (Table 1). They all had typical intermittent claudication during exercise with a walking distance of 20 -> 300 m. In 5 patients an identical study was performed 3-5 months after reconstructive surgery, when they had all achieved a walking distance of >300 m.

Table I. Age (years), duration of symptoms, absolute claudication distance (m) and arteriographic findings in the patient material. The walking tolerance was studied on the tread mill with a 5° inclination at a speed of 2.5 miles/hr and continued until the pain forced the patient to stop. (+++ = total occlusion, ++ = obliteration > 75 % of the lumen, + = obliteration < 75 % of the lumen).

	Age	Duration of sympt.	Claud. dist.	Angiography			
				Iliaca	Fem.com.	Fem.sup.	Popl.
F.N.	57	1 1/2	90	+++	+++		
H.B.	61	1/2	230	+++		+++	
R.J.	64	1 1/2	90	+++	+	+	+
M.E.	60	1	120	+++			
H.E.	45	1	>300	++			
G.N.	61	6	>300	+	++	+++	+
W.B.	63	10	180	+	+	+++	++
B.A.	48	2	115			+++	++
T.C.	43	2	130	+		+++	+
D.J.	52	14	180	+	+	+++	+
C.K.	50	5	>300	+	+	+++	
N.L.	50	1	230			+++	

The <u>control</u> subjects were five healthy male volunteers (age 40-52 years).

The studies were performed in the morning with the subjects in the postabsorptive state. Teflon catheters were placed percutaneously in a brachial artery and a forearm vein. A third catheter was inserted into a femoral vein and the tip placed 20-25 cm below the inguinal ligament. Many of the subjects in the patient group had bilateral symptoms during exertion but the more diseased leg was catheterized in every case.

Albumin-bound oleic acid-I-^{14}C was given continuously as an i.v. infusion (about 0.6 µCi/min) during a rest period of 20 min and during the following exercise period of 35-55 min. The exercise was performed in the upright position on a bicycle ergometer. The work load chosen in the patient group was the highest predetermined individual level, which the subject was able to tolerate for at least 30 min. The work loads used were 150-400, (mean 250), kpm/min. All controls exercised at 400 kpm/min. Blood samples for analyses of individual FFA and FFA radioactivity, $^{14}CO_2$, oxygen and carbon dioxide content glucose and lactate concentration were drawn repeatedly from the artery and the femoral vein at rest and during exercise.

Determinations of pulmonary oxygen uptake were made at rest and twice during the exercise period.

Labeled material. Oleic acid-I-^{14}C (specific activity 53.7 mCi/mM) was obtained from NEN Chemicals (Germany). It was bound to human serum albumin as described earlier (6).

Analytical methods used for individual plasma FFA, total radioactivity in the FFA fraction and $^{14}CO_2$ in blood have been described earlier (6). Oxygen content was calculated after spectrophotometric determination of hemoglobin concentration and oxygen saturation (8); blood RQ was calculated after analysis of blood samples by the van Slyke technique. Blood lactate was analyzed with an enzymatic method (12). Glucose was analyzed with a commercial glucose oxidase analytical kit (Kabi, Stockholm, Sweden). Expired air was collected in Douglas bags and analyzed according to the Scholander technique.

Calculations

The following variables were calculated:

$$\text{Fractional uptake of FFA } (f) = \frac{^{14}C\text{-}FFA_a - {}^{14}C\text{-}FFA_v}{^{14}C\text{-}FFA_a}$$

The subscripts a and v refer to arterial and venous concentrations respectively.

Uptake of FFA, $(U_{FFA}) = FFA_a \cdot f$

Release of FFA, $(R_{FFA}) = U_{FFA} - (FFA_a - FFA_v)$

$$\text{Fractional oxidation of FFA } (f_{ox}) = \frac{(a-v)\ {}^{14}CO_2 \cdot 100}{(a-v)\ {}^{14}C\text{-}FFA\ (1-Hct)}$$

$$\text{Turnover rate of FFA} = \frac{\text{dpm/min infused FFA}}{\text{FFA specific activity}}$$

$$\text{Oxygen consumption for FFA oxidation} = \frac{24.7 \cdot U_{FFA} \cdot f_{ox}\ (1-Hct)}{(a-v)\ O_2}$$

$$\text{Leg blood flow} = \frac{0.75 \cdot (a-v)\ O_2\ \text{rest} + 0.8\ (V_{O_2}\ \text{exercise} - V_{O_2}\ \text{rest})}{(a-v)\ O_2\ \text{exercise}}$$

All values are given as mean ± SE.

508

RESULTS AND COMMENTS

Preoperative Studies

Oxygen uptake and heart rate (Fig. 1). The mean oxygen uptake in the lungs was somewhat higher in the controls than in the patients. The difference, being probably significant ($p<0.05$) is evidently due to the fact that the controls exercised at a somewhat higher mean working intensity than did the patients. Heart rate during exercise was, however, significantly ($p<0.001$) higher in the patient group than in the controls. This difference in heart rate may indicate differences in physical fitness between the two materials. To some extent it may also be connected with the fact that all the patients complained of fatigue and pain in the thigh and calf muscles during the latter part of the exercise period, while the work was stopped arbitrarily in the controls without any symptoms. As shown by Asmussen et al. (1), muscle pains during submaximum exercise induce an increase in heart rate.

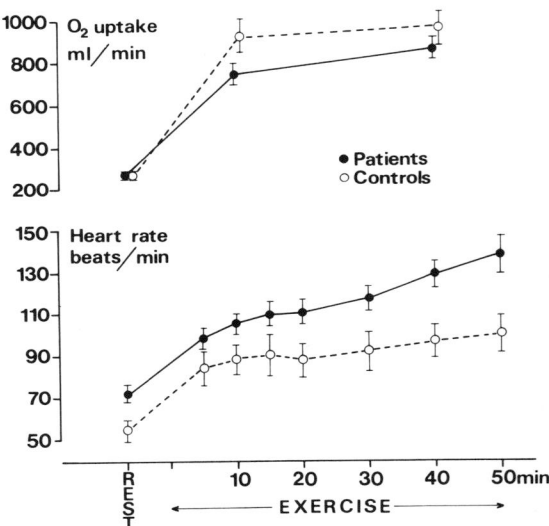

Fig. 1. Oxygen uptake in the lungs and heart rate at rest (supine) and during exercise. Mean values ± SE are given. The difference in oxygen uptake between controls and patients is probably significant ($p<0.05$) while the difference in heart rate is highly significant ($p<0.001$) both at rest and during exercise.

Blood oxygen and lactate (Fig. 2). Oxygen saturation of the femoral venous blood decreased during exercise to significantly ($p<0.01$) lower values in the patients than in the controls. The

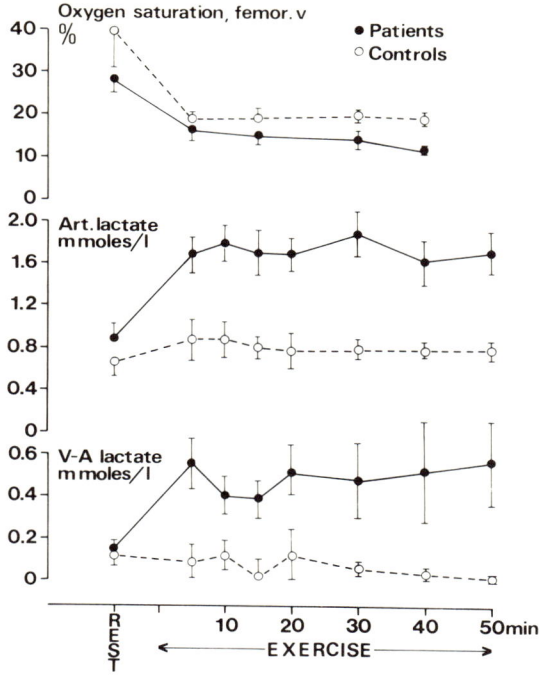

Fig. 2. Femoral venous oxygen saturation, arterial lactate and femoral v - a lactate difference at rest (sitting on the bicycle) and during exercise. The values are given as mean ± SE. The difference between controls and patients is significant in respect of oxygen saturation at the end of exercise ($p < 0.01$) as well as arterial lactate ($p < 0.001$) and v - a lactate ($p < 0.001$) during the whole exercise.

lowest observed individual value for femoral venous oxygen saturation was 3 per cent.

The arterial lactate concentration and the femoral v - a lactate difference at rest were similar in the two materials. During work, however, both arterial and v - a lactate increased to significantly higher values ($p < 0.001$) in the patients than in the controls.

Uptake and release of FFA (Fig. 3). A net uptake of FFA in the leg muscles was observed in all control subjects and in 9 of 12 patients. When the position was changed from supine to sitting both uptake and release of FFA fell considerably, release more than

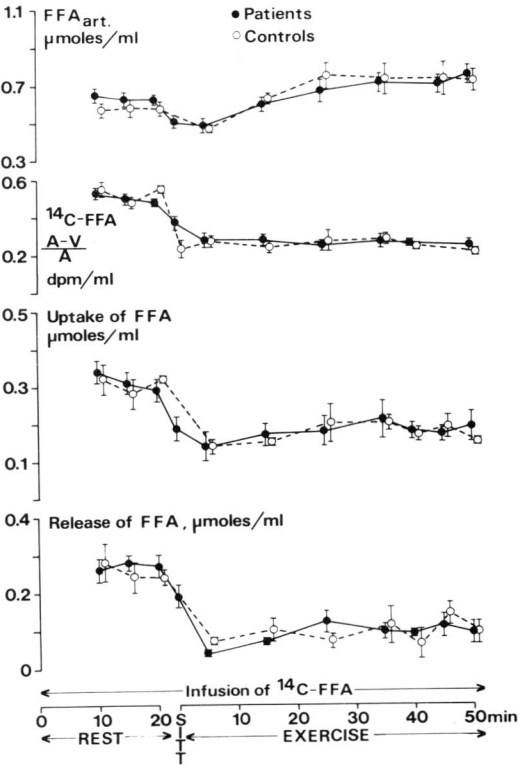

Fig. 3. Arterial concentration, fractional uptake, uptake and release of FFA at rest (supine and sitting) and during exercise. The values are given as mean ± SE. In no respect are there any significant differences between controls and patients ($p > 0.05$).

uptake, which might explain the decrease in arterial FFA concentration in connection with the change in body position at rest. The specific activity of FFA was higher in all subjects in the sitting than in the supine position.

During exercise both the arterial and the venous FFA concentration rose with an almost unchanged a - v difference. Specific activity was in all cases lower during exercise than at rest. A considerable a - v ^{14}C-FFA difference was observed, greater at rest than during exercise. Calculated uptake and release of FFA expressed per 1 blood decreased at the beginning of exercise but then remained almost unchanged during continued work.

No significant differences were found between patients and

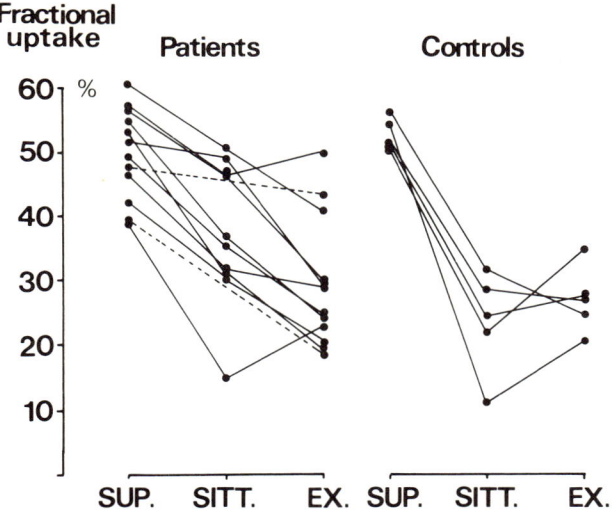

Fig. 4. Individual values for fractional uptake of FFA at rest (supine and sitting) and during exercise.

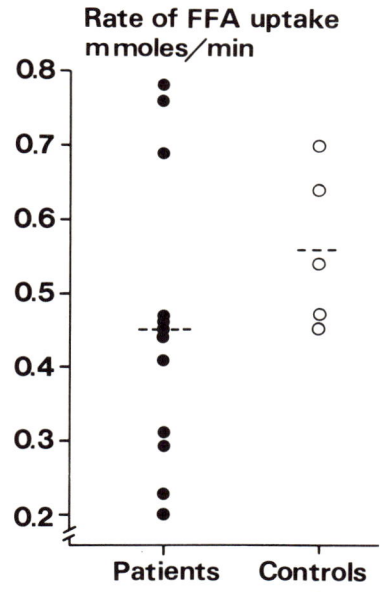

Fig. 5. Calculated rate of FFA uptake (U_{FFA} × plasma flow) during exercise. The difference between controls and patients is probably significant ($p < 0.05$).

controls as regards plasma concentration, uptake and release of FFA at rest or during exercise. The interindividual variations were, however, great. Three patients had a higher fractional extraction of FFA than any of the controls (Fig. 4). They all had occlusions in the iliac artery and the lowest walking tolerance in the material.

The amount of FFA taken up per min was lower in the patient group than in the controls during exercise (Fig. 5).

<u>Turnover rate of FFA</u> (Fig. 6). The turnover rate of oleic acid in relation to its arterial concentration at rest was the same for controls and patients. This value increased considerably during exercise in all subjects, the rise being steeper in the control group than in the patients. The values obtained in the controls agree with earlier findings (10). The increase during exercise is probably due to an increased removal of FFA in the working muscles brought about by the increase in blood flow. Consequently the difference in turnover rate between controls and patients during exercise may reflect differences in muscle blood flow.

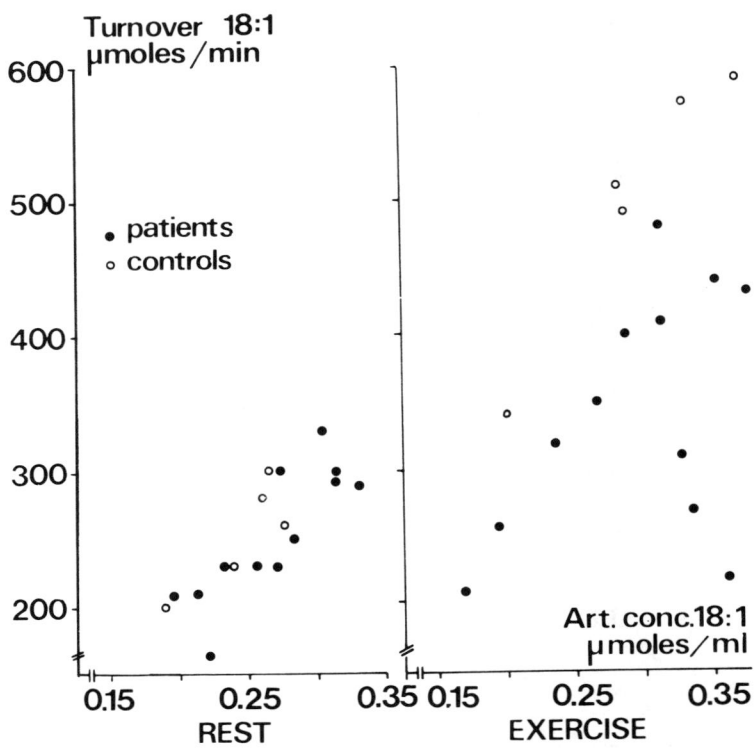

Fig. 6. Turnover rate of oleic acid at rest and during exercise in relation to arterial concentration of oleic acid.

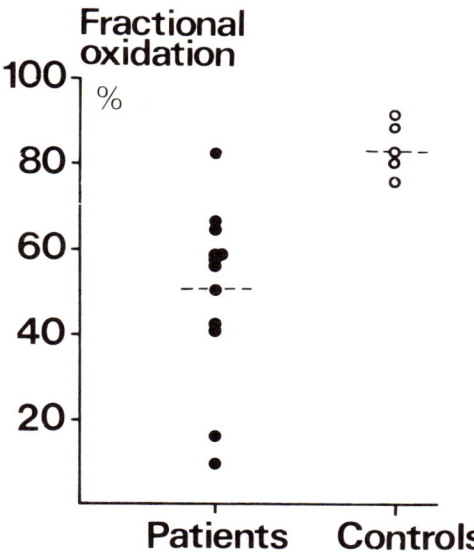

Fig. 7. Fractional oxidation of FFA during exercise. The difference between controls and patients is highly significant ($p < 0.001$).

Fractional oxidation of FFA (Fig. 7). The oxidation of FFA taken up by the leg muscles was studied by simultaneous analyses of the a - v $^{14}CO_2$ and a - v ^{14}C-FFA differences. The release of $^{14}CO_2$ occurred rapidly during exercise and reached a reasonably steady state after about 25 min of exercise. In the controls the fraction of FFA taken up by the muscles and immediately oxidized, calculated as the total recovery of radioactivity from FFA as $^{14}CO_2$ when the a - v $^{14}CO_2$ difference was stabilized, was 76-91 per cent, which agrees fairly well with earlier observations (11). Corresponding values in the patient group were 10-84 per cent. Only case CK among the patients had a fractional oxidation within the range of the controls. He had a distal obliteration and the longest walking tolerance in the patient group. The lowest oxidation values were obtained in the two patients with the most pronounced proximal obliterations and lowest walking tolerances. A linear relation was found between fractional oxidation and arterial lactate concentration (Fig. 8).

In three patients the v - a difference of radioactivity in a perchloric acid blood extract was found to account almost completely (> 90 per cent) for the difference between ^{14}C-FFA uptake and $^{14}CO_2$ production. More than 50 per cent of this water-soluble radio-

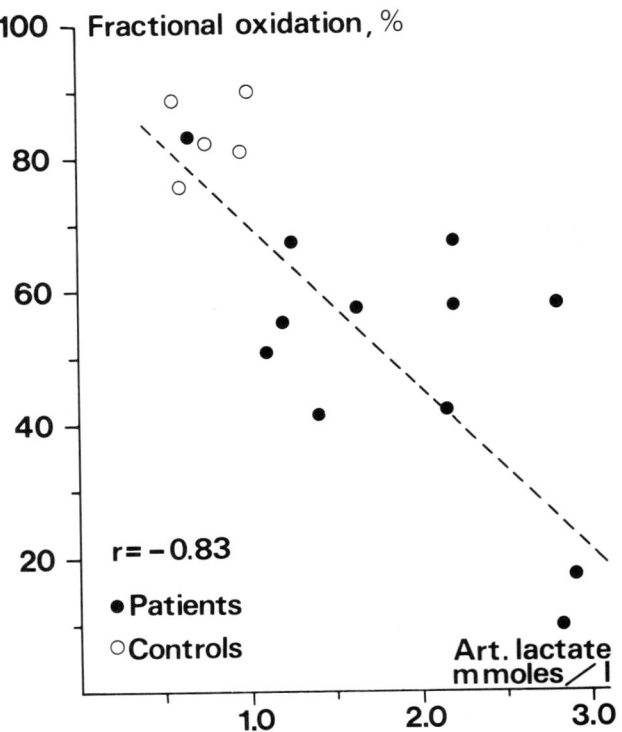

Fig. 8. Fractional oxidation of FFA during exercise in relation to arterial lactate concentration. The correlation is highly significant ($p < 0.001$).

activity has been identified as labeled acetate. The interpretation of this finding has been discussed in a previous communication and the ADP - ATP exchange between cytoplasm and mitochondria has been proposed as the limiting step in the oxidative process (7). In the present patients, with extremely low oxygen saturation in the venous blood, the oxygen supply may have been a further limiting factor.

Oxygen consumed for fat and carbohydrate oxidation (Fig. 9). From the RQ and (a - v) oxygen difference of the leg it was calculated that during exercise, fat combustion accounted for about 80 per cent of the oxygen consumption in the controls, while the corresponding value for the patients was about 40 per cent. Taking the fractional oxidation into consideration, no difference was found between the controls and patients concerning the fraction of FFA in the total fat oxidation (36 and 42 per cent respectively). However, since the fractional oxidation of FFA was considerably lower

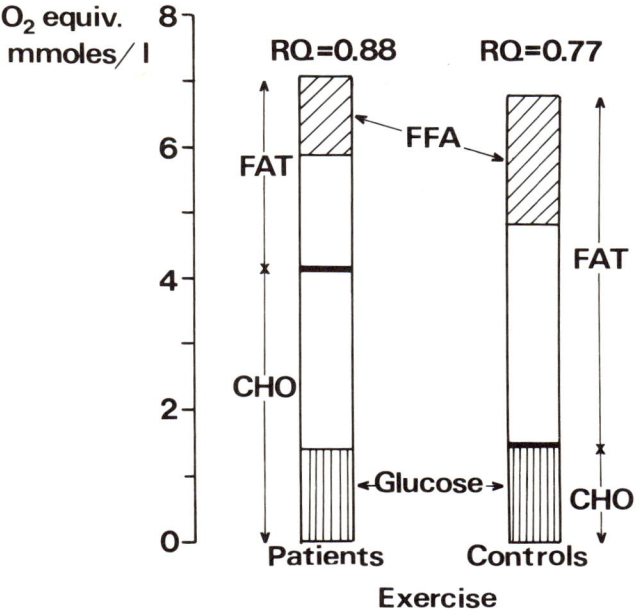

Fig. 9. The proportions of fat and carbohydrates (CHO) as substrates for the leg muscles during exercise, calculated from leg a - v oxygen difference and RQ and expressed as oxygen equivalents. The fractional oxidation values are included in the calculation of oxygen consumed for FFA oxidation. Data for glucose oxidation are based on glucose uptake minus lactate production. For further explanation, see text.

in the patients than in the controls at identical uptake values, it follows that the amount of oxygen consumed for FFA oxidation was lower in the patients than in the controls (mean values 17 and 32 per cent respectively of the (a - v) oxygen difference). The value obtained in the controls agree well with earlier results for leg exercise at comparable work intensity, indicating that 27 per cent of expired CO_2 was derived from FFA (9). Similarly Hagenfeldt and Wahren (4) found that in the forearm, the uptake of palmitic, oleic and linoleic acid accounted for 26 per cent of the oxygen consumption.

The oxygen used for glucose oxidation, assuming complete oxidation of all glucose taken up and not converted to lactate, was similar in the controls and the patients (mean values 1.7 and 1.4 mmoles/l respectively). The caloric expenditure derived from carbohydrate metabolism, as calculated from RQ, was completely covered

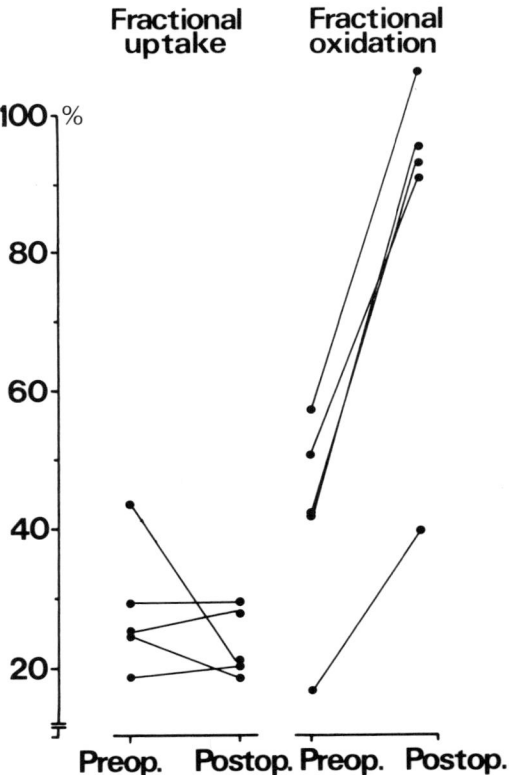

Fig. 10. Fractional uptake and oxidation of FFA during exercise before and 3-5 months after reconstructive surgery.

by glucose oxidation in the controls. In the patients, however, glucose accounted for only 35 per cent of the total carbohydrate combustion.

Postoperative Studies

When identical studies were repeated 3-5 months after reconstructive surgery (5 patients), no significant difference was observed in oxygen uptake, heart rate, arterial concentration of FFA, fractional uptake or release of FFA at rest and during exercise. Fractional oxidation was, however, considerably higher postoperatively, with a complete normalization in four patients and an increase, though not to normal values, in one patient (Fig. 10). The turnover rate of oleic acid during exercise was higher in the two patients with the lowest preoperative values and almost unchanged in the others, who had had only slightly decreased turnover rates preoperatively (Fig. 11).

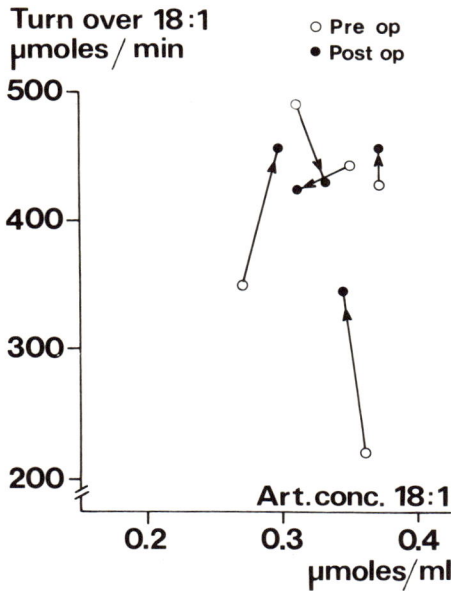

Fig. 11. Turnover rate of oleic acid during exercise in relation to arterial concentration of oleic acid pre- and postoperatively.

SUMMARY AND CONCLUSION

Uptake, release and oxidation of FFA were studied in 5 healthy subjects and 12 patients with decreased walking tolerance due to arterial occlusions of the leg arteries. The metabolic variables were estimated from arterial and femoral venous blood during a continuous intravenous infusion of ^{14}C-FFA at rest and during a 35-55 min exercise period on a work load which gave mild to moderate symptoms of claudication at the end of the exercise period. No intergroup difference was found in oxygen uptake and (a - v) oxygen difference over the leg during exercise, while heart rate and blood lactate was significantly higher in the patient group. No difference was found in the arterial FFA concentration, release or fractional uptake of FFA, while the rate of uptake was lower in the patient group. The turnover rate and the fractional oxidation of FFA taken up by the muscles were likewise significantly lower among the patients as compared to the controls. The fraction of the (a - v) oxygen difference utilized for FFA oxidation was lower in the patients. The radioactivity not recovered as $^{14}CO_2$ was almost completely found in a perchloric acid blood extract. More than 50 per cent of this water-soluble radioactivity was identified as labeled acetate. When identical studies were repeated 3-5 months after reconstructive surgery (5 patients), no difference was observed in fractional uptake, while fractional oxidation of FFA was normalized.

It is concluded that an impairment of the blood flow capacity of the leg brings about not only a decrease in the rate of inflow of metabolities contributing to the oxidative metabolism in the leg but also a reduced capacity to oxidize the FFA taken up by the muscles.

ACKNOWLEDGEMENTS

This work was supported by grants from Karolinska Institutet and the Swedish National Association against Heart and Chest Disease.

REFERENCES

1. ASMUSSEN, E., E. HOHWÜ CHRISTENSEN, AND M. NIELSEN. Kreislaufgrösse und cortikal-motorische Innervation. Scand. Arch. Physiol. 83: 181-187, 1940.

2. CARLSON, L.A., J. BOBERG, AND B. HÖGSTEDT. Some physiological and clinical implications of lipid mobilization from adipose tissue. In: Handbook of Physiology, Adipose tissue. Washington: Am. Physiol. Soc., 1965, sect. 5, p. 625.

3. DONALD, K.W., P.N. WORMLAND, S.H. TAYLOR, AND J.M. BISHOP. Changes in the oxygen content of femoral venous blood during exercise in relation to cardiac output response. Clin. Sci. 16: 567-591, 1957.

4. HAGENFELDT, L., AND J. WAHREN. Simultaneous uptake and release of individual free fatty acids in human forearm muscle during exercise. Life Sci. 5: 357-364, 1966.

5. HAGENFELDT, L., AND J. WAHREN. Production of B-hydroxybutyrate from FFA in working muscle during anaerobic conditions. Excerpta med. Intern. Congr. Series No. 172S: 218-220, 1967.

6. HAGENFELDT, L., AND J. WAHREN. Human forearm muscle metabolism during exercise. II. Uptake, release and oxidation of individual FFA and glycerol. Scand. J. clin. Lab. Invest. 21: 263-276, 1968.

7. HAGENFELDT, L., AND J. WAHREN. Metabolism of free fatty acids and ketone bodies in skeletal muscle. In: Muscle Metabolism during Exercise. New York: Plenum Publ. Corp., 1970, this volume.

8. HOLMGREN, A., AND B. PERNOW. Spectrophotometri measurement of oxygen saturation of blood in the determination of cardiac output. A comparison with the van Slyke method. Scand. J. clin. Lab. Invest.11: 143-149, 1959.

9. HAVEL, R.J., L.A. CARLSON, L-G. EKELUND, AND A. HOLMGREN. Turnover rate and oxidation of different free fatty acids in man during exercise. J. appl. Physiol. 19: 613-618, 1964.

10. HAVEL, R.J., A. NAIMARK, AND C.F. BORCHGREVINK. Turnover rate and oxidation of free fatty acids of blood plasma in man during exercise: Studies during continuous infusion of palmitate-I-C^{14}. J. clin. Invest.42: 1054-1063, 1963.

11. HAVEL, R.J., B. PERNOW, AND N.L. JONES. Uptake and release of free fatty acids and other metabolites in the legs of exercising men. J. appl. Physiol. 23: 90-96, 1967.

12. WAHREN, J. Quantitative aspects of blood flow and oxygen uptake in the human forearm during rhythmic exercise. Acta physiol. scand.67: Suppl. 269, 1966.

CLINICAL AND METABOLIC ASPECTS ON OBESITY IN CHILDHOOD

Göran Sterky

Department of Pediatrics at St. Göran's Children's Hospital, Box 12500, S-11281 Stockholm, Sweden

The etiology of obesity in childhood is very probably multifactorial. We know that genetic, nutritional, environmental, and hormonal factors play important roles. Their quantitative interrelationships are, however, unknown. In obese patients it is possible to observe a number of differences from the normal variation as regards biochemical, hormonal and physiological patterns. Most of them seem to be adaptation to obesity since they are reversible by weight reduction (10). Some data on an illustrative case, a girl we observed a number of years ago are given in table 1.

To be able to identify a child predisposed to obesity before he accumulates excessive amount of fat, it is necessary to have reliable information about normal growth and development. In this connection changes in body composition and the influence by physical work and training are especially important. Few data obtained on adults can be used on growing individuals, e.g. blood and muscle lactate response to work are lower in children than in adults (9). Due to the great individual variation and the influence of growth it is necessary to have a thorough information about each individual taking part in a therapeutic program in order to be able to evaluate its effect. To minimize differences one should use some expression of biological age instead of chronological one.

The work by Hirsch (18) and others suggest a genetic determination of the number of cells in adipose tissue and a nutritional influence on the size. If one can interpret such observations on animals (11) as relevant to humans it would, however, be possible to influence even the number of cells during the first 9 - 11 months postnatally. Thus both under- and overnutrition early in life could change a child's potential growth pattern.

Table 1. Girl born 1956. Heredity for obesity. Overweight since age 5. Good physical activity. Overconsumption.

Age years	Height cm	Weight kg	Weight level "S.D."	K_G	FFA mmol/l 0'	10'	60'	Aerobic power $\dot{V}_{O_2}$ ml/kg × min
9 2/12	143	54	+4.6	0.9	1.41	1.17	0.43	22
9 7/12	147	45	+1.4	0.8	1.06	1.02	0.43	
10 2/12	153	44	±0	2.6	0.67	0.65	0.67	34

In obesity both number and size or adipose cells are increased. Recent observations by Cheek (5) suggest also that there are sex differences in adipose cellularity in obesity in childhood, with boys in a more unfavourable situation. In most obese children not only bodyfat but lean bodymass is increased (5). The physically active child has more lean bodymass and less fat than the inactive one (13). There are observations that some babies are less active than others already during the first months of life, and that those babies are on the average heavier than others (17). During the last years two studies of clinical material suggest that an early excessive weight gain (7) or excessive weight at one year, will result in higher proportion of obese children at schoolage. So far it has, however, been impossible to prove that a reduction of physical activity causes obesity.

From a clinical point of view it is, however, enough to know that physical inactivity might play a role in the development of obesity. When already obese the habitual physical activity is low. In unselected populations obese children do not have excessive caloric intake (2). Many authors have reported low aerobic power in obese children (3). Up to a certain degree of overweight there seem to be a normal relationship between work capacity and circulatory dimensions, such as heart size and blood volume (20).

In our previous studies (16) we found that the duration of obesity played a more important role than the degree itself in explaining metabolic deviations. In all studies we have attempted to select materials in a similar way, i.e. from the schoolnurse's reports and not from our outpatient clinic (3). However, as pediatricians we also have to treat those patients, who seek our advice. Our experience from previous studies (2) is that a physical training program has more influence on weight and fat reduction than any kind of caloric restriction. We have also seen a relationship between the degree of training and the loss of fat storage. It is, however, not necessary to attempt to reduce bodyweight but to prevent it from increasing. Treatment should start before puberty as weight reduction in adolescence might cause decrease of lean body mass.

In an attempt to study "preobesity", slightly overweight boys were offered a physical training program over one years time. The boys had extra gymnastic sessions twice weekly for one hour each. In the beginning each boy was given careful instructions to overcome his inability to perform even simple exercise, but the intensity was then increased, and group training of circulatory dimensions took place. We now have some preliminary data (20), which are given below.

In spite of a great range in work capacities the individuality the mean heart rate at a standard work load is given in Fig. 1. The decrease are, as one might expect, most pronounced during the short intense training periods. We also observed a decrease of body weight and a reduction of skinfold thickness (Fig. 2). The individual variation was great, but almost every one did reduce in body fatness. In comparison with the expected weight increase over the observation period our obese boys only increased 1/6. The degree of overweight, measured as standard deviations for respective height, was significantly reduced. One year afterwards, without any attempt to influence the boys during this time, we observed that they had kept their lower overweight level (Table 2).

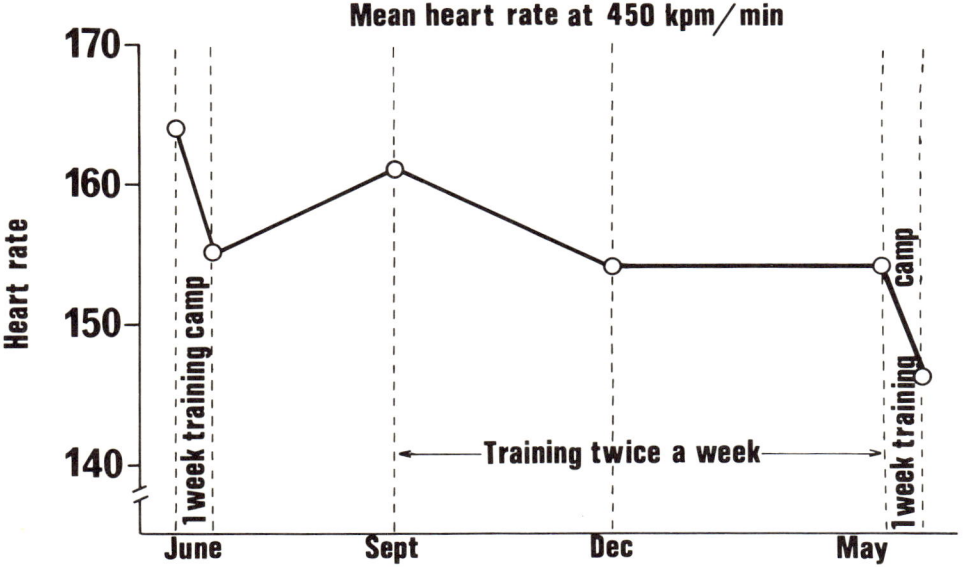

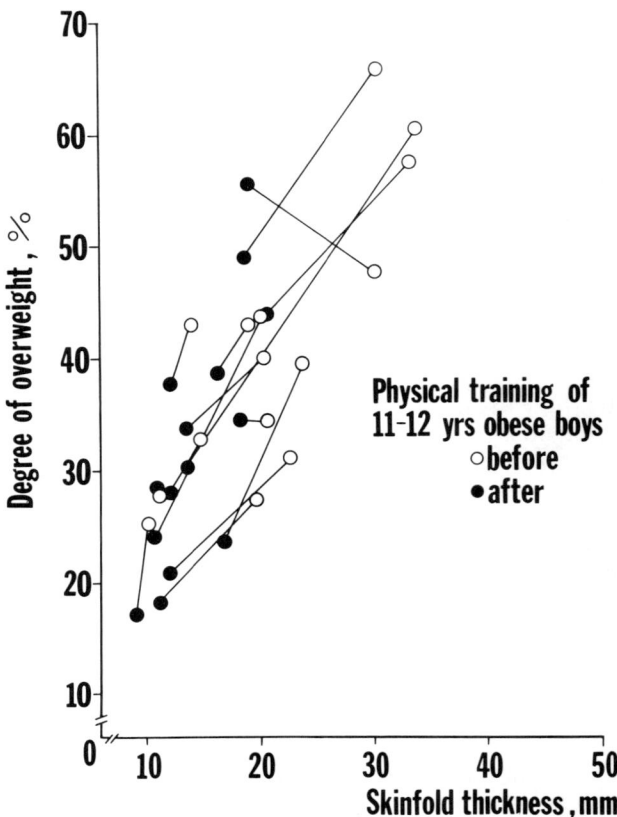

Table 2. Effect of physical training on degree (S.D.) of overweight in 16 preadolescent boys.

Year	Mean	Range
1968	+ 3.7 S.D.	(2.4 - 5.4)
Training program		
1969	+ 2.5 S.D.	(0.2 - 4.0)
No program		
1970	+ 2.6 S.D.	(0.7 - 4.6)

Blood lipids, FFA glycerol and β-hydroxybutyrate were determined in the fasting state and intravenous glucose tolerance test was performed before and after training. The individual variation in lipid response was great but most of the boys showed a decrease of cholesterol and FFA but an increase in triglycerides. One might very well ask, which parameter is most important and life treatening?

As regards hormonal levels others have reported that no obese child has a normal growth hormone response to insulin hypoglycemia and/or to arginine infusion (14). The response to exercise has not been studied in obese children. Obese adults respond with very high levels (19). Normal weighing children are as well capable of increasing there growth hormone level tremendously (8). Insulin levels in obese children are extremely varying. Some investigators report high fasting levels only in those with heredity for diabetes (14), others only in those with long standing obesity (6). The insulin responsiveness is said to be dependent on adipose tissue cellsize but the glucose metabolism related to number of cells (18).

Our finding of glucose-insulin relationships before and after training are reported in Table 3. Both fasting blood glucose and glucose tolerance tended to increase. The most consistent change was an increase in the relative insulin response (21), which is an expression of the glucose insulin interrelationships during the release of the rapidly releasable insulin pool. Factors during exercise such as epinephrine release might selectively block this insulin response (12). Differences against published data might partly be explained by the slight degree of overweight in our cases, and the unknown influence by one years growth.

Table 3. Glucose and insulin before and after 1 yr of physical training.

Case No.	Fasting plasma glucose mg/100 ml		Fasting plasma insulin µU/ml		Glucose tolerance K_G		Relative insulin response RIR	
	b	a	b	a	b	a	b	a
4	69	80	16	6	2.2	2.2	0.08	0.13
6	64	75	12	8	1.4	2.0	0.09	0.14
7	68	64	5	5	1.6	2.4	0.10	0.14
18	77	90	15	6	1.8	1.5	0.02	0.09
20	74	82	5	8	1.3	2.5	0.04	0.12
28	90	80	4	4	1.2	1.5	0.12	0.19

There is no doubt that there are effects by physical training on obese children. The aim of our studies has been to design individual training and nutritional programs, based on physiological and biochemical observations. However, as this is not yet possible, we have decided to use individual as well as group therapy. As you all know there is in the community great psycho-social problems (15) connected to obesity, which has to be born in mind before designing too rigid programs. We will offer the children, the parents and the school a program over one term during which time we will try teaching them to take a more lasting responsibility. We thus hope to be able to handle a greater number of obese children. The hospital will thus try to support the schoolhealth in prevention of obesity as well as treatment. I do hope, it will turn out to be a good application of muscle metabolism during exercise.

REFERENCES

1. ASHER, P. Fat babies and fat children. Arch. Dis. Childh. 41: 672, 1966.
2. BLOMQUIST, B., M. BÖRJESON, Y. LARSSON, B. PERSSON, AND G. STERKY. The effect of physical activity on the body measurements and work capacity in overweight boys. Acta Paed. Scand. 54:566, 1965.
3. BÖRJESON, M. Overweight children. Acta Paed. Scand. Suppl. 132, 1962.
4. CARNELUTTI, M., M.J. DEL GUERCIO, AND G. CHIUMELLO. Influence of growth hormone on the pathogenesis of obesity in children. J. Ped. 77:285, 1970.
5. CHEEK, D.B., R.B. SCHULTZ, A. PARRA, AND R.C. REBA. Overgrowth of lean and adipose tissues in adolescent obesity. Pediat. Res. 4:268, 1970.
6. CHIUMELLO, G., M.J. DEL GUERCIO, M. CARNELUTTI, AND G. BIDONE. Relationship between obesity, chemical diabetes and betapancreatic function in children. Diabetes 18:238, 1969.
7. EID, E.E. Follow-up study of physical growth of children who had excessive weight gain in first six months of life. Brit. Med. J. 2:74, 1970.
8. ERIKSSON, B., AND J. THORELL. The effect of repeated prolonged exercise on plasma HGH-level and blood substrates in boys and adults compared to rest. III. Int. Symp. of Pediatric Work Physiol. (1970) (to be published in Acta Paed. Scand.).
9. ERIKSSON, B., J. KARLSSON, AND B. SALTIN. Muscle metabolites in 13 years old boys during submaximal and maximal work. III. Int. Symp. of Pediatric Work Physiol. (1970) (to be published in Acta Paed. Scand.).

10. FARRANT, P.C., R.W.J. NEVILLE, AND G.A. STEWART. Insulin release in response to oral glucose in obesity - The effect of reduction of bodyweight. Diabetologia 5:198, 1969.
11. HIRSCH, J., AND P.W. HAN. Cellularity of rat adipose tissue: effects of growth, starvation and obesity. J. Lipid Res. 10:77, 1969.
12. LERNER, R.L., AND D. PORTE, JR. Epinephrine and insulin release: A new look. Diabetes 19, Suppl. 1:366, 1970.
13. PARIZKOVÁ, J. Longitudinal study of the development of body composition and body build in boys of various physical activity. Human Biology 40:212, 1968.
14. PAULSEN, E., L. RICHENDORFER, AND F. GINSBERG-FELLNER. Plasma glucose, free fatty acids and immunoreactive insulin in sixty-six obese children. Diabetes 17:261, 1968.
15. PENICK, S.B., AND A.J. STUNKARD. Newer concept of obesity. Med. Clin. N. Am. 54:745, 1970.
16. PERSSON, B., AND G. STERKY. The effect of prolonged fasting and ketogenic diet on blood lipids and ketone bodies in overweight and normal children. Acta Paed. Scand. 55:153, 1966.
17. ROSE, H.E., AND J. MAYER. Activity calorie intake, fat storage, and the energy balance of infants. Pediatrics 41:18, 1968.
18. SALANS, L.B., J.L. KNITTLE, AND J. HIRSCH. The role of adipose cell size and adipose tissue insulin sensitivity in the carbohydrate intolerance of human obesity. J. Clin. Invest. 47:153, 1968.
19. SCHWARZ, F., D.J. TER HAAR, H.G. VAN RIET, AND J.H.H. THIJSSEN. Response of growth hormone (GH), FFA, blood sugar and insulin to exercise in obese patients and normal subjects. Metabolism 18:1013, 1969.
20. STERKY, G., AND C. THORÉN. (unpublished observations)
21. THORELL, J., B. NOSSLIN, AND G. STERKY. Some considerations on the early insulin response (unpublished observations).

METABOLISM DURING EXERCISE IN PATIENTS WITH DIABETES

J. Östman, E. Cerasi, L.-G. Ekelund, R. Luft and
S. Nordlander

Departments of Endocrinology & Metabolism and Clinical
Physiology, Karolinska Hospital, Stockholm, Sweden

Abnormalities in the mobilization of fatty acid during exercise have been described in certain insulin-deficient states, by Issekutz and co-workers (8) as well as by Havel (6) in depancreatized dogs and by Carlström (1, 2) in human diabetes. Present studies were designed to explore the metabolism of endogenous fatty acids in healthy subjects in the prediabetic state, defined according to Cerasi and Luft (4) as a state characterized by poor insulin response to glucose load and normal glucose tolerance.

Experimental

Studies were undertaken in four small groups of subjects: 1) five insulin-dependent, untreated diabetics with non-acidotic ketosis, 2) three diabetics in clinical remission (insulin had been withdrawn 2, 9 and 27 months respectively before the study), 3) six prediabetic subjects, who had as the only abnormality poor early insulin response (k_{i1}) (3) to glucose load, and 4) nine control subjects with normal insulin response to glucose. Although, statistical analysis was not possible because of the limited size of the samples no obvious dissimilarity of age or body-weight was apparent between the four groups.

Since the increases in FFA and growth hormone in plasma (5, 10) are dependent upon the work load, it seemed appropriate to use in all individuals the same relative work. Thus, in all subjects the maximal working capacity was firstly determined on an electrically braked bicycle ergometer according to the technique of Sjöstrand and co-workers (7, 9, 11). The relative work load later used at the experimental exercise amounted to approximately 60-70 per cent of the maximal working capacity.

Results and discussion

Fig. 1 shows that the concentration of arterial plasma FFA decreased during the exercise in the groups of control subjects. From the individual curves of the six prediabetics, it can be seen that four exhibited the normal response, whereas two had more marked rises in plasma FFA, expecially after the period of exercise. The values were then more than 3 S.D. of those of the control group. Our findings that insulin-dependent diabetics exhibited marked rises in plasma FFA (Fig. 2) are in full agreement with those of Carlström (1, 2). One of the diabetic patients in clinical remission had a response similar to that of the ketotic diabetics.

Figures 3 and 4 show that the changes in the concentration of arterial plasma glycerol were closely related to the changes in plasma FFA, as marked rises in plasma glycerol were observed in all insulin-dependent diabetics, two prediabetics and one patient with diabetes in clinical remission, all of whom had augmented rise in plasma FFA.

The findings of elevated concentrations of FFA and glycerol in insulin-dependent diabetics during and shortly after exercise suggest that the mobilization of lipids from adipose tissue is enhanced, although no turnover studies have been performed. Studies by Havel (6) on the rates of turnover of FFA and glycerol in plasma of diabetic dogs indicate that the increased mobilization of FFA from adipose tissue is due to accelerate levels of lipolysis. A diabetes-like response of FFA and glycerol was presently observed in some prediabetics and in one diabetic in remission. Neither the extent of obesity nor that of training seems to be responsible for the abnormal pattern, since the influence of these two factors has been most likely avoided by the design of the study and the selection of the subjects. It is possible, although by no means proven, that the elevated concentrations of plasma FFA and glycerol in diabetics in clinical remission and in prediabetics are the result of those mechanisms operating in insulin-dependent diabetics. Factors possibly contributing to these changes in the lipid mobilization would be 1) abnormalities in the release of insulin, catecholamines, or growth hormone, 2) changes in the sensitivity of adipose tissue to these hormones, and 3) combinations of these factors.

The concentration of plasma insulin did not appreciably change during exercise, nor did a relationship seem to exist between the insulin concentration and the exercise-induced lipid mobilization found in prediabetics and diabetics in cli-

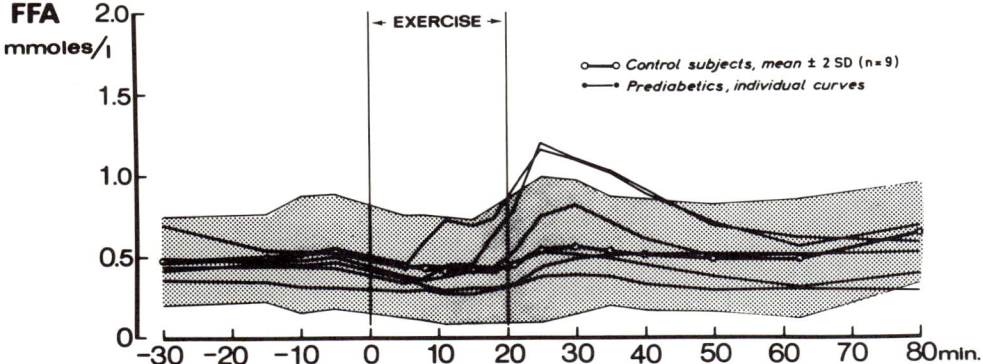

Fig. 1. Effects of exercise in control subjects and prediabetics on the concentration of FFA in arterial plasma.

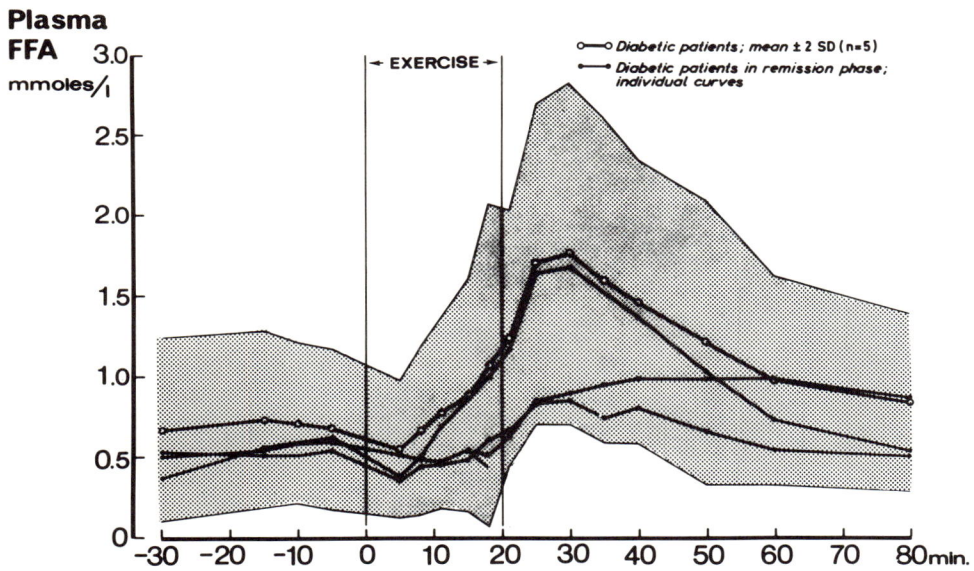

Fig. 2. Effects of exercise in insulin-dependent, untreated diabetics and diabetic patients in remission phase on the concentration of FFA in arterial plasma.

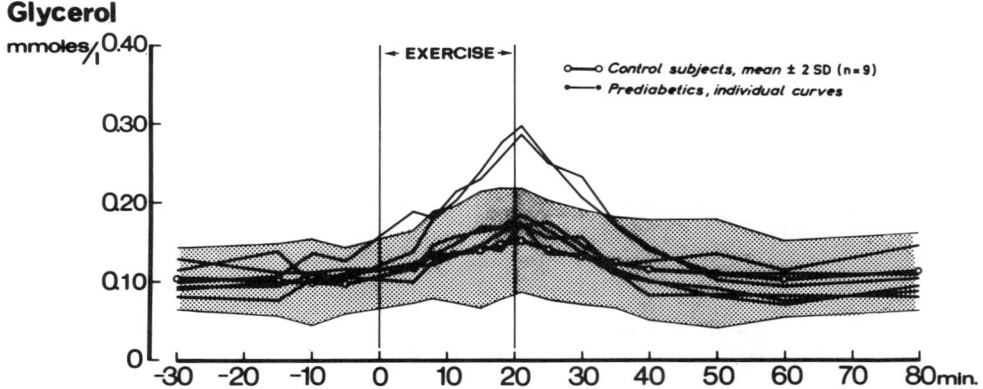

Fig. 3. Effects of exercise in control subjects and prediabetics on the concentration of glycerol in arterial plasma.

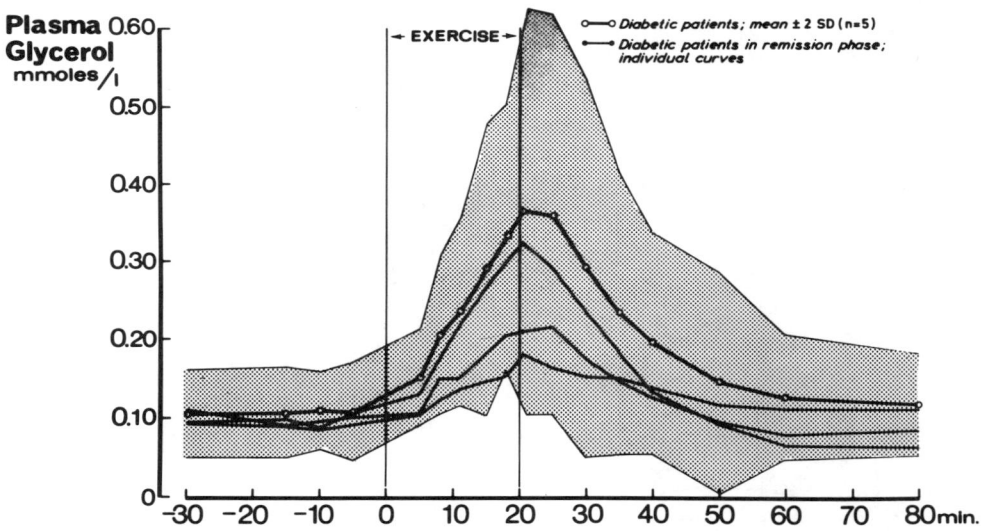

Fig. 4. Effects of exercise in insulin-dependent, untreated diabetics and diabetic patients in remission phase on the concentration of glycerol in arterial plasma.

nical remission. Fig. 5 shows that the increase in pulse rate was similar in all groups. The increase in pulse rate was not more pronounced in the two prediabetics and the remission-diabetic with elevated levels of FFA and glycerol in plasma. Somewhat higher concentrations of arberial lactate during exercise were observed in the control subjects. We have, therefore, tested the possibility that an inverse relationship existed between lactate and either FFA or glycerol. No such correlation was demonstrated. In addition to the catecholamines, growth hormone might have played a role in the excessive mobilization of lipids in the diabetic patients, although previous and present data do not suggest that growth hormone is a major factor in the lipid mobilization in non-diabetic subjects submitted to short-term exercise. Fig. 6 shows tha the mean values ($\pm$ 2 S.D.) of growth hormone found in the group of control subjects increased during the late phase of the exercise period. Furthermore, it can be seen that four insulin-dependent diabetics had elevated pre-exercise levels of growth hormon. In three of those patients there was an expected decrease in the concentration of growth hormone. In one diabetic patient,

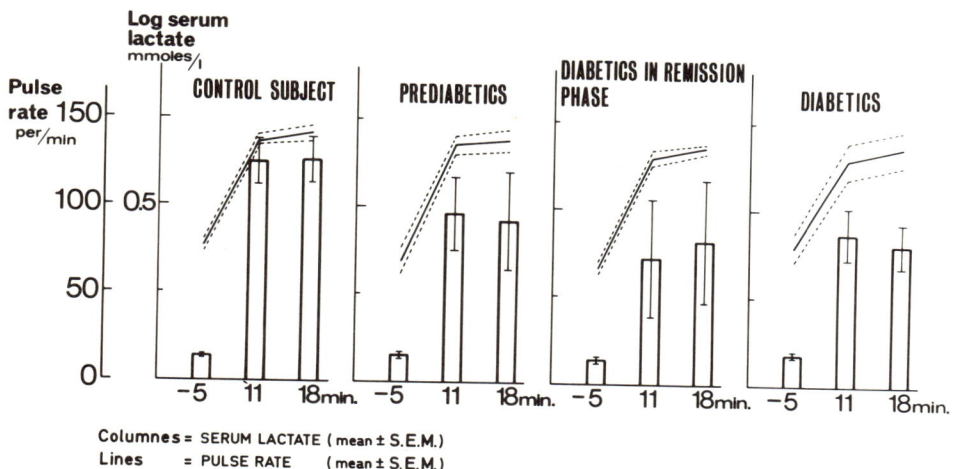

Fig. 5. Effects of exercise in 1) nine control subjects, 2) six prediabetics, 3) three diabetics in remission phase, and 4) five insulin-dependent, untreated diabetic patients on the pulse rate (lines; mean $\pm$ S.E.M.) and the serum lactate (columnes; mean $\pm$ S.E.M.).

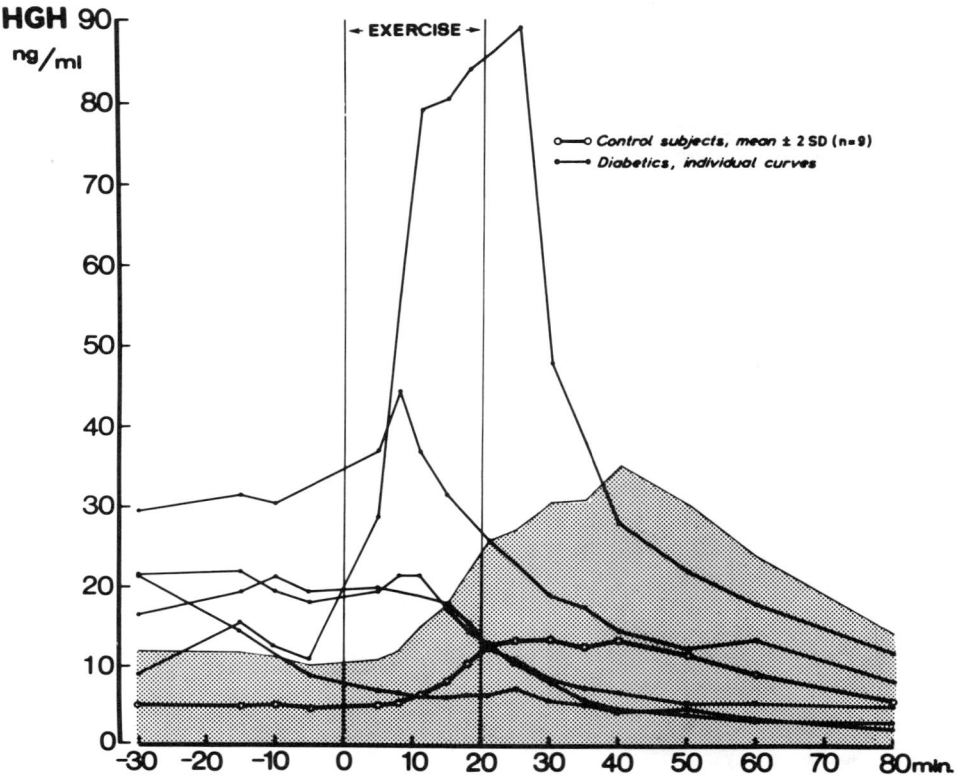

Fig. 6. Effects of exercise in control subjects and five insulin-dependent untreated diabetic patients on the concentration of growth hormone in arterial plasma.

who had a normal baseline of growth hormone, the exercise produced a marked rise in the level of growth hormone, similar to that recently reported by Hansen for juvenile, uncontrolled diabetics (5). One of the prediabetics with an abnormal rise in FFA and glycerol also had a marked elevation in the concentration of growth hormone, whereas all other prediabetics and the three diabetics in clinical remission had a normal response. In general, no association was found between the increases in glycerol and growth hormone produced by the exercise in these two groups of subjects.

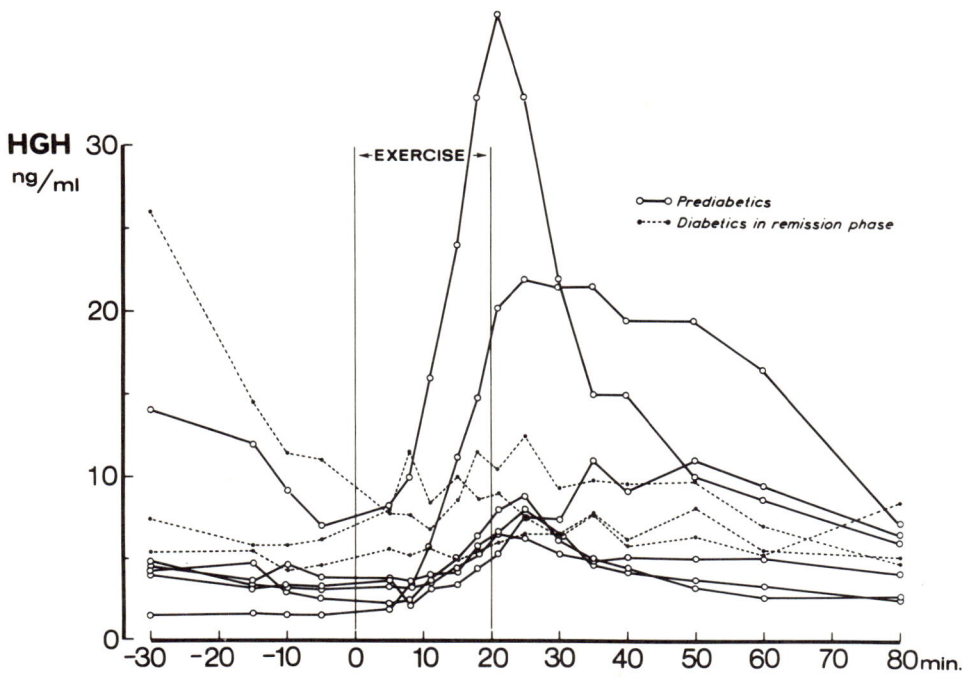

Fig. 7. Effects of exercise in six prediabetics and three diabetics in remission phase on the concentration of growth hormone in arterial plasma.

Summary

Abnormal rises in plasma FFA and glycerol were observed in all insulin-dependent diabetics, one out of three remission-diabetics and two out of six prediabetics when submitted to the same relative work determined by bicycle ergometer. Additional studies as well as clinical follow-up are required to find out whether or not the mechanisms which lead to the abnormal rises of FFA and glycerol in overt diabetes mellitus are the same as those existing in prediabetics and remission-diabetics.

References

1. Carlström, S. & Karlefors, T., Plasma-free-fatty acids during exercise. Lancet i, 331 (1964).

2. Carlström, S., Studies on fatty acid metabolism in male diabetic patients during exercise. Acta univers. Lund. Sectio II, n:o 20 (1967) (Thesis) C.W.K. Gleerup, Sweden.

3. Cerasi, E., An analogue computer model for the insulin response to glucose infusion. Acta endocr. (Kbh.) 55, 163 (1963).

4. Cerasi, E. & Luft, R., "What is inherited – what is added" hypothesis for the pathogenesis of diabetes mellitus. Diabetes 16, 615 (1967).

5. Hansen, A.P., Abnormal serum growth hormone response to exercise in juvenile diabetics. J. clin. invest. 49, 1467 (1970).

6. Havel, R., Some influences of the sympathetic nervous system and insulin on mobilization of fat from adipose tissue: Studies on the turnover rates of free fatty acids and glycerol. Ann. N.Y. Acad. Sci., 131, 91 (1965).

7. Holmgren, A. & Mattsson, K.-H., A new ergometer with constant work load at varying pedalling rate. Scand. J. clin. Lab. invest. 6, 137 (1954).

8. Issekutz, B., Miller, H.J. & Rodahl, K., Effect of exercise on FFA-metabolism of pancreatectomized dogs. Am. J. Physiol. 205, 645 (1963).

9. Sjöstrand, T., Changes in the respiratory organs of workmen at an smelting works. Acta med. scand. 128 (suppl. 196) (1947).

10. Sutton, J.R., Young, J.D., Lazarus, L., Hickie, J.B. & Maksoytis, J., The hormonal response to physical exercise. Austr. Ann. Med. 18, 84 (1969).

11. Wahlund, H., Determination of the physical working capacity. Acta med. scand. 132 (suppl. 215) (1948).

DISCUSSION

(Selected parts of the discussion)

STORAGE, RELEASE, AND METABOLISM OF SUBSTRATES

Lipids

ROWELL: Dr. Fröberg, you have reported a change in the muscle TG concentration in connection with exhaustive exercise. What I have seen regarding this problem shows that the muscle TG concentration either tends to increase slightly with exhaustive exercise over prolonged periods or it stays essentially the same. Masoro and I a few years ago labelled TG and phospholipids in monkey muscle. Using one leg as a resting control we stimulated gastrocnemius muscle in the other leg with electrical impulses three times a second for five hours. We found no change in any of the lipid fractions of the stimulated muscle and concluded that there was no turnover of muscle lipid under these conditions. Thus we saw no sign of disappearance of TG. The amount of TG in most mammalian muscle is really very small and I would like to know why it enters into this discussion as a significant energy source. If its percentage is so very small in muscle, it seems to me that if you consumed all the TG in the muscle tissue, you would account for only a very small fraction of the metabolic rate in the mammalian muscle.

FRÖBERG: In fact the work of Masoro stimulated us to further studies on the metabolism of muscle TG during exercise, especially in the view of Havel's earlier work, where a discrepancy was found between oxidation of fat and rate of utilization of plasma FFA. We started our studies on rat. The gastrocnemius muscle was stimulated through the sciatic nerve. In an earlier study we had found that it was possible to differ between red and white muscle types in this muscle and we also wanted to know if there were any differences between these two muscle types with regard to lipid metabolism during exercise. We stimulated with five impulses/sec for two hours and found a decrease in the TG concentration in the red muscle tissue. Of course, this decrease, which was about 25 per cent, might at least to some extent have been due to increased hydration of the working muscle. The concentration of the normal lipids, cholesterol and phospholipids, did not decrease more than half as much as the TG concentration, e.g. about 10 per cent. We did not see this change either in the resting unstimulated muscle or in the white muscle tissue of the stimulated muscle. I have integrated these results to indicate that the red muscle type, but not the white, oxidized TG fatty acids during the exercise.

ROWELL: If you make a rough calculation of the total amount of energy that was expended by the muscle over this period of time, did it come out as a significant fraction of the total energy expenditure?

FRÖBERG: We have not made such calculations. The rats had an intact flow of blood to the muscle and probably energy was derived also from oxidation of plasma fatty acids and so on. The resting TG concentration was low, not more than 1.5 µM per gram. A net decrease of about 10 - 15 per cent is probably of no major importance in the energy metabolism of the rat during the exercise. It may be of interest in this context that the fractional turnover of plasma FFA in the rat has been reported to be 5 - 6 times higher than in man. In man on the other hand we have much more glycerides in the muscle, about 10 µM/g. From the caloric point of view, muscle TG in man are quite considerable.

PAUL: As we know from a study by Carlson, fat can be accumulated in the muscle and it can also be dispersed (study with norepinephrine). We did some studies on well-trained dogs that were exercised for four hours. At the beginning of the experiment we inhibited the mobilization of free fatty acids by giving nicotinic acid, which also depressed the rate of oxidation of plasma FFA. These dogs were able to exercise for four hours. We analyzed this as intramuscular energy fuel; having calculated the RQ over the muscle, we determined what per cent of energy used inside the muscle came from FFA. It was an appreciable amount.

PERNOW: These studies are truly complicated to evaluate. We did similar studies in man. If we gave enough nicotinic acid to lower the arterial concentration of glycerol and FFA almost to zero, the subjects were still able to perform almost the same amount of work (both short maximal and prolonged submaximal work) as without nicotinic acid. In another series of studies, subjects worked with their legs so that the amount of glycogen in the muscles dropped almost to zero. On the following day the subjects were again exercised but first we gave them

DISCUSSION

nicotinic acid. They were still able to perform work but their performance was reduced by about 50 per cent. RQ analyses indicated a considerable fat combustion, which may at least partly come from local TG stores. These evidently are not blocked by nicotinic acid.

FRÖBERG: We gave nicotinic acid to rats and followed the concentration of TG in both red and white muscle types and also in the myocardium for six hours. In the red muscle type the TG concentration was depressed after two hours and stayed depressed during the rest of the experiment. No significant changes were observed in the TG concentration in the white muscle tissue. In the myocardium, too, the TG concentration was depressed after two hours but in contrast to red muscle tissue not at four and six hours. The decrease in the TG concentration means, as far as I can understand, that nicotinic acid does not block the lipolytic processes in muscle tissue as it does in adipose tissue. Therefore the decrease in the muscle TG concentration is not likely to be due to a reduction of the adipose tissue in muscle tissue, it must be due to a reduction in the muscle fibres.

BOBERG: In the studies discussed by Dr. Pernow, showing that people can go on working after nicotinic acid, other fuels may be available besides those mentioned. I think that the TG circulating in the blood should be kept in mind. We have rather good evidence that nicotinic acid stimulates the lipoprotein lipase system and that nicotinic acid may accelerate the utilization of a fuel like plasma TG. To evaluate this problem, rather complicated studies on the turnover of plasma TG are, however, necessary.

MORGAN: I would like to support what Dr. Fröberg said about the concentration of TG in human muscle. As Rowell knows, our results on human muscles are somewhat different from his findings in the species which he studied. Our average value for the TG concentration in man was 9 μmoles/g wet weight, which is slightly higher than Dr. Fröberg's. I should also emphasize what he has said, that there is an extreme variation in the TG concentration. Even with careful cleaning of biopsy specimens the variation in normal, untrained individuals ranged from 2.6 to approximately 20 μmoles/g wet weight. The effect of longterm exercise training is not appropriate to this discussion.

FRÖBERG: We have studied the TG concentration in muscle of man from 20 up to 60 years of age. With age the concentration of TG does increase. It doubled over this period from 10 to 20 µmoles/g. Helander and I in 1958 found no change in the phospholipid concentration with age. Also the glycogen concentration was unchanged with advancing age. With regard to Dr. Boberg's remark I would like to mention two sets of experiments: one with untrained rats and one with trained rats doing the same amount of work. In the untrained rats we saw no change in the plasma TG concentration but in the trained rats it decreased. This may have been due to increased utilization of plasma TG by the working muscles in trained rats. This suggestion rests on the following hypothesis. The lipoprotein lipase system is responsible for the removal of the plasma TG molecules. The lipoprotein lipase is situated in or in very close relation to the capillary endothelium and during training the functional bed of capillaries in the muscle increases. With increased endothelial surface the amount of lipoprotein lipase also increases, i.e. the capacity to remove TG increases. Finally I would like to mention a study by Carlson and Mossfeldt, who found that the plasma TG concentration decreased quite markedly during prolonged strenuous skiing (Vasaloppet). Later in the same skiing contest we found a decrease both in the muscle TG concentration and in the muscle glycogen concentration. Calculated for the same g of muscle, the net decrease in the TG concentration accounted for about 1.5 more calories than did the decrease in the glycogen concentration. The caloric equivalent from muscle TG was thus greater than from muscle glycogen. This does seem to indicate that muscle TG are of importance for the energy metabolism during exercise.

CAHILL: We are really studying two superimposed states here. One is the metabolic results of exercise. The other is the transition from the fed to the fasted state. We have all emphasized the decrease in muscle glycogen and the decrease in muscle TG with exercise. There then has to be a reconstituting phase for this whole thing to reoccur. It was also pointed out this morning that there is no negative nitrogen balance during the exercise phase. Unless the animal feeds it would not be able to exercise again unless it reconstituted its muscle glycogen, if the glycogen were crucially important for this exercise phase to start

DISCUSSION

with. I would like to ask the investigators if they have looked at any of these prolonged exercise states a second time without an interval for refeeding. In other words, is this possible? Because what we have heard so far today is a total contradiction. You cannot deplete TG, you cannot deplete glycogen and run the whole piece of machinery a second time without having a very crucial transfer of energy from adipose tissue back to muscle to reconstitute the triglyceride. Or does the breakdown of the muscle nitrogen reconstitute the glucose to replenish the muscle glycogen. And I would like to ask Dr. Fröberg, for example, how long it takes to reconstitute muscle TG after it has been depleted by exercise.

FRÖBERG: Unfortunately I have no idea about the time. But we know that if in some other way we produce a lack of external substrates for the TG synthesis the TG concentration decreases to about 50 per cent after two hours. This decrease remains after six hours, so these four hours are not enough to reconstitute the TG level even if the flow of fatty acids is completely unchanged during this time.

CAHILL: This is interesting because I once thought that the hypermobilization of free fatty acids at the termination of exercise was probably a transfer of free fatty acids from adipose tissue to replenish muscle TG. What you are telling me is that there is no evidence for this.

PAUL: We are talking about the glyconeogenesis during exercise and glucose formation but we have not mentioned anything about the glycerol. It is possible that during exercise, when the turnover rate of FFA is tremendously increased, there is enough glycerol to produce a substantial amount of glucose. We did a few studies in Lankenau with starving and found that glycerol during starvation can produce up to 100 g glucose per day. The FFA level was at that time about 1.5 mmoles/l. The same FFA level during exercise would probably have a much bigger turnover of glycerol.

CAHILL: During exercise, where you have rapid mobilization of TG, you might get an adequate amount of glycerol. But if you have a basal caloric expenditure of 2000 Cal/day, there is only 180 g of TG, which would only give you 18-20 g of glycerol. So glycerol turnover basally is very small.

HAGENFELDT: I have a question concerning the TG values in Dr. Grimby's paper (p. 471). In the most active groups of men in the population study, there were slightly lower TG values. You then introduced a group of still more active men in the same age range with much lower TG values. I do not think they matched each other really. Were they so enormously more active?

GRIMBY: Yes, they were, and they had definitely higher maximal oxygen uptake values. I quite agree with you that there are many differences between this group of athletes and the population study, but this was our only chance of getting a large enough sample from what we called activity group IV.

LASSERS: It is known that endurance and hard exercise have an acute effect in reducing serum TG. The group of athletes with hard exercise several days a week may have been sampled close to a period of exercise.

GRIMBY: That is their natural habitual activity and therefore they may have low TG all week.

FRÖBERG: In the study by Carlson and Lindstedt the subjects were divided into three groups: none, moderate, and hard physical work. With increasing physical activity, there was a successive decrease in serum TG.

EDWARDS: Could an important factor in people with high physical activity be the range of diurnal variation? Could you tell us more about the habitual activity in relation to the mealtimes of the people? Were there any significant differences in the diets of these people?

GRIMBY: There were no obvious diet differences between the different activity groups as judged from a short diet interview. I have no answer to your first question.

FRÖBERG: We have followed TG values for 24 hours. The increase in TG concentration after a meal stays for 8-10 hours, so it should be adequate to use fasting morning values.

HOLLOSZY: I think we are not talking about a training effect but an exercise effect. If a trained

DISCUSSION

individual stops exercising for two weeks, there is a tendency for the TG values to increase, but he will not lose the training effect in terms of maximal oxygen uptake. Furthermore, you do not need to be trained to get a reduction in serum TG by exercise. Another point is that exercise will not lower a serum TG level that is already low. In other words, if you are dealing with individuals that have a TG level of 80-100 mg % or lower, it is very difficult to demonstrate any effect of exercise on the serum TG, whereas in individuals with values in the range 150-200 mg % or higher, you will get a marked drop.

Carbohydrates

SALTIN: The table below gives some of our data on the total amount of calories used during prolonged bicycle exercise at 70-75 per cent of max $\dot{V}_{O_2}$ in an untrained (UT; max $\dot{V}_{O_2} \approx 3.0$ l/min) and a well-trained subject (WT; max $\dot{V}_{O_2} \approx 5.0$ l/min).

	Oxygen uptake l/hr	RQ	Calories from carbohydrates g/hr	Muscle glycogen depletion g/kg x hr
WT	210	0.90	165	10
UT	126	0.96	135	12

These results differ from the data presented by Dr. Wahren. He mentioned that the 18 g of glucose that was released from the liver may account for up to 40 per cent of the total energy metabolism. According to our results the overall contribution of glucose may be 10 or at most 15 per cent.

WAHREN: In response to Dr. Saltin I would like to emphasize that my statements concerning the contribution of glucose to the estimated carbohydrate metabolism and to the total oxidative metabolism were based on measurements made after 40 min of exercise at the different work loads. At this time glucose may have contributed 25, 25, and 40 per cent of the total metabolism at the work loads 400, 800, and 1200 kpm/min, respectively, if all glucose taken up was completely oxidized. The data indicate that the glucose contribution is smaller earlier during the exercise period and some preliminary observations after an hour's exercise point to a slightly larger glucose uptake at this time.

Quantitative aspects of glucose utilization in this study are probably best evaluated from the data on splanchnic glucose production since leg blood flow was not measured. The total splanchnic production of glucose during 40 min of exercise at 1200 kpm/min can be calculated to 100 mmoles or 18 g. If the gradually increasing glucose production continued to rise until 60 min of exercise, as the preliminary data indicate, this would entail a total glucose production of approximately 165 mmoles or 30 g. About 10 per cent of this glucose may have been supplied by gluconeogenesis but - in the absence of kidney glucose production - the remaining amount presumably derives from hepatic glycogenolysis. It seems reasonable to assume that the liver even in the post-absorptive state can produce this amount of glucose, particularly if one considers the value for liver glycogen content of 50 g/kg given by Dr. Hultman.

It is clear that the observed rate of glucose production by the liver can be upheld only for a limited period of time. In the absence of experimental data in this study on glucose turnover for exercise beyond the first hour, no extrapolation at this time can be made of the possible metabolic adaptation.

HAVEL: In relation to Dr. Wahren's statement I would like to emphasize that using values obtained at one point in time after the onset of exercise to give the whole picture cannot be done as there is a continuous change in the fractional contribution of glucose, of glycogen and of fat with time. Therefore, one cannot extrapolate and say that 40 per cent of the energy always comes from glucose. It may had occurred at that point at the minute measured.

KARLSSON: We have made experiments similar to those described by Dr. Hermansen with repeated maximal or near maximal activity bursts (5 x 1 min, with 5 min rest between) and included muscle biopsies in the protocol. We found a gradual increase in blood lactate with a tendency to a plateau ($\approx$ 20 mmoles x l^{-1}) after about 15-20 min. The muscle lactate concentration, however, was maximal or near maximal already after the first burst. Thus, after the first burst of activity the muscle lactate concentration averaged 22.5 mmoles x kg^{-1} compared

DISCUSSION

to 23.2 mmoles x kg^{-1} wet muscle after the last one. Another point I would like to make is that Dr. Hermansen discussed the possibility that an increase in muscle pyruvate concentration would influence the lactate formation of the muscle. We have measured pyruvate concentration at rest and after brief exhaustive exercise, but never seen concentrations higher than 0.20 mmoles x kg^{-1} wet muscle. The blood pyruvate concentration at the same time had a range of 0.3-0.6 mmoles x l^{-1} of blood.

HERMANSEN: Still you have a difference between the concentration in the muscle and the blood and this, of course, is most likely since the lactate is produced in the muscle and has to diffuse out in the blood. Our blood lactate values were about 25-30 mmoles and I do not know if in this situation we have minimized the difference between muscle and blood lactate concentration.

LASSERS: I wonder if it is possible to compare muscle lactate concentration in mmoles x kg^{-1} wet weight with concentrations in the blood (mmoles/l) and say that the muscle concentration is higher than the concentration in the blood. Are these two really comparable?

KARLSSON: Muscle specimens from the quadriceps muscle of man contain 74-79 per cent of water at rest and during exercise according to our measurements of the water content by means of freeze drying. Whole blood contains around 90 per cent of water. Thus, from a practical point of view, figures for lactate concentration per kg wet muscle and per l of blood can be used for estimating the concentration differences on a per l of water basis. It should be emphasized that even during maximal exercise the water content of muscle does not differ by more than around 5 per cent as compared with rest.

HULTMAN: Can I give a few figures on this. We have measured lactate content during continuous exercise in mmoles/100 g dried tissue. At the beginning it was 0.60, after 5 min 12.0 mmoles/100 g, and at the end 10.8. Recalculating this to mmole intracellular muscle water, based on the assumption of an unchanged chloride space, gives 1.9 mmoles/l, 32.2 and 29.2. The water content in the cell was in these situations 306, 365, and 350 ml/100 g dried tissue, respectively. Just calculating on the wet weight without taking the changes in water content with exercising into account may give a slightly wrong figure for the intracellular concentration of lactate.

PHOSPHAGENS AND MUSCLE CONTRACTION

OLSON: Dr. Davies, I remember very well your most provocative paper on the mechanism of muscular contraction published in Nature in 1963 (Nature, Lond. 199, 1068, 1963) to which you alluded in your presentation. This paper was, of course, published before the discovery of the calcium regulatory protein of muscle, troponin, by Ebashi (J. Biochem, Japan 55, 604, 1964). In one of your slides, however, you presented the view advanced in 1963 that calcium ions link actin to myosin. Since it is now known that in the absence of troponin, contraction by synthetic actomyosins can occur, which is not sensitive to calcium, do you feel that your view needs some modification?

DAVIES: I agree with this. Troponin was not discovered in 1963 but it was posited here. All the evidence about troponin can be fitted in two ways, either that it releases an inhibition or that it confers specificity. Either of these will give the same answer. Without troponin magnesium will fit, with troponin only calcium will fit. It is known that calcium binds to troponin. I would still believe that the formal link itself involves the molecule of ATP, probably of calcium and of ADP, and that troponin fits there and allows only calcium to fit. With no troponin, magnesium will do. This allows the muscle to be regulated, turning the muscle on by adding calcium and turning it off by the calcium pumps. Nobody knows where calcium goes on the thin filament except that it attaches to troponin in more primitive systems. It is very likely that the calcium attaches to the troponin itself, being bound by troponin-myosin to the active filament. That does not mean that it is not available to attach to the ATP.

I think the finding that muscles can develop a force without reduction in ATP means that the filaments must be able to do one cycle of interaction without ATP being split. There seems to be no escaping the fact that ATP is split somewhere in the cycle. If you break down a single twitch into, say, ten different steps during the time the muscle is contracting, the ATP is being used all the time the work is being done. This finding would suggest that the interactions of the crossbridge are based on the mechanism in which the crossbridge is already charged. They can do one given cycle of whatever the interaction is that causes the movement and this must be restored by ATP. I do not see any way out of that. I suggested the formation of an alpha-helix and it has the right sort of properties but is peculiarly difficult to prove. Just as twenty years ago it was impossible to prove that ATP had anything to do with contraction in the living muscle.

ASMUSSEN: We made some determinations some years ago on negative exercise bicycling down hill and braking the movements and we noticed that the oxygen uptake at the same nominal rate of work was much less in negative exercise than in positive exercise. And further we noticed that increasing the velocity of the movement increased this difference, so that, at the highest possible velocity on the bicycle the ratio between oxygen uptake in positive work to negative work at the same nominal value was 1:125. This means that you did a lot of negative work practically without any increase in energy output. Could that have anything to do with the findings presented by Dr. Davies (p. 327)?

DAVIES: In fact that paper of yours stimulated me to do these experiments. It was really very impressive how e.g. a small woman pedalling backwards on the tandem could tire out a strong man who was pedalling forwards. The woman absorbed all the work that the man did and she was fresh at the end of the experiment and the man was very tired. I suppose that just the finding that walking down a hill is easier than walking up a hill shows that the muscle must be able to conserve this energy. But this is very strange since muscle could be one of three types of motor; like a rocket motor, like an automobile or like a street car. When the human lunar excursion landed on the moon the rocket motor took as many pounds of rocket fuel to land on the moon softly as to get back into orbit from the moon. The muscle clearly is not that sort of machine or it would need as much energy to do negative work. A street car with an electric motor can go down hill and use the gravitational energy to drive electricity back into the power line; it is really a reversible machine. The muscle is, however, not a reversible machine and cannot resynthesize ATP. What you can do is to arrange for it to act like an automobile that can go down hill with the brakes on. An automobile going down hill does not need energy to absorb the work of the difference in gravitational potential. You just put your foot on the brake and turn all the energy into heat.

FACTORS LIMITING OXYGEN TRANSPORT

HOLLOSZY: There are two subjects that seem to have aroused considerable interest. One of these is, what happens during the early phase of exercise while the oxygen deficit is developing. The other has already been discussed this afternoon. What limits oxygen uptake; is it the muscle or the circulatory system. Regarding the second point I personally avoid taking sides because we are dealing with a very complicated system and I cannot imagine that it is an all-or-none sort of thing - that one or the other is limiting under all kinds of circumstances.

KAIJSER: Your point about what limits oxygen transport is very important. I very much doubt that the circulatory system is limiting for oxygen transport. In studies on the adaptation of the capacity for work we increased the oxygen content of arterial blood by oxygen breathing at increased atmospheric pressure. By this means we were unable to increase the performance of the individual, either for work with small muscles like those in the forearm or for exercise with large muscles as in the legs, where there is probably a considerable demand on the central circulation. Nor did we see an increase in the a-v oxygen difference across for example the forearm when we increased the oxygen content. There thus seems to be quite a lot of evidence against the circulatory system as limiting. On the other hand the oxygen content of the blood can be decreased only a very little before performance starts to decline. So obviously even if we state that the circulation is not the limiting factor, it has very little margin for an adequate oxygen supply during exercise.

DI PRAMPERO: Could it not be that some oxidative enzyme was damaged by this high oxygen pressure in your experiments and that therefore oxygen uptake did not increase more just because of this?

KAIJSER: If there were toxic effects on muscle metabolism from breathing high-pressure oxygen, the performance of the muscles, for example in the forearm, would decrease with the duration of oxygen breathing.

DISCUSSION

I have performed experiments with short as well as long periods of oxygen breathing and was unable to demonstrate any decrease in performance time.

GOLLNICK: I think that without a doubt you ultimately come to the cardiovascular system as being the limiting factor. The muscles cannot consume more oxygen than is brought to them.

HOLLOSZY: Or put in another way, it does not matter how much oxygen you bring to the muscle if the muscle does not have the machinery to use it.

GOLLNICK: That is very true, but I do not think that in a normal healthy animal we have ever demonstrated that the capacity to consume oxygen in the muscle cell is really limiting.

PERNOW: I agree with Dr. Holloszy that it is almost impossible to claim just one factor that limits oxygen transport. It is necessary to define what type of work you are discussing. Thus different factors might limit maximal work of short duration on the one hand and work lasting hours (as for instance the long-distance skiing mentioned earlier) on the other. Furthermore, you have to define the size of the muscle groups that are involved in the work. For instance, when forearm muscles work to exhaustion, there is still about 35 per cent oxygen saturation in the venous blood draining the working muscles and fairly low lactate concentration. So evidently in this case the transport of oxygen is not the limiting factor. On the other hand, at the end of maximal work performed with the leg muscles on a bicycle ergometer, the venous oxygen saturation of the femoral venous blood is below 10 per cent even in normals and the lactate concentration is very high. In this situation the oxygen transport might be an important limiting factor.

CHRISTENSEN: In this connection it is also important to see how the capacity of the circulatory system is used at the given moment. And I think some typical experiments done by Saltin some years ago about the effect of dehydration are very clear. You have a marked decrease in endurance capacity, but an unchanged maximal oxygen transport capacity. This means that you can transport the same amount of oxygen during a few minutes, you have an unchanged power but a decreased capacity and still you have the same individual.

Nor do I think one should forget some old experiments with increased oxygen tension in the expired air. If you change from 21 per cent normal pressure up to 30 per cent of oxygen, you will get a marked increase in maximal oxygen uptake capacity. You thus really have a higher power for a short work. In this situation one is therefore justified in saying that the oxygen transport is a limiting factor. And, as Kaijser said, if you make a slight reduction in arterial transport capacity you get the same marked reduction in power.

DAVIES: I agree there cannot be a simple answer to the problem of limiting factors unless you define the question better. From the metabolic point of view there is an absolute limit for maximal muscle power and this is the Ca-activated-actomyosin-ATP-ase. Dr. Berony in New York has made some very nice experiments over the last year or two comparing very slow and quick muscles in mice. He found that the maximal velocity of the muscles, which is the time where approximately the maximal power is delivered, is directly related to ATP-ase, the rate at which the muscle can split ATP. Even when the supply of ATP is optimal, what determines the power output is the efficiency of the enzyme itself. And there is a very nice linear correlation over some orders of magnitude between the rate of the enzyme and the maximal power output. For this system it is very clear that the enzyme itself determines the difference in fast and slow muscles in the same animal. The reason why some species are quick are because the ATP-ases are quick. The reason why they may be able to continue for a long time is determined by these enzymes. And we must not forget the leg of a jumping flee, which can work even quicker than the enzyme can operate. The same is true of the flight muscles; they operate this elastic mechanism so that some of the power strokes deliver power more quickly than the enzymes themsélves can produce this power. It is stored elasticly and then delivered by an elastic recall. So the system is very complex and the question: "What is the limiting factor" must have fifteen different answers according to the particular system you are thinking about.

DISCUSSION

DAVIES: The maximal oxygen uptake in the cell is obtained in less than 1 sec when you look at the turnover of cytocrome B or DPNH. The signal that turns on both oxygen uptake and glycolysis is the appearance of ADP and inorganic phosphate, so you must first have the muscle contraction producing the ADP and the inorganic phosphate. This can then turn on the respiration in less than 1 sec. The quick muscle can use energy at a much higher rate than it can be supplied either from oxygen uptake or from glycolysis. The signal that turns on glycolysis includes AMP, which requires the further conversion of ADP into AMP and ATP by myokinase. So there is a delay in all these things. But the delay is extremely short.

DAVIES: The term submaximal work can only be used when speaking of the muscle as a whole. If a single fibre contracts then it produces enough ATP to turn on the respiration maximally in the region of that fibre. If the work load is submaximal it is because lots of the fibres have not been stimulated and they just stay at their resting levels, assuming an "all-or-none" law. If the muscle fibre is stimulated, then the amount of work it develops would depend on the load that is between the ends. Under nearly all circumstances the splitting in that single fibre is enough to activate it maximally. What one sees in the slow rise, especially in submaximal work, is that not all the fibres have been stimulated. There is always a mixed population; some fibres are resting and some are working rather rapidly.

INDEX

Acetylcholine   30

Actomyosin   335, 546

Adenosine diphosphate (ADP)   59, 303, 335, 431, 434

Adenosine monophosphate (AMP)   303

Adenosine triphosphatase   55, 332

Adenosine triphosphate (ATP)
    from glycolysis   55, 347
    from oxidative phosphorylation   55
    hydrolysis in muscular contraction   38, 329, 341, 371
    utilization   303, 327, 341, 345, 357, 371, 401, 419, 431, 434, 546, 550

Adenylate kinase   35

Adipose tissue   55, 103, 111, 249

Adrenaline   113, 115, 119

Adrenergic nerves   111, 115, 121

Alanine   106, 205, 319

Amino acids   103, 193, 199, 205, 208, 319
    transport, insulin effect   106

α-amino nitrogen   199

Anaerobic metabolism
    anaerobic glycolysis   371
    ATP depletion   329, 342, 371
    CP depletion   342, 371
    lactate formation   371, 509

Carbohydrate   see Glycogen or Glucose

Carnitine   89

Catecholamines   111, 119

Cholesterol
    artherosclerosis   497
    blood content of   470, 474, 497, 501

Creatine phosphate (CP)
    ATP yield 343, 371
    breakdown 341, 371, 401, 419, 431

$3',5'$-cyclic AMP 257, 263, 266

Cross innervation 21

Cytochrome oxidase 54

Diabetes 529

Diet 143, 171, 275, 296

Enzyme adaptation 51, 54

Epinephrine see Adrenaline

Exhaustion 215, 319, 357, 388, 405

Fat
    body content 497, 500
    storage 237, 436, 535

Free fatty acids
    ATP yield 104, 153
    blood concentration 112, 155, 229, 239, 241, 310, 529
    metabolism 89, 154, 225, 232, 235, 245, 315, 505
    release 55, 112, 229, 245, 249, 315, 505
    uptake 153, 225, 315, 436, 505

Fructose-1-6-diphosphate 303, 349

Fructose-6-phosphate 303, 349, 436

Gluconeogenesis 104, 275, 363

Glucose
    exercise and 130, 167, 180, 191, 215, 225, 239, 242, 319, 543
    glycogen and 281, 345
    insulin and 170, 238
    in blood from liver 130, 198, 239, 318
    in ischemic heart disease 498
    phosphorylation by kinases 284
    transport into the cell 261, 285

Glucose-6-phosphate 64, 257, 261, 282, 303, 346, 436

Glycerol 114, 200, 249, 529, 537

Glycogen
    content in liver   143
    content in muscle   13, 51, 143, 181, 257, 273, 289,
        301, 307, 436, 543
    glucose from   193
    synthesis   257, 273
    utilization   297
        effect of propranolol   301
        in exercise   51, 55, 143, 180, 242, 273, 289,
            419, 543
        in starvation   104, 147, 274

Glycogen storage
    diet and   143, 274
    in liver   146
    in muscle   87, 143, 273, 436
    metabolic regulation   143, 257, 260, 267, 319

Glycogen synthetase   88, 257, 260, 264, 278

Glycolysis   319, 349, 421

Glycolytic enzymes   33, 34, 56, 88, 349

Growth hormone   317

Heart muscle
    FFA oxidation in   63, 66, 436, 449, 459
    glucose oxidation in   436, 449, 459, 487
    ischemic heart disease   74, 483
    lactate oxidation in   63, 449 459, 484
    mitochondria in   63, 72, 440

Hexokinase   39, 56, 173, 193

Hypoxia   74

Inorganic phosphate (P_i)   59, 220, 327

Insulin   166, 238, 495, 525, 530
    effect on adipose tissue   317
    glucose transport   239, 488
    in ischemic heart disease   495

Ischemic heart disease
    myocardial lactate   484
    plasma insulin   488, 495
    plasma lipids   486

Ketones   104, 153, 158

Lactate
- determination of ^{14}C-L(+)-lactate in blood   409
- formation   52, 58, 130, 208, 215, 225, 289, 302, 345, 361, 371, 401, 415, 425, 509
- from pyruvate   302, 347, 389, 540
- glycolysis and   218, 292, 304, 347, 357, 385, 404, 414, 420, 427
- ischemic heart disease   483
- lactate-pyruvate ratio   253, 302, 346, 362
- oxidation   413
- turn-over   413

Lactate dehydrogenase (LDH)   see Glycolytic enzymes

Lipid   see Fat, Free fatty acids

Lipoprotein lipase   539

Liver
- alanine-glucose cycle   205
- amino acid metabolism   209
- blood flow   127, 144, 198, 211
- fatty acid metabolism   131
- gluconeogenesis in   133, 137
- glucose metabolism   130, 131, 198, 211, 544
- glycogen content   127, 143
- hypothermia   135
- insulin effect   239
- lactate metabolism   130, 131, 133, 145, 200
- lactate-pyruvate ratio   132, 200
- lipid metabolism   127, 131
- oxygen uptake   133

Malate   303, 345, 348

McArdle's syndrome   319, 320

Membranes   443

Mitochondria
- enzymes   54
- membranes   63, 89
- morphology   54, 72, 74, 92, 97
- oxidative phosphorylation   54
- physical conditioning and   73, 91

Muscle fibers, red and white   13, 21, 38, 215, 316

Myocardium   see Heart muscle

Myoglobin 55

Negative work 547

Nicotinic acid 321, 538, 539

Noradrenaline 111, 113, 119, 317

Obesity
    exercise and 500
    in children 521

Occlusive arterial disease 505
    free fatty acids 509
    lactate 508

Oxygen debt 344, 361, 371

Oxygen deficit 384, 388
    lactate formation and 363, 376, 388
    phosphagen depletion and 344, 376, 388

Oxygen transport
    limiting factors 548, 549, 550

Oxygen uptake
    metabolic rate and 216, 232, 238, 292, 402, 508
    regulation of 219, 441

Palmitate
    oxidation of 228, 237, 440

Phosphagen   see ATP, CP

Phosphofructokinase 56, 352

Phospholipids 90, 307, 436

Phosphorylase 15, 33, 56, 88, 351

Physical conditioning
    glycogen breakdown and 293
    lactate formation and 51, 396
    phosphagen depletion 395

Protein 54, 74, 90, 106

Pyruvate 104, 134, 200, 207, 345, 362

Respiration quotient (RQ)   274, 307, 320

Starvation   104

Succinate dehydrogenase   54

Triglycerides
   in adipose tissue   103
   in blood   469, 474, 497, 501
   in heart muscle   439
   in liver   132
   in skeletal muscle   89, 307, 537
   utilization   103, 307, 318, 537, 538, 539, 540, 541, 542, 543

UDP glucose   257

CANISIUS COLLEGE LIBRARY
BUFFALO, N.Y.

3 5084 00368 6113

QP301 .M753 1970
c.2
Muscle metabolism during
exercise : proceedings of a
Karolinska institutet
symposium held in Stockholm
Sweden, September 6-

DATE DUE

GAYLORD — PRINTED IN U.S.A.